W9-AZF-575

TOUR OF DUTY

★

ALSO BY DOUGLAS BRINKLEY

Wheels for the World: Henry Ford, His Company, and a Century of Progress, 1903–2003

The Mississippi and the Making of a Nation (with Stephen E. Ambrose)

American Heritage History of the United States

The Western Paradox: The Bernard DeVoto Reader (editor, with Patricia Limerick)

Rosa Parks

The Unfinished Presidency: Jimmy Carter's Journey Beyond the White House

John F. Kennedy and Europe (editor)

Witness to America: An Illustrated Documentary History of the United States from the Revolution to Today (editor, with Stephen E. Ambrose, Allan Nevins, and Henry Steele Commager)

Rise to Globalism: American Foreign Policy Since 1939 Eighth edition (with Stephen E. Ambrose)

Dean Acheson and the Making of U.S. Foreign Policy (editor)

Jean Monnet: The Path to European Unity (editor)

The Majic Bus: An American Odyssey

Dean Acheson: The Cold War Years, 1953–1971

Driven Patriot: The Life and Times of James Forrestal (with Townsend Hoopes)

FDR and the Creation of the U.N.

TOUR ☆ OF ☆ DUTY

JOHN KERRY AND THE VIETNAM WAR

DOUGLAS BRINKLEY

wm
WILLIAM MORROW
An Imprint of HarperCollins*Publishers*

HarperCollins books may be purchased for educational, business, or sales promotional use. For information please write: Special Markets Department, HarperCollins Publishers Inc., 10 East 53rd Street, New York, NY 10022.

FIRST EDITION

Designed by Renato Stanisic

Printed on acid-free paper

Library of Congress Cataloging-in-Publication Data has been applied for.

ISBN 0-06-056523-3

04 05 06 07 08 WBC/RRD 10 9 8 7 6 5 4 3 2 1

To Anne Goldman Brinkley

With love and devotion

Labour to keep alive in your breast that little spark of celestial fire, called conscience.

—George Washington

Life, hope, they conquer death, generally always; and if the steamroller goes over the flower, the flower dies.

—Robert Lowell

I will set my ear to catch the moral of the story and tell on the harp how I read the riddle.

—Psalm 49

Seven letters—that's all it takes to make the word Vietnam. But it is much more than a word. More than the name of a country. It is a period in time—it is a one-word encapsulation of history—a one-word summary of a war gone wrong even as young Americans in uniform sought to do what was right, of families divided yet united by love, generations divided, a nation divided yet in a deeper sense united by its ideals. Vietnam, it carries in its seven letters all the confusion, bitterness, love, sacrifice and nobility of America's longest war. It is a one-word, all-encompassing answer to questions: What happened to him? Where was he injured? When did he change? Say the word Vietnam to a veteran and he or she can smell the wood-burning fires, hear the AK-47s and B-52s, see the pajama-clad Viet Cong and the helicopters darting across the sky—you can feel all the emotions of young men and women who in the end were fighting as much for their love of each other as for the love of country that brought them there in the first place.

—John Kerry

CONTENTS

★

AUTHOR'S NOTE

This book tells the story of one young American's Vietnam War odyssey. It is not meant to be a biography of John Kerry or an authoritative history of that era. Everybody who fought in Southeast Asia has his or her own highly personalized story to tell. The same is true of the concerned citizens who took to the streets to protest the war. While hundreds of primary and secondary sources were drawn upon and more than one hundred people were interviewed for this book, the narrative is based largely on journals and correspondence Kerry kept while on his tours of duty. He provided me complete and unrestricted access to these documents and permission to use them as I saw fit. He also granted me twelve hours' worth of interviews at his home in Boston and office in Washington, D.C. He, however, exerted *no* editorial control on the manuscript.

Prologue
April 22, 1971 (Washington, D.C.)

The only unforgivable sin in war is not doing your duty.

—Dwight D. Eisenhower

Every public life has its point of origin. For the twenty-seven-year-old John Kerry, that dramatic moment came two years after he shipped out of the rivers of South Vietnam, in a committee hearing room in the Dirksen Senate Office Building in Washington, D.C. For it was in that grand, high-ceilinged chamber that the young Navy veteran posed what he saw as the fundamental question about the Vietnam War to the senators on the Foreign Relations Committee and beyond them to the nation at large: "How do you ask a man to be the last man to die for a mistake?"

And John Kerry was among the lucky ones who did not die, and who came back whole. The 1966 graduate of Yale University had enlisted in the Navy because it had seemed the right thing to do at the time. Following Officer Candidate School in Newport, Rhode Island, and specialized training in San Francisco and San Diego, Kerry had been assigned to the guided-missile frigate U.S.S. *Gridley*, aboard which he visited Vietnam for the first time in March 1968. His job as ensign was to make sure the *Gridley* stayed perfectly maintained, but it was tedious duty and Kerry was in quest of more adventure, independence, and responsibility. He wanted to get out of the "Black Shoe Navy" and have a command of his own. So he applied for Swift boat school at the Navy's training facility in Coronado, California, and after it, he returned to Vietnam in November 1968, this time as a lieutenant (junior grade) in charge of his own patrol craft fast (PCF). These boats, each manned by a junior officer and five enlisted men, had become the vanguard in the gambit of Elmo R. Zumwalt Jr., the new commander of U.S. Naval Forces, Vietnam, to forestall Viet Cong infiltration into South Vietnam's Mekong Delta and Ca Mau Peninsula.

Between December 1968 and March 1969, Kerry led PCF-44—and when that boat got shot up, another Swift, PCF-94—on scores of dangerous raids up the rivers and canals of South Vietnam's Mekong Delta, including the dangerous territories along the Cambodian border. In the process the young skipper was wounded three times. Between his second and third Purple Hearts, on February 28, 1969, Kerry beached his boat in the center of an ambush and killed a Viet Cong sniper armed with a B-40 rocket-propelled grenade launcher, thereby winning the prestigious Silver Star for valor in combat. Two weeks later, despite an injury to his right arm, the young lieutenant went back to save a drowning colleague under fire, this time earning a Bronze Star for bravery as well as his last Purple Heart. Any way one looked at it, John Kerry returned home in April 1969 a genuine war hero. "But I had my arms and legs," he pointed out. "Many of them I was speaking for did not."

Kerry was one of more than a thousand American veterans who descended upon Washington for five days of antiwar protests late in April 1971. The demonstration was dubbed "Dewey Canyon III," in mocking reference to the Nixon administration's Operation Dewey Canyon I and II efforts against Hanoi's supply lines from Laos in 1969 and 1971. Most of the protesters belonged to Vietnam Veterans Against the War (VVAW), an organization headquartered in New York that claimed upward of eleven thousand members nationwide. Those veterans who made it to the nation's capital for the rally were bitter. None had tried to dodge the draft, filed for conscientious-objector status, or joined the National Guard. No one could question their patriotism. Phillip Lavoie of North Dighton, Massachusetts, for example, arrived wearing his olive-drab fatigues and forearm crutches; he had lost both his legs in Vietnam to a land mine that exploded beneath him while he was on a reconnaissance mission. Lavoie would tell Associated Press reporter Brooks Jackson that he had come to participate in the mass protest against U.S. policies in Vietnam, Laos, and Cambodia because he cared about America. "I love this country, man," Lavoie had explained. "Like, wow, it's really beautiful. But we're not fighting for democracy over there. We're fighting so some people in this country can have more money."

Robert Muller, a former Marine Corps first lieutenant, likewise had come to the nonviolent protest in his wheelchair out of love for his country. Like all the veterans congregating in Washington, D.C., he was a survivor, an "antiwar warrior" in the apt phrase of Robert Jay Lifton, the Yale University professor of psychiatry. The Nixon White House was afraid of Muller and the other veterans because they refused to have amnesia about what had occurred in the rice paddies, elephant grass, mangrove thickets, and murky rivers of Vietnam. "I got shot through the chest," Muller explained. "The bullet went through both lungs and severed the spinal cord. And I was immediately rendered paraplegic, from the fifth thoracic vertebra down. I was conscious for maybe ten seconds after I was hit, and my first thought was, 'I'm hit. I don't fucking believe it. I'm hit!' That was the first thing that went through my head. The second thing was, 'My girl. And my family.' "

Rushed to a hospital ship in the Gulf of Tonkin, Muller had awoken to find seven tubes in his body and no hope of ever again moving his legs. Instead of feeling sorry for himself, he soon felt liberated to speak out for his many friends who came back in bags. "And that's why I can't complain," Muller explained. "I am bitter, not because I was shot in Vietnam. I'm bitter because I put my faith, my allegiance, in my government. I did so with the best, most honest intentions in the world, believing that I was doing right, because [that] my government would lie to me or lead me astray was inconceivable. But having been there, and recognizing what we've done over there, and not being able to justify the death of any of my friends, that's why I'm bitter. I'm bitter because I gave to my country myself, one hundred percent, and they used me."

Many Vietnam veterans who felt the same way had come to Washington that April to make their feelings known. The sentiment among the veterans was that either you had been to "Nam" and understood or you had been brainwashed by the war propaganda machines of the Johnson and Nixon administrations. Nobody who encountered these vets, dressed in their military fatigues, most with shoulder-length hair and angry eyes, could escape the conclusion that at the very least they had been tainted by their combat experiences. "All over Washington veterans were camped out," recalled *Boston Globe* reporter Thomas Oliphant of the two-mile expanse from the Washington Monument to the Capitol. "Many were disabled. They had illegally taken over the Mall. There were constant threats of police raids. The attorney general, John Mitchell, wanted to strong-arm them, and Kerry had the difficult job of keeping the veterans in check. When they got together, emotions ran high. Kerry's job that entire week was Herculean. He responded to it with incredible political skill and grace."

With Oliphant at his side that Thursday, the Vietnam veteran from Waltham, Massachusetts, who had been chosen to speak for the group was running late. John Kerry hustled across Capitol Hill having just helped defuse a dispute between the local police and a few dozen antiwar protesters over whether the latter were permitted to congregate on the steps of the U.S. Supreme Court. He had negotiated a compromise whereby the group would be allowed to demonstrate on the sidewalk in front of the Court steps. Not all their contentious peers had the advantage of Kerry's immediate offices that morning; 108 of the antiwar veterans would be arrested in Washington that day, most freed on ten dollars' bail a head. And not all of the protesters, the majority of whom had been enlisted men in the war, were all that enamored of this Brahmin ex-officer representing them. "A lot of the guys were blue-collar enlisted men, and distrusted officers," former Lieutenant William "Bill" Crandell of VVAW recalled. "Kerry had to convince them he was for real."

Kerry had spent the day before rushing around Washington, D.C., talking with veterans on the Mall and working on his speech. He originally called it his "Letter to America." Some of the language came directly from antiwar addresses he had delivered earlier at the Valley Forge National Battlefield in Pennsylvania and Dedham High School in Massachusetts. "When I began it—in the previous summer—it was in the form of a letter," Kerry recalled. "Then I transitioned it into a speech. It didn't quite work as a letter." Cobbling his best lines from these, together with some late-night-telephoned suggestions from Adam Walinsky, who'd been a speechwriter for Robert F. Kennedy, Kerry crafted a powerful statement declaring why the United States had to end its military occupation of Vietnam. "The pace of things in Washington, D.C., was ferocious that week," he recalled of the experience. "I remember staying up much of the night, working on organizing my thoughts and collecting the testimonies of fellow veterans. I wrote directly from the heart, developed a collection of pages on yellow legal pads, and then joined them all together. The final product represented the best of my thinking, an emotional plea to Congress to stop funding the war."

Waking up at dawn on the Mall, Kerry rushed over to VVAW headquarters on Vermont Avenue, shaved, and then headed back to the Lincoln Memorial; soon thereafter he moved on to the Supreme Court. Tom

Oliphant, who had intuited that the Massachusetts Swift boat veteran was the best antiwar story of Dewey Canyon III, stayed close to Kerry that morning as he moved from location to location. Together they hurried down Constitution Avenue to the Dirksen Senate Office Building with less than five minutes to spare before his appearance. When they got there, a strange calm prevailed around the building, considering approximately 1,000 veterans had flocked to town. "We climbed up the stairs, two steps at a time," Oliphant remembered. "The clock was ticking."

Kerry had been asked to testify only two nights before by Arkansan J. William Fulbright, whom he had met at a fund-raiser hosted by a fellow Democratic senator, Philip Hart of Michigan. Known as a tenacious foreign policy analyst, Fulbright had been elected to the Senate in 1944 and became one of its best-known and most influential members. When it came to promoting international student exchanges, Fulbright had become *the* respected brand name. But it was as a Vietnam dove that Fulbright was garnering headlines. Starting in 1966, when under his chairmanship the Senate Foreign Relations Committee held hearings on Vietnam, he became a lightning rod for dissent. The title of his 1967 book, *The Arrogance of Power*, soon was a catchphrase for those criticizing the U.S. government. "The role he plays in Washington is an indispensable role," Walter Lippman wrote of Fulbright in 1963. "There is no one else who is so powerful and also so wise, and if there were any question of removing him from public life, it would be a national calamity."

At Hart's party Kerry had learned that Chief Justice Warren E. Burger had led the Supreme Court to reverse a District Court order, originally sought by the White House, keeping the veterans from camping on the Mall. A virulent animosity had grown between the Nixon administration and the antiwar movement, including these Vietnam veterans. Vietnam policy critics like Fulbright could certainly see the appeal of a well-spoken, highly decorated, Ivy League–educated Navy officer denouncing the war before his committee.

Kerry had already garnered some attention as perhaps the most articulate antiwar veteran on the public-speaking circuit. His television appearances on both ABC's late-night *The Dick Cavett Show* and NBC News' Sunday morning *Meet the Press* had even brought him to the far outer edges of celebrity. But testifying before the Senate Foreign Relations Committee—even if it was chaired by the fully sympathetic Fulbright—with journalists from around the world reporting his every word was at a different level entirely. The pressure was on. And the opportunity was immense. He felt it his duty to articulate the fears and laments and anguish of an entire generation—the Vietnam generation, those who had actually been there—in his two hours of testimony. After all, he would be the only veteran to testify. "What I wanted to do was give voice to our concerns," Kerry explained, looking back, "to put a stop to the charade."

Scrambling to collect his thoughts, he was ushered into a fourth-floor office adjacent to the committee room. Oliphant waited with him as he went over his prepared remarks. Kerry pried the door slightly ajar and took a peek inside the hearing room. It was packed, with TV cameras and lights everywhere. He turned to Oliphant and whispered: "Oh, shit." He stepped back and took a deep breath. The *Globe* reporter tried to joke away Kerry's stage fright. "Go ahead and be famous," Oliphant jibed. "See if I care." Suddenly, Kerry remembered, "somebody rushed up in to see me and shouted, 'They're waiting for you! They're waiting for you! Get in there quickly! You're late!' " So, clutching his folder, John Kerry headed through the large doors into the committee chamber.

As Oliphant sidled to a spot against the wall, Kerry strode toward the witness table. His wife, Julia, was already there, in the back, while his sister Peggy was anxiously awaiting his arrival in one of the front rows. The solidarity that Kerry felt toward the veterans in the chamber—"brothers," as he called them—was palpable. "I walked into the room and it was just packed: a standing-room-only crowd; lights blazing," he recalled. "It was a media event. Five senators were sitting up there, waiting. I walked up to the table and apologized profusely for being late. I had no idea that I was going to be the only one to testify. Once I was seated, a few of the senators made introductory comments." Behind Kerry sat his college friend and fellow VVAW organizer George Butler, stroking his newly grown beard. Butler was similarly dazzled by the scene. "I counted," he marveled. "There were seventeen cameras in the room. Even the BBC and Russian television were present."

Kerry made a strong impression even before he spoke. He was long-jawed and patrician-featured beneath his dark Beatle mop. His lean six-foot-four-inch frame was clad in neatly pressed military fatigues with rows of colorful ribbons festooning the shirtfront. When he began, it was in a low clear voice, calm and unhesitating. He often looked up from his notes and straight ahead at the five senators on the committee: Chairman Fulbright, his fellow Democrats Claiborne Pell of Rhode Island and Stuart Symington of Missouri, and Republicans Clifford Case of New Jersey and Jacob Javits of New York.* Kerry's testimony proved unflinching. From the outset he took control of the media spectacle, using the limelight to lambaste America's foreign policy leadership and to challenge Congress to end what he described as an immoral war. Kerry's testimony indicted not just Presidents Lyndon Johnson and Richard Nixon and their administrations, but the entire U.S. foreign policy establishment since the Gulf of Tonkin Resolution passed in 1964.

Sitting a dozen rows behind Kerry and quietly urging him on were some 100 more green-fatigued VVAW members whose shared frustrations and aspirations were finding their outlet in his voice. Although security at the Dirksen Building had been tight, Massachusetts Democratic Senator Edward M. Kennedy had managed to finagle passes to the hearing for the veterans. "I would like to say for the record, and also for the men behind me who are also wearing the uniform and their medals, that my sitting here is really symbolic," their decorated representative intoned. "I am not here as John Kerry. I am here as one member of the group of a thousand, which is a small representation of a very much larger group of veterans in this country, and were it possible for all of them to sit at this table they would be here and have the same kind of testimony."

Then the young naval reserve officer launched into an impassioned denunciation of the Vietnam War from a veteran's perspective. Acknowledging his disadvantage compared with senators when it came to formulating foreign policy, Kerry nevertheless suggested they had been getting the big picture wrong. He described what he had seen in Vietnam as a "civil war," a liberation movement aimed at shedding both the remnants of French colonialism and the current burden of American imperialism. "We found that most people didn't even know the difference between communism and democracy," he explained of the South Vietnamese, with the con-

---

**The other members of the Foreign Relations Committee were Democrats John Sparkman (Ala.), Mike Mansfield (Mont.), Frank Church (Idaho), Gale McGee (Wyo.), Edmund Muskie (Maine), and William Spong (Va.), and Republicans George Aiken (Vt.), Karl Mundt (S. Dak.), John Sherman Cooper (Ky.), Hugh Scott (Pa.), and James Pearson (Kans.).*

viction of one who had been in their country to ask them. "They only wanted to work in rice paddies without helicopters strafing them and bombs with napalm burning their villages and tearing their country apart."

While his quiet assessment was moving, in essence Kerry wasn't saying anything that Senators George McGovern, Eugene McCarthy, and others hadn't already told their colleagues. And since the mid-1960s even military leaders had publicly criticized U.S. foreign policy in Southeast Asia. General James Gavin, for example, had called the war "militarily preposterous." General David M. Shoup, former commandant of the Marine Corps, had denounced the war on moral grounds. At the state capitol grounds in Madison, Wisconsin, Brigadier General Robert L. Hughes of the Army Reserve had offered a haunting dissent at a 1967 Memorial Day event. Kerry was in some ways just another military man criticizing war policy. "Yet his voice added a new dimension to the criticism of the Nixon administration," McGovern reflected. "For in a clever oratorical move Kerry, after explaining how the South Vietnamese peasants were victims in the war, shifted gears claiming that the American GIs were also victims—that taxpayers' money was going to support corrupt local dictators throughout Southeast Asia."

What quickly became clear was that Kerry was accusing the U.S. government of war crimes, as ordained through such policies as free fire zones, harassment-and-interdiction fire, search-and-destroy missions, carpet bombings, and the torture and execution of prisoners. He scorned the rationale that one had to destroy a village in order to save it, and excoriated the effects of that policy on the hearts and minds of the Vietnamese people. As Kerry pointed out, they understandably saw U.S. soldiers not as liberators like the GIs of World War II, but as colonialist intruders even worse than the French before them. "We saw America lose her sense of morality as she accepted very coolly a My Lai, and refused to give up the image of American soldiers who hand out chocolate bars and chewing gum," Kerry lamented to his elders. "We learned the meaning of free fire zones, shooting anything that moves, and we watched while America placed cheapness on the lives of Orientals."

Thus the Navy man raised the grim specter of atrocities committed by Americans in Vietnam. He referred to the "winter soldier investigation" that VVAW had held earlier that year in Detroit, at which more than a hundred veterans had described heinous acts they had committed in Southeast Asia.

"They gave me a Bronze Star . . . and they put me up for the Silver Star," one veteran there recalled. "But I said you can shove it up your ass. . . . I threw all the others away. The only thing I kept was the Purple Heart because I still think I was wounded." Other veterans, in gory detail, had told gruesome stories of sadistically torturing and raping Vietnamese women. Burning villages and machine-gunning peasants who were thought to be Viet Cong had become a part of U.S. policy.

Kerry was placing the blame on the U.S. government for instituting such immoral policies. Offering his own experiences as a Swift boat skipper as an example, Kerry detailed how he had been instructed to shoot anything that moved on the Mekong Delta rivers during the U.S.-set curfew hours. He noted that this had led to some unfortunate incidents. Now, he added, veterans across the nation were coming forward to confess to war crimes. And, due in part to the lobbying efforts of VVAW, the next day Senators McGovern and Hart were going to hold hearings investigating atrocities committed by U.S. soldiers in Vietnam. Then Kerry graphically laid out why such an investigation was called for. That winter in Detroit, decorated veterans had "told stories at times that they had personally raped, cut off ears, cut off heads, taped wires from portable telephones to human genitals and turned up the power, cut off limbs, blown up bodies, randomly shot at civilians, razed villages in a fashion reminiscent of Genghis Khan, shot cattle and dogs for fun, poisoned food stocks, and generally ravaged the countryside of South Vietnam, in addition to the normal ravage of war." After this breathless recounting of U.S. violations of the Geneva Conventions, Kerry remarked that the foregoing indictment didn't even include the tens of thousands of Vietnamese deaths caused by the Johnson and Nixon administrations' bombing campaigns.

Now that he had the senators' rapt attention on what was so desperately wrong with the war, Kerry went after the denigrators of those who opposed it. He took direct aim at Vice President Spiro Agnew, who had delivered a gloves-off speech at West Point the year before in which he declared that "some glamorize the criminal misfits of society while our best men die in Asian rice paddies to preserve the freedom which most of those misfits abuse." Kerry glared right at the five senators and charged Agnew with distorting the antiwar view and those who held it. After all, he was no drugged-out counterculture wandering hippie guru, and neither were the other antiwar veterans sitting behind him. They were not the "summer soldiers and sunshine patriots" Thomas Paine had sneered at in December 1776, but true winter soldiers, American patriots who had put their lives on the line in the Cold War battle against Soviet and Chinese Communist expansionism. They were the fallen tiles of the domino theory, many now paraplegics and quadriplegics and amputees left to rot away ignored in poorly run and underfunded Veterans Administration hospitals. As these brave veterans tried to readjust to an unwelcoming nation, bitterness and guilt consumed many of them. They had seen with their own eyes that no "mystical war" against communism was necessary. "In our opinion, and from our experience, there is nothing in South Vietnam, nothing which could happen that realistically threatens the United States of America," Kerry stated matter-of-factly. "And to attempt to justify the loss of one American life in Vietnam, Cambodia, or Laos by linking such loss to the preservation of freedom, which those misfits supposedly abuse, is to us the height of criminal hypocrisy, and it is that kind of hypocrisy which we feel has torn this country apart."

That mention of "hypocrisy" hung over the Senate Foreign Relations gallery like a shroud. Kerry's utterance of the highly charged word—twice in one sentence—got everybody's attention. "Kerry gave me the chills," Symington remembered. "Right out the gate it became clear that this was going to be one of those moments frozen in time. Outside the Capitol thousands of veterans were chanting for peace while Kerry was accusing Nixon's administration, Agnew in particular, of hypocrisy. A worm was turning."

But, after charging the government with immorality, hypocrisy, and war crimes, Kerry was still not finished. Indeed, he was there, under the bright lights with all the cameras pointed at him, to advance the cause of veterans' rights. Recounting the heroism American troops had displayed at Hamburger Hill and Khe Sanh, at Hill 8815 and Fire Base 65, he choked on the very concept of Vietnamization. "Now we are told that the men who fought there must watch quietly while American lives are lost so that we can exercise the incredible arrogance of Vietnamizing the Vietnamese," Kerry declared, his voice quivering with emotion. "Each day—" he started, but he couldn't finish the thought as the gallery erupted into thunderous applause, most of it from his fellow veterans. (Kerry would be interrupted a dozen times over the course of the hours he testified.) Trying to maintain order,

Chairman Fulbright grabbed his microphone and addressed the cheering veterans: "I hope you won't interrupt. He is making a very significant statement. Let him proceed."

Without hesitation Kerry continued to take direct aim at Nixon, accusing the President of letting the war in Southeast Asia drag on just because he refused to go down in history as the first U.S. chief executive to lose a war. Then, in his most famous utterance, the decorated veteran posed a pair of tough questions to the senators. "How do you ask a man to be the last man to die in Vietnam?" Kerry inquired. "How do you ask a man to be the last man to die for a mistake?" An eerie silence fell over the chamber. Then muffled murmurs could be heard. It was a poignant question, and had no quick and easy answer. The reporters covering the hearing knew they had their lead. Kerry's second question would be the next day's headline; they made sure of it.

Maintaining a dignified calm throughout his testimony, even when his voice betrayed some lingering bitterness, Kerry personalized what it was like to join the U.S. military and be sent to Vietnam. He imagined an all-American teenager who one day sees a poster with Uncle Sam pointing at him under the entreaty "I Want YOU." Inspired by what he has been taught the star-spangled figure represents, the young man signs up to serve his country. He endures basic training and then finds himself in Vietnam. Like any good soldier, he follows orders. He kills as many enemy "gooks" as he can, loses a limb, and wins a medal. When he returns home there is no ticker-tape parade to greet him, only contempt from the doves for having killed civilians and resentment from the hawks for having failed to win the war. According to Kerry, employment opportunities for Vietnam veterans barely existed, particularly for those who came back physically or psychologically impaired. He offered the statistic that one out of every ten unemployed Americans was a Vietnam veteran, with the percentage much higher among African Americans. "The hospitals across the country won't or can't meet their demands," Kerry informed the senators of those too disabled even to seek work. "It is not a question of not trying; they haven't got the appropriations. A man recently died after he had a tracheotomy in California, not because of the operation but because there weren't enough personnel to clean the mucus out of the tube, and he suffocated to death."

He went on to tell of other real veterans suffering from their experiences in Vietnam. The My Lai incident—still a point of national contention—was an inferred backdrop to this round of remarks. Perhaps most effective was his anecdote about a Native American in California. "He told me how as a boy on an Indian reservation he had watched television and he used to cheer the cowboys when they came in and shot the Indians, and then suddenly one day he stopped in Vietnam and he said: 'My God, I am doing to these people the very same thing that was done to my people,' and he stopped. And that is what we are trying to say: that we think this thing has to end."

Had his testimony ended there, with the applause that shook the chamber, John Kerry would have been noticed in the next day's newspapers for having delivered an eloquent account of the returning winter soldier's perspective on the Vietnam War. He went on to deliver his stinging indictment of the entire U.S. foreign policy establishment. Specifically, he accused the Johnson years' Robert McNamara, Walt Rostow, McGeorge Bundy, and Roswell Gilpatric—by name—of having "deserted their troops," adding "that there is no more serious crime in the law of war." Pointing out the U.S. infantry's righteous boast that they never left their wounded comrades behind, Kerry excoriated the previous administration's so-called wise men for having done just that. "These men have left all the casualties and retreated behind a pious shield of public rectitude," he proclaimed. "They have left the real stuff of their reputations bleaching behind them in the sun in this country."

After spelling out such a bleak assessment of everything involving the United States and Vietnam policy, Kerry ended his testimony on a hopeful note. "Thirty years from now, when a man is walking down the street without an arm, or a face, or a leg, and a little boy asks him why, he will have to say 'Vietnam,' " Kerry speculated. "And mean not a desert—not an obscene memory—but mean instead a place where America finally turned and soldiers like us helped in the turning."

Upon that, the veterans in the back of the gallery sprang to their feet and gave their spokesman a standing ovation. The senators just looked at one another and smiled. They knew a political star had been born. They let the applause continue to rumble on, according the impassioned young officer his moment in the limelight. Finally, Senator Symington, who had served as the first secretary of the Air Force, asked the witness a question: "You have a Silver Star?"

"Yes, sir," replied Kerry, who was wearing the Navy's third-highest award for combat at the top of his three rows of campaign ribbons.

"You have a Purple Heart with two clusters?" Symington continued.

"Yes, sir," came the answer.

"I have no further questions," concluded the Missouri senator, having established for the record the caliber of the witness.

"Credentials are something we always think about," Senator Javits chimed in. "Your credentials couldn't be higher."

What seemed to impress the Foreign Relations Committee members most during the ensuing question-and-answer session was how the twenty-seven-year-old before them articulated the veterans' pragmatic rationale for getting out of Vietnam in such effectively measured paragraphs. His eloquence was simply expected by those who knew him. The transcript of the committee's questioning of him shows just how well prepared the careful student of Vietnam had come to Capitol Hill. When Chairman Fulbright asked him how Congress should proceed to extricate the United States from Southeast Asia, Kerry had a ready answer. "If we can talk about filibusters for pork-barrel projects, we should talk about filibusters to save lives," he rejoindered. "It's an extraordinary enough question, so it requires an extraordinary response." When Senator Case, a courageous antiwar Republican, quizzed Kerry on why the White House's "peace with honor" approach wasn't the best policy, the VVAW spokesman answered: "As a man who fought in the war, I know this policy has no chance of bringing peace if it arms people of another country and tells them to go on fighting. It would be criminal if the fighting continued and if large numbers of South Vietnamese tried to stand up for something they can't. [It] would place all of their lives on our conscience, along with all the others."

As the hearing wound down, Chairman Fulbright ended the questioning with praise for the witness. Impressed with Kerry's manners and intellect, the Arkansas Democrat lauded the young activist for trying to effect changes in policy by working within the system. Dubbing him a leader of the Vietnam generation, Fulbright implored him not to lose faith in the capacity of Congress to respond to what he had said that day. "I don't think I'd be here if I didn't believe," Kerry quipped. "I won't quit, Senator, but unless the country can respond on the war, how can it respond to poverty and all the other problems? I'll keep trying, and I see no other broad system than democracy, but democracy must remain responsive or there will be pressure for other systems, and that is beginning to happen in this country." Hearing that answer, the renowned pacifist magazine writer I. F. Stone leaped to his feet in the gallery to set off yet another standing ovation for John Kerry.

Kerry left Capitol Hill that day a changed man. In the space of two hours he had become a bona fide celebrity, the household name and face of the Vietnam veterans' antiwar protest. All three nightly TV newscasts aired long clips of his testimony that evening. Walter Cronkite of CBS called Kerry a "force" in the antiwar movement. The print media followed suit, with his photograph published in leading periodicals such as *Time* and *Newsweek*. The *New York Times* ran a profile of him headlined "Angry War Veteran." His new friend Tom Oliphant saluted Kerry in the *Boston Globe* for displaying "a deeply disturbed conscience." Oliphant also persuaded his editor, Thomas Winship, to print the entire transcript of Kerry's testimony in the fiercely antiwar *Globe*; from there, his statements were picked up by dozens of other newspapers nationwide. Reporter Peter Lisagor of the *Chicago Daily News* extolled Kerry's presentation as "so eloquent and moving that a couple of grizzled senators bit their lips. . . . Cool print cannot convey the cool anguish of this tall young man with a handsome face." Robert Jay Lifton, who had been working on Vietnam veterans' issues for the preceding few years, including his landmark insights into what constituted an "atrocity-producing situation," deemed the speech "an exquisite moment." Hundreds of other antiwar intellectuals felt the same way.

Even President Nixon, who had tried to shut down the VVAW march, couldn't help but be impressed by this bright young troublemaker with the impeccable credentials. In an Oval Office meeting the next day, the President noted Kerry's distinctiveness from the other "bearded weirdos." He had been the "real star" of the hearing, Nixon told his chief of staff, H. R. Haldeman, and his national security advisor, Henry Kissinger.

"He did a hell of a great job," Haldeman said.

"He was extremely effective," Nixon agreed.

"He did a superb job on it at the Foreign Relations Committee yesterday," Haldeman repeated. "A Kennedy-type guy; he looks like a Kennedy, and he talks exactly like a Kennedy."

Decades later, in 2003, Leslie Gelb, a former National Security Council official during the Nixon administration, recalled the shock wave that Dewey Canyon III—and John Kerry, in particular—triggered throughout Washington, D.C. A senior fellow of the Brookings Institution at the time of the march, Gelb, who helped compile the Pentagon Papers, remarked how deeply insular and paranoid the White House team of Haldeman, John Ehrlichman, Charles Colson, and Egil "Bud" Krogh were. "It was always the Democrats, liberals, press, and Jews that were trying to get them," Gelb said. "But now suddenly veterans like John Kerry were coming home, and they didn't necessarily fit those categories. They were time bombs coming back from Vietnam saying, 'What the hell are you guys talking about?' These Vietnam veterans weren't containment-doctrine people, they didn't buy it. The questions Kerry and the others posed were so simple, so poignant, so factual, that it was revolutionary. I wasn't in the bunker with Nixon and those guys but Kerry's testimony was seen as a direct threat. There he was, in fatigues, a war hero trying to undermine their foreign policy. Although Kerry went on to have a career in the U.S. Senate, I'll always remember him most for that singular moment."

Equally unbeknownst to Kerry, a more propitious good opinion of him was being formed in the mind of a young Georgia state senator. Former U.S. Army Captain Max Cleland had lost both his legs and his right arm on April 8, 1968, during the siege of Khe Sanh, when a grenade exploded next to him as he stepped off a helicopter. Twenty-five years old when it happened, Cleland spent eight months recuperating at the Walter Reed Army Medical Center in suburban Washington, D.C. When he was finally released from the hospital, an old girlfriend pushed him in his new wheelchair around the nation's capital. Outside the White House, the chair hit a curb and Cleland tumbled forward out of it. "I remember trying to lift myself out of the dirt," he reflected. "There were cigarette butts and trash in the street right alongside of me."

That awful experience, combined with his deep-seated anger about the

Vietnam War, and the disgraceful lack of proper health care for its veterans, set Cleland to seeking elective office. "I figured this [was] a good time to run for the state senate," he recalled. "And politics became my therapy, forcing me to get out of the house and be seen."

Once in the Georgia legislature, Cleland began promoting an immediate peace plan based on a U.S. withdrawal from Vietnam in exchange for the release of any and all American prisoners of war. "I knew that it was the time to end the war," Cleland explained. "But I felt isolated in Georgia. Then all of a sudden a fellow Silver Star winner named John Kerry was on TV saying what I was saying—only better. It took a lot of guts to testify like he did in front of the Fulbright Committee. At that time [the] Veterans of Foreign Wars was threatening me, claiming that I was finished in politics because I was opposed to Nixon's plan. Kerry gave me hope; he renewed my faith. He made us proud to be veterans just like the guys from World War II felt. . . . This guy was speaking for me deeper than I even knew I felt at the time."

Among the myriad gaping differences between veterans of World War II and the Vietnam War was the length of their service. Soldiers and sailors had fought the Good War to its end in Europe and the Pacific; their duty ended when the war did. Vietnam was different; enlisted men were supposed to serve a 365-day tour of duty. But as it turned out, for Vietnam veterans like Cleland, Muller, Lavoie, Kerry, and thousands of others, the tours would last the rest of their lives: after they completed their service in Khe Sanh, the DMZ, the Mekong Delta, or Danang, they had to come home and serve their consciences for having participated in what they saw as an immoral endeavor. Unlike World War II, the Vietnam conflict was filled with ambiguity. As CBS News correspondent Edward R. Murrow is said to have remarked of Vietnam, anyone who wasn't confused didn't really understand the situation there. The war's veterans came back from Southeast Asia having to face new battles—political and social ones they were even less equipped to take on than the firefights in-country. In addition to their noticeable wounds, the Vietnam vets had to confront post-traumatic stress disorder, inadequate VA medical care, forgotten POWs, and the U.S. government's illegal incursions into Cambodia and Laos. It had now become the duty of the war's winter soldiers to put a stop to the horrific errors of the Johnson and Nixon administrations.

Decades later, Cleland, who would go on to be elected a U.S. senator, still vividly remembered April 22, 1971. "It was a day that changed my life," he explained. "Kerry's words still bolt right through what's left of my body. . . . When he said, 'How do you ask a man to be the last man to die for a mistake?'—that was gutsy. Kerry understood that like me, someone seriously wounded, our tour of duty—our generational duty—had just begun."

Chapter One

Up from Denver

The sun was glaring through the windshield of Richard J. Kerry's single-engine light aircraft as he prepared for takeoff from a runway in northern Virginia on February 27, 1954. Mild, with temperatures in the midfifties, no clouds in sight, it was a perfect day to fly. During World War II Kerry had served the United States government as a pilot in the Army Air Corps, flying DC-3s and B-29s. Now he was based in Washington, D.C., serving as an attorney for the State Department's Bureau of United Nations Affairs. This was, however, to be his final flight. With his eleven-year-old son John sitting in the rear seat, Kerry, now a civilian, started the engine and checked his navigational charts. Everything was in working order. "Don't touch the stick," he cautioned his son before takeoff. "Not until you're older."*

Anybody who knew the austere and hardworking Kerry well thought of him as a man with an intense, careful disposition, a pilot whose logbook was as tidy as an accountant's ledger. This particular book, beige in color and three-quarters full, had been kept since 1940. During World War II he had crisscrossed America numerous times, including long stints in Alabama, Ohio, California, and Colorado. Today was no different from any other flight day: he carefully scrawled "Alexandria Local Aeronca" in his book. He was hoping to give his son an aerial view of metropolitan Washington sites. Usually Kerry never editorialized in his log: just the no-nonsense facts. But on this last flight he made an exception, writing something personal: "Flight

---

**In 1974 John Kerry was able to fly his father to the Harvard-Yale game in a Piper Aztec. His dad recalled and marveled at the new instrument panels on this flight.*

over Mt. Vernon with Johnny." The flight lasted for only a brief forty minutes. But forty years later he sent the logbook and wings to his son with a note on his law firm stationery: "Is this last entry prophetic?" Richard Kerry was probably referring to his son's passion for flying, but the flight over Mt. Vernon may inadvertently touched a different prophecy.

Even when he was an eleven-year-old boy, there was a feeling that John Forbes Kerry was touched with destiny—or, more accurately, that public service was instilled in him by his parents. There was, however, a touch of the parvenu in all of this, a fierce family belief, not unlike that which Joseph Kennedy imposed on his four sons, that the Kerry boys—John and Cameron—could accomplish any feat, no matter how difficult. But to do so would take discipline. A touch of old-fashioned chauvinism, however, prohibited Richard Kerry from fully instilling the same attitude in his two daughters, Margaret (Peggy) and Diana. What was important was that his two sons were not slouches. Concepts like diligence, duty, and loyalty were instilled in them, with tenderness usually coming last. Like the fathers in so many second-generation immigrant families, Richard Kerry believed his boys could accomplish anything in America, even following in the oversized footsteps of George Washington, making it all the way to the White House. "Excelling was the Kerry family ethic" is the way *Washington Post* reporter Laura Blumenfeld explained it. She gave an example as a case in point: Richard Kerry taught his sons how to steer a boat under a blanket, so they would learn to navigate in the fog. "He definitely promoted tough love," Peggy recalled. "He wanted us to be equipped with the harsh realities of the real world."

The story of Richard Kerry's rise is one of overcoming obstacles. Born in 1915 in Brookline, Massachusetts—the same Boston suburb where John F. Kennedy was born two years later—Richard Kerry was a handsome, erudite boy, always fighting against the odds. His father, Fredrick A. Kerry, was actually a Czech Jew named Fritz Kohn who had fled the aggressive Austro-Hungarian Empire in 1905, brutalized by anti-Semitism. Three years before his arrival in America he married Ida Lowe, a beautiful Jewish musician from Budapest. According to the *Boston Globe*, the young couple simply studied a map of Europe, found County Kerry in Ireland, and chose it as their last name. Baptized as Catholics, they moved to Chicago with their young son Eric, where Fredrick (or Fred as he was called) earned a living as a business manager. Eventually they moved to Brookline, known as the "town of millionaires" in the early 1900s, had two additional children, Richard and Mildred, and earned a reputation as good neighbors. The local newspaper deemed Fredrick "a prominent man in the shoe business"; his shop was located at 487 Boylston Street in the Back Bay neighborhood of Boston. He seldom missed attending Catholic church services on Sunday. (He kept it secret that he was of Jewish descent.)* With a two-story, Arts and Crafts–style house in Brookline—designed by John C. Spofford—located at 10 Downing Road, a black Cadillac parked in front and three healthy children running happily about, it seemed, to the outside world, that the Kerry family exemplified the American dream.

That notion was brutally dispelled on November 23, 1921, when a depressed Fred Kerry, wandered into the Copley Plaza Hotel in Boston, walked into the men's room, and shot himself in the head. The *Boston Globe* published a short story about the suicide, which took place at 11:30 A.M., claiming he had died instantly. "Kerry had been ill for some time, and he became despondent as a result," the obituary read. "He left his home about the usual hour this morning, and his spirits seemed to be low. After going to his place of business he came out and went to the hotel where he took his life."

It's hard to fully understand how such a grisly death affects a six-year-old boy, but Richard seemed to internalize the suicide. Thinking of it as a badge of shame, he coped with the loss of his father by ignoring it. Years later, Richard, just as he was to begin his foreign service career, suffered another tragic loss when his sister Mildred, a polio victim, died of lung cancer as a young woman. Stoically, in the great tradition of New England, he pressed on. He spent 1930 to 1933 at Andover, a first-rate Massachusetts boarding school. His Andover grades were good enough to get him into Yale University. He graduated with a B.A. in 1937. He grew to be six feet three, an able athlete in every sport he tried. At times genial, but often private, Richard especially enjoyed sailing and mountain climbing.

**Like many assimilated Jewish immigrants, Fred Kerry hid his religious background and never mentioned his European past. Senator John Kerry, his grandson, in fact, never knew he was 50 percent Jewish until 2003, when* Boston Globe *investigative reporter Michael Kranish informed him of his roots.*

Besides the outdoors Richard Kerry was enamored of the legal teachings of Oliver Wendell Holmes and Louis Brandeis. He was fascinated by such arcane legal issues as U.S.-Canadian fishing rights disputes and who owned the petroleum of Antarctica. He went off to Harvard Law School, where he graduated with a J.D. degree in 1940. More cerebral than ambitious, Kerry longed to better understand the world. Infatuated by William James and Søren Kierkegaard, midway through law school he decided to spend the summer in Europe in a search for history's footprints. It was 1938, and German Führer Adolf Hitler was gearing up for war. After seeing the sites of London and Paris, Kerry, in footloose mode, drifted to the French market town of St. Briac in the Brittany region. Here—on a dramatic promontory between the ravines of the Gouët River and its tributary the Gouëdic, not far from the English Channel—he enrolled in sculpture classes. Since Richard was eight or nine, he had been interested in making carefully crafted wooden ships. He hoped to perfect the skill in St. Briac. It was there, while whittling, that he met his future wife—Rosemary Forbes.

The Forbes family had deep New England roots dating back to the colonial era. Shipping was an important family enterprise and documents exist pertaining to their entrepreneurial ventures in such far-flung places as Rio de Janeiro and Canton, China. One of John Kerry's distant cousins, in fact, Robert Bennett Forbes, was a Massachusetts leader in maritime safety, his lobbying leading to the first iron-hulled tugboats. When the potato famine of 1847 struck Ireland, it was a Forbes who petitioned Congress to send a relief ship full of food across the Atlantic as a humanitarian gesture. Based on Cape Cod, the Forbes family represented the so-called Brahmin class of old New England.

But one of the clan—James Grant Forbes, a Harvard graduate—had bolted Massachusetts, making his home in England. In 1928, he also purchased a summer home in the Breton countryside just outside St. Briac. His wife (Rosemary's mother) was Margaret Winthrop, a direct descendant of John Winthrop, the first governor of Massachusetts. His Breton estate was known as Les Essarts. His daughter Rosemary, one of eleven children, was studying to become a nurse. Kind, considerate, and a natural nurturer, Rosemary never exuded more personal ambition than caretaking and gardening; "small and self-effacing, but dogged," is the way her son Cameron Kerry described her. Whether she fell instantly in love with Richard is unclear. But they did start a regular correspondence, talking, in part, about spending their lives together—in a few years because Kerry had to finish up at Harvard Law School. In the meantime he did manage to spend the summer of 1939 back in France, at Rosemary's side.

By the time Richard Kerry graduated in June 1940, he could not go to France to live as planned. With war looming, he enlisted in the Army Air Corps. He hoped Rosemary would join him in America. But she was working in Paris, taking care of refugees. The Nazis had invaded France on May 10, and Paris was in chaos. "She escaped on bicycle and foot with her sister and friend," John Kerry recalled of his mother's flight from Paris as the Germans came in, accompanied by her sister Eileen and Eileen's new husband, Francis Tailleux, a painter. "It was literally the last day before the Nazi occupation. So she dodged Germans and lived by foraging. Struggling to stay alive, they slept in dilapidated barns and in underground wine cellars." Rosemary wrote Richard a hurried letter about her family's last desperate moments before the Nazis arrived in France. "Dick Dearest," she wrote on July 14. "We left Thursday June 13 at 8:30 P.M. just after the gas and electricity had been shut off and explosions were going off where they were blowing up gasoline tanks. . . . At dawn the Germans entered Paris. Next day, we pushed on towards Orleans, missed being bombed . . . by taking a longer route though we saw the planes going on the mission of death and had to duck their machine guns. . . . I am so scared of coming to America but with you I know everything will be all right."

Eventually the group made it all the way to Po in southern France and then down the coast of Portugal. The resourceful Rosemary boarded a ship in Lisbon and headed to Boston, where Richard Kerry planned to meet her. She was going to stay with her sister Angela in Hamilton, Massachusetts. (Two of her brothers—James and Jack—had already moved to America as well.) While she was fleeing Europe, Richard was stationed in Montgomery, Alabama, training to be a test pilot. It was not until she reached Massachusetts soil that she learned the Nazis had also swarmed Les Essarts, capturing the family home to use as an Atlantic fortress for the duration of the war. As for James Grant Forbes, he fled to London with his wife, remaining engaged in international business.

Once in America, Rosemary accepted Richard Kerry's proposal of marriage. The Army Air Corps had sent Richard to Maxwell Air Force Base, which since the early 1930s had been the home base of the Army Air Corps Tactical School, to learn how to test-fly DC-3 paratroop planes. "He was a cadet in Montgomery and had quickly married my mother, in February 1941, in a civil ceremony," John's brother, Cameron Kerry, recalled. "A few weeks later, after his training, they got married again in a church."

From Montgomery, Kerry was assigned to Wright and Patterson Fields near Dayton, Ohio. His main job was to test B-29s for increased propulsion power and efficiency. It was in Dayton that Richard and Rosemary's first child, Margaret, was born on November 11, 1941.

Like the hot-shot pilots Tom Wolfe celebrated in *The Right Stuff*, Kerry's specialty was testing new aircraft at high altitudes. The Army Air Corps, in fact, soon sent him to Camp Cooke, a base built on 86,000 acres on the central coast of California north of Santa Barbara, to do just that. Camp Cooke trained armored and infantry troops for war in Europe and the Pacific at a turnstile pace. Experiments with new missile systems were also conducted there. All things considered, it wasn't a bad assignment. Living in Pismo Beach, the Kerrys often took joyrides down Highway 101, followed by late-night swims on Shell Beach. Then, just as they were settling into paradise, Kerry contracted tuberculosis (TB) and the Army Air Corps shipped him to Denver, to recuperate in the dry Rocky Mountain air.

Spread by infected airborne droplets, the dreaded lung disease TB was—and is—highly contagious. Doctors would often send patients to Denver hoping these "lungers" would find comfort and cures. The decline in Kerry's health was dramatic. One week he was in Camp Cooke flying planes, the next week he was coughing up blood, breaking into night sweats, struggling just to breathe, and constantly harboring a low-grade fever. During World War I soldiers who had contracted tuberculosis in European trenches were often sent to Colorado for recovery; many died there in forlorn hospital beds. There were tense moments when it looked as if Richard Kerry was going to join their ranks. Recognizing the need for better TB treatment facilities, the U.S. Army, however, had built Fitzsimmons Hospital in 1941 to treat chronic lung patients. The doctors there—who prescribed fresh milk and antibiotics—were among the best in the world. It was under such difficult family circumstances, her husband seriously ill, that Rosemary gave birth to their second child, John, on December 11, 1943. "My father got tuberculosis, so he was sent to Denver to Fitzsimmons General Hospital," Kerry recalled about his circumstantial birth in Colorado. "I guess they sent him there to dry out his lungs, and I appeared, in a rush as my mother explained to me. I was there for a very short period of time."

Recovering faster than anticipated, Richard Kerry was released from Fitzsimmons, which he called "Magic Mountain" in reference to the Thomas Mann novel in which the protagonist has TB. He was released from active duty in the Army Air Corps in March 1944 and relocated to Groton, Massachusetts. The following year, he was admitted to the Massachusetts bar, joining the firm of Palmer, Dodge, Chase & Davis. The Kerrys lived not far from where the Nashua and Squannacock Rivers joined.

Although John Kerry spent the first year of his life in Groton, his childhood was one of constant relocating. "Kerry is a man without geographic roots" is the way *Boston Globe* journalist Michael Kranish explained his upbringing. "His youth stretched through a dozen schools across two continents." He was like a character novelist Armistead Maupin described in his 2000 *The Night Listener*, "I had always been on the move, a serial renter leaping from hilltop to hilltop." The one place that Kerry didn't feel like a transient, however, was Millis. Late in 1944, the family purchased a home in that Charles River Valley town fifteen miles southwest of Boston. It had a long driveway lined with rhododendrons. They lived there for almost five years. "We had a great house on a farm in Millis and my dad was then in a law firm, commuting back and forth to Boston," Kerry recalled of his youth. "My memories are of dressing up in a snowsuit in the middle of summer and playing Eskimo with my sister. Also, swimming in a pond. There was a pond on the property and I remember a slithering snake and gaining an early dislike for them. It was a terrific life. We were on a farm so I used to sit in the seat of a John Deere tractor. I remember one day my dad came in and said I had left my tools in the garden and I had to go down and clean them up. I went down to the garden and there was this gleaming red toy tractor, waiting for me in the garden. I also remember accidentally putting my arm through a window, cutting myself pretty badly, getting rushed to see the town doctor."

His years in Millis, more than any other place, gave Kerry a sense of home. And although in so many ways he looked up to his father, it was his mother whom he grew closest to. Listening to Kerry tell stories from those years is like viewing a gallery of Norman Rockwell paintings: dressing up as a pumpkin for Halloween, chopping down a pine tree for Christmas, sledding on New Year's Day, and shooting off fireworks for the Fourth of July. His sister Peggy recalled that one afternoon John played at the farm adjacent to their house. He came back having learned a new curse word. "My father was furious," she said. "He grabbed a bar of soap and washed his mouth out." While the old Millis lace looms and brickyards were no longer in operation, many old-time industrial enterprises were still going strong—for example, Herman Shoes, Safe Pack Mills, and the Cliqout Club ginger ale plant. Walking around Main Street was a real Sinclair Lewis experience, with the Rotary Club and Lion's Club serving as town pillars of local wisdom. "It was idyllic," Kerry later reflected. "It gave me a sense of peace, of belonging to the land."

On April 16, 1947, while living in Millis, the Kerrys had their third child, Diana. With two sisters, and thirty-one first cousins on his mother's side alone, young John was often just one of a large pack of kids. His father, always working, seldom spent quality time with him. "I'm a different father from my father," Kerry offered in a 2003 interview. "Largely because of the things I learned that I had missed or didn't get or would have liked to have had from my father. No doubt about it. Not because he didn't want to, but because he didn't know how, it wasn't part of his experience. He had a very different upbringing because he never knew his father. Never had a father, and his mother was clearly doting and so forth. My father always felt the loss of his sister when she died so horribly young. I think those were big deficits in his life."

The idyllic Millis stint ended in 1949, when Richard Kerry moved his family to Washington, D.C. He had gotten a job with the Office of General Counsel for the Navy. At first the Kerrys lived in a Georgetown town house just blocks from the Potomac River. Since it was not big enough for a family of five, Richard then bought a small colonial house in the suburb of Chevy Chase.

From 1951 to 1954, Kerry worked as an attorney for the Bureau of United Nations Affairs in the State Department. Eager to spend time in Europe, as was his wife, in late 1954 he accepted the post as legal advisor at the U.S. Mission to Berlin and U.S. attorney for Berlin. In West Berlin he helped implement the containment policy of Secretary of State John Foster Dulles. He became well known as a multilingual friend of NATO, meeting with diplomats in London, Bonn, Paris, and The Hague on a regular basis.

He spent 1958 at the NATO Defense College in Paris. When Walter F. George, former chairman of the Senate Foreign Relations Committee, was named President Eisenhower's special ambassador to NATO, Kerry signed on as George's special assistant, drafting memos and policy statements to promote transatlantic relations. By 1958 Kerry was an experienced American diplomat in Europe. Although he never stayed in the Foreign Service long enough to be appointed an ambassador, a job he coveted, he succeeded in becoming chief of the political section of the American embassy in Oslo—a post he held for four years.

It's hard not to be moved by young John Kerry's stories about arriving in Berlin at age eleven. The tension in the fabled German city in 1954 was palpable. "Buildings all over Berlin were bombed out, some practically still smoking," Kerry recalled. "World War II had only ended nine years ago. And now the city was divided into sections." Refugees were everywhere. In fact, between 1950 and 1961 approximately 140,000 Germans escaped the Soviet occupation zone annually to live in the West. Kerry was fascinated riding the night train from Berlin to Frankfurt, armed with a six-pack of Coca-Cola and Marvel comic books, peering out the blinds at the Russian army occupying East Germany. "At times I was shocked by so much dislocation," Kerry recalled. "I looked for escapes from the dreariness. My fondest memories of Berlin were riding my bicycle through the forests on the edge of town. It was a storybooklike experience."

Unlike grim Berlin, Oslo was a vast playground for John Kerry. Always wanting to explore, Kerry made the nearby fjords his Mississippi River, a place where Huck Finn fantasies could be played out. He would camp in Vigeland and sail down the vast fjords that stretched out of Oslo. "I learned a lot on the water in both Cape Cod and Norway," Kerry recalled. "In part that's why I went into the Navy. The Navy's sort of in my blood, partly because I had been raised summers on the ocean, in the water. Every aspect of it: digging clams, getting dragged behind the boat, sailing, swimming—doing all those types of things." And young John, with his father's reluctant approval, was given permission at age sixteen to take a bicycle trip by himself in England. "My dad wasn't so sure it was a great idea, but I went and did it. I hopped on the ferry and went across the North Sea and then hopped on my bike. My goal was to sleep under the trees in Sherwood Forest like Robin Hood, and meet people wherever I went."

But the one place in Europe that took his breath away was the American cemetery in Normandy. While the family stayed at his grandparents' home in Brittany, John would accompany his father there and they would wander among the white crosses discussing what had happened on June 6, 1944, D-Day. "I think it's the most tranquil, beautiful, moving, emotional place I've ever been to," Kerry explained. "The assault beaches—Utah, Omaha, Gold, Juno, and Sword—are stunning and breathtaking. When I was a boy, my father explained how in eighteen difficult hours the walls of Adolf Hitler's Fortress Europe had been breached."

Supplemented by newsreels he watched, the young Kerry imagined what it must have been like for members of the 101st Airborne Division to scale the cliffs while U.S. Navy destroyers sat in the Channel blasting away, point-blank, at the German fortifications. "The scope of D-Day was just monumental," Kerry reflected. "It was a confrontation of Germans versus Americans, tyranny versus freedom, evil versus good. I used to try and imagine what it was like for a young GI in a Higgins boat, wading to shore, knowing you were going to be met with a hail of bullets. My dad took me to see these rusty hulks and burned-out tanks that were scattered all about. Everywhere you went you saw reminders. You saw bullet holes in the sides of buildings and you saw bunkers. In St. Briac we even discovered a mine buried in the driveway to our house. There were German bunkers less than a mile from our house, so I grew up with a sense of World War II. I remember having had a constant sense of wonder, thinking Wow, this is a piece of history and it happened right here. It was very moving."

Although the Kerrys lived in St. Briac, Berlin, and Oslo, young John's real base for two years was Switzerland—at boarding schools. "The first school I went to I was plunked down," the long-jawed Kerry recalled. "I remember arriving in Europe and my mom and dad just dropped me off. It was in the fall of 1954, we left in the late summer. Suddenly we just left Washington, D.C., went over to Europe and boom, I was in a Swiss boarding school. It was a place called Zug, right near Zurich. Our school was up on a hill, right up on a mountain. I just didn't know where the hell I was and it was strange as hell. I'll never forget the empty, sinking feeling when we said good-bye. I've always been bad at good-byes, all my life, probably ever since because I just learned to hate them. That feeling of going back to school, ending vacation, whatever. But I remember my parents getting into that car and driving off, and boy, I tell you I think I cried for about three weeks. I was one homesick puppy." His sister Peggy was dropped off at another boarding school.

Kerry harbored mixed emotions about boarding school in Switzerland. His second year, he contracted scarlet fever and was sad that his parents didn't journey down from Berlin to help lift his spirits. "They just weren't able to get there then," Kerry recalled. "There were a group of us who had contracted it and it was contagious. I think about eight of us were put in a dorm for quarantine and so that was as sick as I have ever been."

But yet for the rest of his life whenever he grew depressed, he thought of the Swiss Alps for inspiration. Kerry wrote his close college friend David Thorne years later while serving in the Navy.

Unless you have been in the Swiss hills during springtime it is almost impossible to understand the spirit that captures the mind and body. It excites me even to think about it now. Each day, imprisoned in a classroom, frozen before a blackboard and unchanging numerals designed to confound the roaming mind of an imaginative schoolboy, the fields and the flowers which grew wildly in them, would pour into our small classroom drawing all semblance of learning far from the mathematical equations that were drearily standing like a guard in front of me. Outside the castle and the princess, inside the dungeon and the jailer. I think I should have kept a diary of the excuses that I used to go anywhere to escape the imprisonment of that room, it drove me out of my mind—both for the feeling of claustrophobia it gave me and for the havoc it wreaked with the naturally imaginative and restless hormones of a young man.

Catholicism also played an important part in Kerry's growing-up years in Switzerland. Often, sitting alone in a back pew, staring at the altar, or lighting a candle, Kerry found his religion brought solace. "I thought of being a priest," he recalled. "I was very religious while at school in Switzerland. I was an altar boy and prayed all the time. I was very centered around the mass and the church." When asked what Bible passages moved him the most in a 2003 interview, he had a ready answer: "The letters of Paul taught me not to feel sorry for myself."

Whatever triggered John Kerry's interest in religion, it was not his father. For while Richard Kerry was certainly a fine Air Force pilot and Massachusetts attorney, he also possessed a brooding intellect and the Bible was not central to his thinking. "One of my father's friends once commented that Richard Kerry was known for two things," Cameron Kerry recalled. "He was a genius at navigation and a natural linguist who wherever he traveled in the world spoke the language and spoke it well." By all accounts, the sea truly was his great love. He sailed across the Atlantic several times, once in a 37-foot sloop. When unnerved about something he would stare out at the ocean for hours, engulfed in a deep contemplative mood no one wanted to disturb. Lined up on a shelf behind his desk, whether in Boston, Washington, D.C., Oslo, Berlin, or Cape Cod, there were his indispensable books: Kennan on Russia, Commager on Jefferson, Clausewitz on war, and Bowditch on navigation. "He was very thoughtful, sentimental, romantic, intelligent, and somewhat unrealistic about the world in some ways—high expectations and not a sense of reality that, at certain times, I think he became a little cynical as his idealism got stomped on," John Kerry reflected after his father's death in 2001. "He went into the State Department thinking they were officers who could change the world, and found instead a bureaucracy. They'd write great memos and no one would read them and nobody would do anything. Finally, he got so exasperated about always being questioned from higher up that he wrote his book *The Star-Spangled Mirror.** I think it's fair to say that he left the foreign service a little bit disillusioned and disappointed by it."

To understand Richard Kerry's global views during the Truman-Eisenhower era, all one needs to do is read his book, a revisionist critique of the Cold War in the tradition of William Appleman Williams. His belief in a commonality of purpose between America and Europe was deep-rooted. Yet he was not merely an Atlantic Alliance booster. He was always a bit of a Cassandra, like George Kennan, worrying that the Eisenhower administration was too often bullying Europe instead of consulting. He frowned on the simplistic Cold War view that the world was divided into two camps—good and evil. Uncomfortable with Dulles's Presbyterian moralism, Kerry, like Chicago attorney George Ball, a high-ranking State Department official during the Kennedy and Johnson administrations, became respected as

---

**Richard J. Kerry,* The Star-Spangled Mirror *(New York: Rowman & Littlefield, 1990).*

an intelligent inside critic of U.S. military intervention. He believed that there was a diplomatic solution to most every global problem. He opposed the Vietnam War from the start. "Casting issues in the form of polar choices (for example: isolationism vs. interventionism) readily leads to the conclusion that if one is wrong, the other must be right," Richard Kerry wrote. "In a more relative view of the issue, both are likely to be wrong."

Believing that self-reliance was the main virtue of character, Richard Kerry, with financial help from his wife's family, decided to send his son John to school back in America. In 1957, when John Kerry was thirteen, his father sent him to the Fessenden School in West Newton, Massachusetts. The motto of the boys' boarding school was "*labor omnia vincit*" (work conquers all). With jacket and tie mandatory, Fessenden was a place where lifelong friendships were formed. With faculty living right on campus and "dorm parents" providing adult supervision, it was the ideal place for the Kerrys to enroll John. While at Fessenden, John Kerry met a tall, very athletic jokester named Richard Pershing. With his sandy blond hair, blue-gray eyes, and incurable penchant for good times, Dick Pershing was a cutup extraordinaire.. By the time he turned fourteen, he was already a prep school legend throughout New England. "His mother was called the Grand Dragon, [and] he would mimic her," Kerry recalled. "He just had a gift. He was one of those guys that often when you listen to you're always in fits of laughter and getting into trouble."

Born in 1943 in New York City, Pershing was the grandson of General John J. "Black Jack" Pershing. Besides playing soccer together, Pershing and Kerry developed a sort of comedy routine that always brought howls of laughter from their peers. "They were like Butch and Sundance or Mutt and Jeff," their mutual friend George Butler later recalled. "They just had this wonderful rap together." Their best humorous bit was using fake English accents, imitating Oxford aristocrats in debate or exuding an overexaggerated caricature of a gaggle of fuddy-duddy lords worrying about the demise of the British Empire. The year they spent together at Fessenden forged an unbreakable friendship between the two. Once Kerry moved on to St. Paul's School, his seventh school by ninth grade, and Pershing enrolled in Phillips Exeter Academy, they stayed in touch, exchanging letters and crossing paths at school and social events.

* * *

Without question, St. Paul's School was where Kerry first engaged in politics. Located in Concord, New Hampshire, an hour's drive from Boston, the boarding school was among the best in America. Spread out over 2,000 manicured acres, the academy was known not only for its high academic standards but also for its competitive sports. Kerry, who, like everybody else, lived in the dormitory on campus, was an avid ice hockey player. Whenever Turkey Pond froze, there would be Kerry skating away. In the mid-1890s St. Paul's, in fact, was the first school in the country to embrace ice hockey as a sport (not a club), playing against Ivy League colleges. Besides promoting athletics, St. Paul's believed strongly in the rigors of scholarship and personal appearance. There was a dress code—ties were mandatory—and sloppiness of any kind was forbidden. Punctuality for *everything* from dawn to dusk was mandatory. "It was pretty stiff," John Shattuck, who was one class ahead of Kerry, recalled. "A typical 1950s New England boys' school, in the best sense."

One of Kerry's closest friends at St. Paul's was a tall, skinny, brown-eyed Italian American classmate from Far Rockaway, New York. They bonded in the final two years before graduation. Danny Barbiero was the son of Italian immigrants from Naples and Bari. His father was a Nassau County assemblyman who later became a district court judge. "He was a larger-than-life character, everybody loved him, just hugely popular in his home area," Barbiero recalled. "We had a fabulous upbringing, but St. Paul's was an alien place for my dad." The bond forged between Kerry and Barbiero would last a lifetime. "John was always talking about global issues," Barbiero said. "He was only eighteen years old and he knew just everything about politics, particularly civil rights. That annoyed some people. No doubt about it."

The catalyst for Kerry's interest in the burgeoning civil rights movement while at St. Paul's came from the Reverend John Walker, the first African American to join the school's faculty. Born in 1925 in Barnesville, Georgia, the son of an A.M.E. minister, Walker grew up in Detroit with the ambition to become an Episcopalian minister. Upon graduation from Wayne State University in 1951, Walker became the first black pastor at St. Mary's Church in Detroit. Eager to spend time on the eastern seaboard, he joined the faculty of St. Paul's in 1957. It was a bold hire by the rector of St. Paul's, the Reverend Matthew M. Warren, signaling a new era of integration. From

1957 to 1986, Walker was involved with all aspects of life at St. Paul's, participating in the school's Chapel Program and serving in the dormitories. Often, during the summer months, he would travel to impoverished places like Nicaragua, Costa Rica, or Uganda to minister to the downtrodden. Kerry and Barbiero used to spend their evenings listening to Reverend Walker discuss issues like *Brown v. Board of Education of Topeka*, the Montgomery bus boycott, and the Little Rock Nine integration controversy. He inspired the fifteen-year-old Kerry to write and deliver an NAACP-like speech titled "The Plight of the Negro." "Bishop Walker wasn't a radical," Barbiero explained about his mentor. "But he had an intense sense of social justice. And John used him as a sounding board."

Tucked away in Kerry's Boston archive is a musical relic from his days at St. Paul's: an album of the Electras, taped locally and cut in New York, in which he played bass guitar. And there, in a group photo, he clutches his instrument, resembling a member of Herman's Hermits or the Turtles. The Electras played honky-tonk rock 'n' roll of the Sun Records variety. "We were part of the Elvis craze," Kerry recalled. "Elvis came in 1955 and look out—it was wild. I used to do Elvis imitations. Curl my lip and throw up collars and sing 'Hound Dog' or 'Jailhouse Rock.' " The liner notes to the Electras album proclaimed that Kerry was "producer of a pulsating rhythm that lends tremendous force to all the numbers." Every day he would practice his bass guitar for an hour. "It amazed me that John played bass guitar," Barbiero recalled. "I never thought of him as a musician because he's so serious. But he applied himself and was pretty good." Aside from a few paid gigs, they played at mixers with girls' schools, with Larry Rand's Les Paul–like guitar licks the highlight.

St. Paul's School, esteemed in establishment tradition, harbored a pronounced social dichotomy that often worked against Kerry. It was a hierarchical preppy kingdom with insiders and outsiders—Kerry was among the latter. At St. Paul's, unless you had a lot of money *and* wore the right clothes *and* had parents who belonged to the right clubs, you could be made to feel inadequate, born on the wrong side of the tracks. By the standards of the upper-class children of multimillionaires, Kerry's father, Richard Kerry, was a mid-level employee of the U.S. diplomatic corps. Danny Barbiero recalled, "At St. Paul's John and I were considered poor kids." Although Kerry's peripatetic father had a 52-foot-long sailboat, for example, the patrician parents of dozens of other classmates had yachts. The difference, while subtle and superficial, was nevertheless real. Kerry explained his family's financial situation: "We were comfortable. My mother had independent money but by St. Paul's standards I would never have called myself rich. I didn't *have* to work, in terms of parental support. But my parents were good parents in that they taught me the value of labor. They weren't saying, 'Johnny, where do you want to go? Okay, here is the check.' They wanted me to learn the value of work. And I thought it was important. I wanted to do it myself. I wanted to make my own way. We were always comfortable, I wouldn't have to worry about where a meal was coming from or have to worry about how or where to live."

Given this class-consciousness of St. Paul's School, Kerry felt especially compelled to prove himself. His life would have been simpler, in fact, if he had been an African American from Atlanta or an Okie from Tulsa. Such clear anomalies at St. Paul's would have been accepted as legitimate outsiders, intelligent flukes of nature trying against ungodly odds to join the Eastern establishment. But Kerry was held to higher standards. He was, after all, from the Boston Brahmin DNA pool and had lived briefly in Europe. His middle name was Forbes, and his family was related to the Winthrops. Given such blue-blood credentials, he was not supposed to be overtly ambitious—but he was. To be fully accepted in the world of St. Paul's, you were supposed to be slightly cynical, laid-back, condescending and arch—in a regal and snobbish way. Kerry embodied none of these attributes. Instead he acted like Horatio Alger on the make, believing that the social order should be based on temerity and merit. Or, put another way, he was driven like a Kennedy.

Instead of hiding his ambition under a rock, Kerry embraced it, creating the John Winant Society debate club, discussing quite frankly his hope of going into public life and perhaps being a congressman or senator someday. He was greatly motivated by Congressman John V. Lindsay, a St. Paul's graduate who visited the school and impressed upon Kerry the importance of public service. Such raw ambition rubbed some schoolmates the wrong way. For every three friends Kerry made at St. Paul's—like Danny Barbiero, Stephen Kelsey, or Pete Johnson—there was an enemy who even decades after graduation remained bold in declaring his committed animosity toward Kerry. It was at St. Paul's School that a cult of envy emerged toward

Kerry, one that would follow him throughout his entire political career. The sentiment was that anybody who excelled at *everything* he tried had to be a phony. But one thing nobody at St. Paul's—friends and foes alike—disputed was that John Kerry was on a forward trajectory, which spelled professional success of some kind or another. "His interest in politics was very real," Shattuck recollected. "We both were strongly for John Kennedy and we weren't shy about letting people know."

There is no record that Kerry had any strong opinions about a faraway place called Vietnam while at St. Paul's School. He did know about the French defeat at Dien Bien Phu in 1954 and the Geneva Accords. He learned about the strange career of Ho Chi Minh. He knew that the French pulled out of Vietnam in 1956 and that the Viet Cong were determined to wage guerrilla warfare. In a history class he learned about Vietnam (or Annam) in about A.D. 800. While at St. Paul's he did debate whether the United States should establish diplomatic relations with so-called Red China (he was for it). But place-names like Danang, Saigon, Hanoi, and the Mekong Delta meant nothing to him. He did not know that young soldiers with the U.S. Military Assistance Program like Major Dale R. Buis of Texas and Master Sergeant Chester M. Ovnard of California were killed by a sneak Viet Cong attack at Bien Hoa Air Base in 1959. (They had been gunned down while watching a movie in a mess hall. Three VC with automatic weapons appeared at the windows and started firing away.) Why should he have? The news of their gruesome deaths was buried in the back pages of America's newspapers. Nobody could have imagined in 1959 that 3 million Americans would eventually serve in Vietnam and that more than 58,000 of them would be killed. President Dwight Eisenhower, in fact, had summed up his administration's position vis à vis Southeast Asia as follows: "No one could be more bitterly opposed to ever getting the United States involved in a hot war in that region than I am." It was decades later, when Kerry, by then a senator, located the names of Buis and Ovnard on "The Wall" in Washington, D.C., that he made the correlation that while he was digging clams on Cape Cod and playing lacrosse in Concord, fellow citizens were being blown away to stop a Communist domino from toppling the non-Communist states of South Vietnam, Cambodia, Laos, Thailand, Burma, and Malaysia.

* * *

An important moment of reckoning came to Kerry in early November 1960. He was a sixteen-year-old student fascinated by the junior senator from Massachusetts, John F. Kennedy. Kerry is remembered at St. Paul's for his zealous promotion of Kennedy, once he became the Democratic nominee for president, running against Republican Richard M. Nixon. Not only did Kerry share initials with his political hero—JFK—but he spoke with a somewhat similar inflection. A debate was set up at St. Paul's with Kerry cast as Kennedy and Lloyd McDonald representing Nixon. On the eve of the debate, Kerry took a train from Concord to Boston. Near North Station in Boston people were congregating to hear Kennedy speak at Boston Garden, wearing hats and waving placards with great anticipation. Chants of "JFK" swirled all about. "What's going on?" Kerry asked. Somebody told him, "Oh, Senator Kennedy's coming in for the last speech before he goes to the Cape for election day."

So an excited Kerry wandered into the rally, watching the political theater from a standing position near the exit. Missing his scheduled train back to Concord, Kerry eagerly waited Kennedy's arrival. When Kennedy arrived onstage smiling, throngs of people cheering wildly when he looked in their direction, Kerry got caught up in the emotion. The natural bond he had felt with Kennedy had been heightened. "Suddenly I felt as if I were part of his campaign," Kerry recalled. "It was exhilarating." After the speech he collected pro-Kennedy pamphlets and headed back to New Hampshire. "I got all these materials I was able to use and worked on the train the whole way back to Concord," Kerry recalled. "They helped me prepare for my 'Why Kennedy Should Be President' argument the next day. Unfortunately the school was predominantly Republican and McDonald, defending Nixon, won the votes. But I took comfort—my man, Kennedy, won the real contest."

Kerry, of course, was not unique in being moved by John F. Kennedy. Blessed with charm and humor, the handsome president, the youngest ever elected, had a mesmerizing effect on an entire generation. Everything about him exuded what Ernest Hemingway called "grace under pressure," and he was seemingly apt for every occasion. Young people, in particular, were captivated by his call for national duty. Concepts like the Peace Corps and the Space Race were designed to capture the imaginations of a new generation of Americans who were suddenly saddled with global responsibility. Not since Theodore Roosevelt mounted his pulpit had a U.S. president galvanized young people to stand up and be accounted for.

Kerry was intrigued by the energy and idealism of President Kennedy and his family. They made American public life exciting. He was, like many, caught up in the enthusiasm of Camelot. When Edward Kennedy, the President's thirty-year-old brother, ran for the Senate from Massachusetts in 1962, Kerry jumped on the bandwagon, handing out pamphlets and blaring endorsements from the loudspeaker attached to the roof of his VW Bug. "Teddy Kennedy probably didn't know me from Adam back then," Kerry recalled in 2003. "I suspect that he might have known me visually as one of the kids he'd see working on the campaign. A lot of the people around him who did that, I still know to this day. But I don't think I shook Teddy's hand more than a couple of times during the campaign."

Besides grassroots organizing, Kerry had a more tangible connection to that clan. During his last year at St. Paul's, he dated Janet Auchincloss, the half sister of First Lady Jacqueline Kennedy. Stunningly beautiful, with a regal bone structure and elegant smile, Auchincloss was swooned over by dozens of New England prep school boys. The Kerry-Auchinchloss romance was, for the most part, platonic in nature, but they did truly enjoy each other's company and for a while they were an item. In August 1962, she invited Kerry to spend a few days at Hammersmith Farm in Newport, Rhode Island, Jackie Kennedy's childhood summer home. Without question Hammersmith Farm, in working operation since 1639, was one of the more scenic outposts on the Atlantic coast. Nestled atop fifty magnificent acres, with gardens designed by Frederick Law Olmsted, the twenty-eight-room "summer cottage" was built in 1887. It had been at this Auchincloss estate on the morning of September 12, 1953, that 1,200 guests arrived for the wedding reception of John F. Kennedy and Jacqueline Bouvier.

So it was with a sense of awe that the eighteen-year-old Kerry drove his father's car to the estate to spend a weekend before his freshman year at Yale with Janet. In order to earn pocket money, Kerry was working that summer at First National Stores, loading trucks in Somerville, Massachusetts. It was the second summer he had done this. "I joined the Teamsters, punched my card every single day, ran an electric cart around a warehouse, selecting all the food, and putting it on the pallets and loading the pallets onto the trucks," he recalled. "A great job, I thought, because it paid pretty well."

At noon on Friday, he punched out of First National Stores, stopped for gas, and telephoned Janet, to let her know he was a "little delayed." To his surprise, she snapped, "Well, hurry up, because the President's here and he wants to go off sailing." Kerry could hardly believe his ears. The President? His president—John F. Kennedy? Like a bat out of hell he raced the back roads of Massachusetts and Rhode Island, navigating a thick rush of weekend getaway traffic to appear at the estate's front door in Barney Oldfield time. "When I got there, only one little guard was posted in front," Kerry recalled. "I said, 'Hi, I'm John Kerry, I'm here to visit the Auchinclosses.' And they said: 'Oh yeah, great, go ahead.' So I drive in, drive right up to the front door and there's only one guy in front of the house. I walk up, open the door and there's nobody inside. I can't see anybody. But there's this guy standing there peering out a window at the bay. He turns around and it's the President of the United States. He walks over to me and I said, 'Hi, I'm John Kerry.' I was so unsavvy, so unschooled. I didn't know what you called him, 'Mr. President' or 'sir' or 'Mr. Kennedy' or whatever. So he said, 'What are you doing with yourself?' and I said, 'Well, I'm about to go to Yale.' I grimaced because he was a Harvard man and I said, 'Sorry.' And he said, 'Oh no, I'm a Yale man too now!' You see this was the summer he'd gotten his honorary Yale degree and made the quip 'I have the best of two worlds, a Harvard education and a Yale degree.' At that time I was volunteering for his brother Teddy, who was running for the U.S. Senate. We joked and chatted about that for a bit. It was more than a kid my age deserved or would have expected." That afternoon, Kerry, dressed in a polo shirt, joined the President on the 60-foot Coast Guard yawl *Manitou* for a relaxing cruise in Narragansett Bay.

A few weeks later, Janet Auchincloss again invited Kerry to Hammersmith Farm, this time to watch the America's Cup race from the U.S.S. *Joseph P. Kennedy* along with President and Mrs. Kennedy. Lunch was served, followed by a few hours soaking up some sun and watching with binoculars the flotilla of sailboats that had come to watch or participate in the America's Cup races. The Kennedy entourage—including Kerry—was pulling for US-17 *Weatherly*, constructed of African mahogany, white oak, and bronze. Skippered by Emil "Bus" Mosbacher, it defeated Australia's *Gretel*. Later that day, Kerry had a brief private conversation with the President. "Nothing incredible," he recalled. "But, boy, it was memorable." Kerry treasures his three photographs as memories of those afternoons in Narragansett Bay. More than one journalist eyeing these photographs has coughed up the easy punch line that Kerry is like Woody Allen's Zelig or Winston Groom's Forrest Gump. Taken with Kodak color film, the photos have that faded, long-ago Polaroid look, but they reveal much about the clean-cut Kerry's demeanor. Clutching his binoculars, the tall youth wore a dark blue suit with a striped Brooks Brothers tie, yet there is nothing presumptuous about him. While Kennedy was wearing dark sunglasses, swapping boisterous jokes with friends, Kerry had the opportunity to meet Adlai Stevenson and Kennedy aides Red Fay and Larry O'Brien. It was heady stuff for an eighteen-year-old and Kerry was absorbed with everything about President Kennedy: from the way he told jokes to the careful way he watched the race. "Having met you several times this summer at Hammersmith Farm, and having worked for your brother in Massachusetts during the same time, I am to say the least an ardent Kennedy supporter," Kerry wrote the President in late September. He went on to thank him "for a very unforgettable and exciting time the weekend of the America's Cup races."

Spending those few hours on two different occasions with President Kennedy was the highlight of his summer. Nothing else came close. He had come from defending Kennedy in November 1960 at a St. Paul's debate to being with him on a yacht docked at Hammersmith Farm in August 1962. All he could do was let it sink in—and keep striving for more. He had a new world to conquer: Yale University.

Chapter Two

The Yale Years

It was an unusually hot September, by Connecticut standards, when John Kerry followed his father's footsteps to Yale University in 1962. The grand old New Haven campus fairly glittered with possibilities in an eager freshman's eyes. Danny Barbiero found himself stunned that first semester by the volume of historical detail his former prep school and now college roommate had absorbed about the place. "John was just so excited to be at Yale," Barbiero recalled. "He was absolutely enthralled. Just beside himself with glee."

Most freshmen knew the basics, of course: founded in 1701, Yale had amassed a distinguished and deserved reputation as a breeding ground for great Americans, its accomplished alumni spanning from eighteenth-century theologian Jonathan Edwards to cotton gin inventor Eli Whitney, from William Howard Taft, the twenty-seventh president, to Dean Acheson, Harry Truman's secretary of state. Yale's august legacy was, indeed, why every year's bright crop was there. John Kerry just seemed more excited to be part of the tradition, pointing out to Barbiero various notable neo-Gothic buildings and ivy-clambered walls. Kerry explained how the U.S. Forest Service had been hatched in the red-stone mansion known as Marsh Hall and related that the campus's Peabody Museum housed some of the most valuable dinosaur bones in the world. "The thing that amazed me was that he would walk down the street the first week we were there and say, 'Dan, look where we are; this is Vanderbilt Hall,' and he'd recite chapter and verse of the entire history of the building. We walked by a statue of Nathan Hale and off he'd go. He just knew the school, knew the traditions. He was a man who always had a sense of the history of things, of his place in history as a young man, which I did not possess."

Of particularly enticing interest to Kerry were the gargoyles perched high atop Harkness Tower, an impressive 216-foot structure on New Haven's High Street. Engraved above the stone totems were the four traditional qualities Yalies strove to achieve: Pen Wielding, Proficient Athlete, Tea-Drinking Socialite, and Diligent Student Scholar. Kerry made a pact with himself to embrace all of them.

The two ex–St. Paulies roomed together in a suite on the freshman campus in Bingham Hall. About midway through their first semester, Kerry and Barbiero went to the New Haven Green to see President Kennedy address a rally next to campus. It was at that October event, around the time of the Cuban Missile Crisis, that Barbiero introduced Kerry to fellow classmate Harvey Bundy. "I was less excited about Kennedy than probably one might have expected given the family connections," the always honest Bundy remembered. "Kennedy gave a political-hack speech. Kerry was all enthused about it; I wasn't. We agreed to disagree."

What rang odd in Harvey Bundy's remark was that at the time two of his uncles ranked among the leading foreign policy hands in the Kennedy administration. McGeorge Bundy served in the White House as a national security advisor, and William P. Bundy was assistant secretary of state for Near Eastern affairs. They were two of the top New Frontiersmen, and as such heroes to the politically fascinated young John Kerry. Harvey Bundy's blasé view of his powerful family connections interested and surprised Kerry. While his other friends seemed mighty impressed that Kerry had enjoyed "face time" with President Kennedy in Rhode Island over the summer, and had the pictures to prove it, Bundy shrugged his peer's encounters off as run-of-the-mill stuff. Kerry and Bundy overcame their differences over Kennedy and became good friends. Later, Bundy even opened an unexpected door. "It was through Harvey that I actually visited the White House," Kerry explained. "We went down to Washington, D.C., together, and McGeorge Bundy actually gave us a guided tour. Just the two of us. He took us into the Oval Office and the Cabinet Room. It was terrific—just a wonderful, detailed tour."

Throughout the rest of their freshman year Kerry and Bundy bonded. At the end of the term, Kerry, Barbiero, and Bundy arranged to room together their sophomore year at Jonathan Edwards College. They shared a suite—a living room and two bedrooms. "One thing that attracted me to John was that he had a car illegally," Bundy recalled. "Freshmen at Yale weren't allowed cars. He couldn't figure out what to do with it. He wanted it on weekends, but he didn't want it during the week. He couldn't figure where to get rid of it, and so what happened eventually is that my girlfriend, Blakely Fetridge, who needed a car to come up every weekend from Wheaton, took John's car for the week and then brought it back on the weekends."

Fetridge, who would marry Bundy the summer after he graduated, kept a diary during those years that remains the best primary source of information on Kerry at Yale. A devotee of Ayn Rand and the daughter of a prominent Illinois Republican, Fetridge was, for all intents and purposes, part of the cabal. As such she was privy to their dorm room bull sessions and recorded the gossipy details in her diary. "Harvey drove down in John's car to pick Sand, Kath and me up for the weekend," Fetridge wrote on May 10, 1963. "We dropped S and K off and picked up John Kerry and drove to Farmington to get his date—Janet Auchincloss. By this time, Harvey was in a good mood and we all had a wonderful time—really hit it off! After picking up Janet, we drove back, went to a band concert, then a play at J.E., *Damn Yankees*, then to Morse to dance, and finally to WYBC. We did an all nighter. . . . [Janet left at 2] and I was the only girl." A few days later she wrote: "John took me on a motorcycle ride—such fun!"

Fetridge noted that, besides partying, Kerry relished holding court on every policy issue before the Kennedy administration, particularly James Meredith's seminal enrollment as an African American at the University of Mississippi and the building of the Berlin Wall. At the time, in 1962, Vietnam had not yet become a hot discussion topic; after all, when Kennedy had been elected in November 1960, only 900 U.S. military personnel were stationed in Vietnam. By 1963, however, there were some 16,000 American troops there.*

**On May 11, 1961, President Kennedy had issued National Security Administration Memo 52, committing the United States to the prevention of the Communist domination of South Vietnam. The President increased the number of U.S. personnel, designated an additional $42 million per year in financial support for the government at Ngo Dinh Diem, and approved the CIA's plans to carry out commando raids in North Vietnam.*

The Kennedy administration went on to condone the overthrow of South Vietnamese President Ngo Dinh Diem, to dispatch the U.S. Army Green Berets to Southeast Asia, and to commit American-manned helicopters and tactical aircraft to help defeat North Vietnam's Ho Chi Minh. John Kerry, meanwhile, was getting an education about the war, from the U.S. policy perspective, from a pretty good source: his roommate's uncle. "Bill Bundy used to come up to New Haven occasionally," Kerry recalled. "And I remember one night sitting on the floor of our room late into the night drinking beer with the then assistant secretary of state. I was all ears, a sponge. We listened to him talk about Vietnam in depth, why it was important, and what our generation would have to do."

While Harvey Bundy, a straitlaced Republican, shared Kerry's sense of academic responsibility, Danny Barbiero had begun to sow some wild oats in New York City. Barbiero often hung out in Greenwich Village with his cousin Felix Pappalardi, a classically trained musician who would become a rock impresario, signing the Rascals, writing and producing songs for Cream, and working with the Youngbloods and producing their 1967 anthem "Get Together." After that he would form Mountain, with Leslie West on lead guitar and himself on bass, an instrument Kerry had dabbled in. "Remember," noted Barbiero, "John had played bass guitar in the Electras at St. Paul's. So Felix, who had mastered the thing called the guitaron, which is a huge oversize bass, probably intrigued him."

In 1962, Pappalardi was eking out a living playing bass for ragtime legend Max Morath. On a few occasions Kerry accompanied Barbiero to Greenwich Village to meet his cousin and hear folkies such as Ramblin' Jack Elliott and Dave Van Ronk. The bohemian coffeehouse scene fascinated Kerry, but he never fit in or felt a part of it in any meaningful way. That said, he did think about making a short documentary on the so-called beatnik lifestyle. "There is a piece of me that has enjoyed dabbling in film," Kerry admitted. "I did that at Yale—filming on my 8-millimeter camera, going to Washington Square in New York City to absorb the happenings. It was just sort of fun, taking in images with my camera. I enjoyed observing and capturing the things around me. That's exactly what I did in 1963, 1964. I shot film of weird folks and surreal images of different people. I guess I was trying to be Fellini Jr. or something."

* * *

It was in October of his freshman year that John Kerry also met David Thorne. An energetic, high-strung athletic six-footer with keen hazel eyes and neatly groomed, light brown hair, the charming Thorne lived in Yale's Vanderbilt Hall and majored in American literature. Harvey Bundy introduced Kerry and Thorne one evening at My Brother's Place, a greasy-spoon diner near campus on Chapel Street. Within minutes of meeting, Kerry and Thorne plunged headlong into a wide-ranging dialogue in which each seemed to be trying to outline the contours of his life in blurted sound bites. When their conversation turned to Europe, where both Kerry and Thorne had spent part of their young lives, the bond was sealed. To their surprise, they discovered they had something else in common: both had been dating Janet Auchincloss. "That came as a shock to both of us," Thorne laughed decades later. "Believe me, that piqued a lot of interest. So we talked warily for a couple of hours and realized there was a deep commonality in our backgrounds."

Throughout the rest of their first year at Yale, Kerry and Thorne remained inseparable, eagerly discussing every freshman passion, from Hemingway novels to Wellesley girls, over endless mugs of beer. When it came time for spring break, Kerry and Thorne decided to go to Florida together—a road trip inspired by the popular 1960 movie *Where the Boys Are*.

Thus motivated, the Yale duo made arrangements to stay in Miami Beach, Palm Beach, and Hobe Sound. As soon as classes ended, they headed south from New Haven straight down Route 1, in Thorne's pale blue two-year-old Volkswagen Beetle, from the back of which stared a pair of cartoonish eyes from a sign that read "Moon Equipped." The car set neighboring motorists to pointing at them in gales of laughter. The good times continued through pit stops in New York, Washington, D.C., and points south. On the drive through the Deep South, however, Jim Crow's whites-only signs hanging from diners and gas stations shocked Kerry. But youthful high spirits prevailed as they bought fireworks and pressed on to the free orange juice stand at the Florida border and beyond. The pair spent the next week cruising from Miami's swank Fontainebleau Hotel to their sterile Holiday Inn in Valdosta in search of those *Where the Boys Are* girls. "We did typical prankish stunts," Thorne confessed. "The worst was when we tried to streak Route 1. I only got halfway across and then got scared and turned back. Johnny completed the mission."

Kerry and Thorne shared an interest in other sports as well. Both played on Yale's soccer team (Kerry at right wing), enjoyed hockey, and went skiing at the Sugarbush resort in Vermont's Green Mountains, even in poor conditions. Kerry, an accomplished and agile skier, particularly enjoyed carving through the bumps and racing giant slalom. It was on these Green Mountain slopes that Thorne took to taunting Kerry about his beautiful twin sister, who owned an autographed picture of John and Jackie Kennedy she had been given by Letitia Baldrige, the First Lady's personal secretary. Kerry couldn't wait to be properly introduced to this bewitching Julia Thorne, even if she did live far away in Italy.

Kerry also reunited with his old Fessenden cohort Dick Pershing at Yale. Occasionally, on weekends, Kerry would stay with Pershing at his Park Avenue penthouse or Long Island mansion. Pershing's parents had another home in Jamaica, just outside of Montego Bay. It was on that island that Pershing had befriended George Butler, who was about as old-stock American as one could get. While Pershing had the battlefield heroics of his grandfather to boast about, Butler could claim that his ancestor Sir Richard Saltonstall came to the New World in 1630. Butler's father was a retired Anglo-Irish officer who had gone to Sandhurst, served with the Royal Welch Fusiliers (which was the regiment of poets Robert Graves and Siegfried Sassoon) and had served in India and Africa, and finally retired to Jamaica. "The bond Dick and I had was soccer," Butler recalled. "And once you met Dick he stayed with you. He had a rubbery kind of face and always made these outrageous contortions with it for laughs. The women were crazy about him. But he was *not* always a serious person."

Butler first met Kerry in June 1964 at Manchester, Massachusetts. It was Harvey Bundy, a prep school classmate, who introduced them. Although Butler had a *Mayflower*-era ancestry, he had turned his back on it. Riveted by the novels of Thomas Wolfe, particularly *Look Homeward, Angel,* Butler had rejected the Ivy League for the University of North Carolina at Chapel Hill. That summer Butler was getting ready to move to Dallas to sell dictionaries door to door, while Kerry was gearing up to do the same throughout New England, only with encyclopedias as his product. "The reason we both went the door-to-door route is that you could make a lot of money on commissions," Butler recalled. "Remember, John—nor I—was really wealthy. We could use all the money we could get."

* * *

In the summer of 1963, when the Kennedy administration was just starting to send U.S. soldiers to Vietnam, Kerry went to Europe with Harvey Bundy. "On the plane over, John and I played gin (I won, of course), drank champagne (John won), and talked to the girls in the seat behind us (we won)," Bundy wrote his girlfriend, Blakely Fetridge. "We landed at 9 A.M. and spent five and a half hours getting the car. Finally, we succeeded only to find we couldn't go over 45 mph till we had gone 500 miles and had a two-day checkup. Then it started to leak gas and we had to have it fixed. Finally all systems were go and we visited Oxford, Cambridge, and Windsor Castle, all at the speed of 45 mph. At night we went to two plays—*Oliver* and *How to Succeed in Business Without Really Trying*."

The car Bundy was writing about was an Austin-Healey 3000 Mark II he had ordered in London for them to tour in. "We did everything you do in London," Bundy remembered. "We visited castles and museums. I remember driving through the beautiful countryside. We had the top down, and John was singing Broadway tunes." After one afternoon of sightseeing, Kerry posed in front of Westminster Abbey, Big Ben in the background, pointing his forefinger skyward à la Winston Churchill intoning we would fight them on the beaches.

Acting the orator was a fantasy Kerry occasionally played out in lighthearted moments. One such remains the deepest ingrained memory Bundy took from their three weeks together in Britain that summer: an afternoon spent at Hyde Park Corner near London's Marble Arch. On the flight over, Kerry had remarked that he would count the summer a success if he got to speak on the famous corner where the likes of Winston Churchill, Marcus Garvey, and George Orwell had once declaimed. "There was only one thing John *had* to do in London, and that was go to Hyde Park Corner and make a speech," Bundy remembered fondly. "He stood up on a soapbox, and off he went."

After London, Bundy and Kerry took the ferry from Dover to Le Havre and then drove to Paris, where they put up at a cheap Left Bank hotel and celebrated Bastille Day with the locals. Fodor's guide in hand, they hit the typical tourist highlights. "All we do is talk about you and Janet [Auchincloss]," Bundy wrote his girlfriend from the French capital, "and how much we love and miss you." The pair next headed to Zurich, arriving at midnight. Everything proved tidily Swiss: manicured, sedate, boring, and closed. So

Kerry came up with an inspired plan. "His brainstorm was: Let's go to Austria. This is typical Kerry—I mean, you're in Zurich, so let's go to Austria; it's only a five-hour drive," Bundy continued. "There was a ski village that he loved and he wanted to look up his old ski instructor, so we took off from Switzerland across Liechtenstein, where we woke up the border guards so they would stamp our passports. They were not happy with that [but] we wanted to show people that we had been to Liechtenstein. Eventually we arrived on the outskirts of this little alpine village in Austria where John used to ski. It was about five in the morning. We both agreed that it was way too early to look up his ski instructor. There is a mountain outside of town, and John says, 'Let's climb it.' Typical. Classic John. He wanted a race to see who could get to the top first. So we're both panting, out of breath on top of the mountain."

From Austria, Kerry and Bundy headed south and vacationed along the French Riviera and then headed to Brittany. At Kerry's family place, the two Yalies enjoyed afternoon swims in the frigid Gulf of Saint-Malo and bottles of wine at night. And then a near disaster struck. Neither Kerry nor Bundy likes to talk about it. One evening Kerry, with another Yalie, Pete Kornblum, as his passenger, was driving Bundy's Austin-Healey along a country road between the northern coastal town of Dinard and Dinan, fifteen miles inland. As they came around a corner, another car suddenly appeared in the middle of the road heading straight for them. Kerry instantly turned the wheel, running off the road into a ditch so hard that the little sportster flipped over and bounced back onto the road, landing upside down. Shattered glass and forlorn bits of metal lay everywhere. That Kerry and Kornblum survived belies the brutality of the wreck. "How they didn't get their heads cut off I don't know," wondered Harvey Bundy. "John still has a scar on his chin from the accident," pointed out David Thorne. Bundy, who had gotten the police call and rushed to the accident scene, said, "John was bleeding all over the place. He had cuts between his elbows and hands." Even worse, "he was shocked that he had totaled the car."

With the Austin-Healey gone and autumn approaching, Kerry's grand tour of Europe came crashing to an end and he returned to Yale having decided to major in political science. His interest, however, in public service remained intact. Decades later he wished he had chosen otherwise. "I regret being a political science major," Kerry lamented. "I wish I had a stronger mentor/guide in my early choices. I was a capable student, but not a very dedicated one."

The one course that did excite Kerry was historian Gaddis Smith's American diplomacy. (He also enjoyed studying the modern presidency with historian John Morton Blum.) An expert on World War II and the Cold War, Smith peppered his lively lectures with engrossing anecdotes from all of America's past. Whether his topic was the XYZ Affair, the Gadsden Purchase of 1853, or that July's signing of the nuclear test ban treaty in Moscow, Professor Smith made history come alive. Many of the students he taught in the 1960s would in fact take on a range of national roles as soon as a decade later, from Bob Woodward to David Gergen and George W. Bush. "Gaddis was a great, great lecturer," Kerry attested. "The lectures were all full of energy and knowledge. I confess that I was . . . I don't know, maybe I suffered from attention-deficit disorder . . . never really paying full attention in [any other] class. I just had so many extracurricular things going on. But I absorbed every word uttered in Gaddis's class."

Unusually for that era, Professor Smith turned his back on traditional textbooks like Samuel Flagg Bemis's standard *American Foreign Policy and Diplomacy*. Instead he would hand out several hundred mimeographed pages of primary-source texts, so that his students could study the real documents of history, such as James K. Polk's declaration of war against Mexico, Abraham Lincoln's Emancipation Proclamation, and Dwight Eisenhower's "military-industrial complex" farewell address. Thus having prepared his students, Smith then would have them on the edges of their chairs with descriptions so vivid they could practically see Woodrow Wilson at Versailles or Franklin D. Roosevelt at Yalta. Perhaps the hottest topic in the class, and destined to grow much more so in the ensuing years, of course, was the Vietnam conflict. "I tried to offer a full spectrum of opinions on Vietnam," Smith recalled, especially after the doubling of the draft call in July 1965. "There was the Lyndon Johnson interpretation that Vietnam was part of a Cold War necessity like the Truman Doctrine, the creation of NATO, and the Berlin crisis. At that point Dean Acheson hadn't turned against the war, and I knew from a White House friend that he was advising Johnson to stay the course in Vietnam just as Truman had done in Korea. We discussed the antiwar opinions of [then Senators] Wayne Morse, Ernest Gruening, Mike Mansfield, and J. William Fulbright—the whole 'arrogance of power' view. Then there was the view, quite prevalent in the academy, that our biggest problem regarding Vietnam was that we were ignorant of Asian culture. Finally, there was the radical-left interpretation that Vietnam was a typical American imperialistic endeavor, just another ugly chapter in our long history as conquerors."

While Smith remained open-minded and careful to present the "full spectrum of opinions" on Vietnam to his students, personally he definitely sided with the doves. When he reflected on the many students who took his course in the mid-1960s, one haunted the professor years later: Army Lieutenant Don Masters, who had written a paper for Smith on the history of U.S. Special Forces, from the British-American Rogers' Rangers of the French and Indian War to the Green Berets of Vietnam. What stayed with Smith was what Lieutenant Masters once told him about how the American soldier was trained to think in Vietnam: "If it runs, it's VC; waste it. If it hides, it's VC; waste it. If it's dead, it's VC; count it and wait for your promotion."

John Kerry's life changed forever at the start of his sophomore year at Yale. One crisp September Saturday morning, he drove his blue VW Beetle to Bay Shore, New York, to spend the weekend with David Thorne at his grandparents' sprawling estate on Long Island's Great South Bay. Shortly after arriving, he finally laid eyes on David's free-spirited twin, Julia. A five-foot-eight nineteen-year-old, with shoulder-skimming black hair and lively dark eyes, Julia was a real head-turner, puffing Marlboros because that's what teenagers in Italy were doing. Demonstrative and spontaneous, she would burst out in song or crack a joke at the unlikeliest moments. "Black-eyed Susan," her governess called her. "John was totally mesmerized by Julia at first sight," David Thorne reminisced. "She was everything he wanted: good-looking, spoke Italian and French, totally captivating. She was full of energy, very international and cosmopolitan."

Julia and David Thorne had been born in New York on September 16, 1944. Their parents moved to Italy when the twins were eight, and Rome became the family's home. In 1959 Julia was sent to Foxcroft School in Middleburg, Virginia. She loathed every minute of the three years she attended the horse-country boarding school. The girls slept on porches and were required to drill with the U.S. Marines twice a week. Thorne had little in common with her classmates, preferring to speak Italian and not even feigning interest in late-1950s American teenage pop fare like Elvis Presley's latest record or Sandra Dee's new *Gidget* movie. Old Europe remained her home, and her culture. "I sort of commuted between boarding school in Virginia and home in Italy," she recalled decades later. "I had been raised by Scot nannies and Swiss governesses."

When John Kerry eyed her in a bikini that September afternoon in 1963, he was immediately captivated. They dated for the next few months, with Kerry clearly unrelenting in pursuit of her affections. "A bond was created very quickly," Thorne remembered. "I was totally fluent in three languages, and considered myself European. Unfortunately, I had been thrown into an American lifestyle that I didn't understand at all. I had just grown up in this world of palaces and princesses and privilege. John didn't live that way at all."

Many families have genealogical charts with ancestors of merit, but Julia Thorne's was chock-full of some of the most distinguished founders in America. The whole history of early New England, in fact, was populated with Thornes and Stocktons. Among her maternal grandmother's forebears was William Bradford, who served as President George Washington's attorney general and ended the Whiskey Rebellion. Julia's great-great uncle was Henry L. Stimson, President Herbert Hoover's secretary of state and President Franklin D. Roosevelt's secretary of war. A great-uncle was Alfred Lee Loomis, the philanthropist who established the famed scientific laboratory at Tuxedo Park, New York. Julia's grandfather owned three-quarters of Hilton Head, which he sold to entrepreneurs who built the world-class resort off the coast of South Carolina. Between the two of them, John Kerry and Julia Thorne constituted a virtual storehouse of America's most productive and distinguished bloodlines.

Julia Thorne, called Judy by her family and friends, found herself amused and fascinated by the difference in the background of her avid young suitor. There seemed to be less joy in his immediate family, and much belief in the ethic of public service and duty to country. John's father, Richard Kerry, in particular, could be cold and overly serious, even if he possessed great charm and enjoyed telling a good story. "He did not let his emotions out easily," Kerry recalled of his father. "Yet, he was also quite a romantic."

When John visited Julia's grandmother's apartment on Park Avenue in New York City, his new girlfriend played classical LPs, while his own tastes ran more to Peter, Paul and Mary and his besotted classmates' raucous renditions of "Louie, Louie." He spoke French near fluently, though to Julia Thorne's ears his pronunciation made him sound like a Norman truck driver.

Unlike Kerry's old girlfriend Janet Auchincloss, Julia Thorne made no pretense to anything beyond haute couture and the lifestyle of those who could afford it. She was photographed for *Women's Wear Daily*. She moved in a faster, more international, and even cynical world and introduced him to all its excess and style. Nonetheless, she grew impressed with Kerry's love of poetry, particularly the verse of Rudyard Kipling and Rupert Brooke. Another favorite was Robert Frost, whom he had read at St. Paul's and who represented a different inspiration more telling in the young man. "I loved him as a poet from New England," Kerry explained. "Some of his stuff is so dry; there is always a piece of New England in him. There is a great simplicity, directness, honesty, openness—opened and closed at the same time, which is very New England."

Kerry had been introduced to Frost as a student at St. Paul's School. He liked him then but was reintroduced in an even more special way in January 1961. John F. Kennedy had elevated his own inauguration by enlisting the esteemed poet to read something uplifting that afternoon. Frost, as one commentator noted, had been picked to play Virgil in the court of the modern Augustus. Decades later, Kerry could still recall every detail of Frost's appearance at Kennedy's inauguration. It was bitterly cold and dazzlingly bright, but as Frost approached the lectern a howling gust nearly bowled him over, ruffling his thinning gray hair and flapping his prepared pages uncontrollably in the wind under a relentlessly blinding sun.

Two years later, in 1963, Kerry was a college sophomore whose appreciation for poetry barely extended past Frost, Brooke, Kipling, and a few others. Julia Thorne determined that she needed to introduce her American boyfriend to figures of more contemporary cultural relevance such as T. S. Eliot and Robert Graves. Whatever instinctive enthusiasms Kerry evinced for popular culture—and he adored Broadway musicals and Elvis Presley albums and James Bond movies—Thorne tried to cure them. "He was essentially a Romantic in his tastes," she lamented. "He really liked Paganini and Rachmaninoff, while my passion was Bach and Mozart. I had played them on the piano. But John preferred very Romantic, dramatic, and emotive music."

As it did for so many Americans of his generation, John Kerry's youth came to a cruel end on November 22, 1963, with the assassination of President John F. Kennedy in Dallas. The news reached Kerry that Friday afternoon on a soccer field at Yale during the annual game against Harvard. "Back then the soccer team played on Friday afternoon and the football game was the next day," recalled Kerry, who played varsity soccer for three years at Yale. "I remember the ripple going through the crowd and wondering what was going on because people were clearly distracted from the game. And then we heard the President had been shot in Dallas. Somehow we played the game, and to this day I can't tell you who won. When we finished the game, we sort of huddled up and hugged each other."

After walking off the soccer field Kerry burst into tears, choking out over and over again that it just couldn't be true that the President was dead. David Thorne, who had been with his teammate on the bench when the horrific news was announced, spent all weekend trying to console his friend. "My parents were staunch Republicans, so we hadn't invested in the whole Kennedy thing," Thorne remembered. "Don't get me wrong—I was upset—but Johnny was devastated." Even forty years later it pained Harvey Bundy to think of the completely distracted Kerry sitting inches from their dormitory television, obsessively absorbing every news update for days. "We couldn't get him away from the TV set," Bundy recalled. "The entire Kennedy assassination, as I remember it, boils down to John staring at that set, numb."

Practically everybody who knew John Kerry at Yale seems to have a story illuminating how upset—and engaged—he was by the national tragedy. Danny Barbiero, for one, remained stunned that his roommate knew the name of every political figure who appeared on TV news reports, from some Nebraska congressman to an assistant secretary of commerce. "What sophomore in college today can look at the television and know every Cabinet member, [and] who [were the] Senate committee heads?" Barbiero marveled. "He just knew it. He was so immersed in government already at that age it just amazed me."

His cousin Serita Winthrop came straight up to New Haven from New York to be with John after hearing the news. Anxious to pry him away from the TV, Serita finally persuaded her cousin to take a break and join her on a long, therapeutic walk. "We were walking about the campus at Yale University," Winthrop remembered, "and he talked at length about how much he admired Kennedy. He had a kind of grief [over the President's] death that I didn't want to intrude upon. We walked around the campus at night for hours, solemn. It was pitch-black outside. He had so clearly invested in the President. It was more than just sharing the same initials—it was as if a family member had been killed. I just didn't know John well enough to comfort him," Winthrop continued. "I wish I would have consoled him more."

Winthrop and Kerry interrupted their walk all over New Haven to pray for his fallen hero at St. Thomas More Chapel. "It was a huge stunner to me," Kerry said about that tragic day. "Just really pulled the rug out from under me. It was incomprehensible, enormous. It sort of robbed us of our moment, our youth, all the enthusiasm we had for a while. And then I watched everything on television. . . . I watched Lee Harvey Oswald getting shot on TV. Like millions of Americans, I just sat there through this incredibly sad drama. Watched John-John salute. I'll never forget." Blakely Fetridge recorded the drama in her diary: "We watched the ceremony of placing Kennedy's body in the Capitol to lie in State," she wrote on November 24. "So very touching and between H and Johnny—they knew so many people there." The following afternoon she simply wrote of the funeral: "A sad, sad day."

Julia Thorne echoed other friends' observations of how distraught Kerry was. She was at home in Manhattan when John called from New Haven to deliver the awful news, which she had already heard. "John just loved Kennedy," she noted. "With the President gone he felt cheated." Kerry, however, would soon feel more directly abandoned. While John Kerry was at Yale that semester, Julia was working at the Tiffany store on Fifth Avenue, selling crystal and china on the third floor. She had succumbed to the gainful employment only at the insistence of her father, who had wanted her to know what it meant to earn one's way through life. "I left Tiffany before Christmas," Thorne laughed, "and moved back to Rome. No way was I going to stick around and do any of that return-business stuff." She also left John Kerry behind.

She did take the occasional break from jet-setting to aid her father at his behest: for example, she fund-raised in Florence and Milan supporting the presidential candidacy of Barry Goldwater, the right-wing Republican senator whose European campaign operation the elder Thorne spearheaded. She also found time to date a few handsome Italians, but John Kerry never faded from her mind. They had become too close; what's more, he simply wouldn't let her be. "John was nothing if not persistent," Thorne related. "He is a fighter."

One of the most significant days of Kerry's college career occurred on May 6, 1964, when he was elected president of the Yale Political Union, an undergraduate organization aimed at promoting public policy dialogue on campus. "The Union can be of undoubted value to the nation as well as the University, provided it maintains independence and voices the true thoughts of those participating," Franklin Roosevelt had proclaimed about the venerable Yale forum, which was started in 1935. "Honest debates will help in the search for truthful answers." As president of the Union, Kerry was following in the esteemed footsteps of McGeorge Bundy, William F. Buckley Jr., William Scranton, William P. Bundy, and John Lindsay. Every month the Union held debates and brought in guest speakers. In recent years Eleanor Roosevelt, Robert Taft, and Barry Goldwater had spoken at the Union. Kerry immediately announced he would expand that tradition. "At dinner we gave Johnny a cake—red, white, and blue, that said 'Yippee!!' on it," Blakely Fetridge wrote in her diary. "We decided that for all the further successes—especially when he's elected President of the U.S.—that we'll send him a 'Yippee!' cake. He was really touched by both the cake and the telegram—he was just like a little boy and really didn't know what to say or how to say it."

The Yale Political Union immersed Kerry in politics more than ever. Although he could never warm up to the notion of Lyndon Johnson as president, he thought the prospect of Goldwater in the White House would be a disaster for America. While he championed LBJ's Great Society programs, which he credited John F. Kennedy with initiating, it was Martin Luther King Jr. whom he emotionally embraced. Often, after dinner, he would call Reverend Walker at St. Paul's to get his take on civil rights legislation like the 1964 act, which forbade job discrimination and the segregation of public accommodations, and the 1965 law, which guaranteed African-American voting rights. When Congressman Allard Lowenstein visited Yale, Kerry went to listen to him speak. "He was amazing," Kerry recalled. "He personally inspired me more than any other speaker I've ever heard. He galvanized my emotions about the civil rights movement." And when in early August 1964 the U.S. destroyers *Maddox* and *Turner Joy* were fired on (supposedly) by North Vietnamese torpedo boats, Kerry like most Americans was outraged. As President Johnson rushed the Gulf of Tonkin Resolution through Congress, accusing the North Vietnamese of "open aggression on the high seas," Kerry agreed that the United States had a right to defend itself. Only years later, when he became friends with sailors on the *Maddox* and *Turner Joy*, did he realize that Johnson had misled both Congress and the American people—the destroyers had not been fired on.

On November 3, Lyndon B. Johnson defeated Barry Goldwater in a landslide victory. Kerry was, of course, ecstatic. But in a heated debate with Danny Barbiero he stood up for Goldwater's belief in tax cuts. "Johnny has really begun to see 'the light,' " Fetridge, a reluctant Goldwater supporter recorded in her diary on election night. "I'll bet eventually Johnny switches over to the GOP, which means quite a bit to H and me, for I'm almost positive that Johnny will run for office and that H (and I) will have a great part and influence upon his campaign. (I can just see in 30 years—President J.F. Kerry, GOP, with Secretary of the Treasury HH Bundy, III, Attorney General D.P. Barbiero, [both Secretary of Health, Education and Welfare and] official White House hostess for the unmarried President, Blakely Fetridge Bundy!!) But even if not that, I'll bet that Johnny will at least be Congressman and probably Senator. . . ."

Besides politics, Kerry continued to be obsessed by Julia Thorne, who came to New Haven for a visit in January 1965. Things went extremely well. "Judy and Johnny are really in love," Fetridge wrote in her diary that month. When Julia returned to Europe, Kerry did his best to hold on to her via a flood of gushy letters addressed to "Darling Bambi," "My Darling Judy," or some variation on his pet name for her, Chipmunk. (Her nickname for him was Pterodactyl, because his long face made him look like a dinosaur.) His letters recounted soccer games and mentioned his dreams of skiing the world's best slopes. But it was ice hockey that, although not his best sport, captured his enthusiasm most: "Bambi, I don't want to sound my horn but I've been playing some 'good' hockey these last few games," Kerry wrote on February 19. "We beat Brown 11–2 last week and I scored 2 goals and got five assists. Today we played Holy Cross and beat them 13–4 and I scored 4 goals and got 4 assists. Our line scored 9 out of 13 goals and has generally been moving well together."

But mostly these letters were overflowing with wistful longings for his girl: wishing they could hike the Appalachians or browse the antique shops of Boston together. Simultaneously he began a habit (which continued into the 1970s) of keeping an occasional journal, its entries often filled with meditations about Julia. "It's a marvelous feeling to communicate with another individual as we have been—openly and frankly—and most importantly with a complete lack of pretensions or insincerity," Kerry wrote her on February 10, 1965. "Some of what we talked about I wrote for inclusion in my [journal, in] which I try to jot any miscellaneous and semi-important ideas that crop up at all times."

As head of Yale's Political Union in his junior year, he got to meet with many of the era's leading political lights; during one week in February, for example, he had private audiences with Florida Congressman Paul Rogers, Senators Hugh Scott Jr. of Pennsylvania and Thomas Kuchel of California, Governor John Chafee of Rhode Island, and civil rights leader James Farmer of the Congress of Racial Equality. "Sounds faster than it is actually, but I imagine that I'll survive," he wrote Thorne about his busy schedule. "With everything that has to be done, I don't see how there'll be time to really languish."

Kerry's growing political involvement at Yale got him invited to be a guest speaker at Choate—the renowned Connecticut prep school John F. Kennedy attended—that same month. It was an unusual honor for a junior in college, and he took it seriously. He arrived at the boarding school accompanied by David Thorne, both in their handsomest suits, and spoke to the thirty-odd high schoolers for nearly an hour. Kerry boldly chose the Vietnam War as his topic, though he declined to offer any proposals for ending the conflict. Instead, he outlined the history of Vietnam, covering everything from French colonialism to the rise of Ho Chi Minh to the lessons of the 1954 Geneva Conference, which had partitioned the country into two uneasy nations. According to an article in the *Choate News*, Kerry claimed that he had originally supported a complete U.S. withdrawal on the grounds that the South Vietnamese government had fallen into disarray, anti-Americanism pervaded Southeast Asia, the Johnson administration's policies were failing, and the domino theory was a myth. That said, Kerry asserted that he had come to realize how important it remained for the United States not to lose face, and that President Johnson thus had to either score a military victory or negotiate a peace. "In the future, the U.S. must fix goals which are tenable," the Yale junior intoned. "These goals should recognize priorities that correspond minutely with our best national interests. We should concern ourselves less with other ideologies and attempt to apply a policy which is both sensitive and compatible with the expressed desires and cultures of the people involved."

The Choate speech, at least by Kerry's account, was a success. "In all honesty, I really believe that it went well and that they were pleased," he wrote to Julia.

I hadn't any time to go over my speech at length before I gave it, and I was afraid that I would be too glued to my notes. But when I got up there, I felt sharper and more confident than I have ever felt before. I rarely looked at my notes and really talked to these guys. The questions went very well also. I really was pleased to pieces and very encouraged by the whole visit. A photographer took pictures for the Choate paper, which they said they would send me, and when they do, I'll send it along to you. It was very strange to have these young, and some not so young [people], calling me "sir" and asking me questions about policy in Vietnam. I got a great kick out of the entire evening, and more important, I proved something to myself—something very important too.

What Kerry had discovered was that he had the ability to be a public person and could translate the complexities of foreign policy into clear analytical platitudes. Kerry found that getting recognition as well as respect for his ideas was gratifying. He couldn't help but tell his girlfriend about how he had heard his name announced over the loudspeaker, how a Hartford radio station had interviewed him, and how he had been given a VIP tour of the school. And that wasn't all. The day after his triumph at Choate, Kerry reported having the privilege of hosting Jonathan B. Bingham, a Democratic congressman from New York who was the featured speaker at the Yale Political Union's thirtieth-anniversary celebration. Bingham, a member of the class of 1936, spoke about China to a packed house at Yale's Woolsy Hall. After his address Kerry, who had introduced the congressman, presented Bingham with the William Benton Bowl. Kerry was mentioned by name in the *Hartford Courant*'s report on the event and was shown with Bingham in a *Yale Daily News* photograph. "The banquet was an unqualified success," he crowed to the distant object of his affections. "Everything went perfectly and both Alumni and members were abundant with their congratulations."

On March 12, 1965, the *Yale News* ran a story announcing that Kerry had won the Ten Eyck Speech Prize, winning $125. The theme of his competitive address—billed as a search for a modern-day Prometheus—was the inherent danger of America's stretching itself too thin in its international commitments. "It is the specter of Western Imperialism that causes more fear among Africans and Asians than communism, and thus it is self-defeating," Kerry orated, even though he was sick with a cold. He complained that the United States had "grossly overextended" itself in "areas where we have no vital primary interest." The speech set out a theme he later expanded on during his class oration at graduation.

Vietnam was only partially on the minds of John Kerry and David Thorne when they traveled around Europe during the summer of 1965, the months when President Lyndon Johnson ordered the deployment of 50,000 additional U.S. troops to Southeast Asia, with a commitment for 50,000 more to follow. Thorne had purchased an English cab they named Baxter, and they treaded all over France and Spain. "It was fun to cruise along at forty miles per hour in this big old black chugga-chugga taxicab with this sundeck chair," Thorne recalled. "English taxis had this open baggage compartment and we put a big comfortable folding canvas chair inside it. So one of us would drive, the other would sit in the chair. It was like an open-air car." They visited Julia at the Thorne summer house along the Tuscan seacoast near Porto Ercole. The highlight of the trip came in Pamplona, Spain, where like so many young men under the spell of Hemingway they decided to run with the bulls.

Throughout his Yale years, Kerry maintained a large green leather scrapbook, which was a gift from his mother.* Among its contents were pictures of himself with John F. Kennedy at Newport, clippings about his athletic feats from the *Yale Daily News*, invitations to various public forums, and incoming correspondence—some of considerable note. These include letters praising the job young Kerry was doing as head of Yale's Political Union from no less than Adlai Stevenson, Clare Boothe Luce, and Clementine Churchill, among others. "You are going forth into a fairly troubled world," William P. Bundy wrote in an April 16, 1966, missive to Kerry on

---

**Kerry continued to save various items in this scrapbook even after he returned from Vietnam—for example, his certificate for having visited "the Top of the World" in Greenland in 1969.*

the obligations of hailing from Yale, "but I have a hunch you will make some contribution to it." For every item on Kerry's athletic accomplishments, such as a November 22, 1965, *Yale Daily News* article, "Yale Booters Bomb Crimson 6–3," the scrapbook boasts two pertaining to his first-rate agility as a debater. Judging by his own yellowed clippings, it appears that Kerry, or the team he was on, defeated every comer at the podium—or at least planned to, as indicated in the May 13, 1965, *Yale Daily News* article "Triangular Ivy Debaters to Discuss Vietnam Policy." "Without question the Yale Political Union exposed me to the attention of a lot of people," Kerry admitted. "I made friends through debating."

It was Kerry's dynamic performance as president of the Yale Political Union that made him known, for example, to Fred Smith, a smart, burly young aviation enthusiast from Memphis, Tennessee, who would go on to found Federal Express. Although born in the Mississippi Delta, Smith spoke in the clipped, businesslike tones of American's ruling class, befitting his childhood in posh central Memphis, attendance at its tony University School, and summers at a pricey camp in the hills of North Carolina. These privileges resulted from the fortune his father had made as the founder of Toddle House, a franchise chain of twenty-four-hour, all-day-breakfast eateries specializing in hash browns and homemade pies. "Every Southerner," Fred Smith proclaimed with filial pride, "has been in a Toddle House." Fred Smith also knew Elvis Presley, who lived just a few blocks from his parents' house. "Elvis was a great guy," Smith demurred years later. "Later, after I got out of the military, I flew him in my plane." One summer young Fred was sent to St. John's Military Academy in Wisconsin, where he met football coach Gib Holgate, a legendary recruiter with connections at Yale University. "He convinced me to go to Yale," Smith remembered. "It was that simple."

Smith recalled that Kerry "was a very, very powerful speaker, very articulate. He was a very visible figure on campus." Their friendship blossomed when both were tapped on April 23, 1965, to join Skull and Bones, the most elite secret society on campus. "I always thought that it was a mistake, that they had asked the wrong Smith," Kerry's pal said. "But of course they hadn't."

The storied Order of Skull and Bones was created in 1832 by Yale student William H. Russell, no doubt for reasons utterly unconnected to the club's suspiciously nefarious image today. It is still housed in a strange mausoleum-style edifice now entwined in ivy, 322 its secret code number in honor of the years of its founding and the death of Athenian statesman Demosthenes. Skull and Bones' reputation as a breeding ground for high-level public servants, rather than the particularly silly frat house it at first appears to be, evolved over the decades because so many of its members hailed from Eastern establishment families (including the Bundys, Bushes, Harrimans, Lords, Lovetts, Phelpses, Rockefellers, Tafts, and Whitneys). That said, at least to the uninitiated the organization has always smacked mostly of the overprivileged yet jejune. Because its members are sworn to lifelong secrecy, no outsider can say for sure what goes on at Skull and Bones, or why. "The initiation of new Knights—Bonesmen are called 'Knights' from the time they are initiated to the time they initiate the next group," Alexandra Robbins surmises in *Secrets of the Tomb: Skull and Bones, the Ivy League, and the Hidden Paths to Power*, "is a ceremony that surely has sophomoric moments, but my understanding is that it is intended to introduce new members to the society's culture, rituals, songs, history, and love in a way that will impress and awe them, not disgust and repel them."

Like most Yalies so honored, John Kerry was thrilled with getting tapped. And it was gratifying that both Dick Pershing and David Thorne were also chosen to join. Since his first day wandering around New Haven informing Danny Barbiero of the grandeur of Eli tradition, Kerry had sworn to exemplify the qualities carved above the gargoyles guarding Harkness Tower. Joining Skull and Bones was another step in what he had come to see as a natural progression. As an old *Scribner's* article had described the invitation to Skull and Bones: "This is the highest honor which a Yale man can receive from his fellows, and because it comes from them he sets it above scholastic distinction."

While going through the Bones initiation process, John Kerry and Fred Smith also began flying together, another Yale tradition. The Yale Aviation Club had been founded in 1915 by sophomore Trubee Davison and was the first naval air reserve unit in the United States. Davison had met the renowned Arctic explorer Admiral Robert Peary and gotten him to help the club acquire a Curtiss Model F seaplane. The Yale Aviation Club really took off during World War I, spawning such leaders of the next war as Robert Lovett, who became assistant secretary of war, and Artemus Gates, who would serve as assistant secretary of the Navy for air. Despite its illustrious history, the club had been dormant for quite some time when Fred Smith decided to resurrect it in 1964. Kerry was one of his first recruits. Among the renascent club's flight instructors was Norwood Russell Hanson, a philosophy of science professor who had served as a Marine fighter pilot in World War II, logging 2,600 flight hours. Nicknamed the Flying Professor, Hanson specialized in aerial stunts. "He had an F8F Grumman Bearcat that he used to buzz the Yale Bowl with during football games, and he was tragically killed in the airplane flying in some weather that he shouldn't have been in," Smith remembered. Another Yale Aviation Club flight teacher was Howard Weaver, the provost of the university.

Escaping New Haven via Piper Comanche became Kerry and Smith's favorite occasional diversion. "I used to fly instead of going to class," Kerry confessed. "I majored in flying my senior year." The pair sometimes took off to Long Island to track down Dick Pershing or David Thorne. Massachusetts also proved a favorite destination, with Kerry and Smith putting down in a town like Amherst or Boston just to attend a party. Kerry sometimes talked about his political aspirations on these flights, casually mentioning he might aim to be elected a congressman or senator someday. "It certainly didn't bother me," Smith recalled. "I didn't take it any differently than somebody who said [he] wanted to be a doctor or a professor or an Arctic explorer, which is what all of our other friends wanted to do. You've got to understand that a large percentage of our class was much more traditional in their view of society, and of their obligations to the country, and so forth. And we were all so heavily influenced by John Kennedy. I mean, Kerry was a Kennedy Democrat from the get-go. You know: bear any burden, pay any price. That was a sort of mantra of our generation and our class. It shifted dramatically between '66 and '68. So I would say we were much more like the class of '56—you know, the classes behind us."

Without question, the draft was on the minds of all high school and college seniors in 1965–66. As the need for more soldiers to serve in Vietnam rose, the Selective Service System was tapping an unprecedented 30,000 draftees a month. Graduation meant the end of a young man's educational deferment, and other deferments were harder to obtain. Some men paid a draft counselor or attorney for advice—a popular slogan that year was "If you got the dough, you don't have to go"—and others applied to graduate school. Recognizing the changing nature of the draft, Kerry signed up in late 1965 for the U.S. Navy, with actual enlistment to come after graduation the following spring. As his last year at Yale passed, he was forced to live with that decision.

A week before graduation, in early June 1966, Kerry and the fourteen other senior Bonesmen convened for a memorable weekend on an island in the St. Lawrence River in upstate New York. Besides beer drinking and gossip, Vietnam was on everybody's mind. In a matter of months, Kerry, Thorne, Smith, and Barbiero would all be entering the U.S. military. Journalist Joe Klein interviewed Kerry about what transpired that long weekend for *The New Yorker* in 2002. At the time Kerry had already written what he deemed a "sophomoric" speech about "life after graduation" to deliver as the class oration. "I decided that I couldn't give that speech," Kerry told Klein. "I couldn't get up there and go through the claptrap. I remember there was no electricity in the cabin. I remember staying up with a candle writing my speech in the wee hours of the night, rewriting and rewriting. It reflected what I felt and what we were all thinking about."

Certainly the greatest honor accorded John Kerry in his four years at Yale came at the end of his time there, when he was chosen to deliver his class oration on Sunday, June 12, 1966. Yale president Kingman Brewster Jr. started the convocation with an admonition for national obligation. It was a boilerplate speech. So it came as something of a surprise when Kerry, towering over all the professors and university officials on the stage, took the microphone and offered a critical analysis of the Johnson administration's foreign policy. "What was an excess of isolationism has become an excess of interventionism," Kerry proclaimed, drawing upon lessons learned in Gaddis Smith's history class. "And this Vietnam War has found our policymakers forcing Americans into a strange corner . . . that if victory escapes us, it would not be the fault of those who led, but of the doubters who stabbed them in the back—notions all too typical of an America that had to find Americans to blame for the takeover in China by the Communists, and then for the takeover in Cuba."

It would be as ridiculous to claim that John Kerry's graduation oration had the slightest impact on his audience as it would be to say the same of any valedictorian's, ever, anywhere. Many members of Yale's class of 1966 profess to remember that Kerry gave a "solid speech," but none can quite recall what it was about. But the day after commencement, the *New Haven*

Journal-Courier did feature the Brewster and Kerry speeches on the front page. As Joe Klein noted in his profile of Kerry, the words the budding statesman uttered that spring afternoon in New Haven would prove prophetic. The address groaned, for example, with nods to John F. Kennedy's call to public duty: "We have not really lost the desire to serve," Kerry pronounced. "We question the very roots of what we are serving."

That "we" marks what mattered in Kerry's speech. It indicated his conscious choice to present himself as a leader of a new generation. He, like others of his social class, wanted it known they were not only willing to serve but eager to serve their country, be it in VISTA, the Peace Corps, or the armed services. By the same token, the more he read about Vietnam the more Kerry began to ask tough questions about U.S. intervention in Vietnam. That June, as Kerry and his classmates were getting fitted for robes, President Johnson rejected the advice of State Department advisor George Ball and others and instead embraced General William Westmoreland's request for more troops to serve in Southeast Asia. The total needed was 431,000. It was a staggering number of young Americans to send into a war zone. "The United States must, I think, bring itself to understand," Kerry avowed, "that the policy of intervention that was right for Western Europe does not and cannot find the same application to the rest of the world." Despite the messages of his oration, as a duty-bound Kennedyite, Kerry was now ready to serve his country.

Kerry's friends Harvey Bundy and Blakely Fetridge were getting married that summer in Chicago. "Everybody stayed at the Ambassador East Hotel," Bundy recalled. We had thirteen groomsmen and seven were from Yale." While most of the gang would be staying in Chicago for the weekend, Kerry was forced to leave right after the ceremony. His number was up: the U.S. Navy was expecting him at boot camp. It was time to face his future, and it lay before him half a world away in the battle zones of Southeast Asia. It saddened Kerry to acknowledge that he and his best pals—Harvey Bundy, David Thorne, Freddie Smith, John Whitman, Dick Pershing, and Danny Barbiero—would never again be together the way they had at Yale. But the ever more serious side of John Kerry understood that growing up meant putting away childish things like spring breaks and even dear friends whose own futures demanded they all go their separate ways.

So Kerry packed his duffel bag solemnly and contemplated the potential realities of the service he was about to enter. He had made an irrevocable decision months earlier; now he had to live with the consequences. "I'll never forget the weird shock of having been at a wedding dinner the night before, and then the next day standing at attention while some other guy blared in my face and shaved my head," Kerry could laugh as a U.S. senator years later. "I thought the world had come to an end. It was awful."

As Kerry peered out the window of the taxi taking him to O'Hare Airport, he scrutinized the proof that Chicago still stood as America's most American city: its Italian pizzerias and German beer halls, side by side with its Jewish synagogues, Polish bakeries, African-American churches, and all the other savory ingredients in that windy microcosm of this nation's boiling melting pot. The streets teemed with people beneath skyscrapers surpassed only in New York, and it occurred to him that Chicago was where the American dream had taken root. He couldn't possibly have guessed that two summers later Carl Sandburg's "City of the Big Shoulders" would turn into the home-front battle zone for his generation, the whole world watching as the clangs of tear-gas canisters and the thuds of billy clubs drowned out the civility Americans expected from their national political conventions.

On his flight to Boston, Kerry realized he had crossed one of those life-changing Rubicons he had read about at St. Paul's School. He was unsettled but excited about what lay before him, and something in him warmed at the notion that when he came to his fork in the road he had shown the independence to opt for the path not taken by most Ivy Leaguers his age: Vietnam. As a child of considerable privilege he could certainly have gone for the easy out, be it joining the National Guard, getting married, or asking a well-placed family friend to finagle him a draft deferment. Although he never considered it, he could even have proclaimed himself a conscientious objector, fled to Canada, or found Quakerism. Instead he had done the honorable thing and enlisted in the Navy, and he was proud of it. While his college friends were still partying at the wedding reception, he was headed back to Boston, where he would quickly pack and drive to Officer Candidate School in Newport, Rhode Island.

"As Yogi Berra said, 'I came to the fork in the road and I took it.' " Kerry chuckled at the mangling of Robert Frost's "The Road Not Taken." "Now that, to me, is real philosophy. That's one of my favorite philosophical statements. But keep in mind, many of my best friends were going into the military—Dick Pershing, Freddie Smith, Danny Barbiero, John Whitman. We never plotted it. We just all thought that was the right thing to do. President Kennedy had called on our generation to 'pay any price' for global freedom, so duty dictated that we enlist."

Chapter Three

California Bound

Newport's Officer Candidate School (OCS) put John Kerry through a regimented sixteen-week program that whipped him into shape and jarred the newfound sense of fashion Julia Thorne had cultivated in him. Every day he stared longingly out at Narragansett Bay, wishing he could suddenly set sail into Rhode Island Sound and visit his beloved Elizabeth Islands up the craggy coast of New England or down to the Florida Keys. Instead, he was a captive in Newport, yelling, "Yes, sir," and dropping for fifty push-ups at the bark of every superior. Living together with an entire unit of men in close quarters also meant that privacy was nonexistent. He didn't even look like himself anymore; the image that now stared back at him from the mirror every morning bore no resemblance to the Kennedyesque Yale debater he had been. John Kerry had been turned into the OCS version of a common grunt. Years later, he could joke about how bad he looked bullet-headed. "Big-time shaved, big-time shock, and big-time ugly!" he laughed. "I was not meant to have no hair. God gave me a strange head shape, and baldness just doesn't work. It's not Bruce Willis's look at all that I had. It was an ungainly sight. I was miserable, glad that we weren't able to leave the base. Going off base would have [had] riot-shock value: cruel and unusual punishment to gazers."

Kerry's memories of OCS center on the grueling physical demands: moving all the time, up early, marching everywhere, close order drills with his fellow recruits. It was not an easygoing environment, to say the least. For four months, Kerry learned what it meant to be part of the proud U.S. Navy tradition. The OCS program was an indoctrination process, teaching its men to exercise and march and salute and conduct themselves according to regulations at all times. The officers-to-be received technical training as well. "We learned the basics of the Navy," Kerry recalled, "the basics of navigation. We learned how to speak in naval terms and learned about American sea power. I learned shipboard communications and Morse code—all that kind of stuff. There was not a lot of free time, but I stayed in touch with my family and Yale friends." The first eight weeks were the easiest, largely because of the mild autumn weather. By November, however, the cold, gray wind blew in from Canada, making the trainees shiver all day long. "Sweat would be pouring out of our bodies, yet our faces were frozen," Kerry remembered. "It was an easy way to get terribly sick."

Even in interviews many years later, men who trained with Kerry at Newport invariably mentioned how "serious minded" and "deeply intellectual" he was. While his barracks mates played cards and traded scuttlebutt, Kerry would often be off in a corner reading William Styron or John O'Hara. Progressive politics became ever more his passion. Like Robert F. Kennedy, for a young white man of privileged background, Kerry always displayed an uncommonly incongruous instinct for siding with the underdog. While others laughed at heavyweight champion Cassius Clay for changing his name to Muhammad Ali, Kerry lauded him for "shedding his slave name." Clearly the civil rights movement had touched him deeply—inspired, in part, by Reverend John Walker. Letters and journals Kerry wrote in the late 1960s show the depth of his interest in racial equality, lauding the brilliance of James Baldwin's 1963 essay collection *The Fire Next Time* and recommending *The Autobiography of Malcolm X*, which he couldn't put down. "Our mother had a great sense of justice," explained his sister Peggy. "So John identified with Martin Luther King Jr."

John Kerry's political and social thinking may have pushed the cutting edge of progressivism, yet there remained something quaint and old-fashioned about him. While he adored the Beatles, the Rolling Stones, Simon and Garfunkel, and Joan Baez, he also strategized how to get tickets to the Broadway opening of Angela Lansbury and Beatrice Arthur in *Mame*. Uninterested in Andy Warhol's pop art, he eagerly attended the new J.M.W. Turner exhibit at New York's Museum of Modern Art. Untempted by *Bewitched* or *The Monkees*, Kerry's TV viewing centered on news programs such as *ABC Scope*, and the CBS and NBC News segments "Vietnam Perspective" and "Vietnam Weekly Review."

Of course, Kerry spent much of his time at Newport pining for Julia Thorne, who was living in Rome and dating other men while he was stuck in the bilge level of the U.S. Navy, sweating and shivering. What he didn't know was how much Thorne longed for him, too. "What I missed was his passion, his focus—this man who knew what he was doing and what he wanted and was not afraid of being successful," she recalled. "I mean, I was very impressed by that. And he had a lot of parts of him, threads of him, that I knew well. He always knew what he wanted to do, and he wasn't afraid to go get it. I had come from this very precious world that was very weighted down, not with stuffiness, but proper ways of behaving. One doesn't do this, one doesn't do that and, you know, it's a very mannered world. Here was John Kerry, who was so focused, and cared so much, that he was a liberating force; he was my way out of the corner I had been painted in. . . . So while I was in Italy, I missed him."

Anxious to get his mind off Julia as well as to reconnect with his college friends, on one of his first liberties Kerry attended the Harvard-Yale football game. Wanting to stay current as a pilot, he rented a Cessna 172 that Saturday to fly himself the ninety-odd miles to New Haven and, as always, even a short hop in the wild blue yonder brought him clarity and calm. After the game, however—won by Yale 24–20—a wave of nausea engulfed him. "I just got sick as hell," Kerry recalled. "I remember collapsing and being whisked to the infirmary at Yale with shivers and chills and fever. It was the sickest I've ever been, except for that bout of scarlet fever as a kid."

He was hospitalized with a diagnosis of pneumonia. His parents were in Groton, his girlfriend was in Rome, his OCS classmates were in Newport, and he felt pretty miserable. Convincing himself he felt up to it, Kerry finagled a premature release from the hospital, returned to Rhode Island, and resumed his regular training duties. He soon suffered a relapse and wound up in the Newport Naval Hospital for almost two weeks. "I kept studying from bed so that I could graduate with my class, and I did," Kerry declared with pride. "I got commissioned around December 11—my birthday—and got to go home to the North Shore for Christmas Eve."

Home had always been a complicated concept for John Kerry. He was a son of Massachusetts, for sure, but he had spent his life bouncing between New England and Europe, with the occasional long layover in Washington, D.C. He couldn't help but feel like an itinerant, a young nomad with no geographic base. He did, however, have the touchstone, of Groton House, the Massachusetts home of his uncle and aunt, Fred and Angela Winthrop. While at boarding school, he often spent Christmas at this 300-acre estate, his "home away from home."

The immediate area boasted more history than most in the United States: the town of Ipswich claimed to be the birthplace of the American Revolution, on the grounds that it had hosted the Reverend John Wise's pioneering 1687 protest refusing "taxation without representation." Two centuries later, in 1903, Frederic Winthrop Sr. bought the Groton House property, and raised a family and many fine horses there at his summer house. A well-connected financier, Winthrop Sr. enjoyed, among many others, the distinction of serving as best man at the celebrated wedding of Alice Roosevelt and Nicholas Longworth. His son, Frederic Winthrop Jr., would keep the Groton House property running during the Great Depression, raising Guernsey cows and cultivating premium hay. An ardent conservationist, Winthrop Jr.—John Kerry's uncle—transformed the family estate into one of the most beautifully landscaped parcels on the North Shore. "Groton House was a special place for all of us," Peggy Kerry averred. "There were seven Winthrop cousins, so when we all got together there was quite a stir. Christmastime was especially wonderful. There was always a great big tree in the window with lots of presents around it. The entire house just bustled with activity and good cheer."

Christmas Day at Groton House in 1966 marked a poignant moment in John Kerry's family life. After a great Christmas lunch, as the snow whirled about the grand old house, the deliberate young ensign embraced each of his relations with an unspoken finality. They knew he wasn't heading to the Riviera this time. Then, he got into the car and was driven off to Boston's Logan Airport by his father, headed for San Francisco's Treasure Island U.S. Naval Training Base. Shortly after they set out, however, Richard Kerry's car broke down. The elder Kerry managed to maneuver the vehicle into a gas station, whereupon his son quickly asked one of the customers if he could drop him off at the airport. He embraced his father and set off for California. "There was a great sense of journey, heading to San Francisco," he reminisced. "I was twenty-three years old and going off to be in the Navy, to duty on a ship. There was a sense of romance about it. It was exciting. I was kind of up for it. My father was more emotional. His kid was going off in a uniform to perhaps fight in a war. He was also more sensitive than I because he was very dubious about Vietnam at that point."

By the end of 1966, San Francisco had turned a giggle into a wild, carnival-like counterculture street theater that would reach its psychedelic apotheosis a few months later in the 1967 Summer of Love. Anarchy was in the air; spelling "Amerika" and putting Chairman Mao's *Little Red Book* of Communist precepts in vogue. It was stylish not to be, and downright hip, to look unkempt. "The signs of youth rebellion were everywhere," Kerry marveled. "Fashion-defying clothes and soapbox radicalism were rampant."

Ensign John F. Kerry arrived at the Treasure Island Training Base to learn "Damage Control Assistance," a duty assigned junior officers on Navy destroyers. He spent the first months of 1967 billeted at the base's bachelor officers' quarters, and generally attended class from 7:30 A.M. to 5:00 P.M., five days a week. Hardly the beehive of activity it had been during the World War II years, Treasure Island had a combined military and civilian population of about 9,000. Many of those personnel proved interesting, however, and despite his polite reserve—and his taste for political debate—Kerry made friends quickly. Among the first was twenty-four-year-old Paul Nace, a Brooklyn native who had earned an M.B.A. at Columbia University. "More than anything else, it was fun," Nace remembered of his time in the Damage Control Assistance program with Kerry at Treasure Island. "We had most evenings off. There was some serious studying to do but neither John nor I found it a very difficult school." On weekends, they would wear civilian clothes and wander around San Francisco, often dining at the Old Spaghetti Factory near Fisherman's Wharf where for $3.95 they could get all the salad and pasta they could pack away—not to mention pitchers of beer at just a dollar apiece. "That restaurant didn't make any money on our table of young naval officers," Nace laughed, "because we could all eat a lot."

Kerry and Nace also assayed a couple of Friday night rock concerts at venues like legendary music promoter Bill Graham's Fillmore Auditorium, where they stood dumbstruck by the pulsating strobe lights, back-projected lava-lamp color blobs, and Day-Glo body-painted dancers. It was jarring to move from a military base to the emerging hippie culture. It was not their scene, to say the least. Kerry harbored a fundamental skepticism for the counterculture in general, and its enthusiasm for illicit drugs in particular.

"I wouldn't have been a part of his life if he had even a touch of the hippie in him," Julia Thorne declared. "The drug culture was not for me." What Kerry disapproved of most about the hippies, however, was their "sloppy thinking." He found their political views simplistic and self-indulgent. "Allen Ginsberg, Ken Kesey, and stuff like that didn't grab me," he emphasized. "It never did—even when I came back from Vietnam. I could never warm up to Abbie Hoffman, Tim Leary, and those type of guys at all. I didn't like their style, their message. I was in a different place. I mean, I was against the war, but I didn't like either their social or cultural or political agenda."

Sometimes, to clear his mind, Kerry would hike up Lombard Street to view the San Francisco waterfront from Coit Tower, and be dazzled by the sight of hundreds of small sailboats plying the choppy blue water. He also wandered along the piers of Fisherman's Wharf, occasionally eyeing a battle-scarred World War II destroyer or a refitted timber barge looking straight out of the nineteenth century. He enjoyed sampling the abalone and stone crabs and oysters from the fishmongers' stalls. By far his favorite diversion, however, came on the occasional Saturday when he could rent a tiny airplane and take Nace up for a short joy flight, setting down on the little airstrips in towns like San Luis Obispo or Crescent City for lunch.

Although both Kerry and Nace seem to enjoy reminiscing about their good times on Treasure Island, the naval training they received there was nothing if not serious. The instructors drilled into every officer's head that men's lives would be in their hands in the event of an emergency. They were taught what to do in a variety of dire circumstances—for example, how to quarantine a ship under biological attack, and how to shut a ship down should nuclear radiation be detected. "If a ship were hit by a torpedo you were in charge of it," Nace summed up, "you know, shifting the load around, saving fuel, flooding compartments to keep the ship balanced and afloat. . . . You're in charge of all nuclear and biological and chemical warfare attacks on the ship. So you're responsible to make sure that the ship can be sealed and pressurized so that nothing can get in if you're under a gas attack. That was the schooling that we did."

Nevertheless, sometimes Kerry and Nace still had a hard time taking their training seriously—seeing that the Navy had manufactured a training boat with so many holes in its hull that *Kon-Tiki* craftsman Thor Heyerdahl himself couldn't have been expected to keep it afloat. "I mean, we were up to our asses in water sometimes, plugging some hole in a fake ship with a mattress as if we were in World War II and someone fired a torpedo at us," Kerry chortled. "It was completely comical, running around from hole to hole."

The high point of Kerry's four months in San Francisco occurred when Julia Thorne flew in from New York for a weeklong visit. She put up at the St. Francis Hotel at Union Square. They traveled around the Bay Area, having the romantic time of their lives. "John took me all around," Thorne recalled. "As always, his enthusiasm for everything was overwhelming. He showed me all around Fisherman's Wharf and the Presidio. We ate in Chinatown, and went shopping for clothes. It was wonderful." No matter how hard she tried, Julia Thorne couldn't get John Kerry out of her system. "There was no question that John and I were in love with each other, but I had consistently left him over those three years. We came back together in San Francisco. I think I had left him three times—at least once a year. It was impetuous, it was spoiled on my part, and it was always for a European man. I never two-timed him, I just left him. At one point he never wanted to see me again. Yeah, I just behaved so badly, his friends didn't want anything to do with me either." But those idyllic days at the St. Francis Hotel brought the couple back together again, and closer than ever. "That trip gave me my first glimmer of a question about my political convictions. The world that my parents had espoused was perhaps not right."

At the conclusion of naval-officer training, Kerry got a short leave, which he spent with family in Massachusetts, and then received his assignment, to the U.S.S. *Gridley*, which was then deployed in the South China Sea and homeported in Long Beach, California. Hoping to join the *Gridley* at sea, he was instead sent for several more months of classes, this time at air intercept controller school in San Diego. In between his studies, Kerry fell in love with the city. "We screwed around a lot," said David Thorne, who was there to attend submarine warfare school at the same time. "Meaning John and I really checked out California. There was a *lot* of time flying. John had gotten his license in New Haven and had gotten very proficient in aerobatics. Meanwhile, I earned my own pilot's license during those San Diego months."

Over time Kerry would earn a reputation among his friends as a qualified risk-taker. Whether behind the wheel of a car or the throttle of an airplane, he took pleasure in learning the discipline of proficiency and in testing himself against the conditions. He enjoyed "extreme" sports before they had the name: downhill racing, surfing, parasailing, full-impact ice hockey, motorcycle riding, you name it. If it was fun and a test of agility and endurance, John Kerry wanted to try it—and usually succeeded in the offing.

An incident that spring illustrates the point. Kerry and David Thorne each got three-day leaves. As soon as they rented a T-34 airplane from the local Navy club, Thorne proposed flying the 406 miles north to San Francisco. Off they went. Clinging to the Pacific coast, the pair buzzed low over San Simeon and Big Sur, marveling at the bright blue waters crashing upon the rugged rocks. An even more spectacular view could, of course, be found flying over the Golden Gate Bridge. "We were just two naval officers looking for some fun," Thorne rationalized, adding: "Johnny loved to show off in the air, doing various tricks." One trick—which Kerry had heard that Yale provost and World War II flying ace Norwood Russell Hanson had once performed—involved looping around a suspension bridge by pulling 360 degrees tightly around the huge span. "Let's fly under the bridge," Kerry shouted at Thorne. "And then he headed straight for it," Thorne exclaimed. "There was no turning back."

As Kerry steered the T-34 toward the Golden Gate Bridge, a seagull suddenly appeared off their wing, affording a moment of comic relief. Then they heard a loud thump—very loud. The little plane shuddered and Kerry immediately pulled up for altitude. They had run into an entire flock of seagulls. "The scene flipped from 'Looney Tunes' to Alfred Hitchcock," as reporter Michael Kranish wrote in his 2003 *Boston Globe* profile of Kerry. "Suck one into an engine and a young pilot's life story could conclude right there: Yale aviator dreamed of being president, killed on joyride." Unflappable even under the circumstances, Kerry deftly turned the plane away from the bridge, then radioed in to the nearest airfield that he had to make an emergency landing. "It was hair-raising," Thorne marveled decades later. "Our fear was that the wing had been so damaged by the birds that it was going to fall off."

One sure way to raise Kerry's dander is to suggest that he may be a tad reckless. When pressed to explain how flying under the Golden Gate Bridge could possibly be seen as anything but pure kicks, Kerry turned defensive.

"It's not the edge factor," he snapped. "People mistake that; I've heard people say that. It has nothing to do with being 'on the edge.' I don't consider it 'on the edge.' I'm simply fascinated by the discipline [and the] freedom of aerial acrobatics—the art connected to the sensation and the visual. That's what it is. I love the sensation and the view.

"People have loved flying since Icarus," he continued, "I appreciate the technique of combined discipline: the instrumentation, the efficiency, the dozen other things you must do to make a flight work properly. I very much feel in control. I don't feel 'on the edge.' . . . I feel safe, in control. I'm tired of people misinterpreting that. For instance, I'm a glider pilot. I love the effect of flying without an engine. Working the air currents and the thermals, the ridgeways, playing the elements. But part of it is the discipline that you can land almost anywhere. You have a spoiler; you have great control; you can put it down in a desert, in a field, on the road. So I consider it safe."

Just about anybody who has sailing or flying in his or her blood and has been there has a warm spot for San Diego. As Richard Henry Dana opined in his 1840 classic *Two Years Before the Mast*, for example, there are a "security and snugness" to San Diego that make it an ideal site for a Navy port. "For landing and taking on board hides, San Diego is decidedly the best place in California," Dana observed. "The harbour is small and land-locked, there is no surf, the vessels lie within a cable's length of all the beach, and the beach itself is smooth hard sand without rocks or stones." With its fishhook-curved bay, San Diego is the only California harbor south of San Francisco that offers sheltered anchorage from the Pacific coast's treacherous storms.

Long before Ensign Kerry got there, San Diego's Naval Training Center (NTC) had established its reputation as the premier facility in the United States for developing sailors' skills and specialties. Conceived in 1916 by Secretary of the Navy Joseph Daniels, along with Assistant Secretary Franklin D. Roosevelt, the NTC received its official commission in 1923. At the NTC's inception, its Camp Paul Jones could house a maximum of 1,500 recruits. By the time John Kerry arrived in 1967, the facility had grown into a huge compound of sailors who were undergoing various kinds of training in anticipation of their assignments, many of them headed to Vietnam.

As he had at Treasure Island, Kerry lived in the bachelor officers' quarters. He recalled, "We trained hard. But then I got a relapse of pneumonia. I spent a couple of weeks in the hospital. That was just about when I was to report to my ship in Long Beach." A short while later, in front of the Century Plaza Hotel in Los Angeles, the two Navy officers witnessed their first antiwar demonstration. "We just happened to stumble upon it," Kerry said. "The police started moving in for arrests and we bolted. But it left a deep impression on me."

The guided-missile frigate (DLG-21) *Gridley* had been first launched July 31, 1961, from the Puget Sound Bridge and Drydock Company. After a shakedown cruise out of San Diego, on April 8, 1964, the *Gridley* had left Long Beach for duty in Pearl Harbor, Australia, the Philippines, Okinawa, and Japan. In the wake of the infamous Gulf of Tonkin "incident," the *Gridley* rushed to the South China Sea to help bolster the American naval presence off the North Vietnamese coast. By the time Ensign Kerry boarded the *Gridley* on June 8, 1967, the so-called Gray Ghost of the South China Sea had racked up quite an admirable record.

In 2003, the U.S.S. *Gridley*'s Web site opened with a photograph of a twenty-four-year-old John Kerry in his Navy whites, hand on hip, hovering over a pair of veteran sailors. The picture was taken while he was in charge of the frigate's First Division. The caption reads, "Ensign Kerry positions himself to run for the Presidency of the United States!" Clearly proud that a bona fide presidential contender had served on their own U.S.S. *Gridley*, which Kerry did from June 1967 until June 1968, the ship's alumni association responsible for the Web site just as clearly meant to tweak the senator for ignoring them on his march toward the White House. "His biographical materials never mention this period even though he was aboard the *Gridley* much longer than he was in Vietnam," the Web site noted. "Between having these two old salts reporting to him and Captain Allen W. Slifer after him all the time, Ensign Kerry was a busy young man!"

Kerry maintains that he never spoke much about the U.S.S. *Gridley* simply because nothing much of note took place during his tour aboard the vessel. But his correspondence and memories from that period reveal a good deal about his development as a Navy officer. "I was what was called a First Lieutenant aboard the ship, which meant I was responsible for what is called First Division," Kerry said, "which is all the deck work: overseeing all seamen on the ship, the appearance of the vessel. . . . Then when you got out to sea, when you deployed, you're the guys who are manning the line and anchor. You're responsible for the seagoing components."

One *Gridley* veteran, Robert E. Jack, quickly developed a lasting impression of Kerry. "When John first came aboard, he (like many other people on board) was given a nickname (behind his back)—The Beatle," Jack wrote. "The reference was to . . . John's haircut, which was longer than usually seen on a naval officer of the time." Jack also recalled that at first Kerry was given a "hard time" for any shortcomings he had, but grew within a few months to be "considered by all as a solid, responsible officer."

David Thorne lived at the time close to the ocean in nearby Belmont Shores. Whenever Kerry could, he trekked over to the two-bedroom bachelor pad his old pal rented with a friend on the second floor of a weathered bungalow. "It was weird living in a military port unless you were drinking all the time," Thorne recalled. "It wasn't a particularly fun place to be." Desperate to get away from the monotony of maintaining the *Gridley*, Kerry often went surfing in the Pacific at dusk, splitting a pepperoni pizza and quarts of beer with his compatriots for dinner, and then crashing on Thorne's thrift-shop sofa. "He really made it quite clear that he wanted to live a life dedicated to public service," Wade Sanders, Thorne's roommate, noted decades later. "He knew what he wanted to do with his life."

Supremely fit and thoroughly trained, his bouts of pneumonia behind him, Kerry felt more than ready for his first Pacific tour of duty. As gung ho as he was in that sense, however, he also could not stop worrying that the Johnson administration was failing in its effort to defeat the Viet Cong and Ho Chi Minh. When Secretary of Defense Robert S. McNamara resigned in November 1967, Kerry saw the writing on the wall and it read, "No light at the end of the tunnel." Yet he also believed that the Johnson administration's basic motive was altruistic: to bring democracy to Southeast Asia. But it seemed to him that the Pentagon's Operation Rolling Thunder—bombs, bombs, and more bombs—was wrongheaded and inadequate, a Dr. Strangelove policy that was accomplishing little. As he continued to read about the war in the national newspapers, Kerry noted the reports that although hundreds of North Vietnamese bridges had been blown up in Rolling Thunder, nearly all of them had already been rebuilt. The U.S. Army boasted about eliminating three quarters of Vietnam's oil reserves, but this success appeared to have no impact; months later, there was still no gasoline shortage in Hanoi, Haiphong, or Cao Bang. And as the United States dropped its bombs day after day, it seemed to John Kerry that the North Vietnamese's resolve only intensified. Meanwhile, every day the chorus of antiwar protesters shouting, "Hey, hey, LBJ, how many kids did you kill today?" grew louder.

Just before the *Gridley* was about to leave Long Beach for its tour of the eastern Pacific, a startling development in Vietnam would turn U.S. military strategy on its head. On January 30–31, 1968, the North Vietnamese Army (NVA) and the Viet Cong attacked Saigon and thirty of South Vietnam's provincial capitals, inflicting heavy casualties. Although the offensive was a clear military failure, the fact that the NVA had so much fight in them shocked the American people. Television brought vivid images of fighting near the U.S. embassy in Saigon into their living rooms. For many, the Tet Offensive was a clear sign that the United States simply could not win this war. In just a few days, this "Tet Offensive" also wiped out the credibility of Johnson's bombing campaign.

Tet was a classic example of the consequences of not remembering the past, as George Santayana so famously pointed out. In 1789—the same year as the French Revolution—Vietnamese leader Tay Son Nguyen Hue drove the invading Chinese out of Vietnam via a surprise attack during Tet, the Vietnamese New Year celebrated over the three days of the first new moon following each January 20. "In January 1789 the Vietnamese defeated a Chinese Army and drove it from Vietnam," wrote military historian Spencer C. Tucker in a seminal article in the February 2003 *Vietnam* magazine. "What might be called the first Tet Offensive is regarded as the greatest military achievement in modern Vietnamese history. Just as the 1904 Japanese strike on Port Arthur foreshadowed their 1941 attack on Pearl Harbor, this 1789 offensive should have been a lesson for the United States that Tet had not always been observed peacefully in Vietnam."

Two months after the Tet offensive, President Johnson declared that he would not seek reelection that fall. If Vietnam was a quagmire by the late winter of 1968, so too was America.

Chapter Four

High Seas Adventures

After sixteen months of basic and specialized U.S. Navy officer training, Ensign John Kerry, was more than ready to go when the U.S.S. *Gridley* was at last deployed on a Pacific tour in February 1968. The youthful sense of invincibility in him craved action and adventure; after all, for fun in his old life he had sailed the Norwegian coast, climbed the Swiss Alps, motored across France, and run with the bulls in Spain. This time, however, the adventure he headed to was war. The *Gridley*'s first stop on this tour of duty was at Hawaii's Pearl Harbor. "My imagination naturally ran rampant with the visions of Japanese planes coming in over the mountains, shooting up Schofield Barracks and then swooping down over the harbor to attack Battleship Row and cripple the pride of the Seventh Fleet," Kerry wrote in a long typed letter to his parents on February 18.

I couldn't think but of how strange it was how many times I have read of Pearl Harbor, seen movies of what happened, asked questions from those who were still there, and still never been struck by the reality of what took place as during those moments that we navigated the channel that those same ships at one time sailed. The channel itself was extremely small and it became patently evident why the Oklahoma *(beg your pardon—*Nevada*) had to beach on the coral lining in order to leave it clear for other ships. We moored opposite Battleship Row and the Memorial of the* Arizona*—which I later visited, and I then had to forget that I was in Pearl Harbor and settle down to get some of the chores done that follow a week at sea and the first moments after docking.*

Kerry's shipboard duties remained the same at sea as in Long Beach, and no less tedious. It was his responsibility to make sure the *Gridley* stayed in tip-top shape—"spic-and-span" was the phrase the ship's captain used. The demands for meticulous maintenance were onerous and exhausting. As he had learned in Long Beach, it was one thing to keep your own gun polished, but quite another to motivate 400 swabbies to keep a 533-foot-long ship glistening. Kerry's many letters home to "Mama and Papa" from the *Gridley* show a definite pattern, generally beginning with gripes about how little sleep he was getting and how ridiculous certain orders seemed. Then, as if catching himself in an unacceptable lack of stoicism, he would switch into travel-brochure mode. In Hawaii, for example, on liberty from the *Gridley*, he rented a car to explore Oahu. At first, he confessed to his parents, Honolulu turned out to be a disappointment. "I was most disappointed in the lack of any really interesting non-tourist-trap places in the downtown area, which one sees advertised so often as the in-place where the in-people go," he wrote. "About the [only] thing these people had to offer was their in-debtedness for the once-in-a-lifetime trip to the tropics that they were all making to escape the excitement of the old-age home and the pension plan. It was really depressing."

The people aside, Hawaii did live up to his image of it as a tropical paradise. "The green goes on forever and the palm trees and the hills seem to beckon to anyone who has a certain amount of leisure-disease in their blood," Kerry opined, adding how it made him feel like James Bond to speed along the treacherous roads with sumptuous tropical flora on one side and breathtaking Pacific views on the other. "Despite the threatening skies and occasional downpour, I was taken by the climate and the countryside," he allowed. His chronological accounts went on to detail everything from attending a hokey luau and a rare bird show to leading some of his crew on tours of the *Arizona* Memorial and visiting a dental clinic to get his wisdom teeth pulled. He marveled at the "big surf" of Waimanalo Beach and the rolling fields growing pineapples that "smelled like fermenting wine." He also mentioned his frequent consumption of rum drinks at the Royal Hawaiian Hotel. "I can't describe the pleasure as we drove slowly through the countryside, stopping where we wanted and doing pretty much as we desired," Kerry enthused to his folks. "In many ways it reminded me of what I had always dreamed Africa would be like."

He also made clear how right his choice of service had been. Indeed, his great love had always been the sea. His letters and journals from his time on the *Gridley* shimmer with Horatio Hornblower images of foghorn-heavy sea mists, lilting waves, and wild monsoons. "There are some lazy moments at sea," he wrote to his parents, "the kind of moments that I, as a sailor, really joined the Navy for. These are the moments after dinner or during watch, or on Sunday during Holiday routine, when you can just stand on the fantail, the fo'c'sle, anywhere, and look at the sea, enjoying it to the fullest. Sunsets and sunrise on the bridge always fill me with the most basic tingles and provide an internal excitement that helps to erase all the pettiness and waste that surrounds so much of the rest of the effort." But then the seemingly happy-go-lucky ensign would end his missives home on a darker, more philosophical note about the war he was about to join. He had serious questions about the conflict, was increasingly certain that U.S. involvement in Vietnam was a terrible mistake, yet never doubted that his duty lay in serving his country. This, of course, left him with a moral conundrum. "I have been thinking a lot about Vietnam and the reasoning of the uncommitted soldiers," Kerry averred. "How one can oppose the war and still fight it."

The plight of the "uncommitted soldier" in Vietnam has been explored in Hollywood movies like *Platoon*, and popular songs like Bob Dylan's "Clean Cut Kid," but the most direct indication of what the uncommitted soldier in Vietnam went through comes from their correspondence home. A case in point is the letters of combat medic Thomas W. Bennett of Morgantown, West Virginia, only the second conscientious objector in history to earn the Congressional Medal of Honor. Refusing to carry a gun, Bennett was instead trained by the Army at Fort Sam Houston to help the wounded. His missives home from Texas teem with antiwar sentiments. But then, in December 1968, after complaining about the futility of U.S. military intervention in Vietnam, he wrote, "If I am called to Nam, I will go. Out of obligation to a country I love I will go and possibly die for a cause I vehemently disagree with. It is my obligation to give service to my country. That's why I'm here—to help provide freedom for dissenting voices. . . . I believe in America."

So did John Kerry. But that conviction was soon sorely tested several days after departing Pearl Harbor. On the afternoon of February 26, 1968, Kerry was on watch on the bridge of his ship. The *Gridley*'s executive officer approached him and asked whether he had a friend named Pershing. There could be only one reason for the question and Kerry did not want to hear it.

His stomach went hollow and he slumped onto a railing for balance. "I knew immediately it was all over, but even when I read the telegram it took moments to sink in," Kerry wrote to his parents of the moment he learned his dear friend Dick Pershing was dead. "Then I just . . . cried a pathetic empty kind of crying that turned into anger and bitterness. I have never felt so void of any feeling at all—so numb."

The dashing Pershing had been killed in combat on February 17, 1968, near the hamlet of Hung Nhon, 400 miles north of Saigon. His platoon had been slogging through mud in search of a lost comrade when the ambush came. "Shift over to the left!" Pershing reportedly had shouted as he tried to wave his men away from the danger. Then, kaboom. A rocket-propelled grenade slammed into a rice dike just a few feet in front of him, hurling Pershing into the air, his entire body sliced through with grenade fragments. He was killed immediately. The charmer of John Kerry's circle had turned into a statistic: another second lieutenant with the 101st Airborne Division gave his life for his country. "The loss of Persh hit me like a ton of bricks," Kerry rasped thirty-five years later. "The Vietnam War suddenly came into focus like it had never before. It wasn't a distant thing. It wasn't a newsreel or news report or newspaper story. It wasn't something at an arm's distance that you were able to grapple with—it was right inside your gut."

Devastated as he felt, Kerry managed to get a telegram off to his fallen friend's parents:

> THERE IS NO WAY TO EXPRESS MY SHOCK AND SORROW FOR THIS LOSS THAT YOU HAVE SUFFERED. LACK OF AIR TRANSPORTATION AND OUR DISTANCE FROM LAND PRECLUDES RETURN. PERSH WAS UNIQUE IN EVERY WAY AND THE FRIENDSHIP HE GAVE AND THE PLEASURE OF HIS COMPANY WILL NEVER BE FORGOTTEN—JOHN KERRY

Pershing's death brought out a profound anger as well as sadness in Kerry. Memories of his liveliest friend kept flashing through his mind, beloved images ruined forever by the knowledge that he would never see their like again: no more of Persh's prankster grin, his wonderful bullshit, his artful skirt-chasing, his uproarious laugh. It was all gone: the boyish mischievousness that bordered on deviance and had so perfectly balanced

Kerry's serious leanings at Yale. Nothing had been off limits to Dick Pershing when it came to the pursuit of fun—and how women, and men, had been susceptible to his charm. Yet when it had come time for him to serve, the life of the party had offered himself unhesitatingly to his country. By the time he got to Vietnam, Pershing had remade himself into the perfect paratrooper, rock solid in body and pure stalwart in spirit. And now he was gone, and for what? "Pershing's death was just one more major-league souring for John of figuring out what the hell Vietnam was all about," explained David Thorne, who was still in California when he got the news. "Why did Dick have to die for this? That's what John wanted to know."

Kerry's anger began to settle on the politics of the war. The very week Pershing was killed, Chairman of the Joint Chiefs of Staff General Earle G. Wheeler had made his eleventh inspection tour of South Vietnam. John Kerry suspected that Wheeler would return with the same message he always had, telling the American people their great nation was winning another war, which he would prove by writing up some overoptimistic reports for the White House's consumption. What Wheeler wouldn't mention was that 543 U.S. troops had been killed the week Pershing died. Nor would he note the 2,457 wounded. And for what? To the mourning Kerry, instead of looking for ways to extract its 525,000 personnel from Vietnam, the Pentagon appeared to be seeking more fresh recruits to shed their blood for the Stars and Stripes. Kerry's friends already "in-country" had begun to complain that the North Vietnamese were now equipped with Chinese-designed 107-mm rockets that weighed only forty-five pounds but could blow a hole in a building six or seven miles away. As the even more uncommitted soldier John Kerry now saw it, he was nothing more than an expendable pawn in a political power struggle between Washington, D.C., and Beijing—a "second wave" sailor, expendable.

During World War I, General "Black Jack" Pershing had visited the grave of the famous Frenchman who had supported the American Revolution. Pershing, contemplating the headstone, is reported to have proclaimed: "Lafayette, we are here." It was a symbolic moment, signifying that the United States was repaying an old war debt. But were the Americans really the liberators of Vietnam? It pleased Kerry to learn that Dick Pershing had been buried next to his legendary grandfather in Arlington National Cemetery. That spoke of the great continuum of duty, honor, and country.

But Kerry also could not help but feel that the Johnson administration was doing its servicemen not just a grave but indeed a lethal injustice by sending a new wave of young people to die in a conflict that the Pentagon itself—as McNamara's resignation in November 1967 showed—did not believe could be won. Kerry had studied geopolitics and he understood that Vietnam—like Korea—was an important Cold War battleground. He believed that both China and the Soviet Union were politically bankrupt totalitarian states. But he had also come to believe that the Johnson administration was mismanaging the effort to deal with global communism.

Once the initial shock of Pershing's death had sunk in, Kerry, as he tended to do, poured out his feelings on paper to his parents. "With the loss of Persh something has gone out of me," he wrote. "Persh was an undefinable spark in all of us and we took for granted that we would always be together—go crashing through life in our own unconquerable fashion as one entity. Now that is gone in an incomprehensible moment. Time will never heal this. It may alleviate—but it will never heal."

Later, in a similar, but even more intense fashion, Kerry poured out his anguish in a letter to Julia Thorne:

Judy Darling,
There are so many ways this letter could become a bitter diatribe and go rambling off into irrational nothings. I don't know really where to begin—everything is so hollow and ridiculous, so stilted and so empty. I have never in my life been so alone with something like this before. I feel so bitter and angry and everywhere around me there is nothing but violence and war and gross insensitivity. I am really very frightened to be honest because when the news sunk in I had no alternatives but to carry on in the face of trivia that forced me to build a horrible protective screen around myself. Something that has never happened to my feelings before. I could not even allow myself the right to think about what was happening as much as everything inside me wanted to. I was standing watch on the bridge when the executive officer called me over and after an ominous pause asked me if I had a friend called Pershing. I just stood there frozen and then read your telegram knowing already in my heart the Godawful wasteful stupid thing that had happened.

Right now everything that is superficial and emotional wants to give up and just feel sorry but I can't. I am involved in something that keeps pushing on regardless of the individual and which even with what has happened must, I know deep, deep down inside me, must be coped with rationally and with strength. I do feel strong and despite emptiness and waste, I still have hope and confidence. There is a beast in me that keeps pushing me on saying Johnny you can't let go because of this—Johnny you find some sense from this—Johnny you are too strong to stop now—something keeps me going harder than before. Judy, if I do nothing else in my life I will never stop trying to bring to people the conviction of how wasteful and asinine is a human expenditure of this kind. I don't mean this in an all-consuming world saving fashion. I just mean that my own effort must be entire and thorough and that it must do what it can to help make this a better world to live in. I have not lost faith—on the contrary—I have gained a conviction and desire greater than ever before—and now, a sense of inevitability—a weighty fatalism that takes worry out of the small actions of late and makes the personal much more important.

The world I'm apart [sic] *of out there is so very different from anything you, I, or our close friends can imagine. It's fitted with primitive survival, with destruction of an endless dying seemingly pointless nature and forces one to grow up in a fast—no holds barred fashion. In the small time I have been gone, does it seem strange to say that I feel as though I have seen several years experience go by. Wherever we go we see B-52s flying overhead going and returning from strikes on the Guam–Vietnam route. Two aircraft carriers are now in port to reload ammunition, rest the crew, and repair airplanes and the talk is of pilots lost and air strikes that were successful for the number of lives taken or unsuccessful for the number of lives lost—both the same and both creating the same hole and sorrow for some unsuspecting person somewhere. Small boats tear around the harbor practicing maneuvers, we train nearly every day for any eventuality. Everything is hot and fast—there is no joking like there was back in California. No matter which one is—no matter what job—you do not and cannot forget that you are at war and that the enemy is ever present—that anyone could at some time for the same stupid irrational something that stole Persh be gone tomorrow.*

Yet as Kerry reflected further on the meaning of Pershing's death, he came to recognize that now, more than ever, he had to perform his own duty to his country, even if it meant dying in the waters off Southeast Asia. Pershing had been the unspoken leader of Kerry's gang at Yale, and anything but a coward. So however disappointed Kerry was of the leadership of the Wheelers and Westmorelands and the other death-dealing hawks Bob Dylan had scorned since 1965 as the "masters of war," he also felt readier than ever to take on the Viet Cong. They had killed his friend, and he was ready to kill them if he had to. For the time being, however, as the *Gridley* headed to Guam, he would keep following orders and try to be the best U.S. Navy officer he could. And that meant keeping his emerging antiwar thoughts to himself or consigned to paper, not out where they might jar the morale of his shipmates.

Kerry sought to steady his uneasy mind by staying focused on his job and spending his spare time reading everything he could get his hands on about World War II—the "Good War," as so many thought it. Like all sailors braving the Pacific Ocean for the first time, Kerry felt humbled before its vastness. From Long Beach to Honolulu alone the *Gridley* had notched 2,267 miles, which put it less than halfway to the Philippines. The long voyage offered time for intellectual improvement. Kerry, at the captain's request, fulfilled his responsibilities as public affairs officer, by sharing the history of their whereabouts with the crew. Every few days while at sea he would write an 800-word vignette about World War II battles that he would then read over the intercom in his best Edward R. Murrow stentorian tones.

Drawing primarily on Samuel Eliot Morison's sweeping multivolume *History of the United States Naval Operations in World War Two*, he paraphrased historically precise war stories from the Battle of Midway to Hiroshima with dramatic flair. All of his "Hear It Now" addresses, delivered at appropriate lulls in shipboard activities, began, "During the night the *Gridley* steamed into some of the most historic waters of World War II"; then Kerry described what had happened at Guam or Wake Island with a palpable excitement at actually being there. A report he broadcast as the *Gridley* approached the Philippines illustrates his style: "During the night and most of today, *Gridley* has been steaming in proximity to the Solomon Islands, where the first offensive moves of World War II by the Allied Forces took place," Kerry said. "It is also the scene of the much publicized adventures of President Kennedy and his PT boat crew. Just before the Battle of Coral Sea, which took place in early May of 1942, the Japanese were occupying Tulagi

Island in the Solomons and they had begun construction of an airfield at Guadalcanal from which land-based planes could have endangered Allied control of the area. This penetration toward Allied lines of communication to Australia and New Zealand represented a threat that had to be stopped at all costs, and prompt steps were initiated to eject the Japanese from the Solomons."

In addition to serving as his shipmates' radio correspondent of the past, Kerry also helped oversee the writing, editing, mimeographing, and mailing of *Gridleygrams*, a weekly newsletter. Full of gossip and tidbits covering news from who the frigate's best backgammon player was to sightings of gray whales, the newsletter provided a way for the crew to communicate with family and friends. Every day some shipmate would lobby Kerry to profile him in print. He tried hard to accommodate every legitimate request. "If you received the *Gridleygrams* then you have in a sense heard from me because I am the talented idiot who has put out that grizzly piece of worded propaganda," he wrote to his family on April 14, 1968. "I suppose that for husbands and wives that are not hearing from their sons and husbands, it is a welcome bit of news."

By the time *Gridley* reached the Philippines, Kerry had established an excellent rapport with his first commanding officer (CO), Captain Allen W. Slifer. A native of Mt. Carroll, Illinois, Slifer had joined the Navy in 1940. He was attached to the *Pittsburgh* during the invasions of Iwo Jima and Okinawa. From the CO's perspective, the eager Kerry made an ideal junior officer, executing even the most menial tasks with determination and diligence. He was also a natural leader. "The days simply did not have enough hours for all they ask us to do," Kerry wrote his sister Peggy. "My men are incredible—they get less sleep than anybody on board and they still keep going in good spirits." Slifer enjoyed Kerry's World War II intercom broadcasts and urged him to expand the format to include tales of Admiral George Dewey's heroic derring-do during the Spanish-American War. "He was extremely excitable on the bridge and would often blow his cool but if you let these moments of seafaring anguish go their way, he was a good instructor and a good commanding officer," Kerry wrote of Captain Slifer. "It is strange, but I think that I managed to get closer to him than many of the officers on board and in doing so garnered a good deal of knowledge that I might not have."

Not long after they became true friends, a grand ceremony was held in the Gulf of Tonkin: Captain Slifer was relieved of his duties aboard the *Gridley* and replaced by a new CO, Captain Wyatt E. Harper Jr. A native of Massachusetts, Harper had served on the U.S.S. *Register* during World War II, surviving a kamikaze attack and helping rescue the survivors of the *Indianapolis*. Balding, with a large nose and darting eyes, Harper believed in upholding rules and in maintaining strict discipline. One veteran sailor who served with Kerry summed Harper up as "grumpy and panicky," and always in a foul mood. "From the moment Captain Wyatt Harper stepped aboard ship . . . life on *Gridley* came to a screeching and abrupt turnaround," Kerry groused to his parents. "This is the most exacting and picayune, most demanding and unfathomable man I have come into contact with since I have been in this brilliant organization."

Kerry was not the only *Gridley* mate who had a hard time dealing with Harper. "Unfortunately for John, me (and the ship), our captain during that period, Captain Harper, was an extremely difficult individual, given to quick, and often incorrect, decisions," Robert E. Jack recalled. "I think John held the record for being 'thrown off the bridge' as officer of the deck by Captain Harper. I was very aware of this, as inevitably, John called me to relieve him. 'What for this time?' became a pretty common question. John was not the only one to be asked the question. But, although it disturbed and troubled him (as well as everyone else), he never developed a bad attitude, which many others did." According to Jack, Kerry did eventually earn Captain Harper's respect because of his clear speaking voice and ship-handling ability. One time the ensign was directing helicopters during an exercise from the Combat Information Center. Jack, who was the watch officer that day, said, "Captain Harper (who had been listening to the radio chatter on the bridge) burst into CIC and asked me who was that person talking to the helos with that great voice. So, I guess the skipper did at least give John a compliment upon one occasion."

On another occasion while the Gridley was steaming through the Gulf of Tonkin, Kerry saw a floating glass jar fifty yards away. Knowing that the captain fancied these souvenirs, he reversed course, carefully maneuvering the ship until it almost touched the bottle that was hoisted on deck. "That evening, in the wardroom at dinner, the captain commended me for ship handling," Kerry recalled. "Coming from him, it meant a lot."

Every day that the *Gridley* patrolled the Gulf of Tonkin an enemy attack was remotely possible. For a few days it became Kerry's job to shuttle sailors and provisions back and forth between the *Gridley* and the aircraft carrier *Kitty Hawk* on a small, motorized whaleboat. The pilot in him envied the naval aviators who got to roar on and off the tiny runways on the decks of the gargantuan carriers. "My own greatest thrill came from the sound and the fury of the jets taking off from the flattop—bird farm—as we call them," he wrote home. "Two by two they would suddenly dart off the ramp and be airborne in a seemingly precarious and unstable fashion. Then they would steady out and you could easily be awed by the rate of climb as they disappeared into the clouds. Every day was cloudy and flying always marginal. Our job was to provide a source of reassurance to the pilots should they crash on takeoff or landing and to lend support of any kind—anti-submarine and anti-air—to the flattop."

After two weeks with the *Kitty Hawk*, the *Gridley* returned to Subic Bay. This time there would be no shopping or carousing; instead, Captain Harper ordered Kerry to spit-shine his ship. The junior officer dutifully obeyed orders, although with some annoyance, since he'd been granted only a few hours of leave time ashore. As the main U.S. naval repair facility in the Pacific, Subic Bay pulsed with activity both on and off duty; not a ship in the Seventh Fleet didn't anchor there for a spell. Kerry was assigned shore patrol in Olongapo City, the deep-water liberty port adjacent to the Subic Bay base. It wasn't a very pleasant experience. He had found Subic Bay a polluted and impoverished ghetto. "The city itself is unbelievable," Kerry wrote his folks. "It is the outgrowth entirely of the presence of the Navy. In fact there are only two things in this part of the Philippines—the Navy base and the city. Behind these is a jungle that is replete with pythons, monkeys, cobras and every other form of wildlife."

Kerry displayed a flair for travelogue in his descriptions of what it was like to be on shore patrol in the fetid port city. In Olongapo, he encountered bloated corpses floating in the river, street urchins begging for coins, shantytowns that made the Great Depression's Hoovervilles look like Palm Beach, and, everywhere, starving mothers struggling to nurse glaze-eyed infants dying from malnutrition. He was hustled by a rogues' gallery of swindlers, pimps, and con men. Olongapo's streets teemed with Jeepneys, modified versions of the U.S. Army's vehicle that had been stretched in the back to carry six or eight passengers. The Jeepneys alone made quite an impression: many of them had psychedelic Day-Glo paint jobs and virtually all of them had constantly blaring horns, cursing drivers, and belching exhaust pipes—it was pure mayhem. Drunken sailors got themselves fleeced with shocking regularity, others popped downers and smoked pot for kicks. Gangs of pick-pockets and corner opium dealers infested the seedy streets. To a well-bred boy, Olongapo was as close to Sodom and Gomorrah as he ever wanted to get. It made the Fillmore look like a Christian Science Reading Room. "From the moment you enter the town there is no cessation to this view of bars and fights," he marveled to his straitlaced parents. "It is raucous. Walking, stumbling, running up and down the streets are sailors interspersed with Filipino girls in tight dresses and little children, ostensibly approaching you to beg some money or something in exchange for a pack of chewing gum but who are really waiting to cleverly slip your watch or wallet off your body."

Kerry was both amused and surprised by the squalid life of this liberty city. His "beat" was the bars and brothels part of town. His mandatory tropical white uniform was a target for abuse. He felt embarrassed by the fat blue band around his right bicep with SP in big yellow letters. What's more, shore patrol was hopeless: crime wasn't a problem in Olongapo, it was the port's way of life. Ten months before Kerry arrived, the city's mayor had been assassinated for trying to clamp down on the vice trade. A crime ring was running the wharfs, and any attempt to try to curb their illegal activities meant certain death. Prostitution was the main racket. Sloppy party girls would fall all over the young officers, purposely distracting the shore patrols from other activities. To his horror, one woman grabbed hold of his hand and tried to pull off the Cartier ring Julia Thorne had given him. On another occasion he came upon a woman passed out on the floor of a bar, a sailor standing above her muttering, "Please don't let her die," over and over. Kerry felt for her pulse and tried to bring her back to consciousness. He succeeded.

Shore patrol was an easy target for mockery, and generally acting like the all-American rube the locals couldn't help but laugh at, the innocent young ensign blanched when a Filipino woman ran up to him and planted a huge wet kiss on his white-clad chest, leaving a red lipstick print that could not be easily scrubbed out. He wrote home about how sad it was to see so many poor women struggling to eke out a meager living.

Back at Yale, Kerry had read Michael Harrington's wrenching study *The Other America*, and came away stunned by the statistics Harrington presented about poverty in rural Appalachia. And during his stay in San Diego he had wandered the grimy back alleys of Tijuana, observing the sad life of the broken-down barrios where people actually dwelled in makeshift homes erected out of plywood and abandoned Chevys. But nothing could have prepared him for the utter squalor of the Philippines, where skinny three-legged dogs lapped unsanitary water out of mud holes and huge, disease-laden rats scurried about in broad daylight. In one letter home Kerry detailed some of his bleaker encounters with Third World poverty:

Here, lined up and down the sidewalk in small little business ventures were old ladies selling cigarettes and more children trying to peddle wares. Where the children weren't selling something, they were curled up on a piece of sackcloth, trying to sleep amidst the noises of the entertainment. More than anything I think this sight of children out so late and in such condition haunted me. I couldn't stop thinking . . . how it must affect them in terms of what they will . . . think about the world when they have a chance to express themselves. Little girls would sometimes walk up the streets with their mothers, huddled against them as though the mother was trying in some way to protect her daughter against the vice and filth. In some of the small shops . . . there were shirtless peasants slaving over sewing machines and shoehorns—trying to make tomorrow's order for some American who that day had walked in and placed a large order. . . . [O]ne sight turned my stomach. On one corner was a small little delicatessen that had a window full of foods. Among the choices was a stew of some kind that was covered—and I mean covered—with flies and moths. As I watched in disbelief just at the number of insects on the food and asked myself who could possibly eat that, a huge ladle lowered into it, scooped up a healthy portion of stew—flies and moths and all—and dumped it in a plate for someone to eat. I walked away in a state of total shock. I imagine these people must have a digestive system that would defy medical history.

Edgy from his dispiriting shore duty and plagued by insomnia, Kerry was actually eager to get to Vietnam. After another month in the Gulf of Tonkin, he got his wish. The U.S.S. *Gridley* was ordered to Danang for briefings, and at last he would see the country that had so preoccupied him since he was a sophomore in college. He planned to try to contact Freddie Smith and Danny Barbiero once he got onshore. He wanted to talk about Dick Pershing's death with his friends. Perhaps, he imagined, they could hold a reunion in the middle of a war zone. It was his way of putting off reality until that was no longer an option. "As the land grew larger in front of us I could sense the tension that was beginning to ride on everyone," Kerry explained in a letter to his parents. "It is strange. One can talk and talk about the meaning of war and the dangers and the horrors and all the sensations that a man has when he gets near the possibility of dying. But until you actually sense them somewhat, you do not really know what you are talking about. And once you have sensed them, you tend not to want to talk about them at all. That is the way I feel now."

Danang—just a couple of hours south of the Demilitarized Zone (DMZ) dividing North from South Vietnam—had become an important port in the late nineteenth century for the country's French colonists, who called it Tourane. The city soon flowered into a Vietnamese hub, second only to Saigon in its cosmopolitan air. But in March 1965 Danang had taken an unexpected turn with the landing of 3,500 Marines, followed shortly thereafter by the construction of an enormous U.S. Air Force base. The Johnson administration had deemed Danang an essential supply depot and the ideal launching pad for its bombing runs into North Vietnam. Soon the once lovely city was catering to the needs of American military personnel, with a proliferation of movie theaters, honky-tonks, whorehouses, and air-conditioned short-term hotels springing up seemingly overnight.

Until Tet, Danang had been a relatively safe port for American sailors. But that momentum-switching assault changed everything. Security became the post-Tet obsession in Danang. Upon entering the harbor, Kerry was astonished to see dozens of American transport ships, many rusted out from too many months at sea. Also at anchor swayed some twenty or thirty freighters of varying registries and size, along with a hospital ship flying the Red Cross flag. Put in charge of anchoring the *Gridley*, Kerry could not help but keep his eyes fixed on the rolling green hills beyond the city, where helicopters filled the sky and the low rumble of artillery echoed. Patiently, he waited his turn to set foot on Vietnamese soil. He described the moment in a letter: "Finally, after the first boat lead went ashore, I was issued my .45 and some ammo and allowed to climb aboard the motor whaleboat to escape the panic and pressure that existed on *Gridley*. It was strange again to strap a .45 automatic around my middle, and I wondered whether I would have to use it."

A gentle wind blew across the harbor as Kerry's whaleboat landed at the base of Monkey Mountain. He disembarked anxious about the prospect of wandering around aimlessly, ten thousand miles from home. Yet he quickly found himself at ease: the echoes of its colonial past made Danang a much less alien landscape to him than the Philippines had been. Overlying the architectural remnants of French colonial rule were the unmistakable marks of the U.S. empire. The handiwork of American construction firms such as Raymond International and Morrison-Knudsen appeared everywhere. It seemed that some military-dull-colored concrete something-or-other had been poured along virtually every block on the wharf. B-52 long-range jet bombers howled across the sky to the northeast, while U.S. Army helicopters were touching down to the west. Pentagon-issue sandbags sat in stacks near the water, in case the tides rose too high. To Kerry's surprise, he saw far more barbed wire than bamboo.

The Americanization of South Vietnam was obvious. But all John Kerry's eyes could focus on was the sight of a 50-foot American-made aluminum patrol craft fast (PCF), commonly called a Swift boat, the acronym for the even clumsier Pentagon moniker "Shallow Water Inshore Fast Tactical Craft." What it was called didn't matter to Kerry; the thing gleamed with the possibilities of a different Navy filled with responsibility and independence. "Tied up to the same wharf we got off at was a small Swift boat, and I thought jealously of my own desires to have one," Kerry admitted in a letter to his parents. "This one apparently had been given to the Vietnamese Navy and they were using it for coastal surveillance," he wrote. He had already applied for Swift boat duty and was waiting for his assignment orders, but seeing the boat confirmed his hope of training at the top-notch Coronado School back in San Diego.

Kerry had little company in his desire to volunteer for Swift boat school. A number of the men on the *Gridley* had already seen enough combat, and preferred to stay *out* of harm's way to the extent possible. David Simons of La Puente, California, for example, had been assigned to the *Gridley* on this Pacific tour. "Kerry was a nice guy," Simons recalled. "When we were in Danang harbor we all saw those Swift boats speeding around.

Others were in repair, some of the skippers on R and R. Kerry wanted one; not me. Those Swift boat guys had guts. They were more gutsy than me." As interim communications electrician aboard the *Gridley*, the handsome, serious-minded Simons wanted nothing more than to return home in one piece to make money in electronics, his passion since the age of six. "You had to be a bit of a cowboy to want a Swift," added Simons. "It meant that you were willing to get shot up all the time."

After his focus on the PCF, however, Kerry honed in on a more immediate objective of his time in Danang: to get to a telephone and reach his buddies Freddie Smith and Danny Barbiero. As he walked along the docks he was startled to see Maoist graffiti spray-painted on walls, and shocked by the gruesome sight of a pile of dead VC awaiting mass burial. "The thoughts of what must have taken place turned my stomach," he wrote to his family. More revelations awaited him on his way to Third Marine headquarters, where he hoped to learn the whereabouts of his Yale buddies. "The ride there was still further education," Kerry continued. "The scene of Vietnamese peasants in their pajama-type costumes and huge slanting straw hats was blatantly juxtaposed to the scene of Army and Marine vehicles moving down the road while they walked slowly at their own pace. I watched as several women moved along with a bar across their shoulders and two loads supported from each end. [They] would walk at a certain speed that got the loads bouncing in rhythm on the end of the bar and . . . would, I am told, take a step every time the load was coming up. That way [they] would lighten the effort of walking."

Looking into the faces of the Vietnamese peasants he encountered, a wave of compassion shot through John Kerry. For two thousand years these people had been warding off invaders—Chinese, French, Japanese—and now he was one of the latest incarnation, one of novelists Eugene Burdick and William Lederer's Ugly Americans. He had read history books about how Vietnam's would-be conquerors had always been vanquished in the end, and it seemed to him inevitable that his own nation was next in line to lose. The pattern was undeniable: Danang's Cham Museum offered the evidence of crates filled with artifacts from the kingdom of Champa, which had ruled the region for more than a thousand years. Like all other comers, Champa too eventually fell to the Vietnamese. That turnaround had taken a millennium; the first U.S. Marines had landed at Danang in 1965, and now less than three years later they were already losing their grip. Kerry stared into the sun-caked fields and grew mesmerized by the peasants matter-of-factly going about their business of cultivating rice as the deadly explosions of U.S. ordnance echoed off the nearby Marble Mountains. "Every so often, there would be an open field where there were a few huts and people working in it with their pant trousers rolled up and their large hats covering up expressionless faces," he wrote to his family. "How could these people really believe we are helping them? It seemed so utterly crazy—the idea of all this modern equipment fighting for an ideal that meant nothing to those of whom the fighting was supposed to be for. . . . I can't help getting the feeling that their faces seemed to say, 'Go away and let us alone.' "

A little sleuthing around at Marine headquarters turned up the news that Freddie Smith was on R and R in Australia, but that Danny Barbiero, who had trained in Quantico, was up north in Quangtri, serving in the headquarters of the Battalion Communications System. For half an hour operators struggled to place Kerry's call there. When they eventually patched him through to Quangtri, the voice on the other end stated that Barbiero had just left the office. Frustrated, Kerry was about to hang up when suddenly he heard a more familiar voice saying "Lieutenant Barbiero speaking." Grinning from ear to ear, the caller shot back, "Ensign John Kerry," and the two were off, trading stories for nearly an hour. "I was standing in this bunker when John called," Barbiero recalled. "We had all these switchboards and wires and suddenly I was talking to him like he lived next door." They spoke, of course, about the loss of Dick Pershing. "It was hard to say good-bye to him, knowing what had happened to Persh," Kerry wrote his parents about Barbiero, who became a platoon commander in the DMZ, "and when I hung up I felt a larger cut than just one of communications."

The main theme in John Kerry's correspondence from Vietnam during that short visit in the spring of 1968 was how disturbing it felt to be an unwelcome soldier in a foreign land. It wasn't that he couldn't stomach being served *thit cho*—dog meat—or barbecued songbird; he could. What annoyed him were the hostile stares at his uniform. He sat in a portside bar drinking pints of *bia hoi* and asking questions about the enemy. A Danang woman who worked for the Red Cross told him that there were more than a thousand VC living among them in Danang. Exaggerated or not, the report made Kerry nervous. "Wherever I went and young Vietnamese men would look at me I grew scared," he confessed to his mother and father on paper. "There really was no way to tell who was who. You could be in a room with one and not know whether he was really a Charlie or not. It became easy to sense the distrust that must exist in the outlying areas. How could one really fight in the fields and know whether at any time the men beside you were not going to turn tail and train their guns on you? Whom did you begin to trust, and where did you draw the line? Another ludicrous aspect of the war."

The *Gridley* left Danang and headed toward Haiphong. Kerry, in his typical fashion, went around canvassing the few crewmates who had gone ashore about what they experienced and learned. The ship's doctor detailed for Kerry every note of the hospital he had visited, to which American casualties were being medevaced to be treated for everything from shrapnel wounds to lost limbs. Kerry spoke openly to his doctor friend about how much he missed Pershing, and about the overall senselessness of the U.S. military involvement in Southeast Asia—but not for too long. He never could forget how often his father had quoted to him one of Dean Acheson's favorite axioms: "Complaints are a bore and a nuisance to all and undermine the security essential for endurance." After all, he served as the U.S.S. *Gridley*'s officer of the deck, in charge of 7,400 tons of ship and 350 men, with only a CO and XO outranking him during those important watch hours. The *Gridley* was now the closest destroyer to North Vietnam's treacherous Haiphong, only forty miles away, so once again he kept his growing doubts stoically to himself for the sake of his command's morale. John Kerry may have been a thoughtfully "uncommitted soldier," but he would be damned if anyone on the *Gridley*, except the doctor, was going to find out about the depths of his skepticism. His government had asked him to perform his patriotic duty, and that was what he meant to do, with as little complaint as he could.

On May 1 Kerry wrote his parents about what it was like ghosting the coast of Vietnam:

The war drags on—air attacks every day and special operations up and down the coast and generally attacks every day. Most of what we are doing and what is going [on] I can't put on paper. . . . The Viet Cong have tremendously increased their counter batteries along the coast and there is not a ship on the shore bombardment that does not encounter opposition. Most of the shore effort is down south—in the I Corps area where Persh was killed. The Tonkin Gulf is like an American bath tub it is so full of American ships and I am never failing to be amazed by the effort that we expend daily for this extravaganza. . . . And so, in three days, we head to Subic Bay for two days, and then off, across the equator, and to Wellington, New Zealand for seven days. This will also mean that we will probably be home early—perhaps the second week in June. Can I possibly tell you how happy I will be to get back to the United States.

One attribute of Kerry's served him particularly well: he was a first-rate listener—asking sincere questions, probing for information, paying attention to the nuances of every crack in a voice. He never acted the know-it-all. As a result, he became something of a ship's therapist, listening to everyone's beefs and thereby earning the clout that comes with the respect of one's peers. He took care to stay open-minded and focused on solving problems rather than dwelling on them, and his temperament and his penchant for aiding the underdog remained intact. He had always taken as a given that the little guy was forever getting screwed by "the system." St. Paul's School, Yale University, Skull and Bones, and all that aside, John Kerry honestly fancied himself more as a liaison between the establishment and the have-nots than as a true member of either. The distance of the outsider afforded him a perspective broader than most enjoy. He had felt the hurt of pecking orders and social stratification, and whenever a related personal problem arose on the *Gridley*, young Kerry would be there to intercede. This journal entry is a case in point:

> *Sometimes a situation plays a game with you that you neither ask for, want, nor enjoy. Sometimes, things seem to gather up into one swell of nonsensical contradictions that make you wonder about all the things—large and small, important and farcical—that you thought had shown some sort of consistency in the past. Today I was asked to investigate the unauthorized absence of one of the* Gridley's *young sailors. After I had asked him to come to my stateroom I had disappeared for a moment to get some silly thing done, and when I came back there was this innocent, almost frightened guy standing outside my stateroom in his dress blues. . . . Obviously, he isn't a fit for the*

military service and has a problem which goes a great deal deeper than any of the petty restrictions the Navy dreams up to keep men thinking in the military way. He needs help in a way that people who are really sick need help and I'm afraid that if he doesn't get it he will only get himself deeper into the trouble. Now he is a man [who] wants his freedom and he is restricted to the ship. Nothing will get him into deeper hell faster than this.

By now the *Gridley* had reached New Zealand. As public affairs officer, Kerry was responsible for making 40,000 "welcome aboard brochures, press kits, pictures, balloons and so on ad infinitum." Part of the *Gridley* mission was to show the American flag. Kerry constantly gave local journalists guided tours of a real U.S. Navy frigate. Kerry had survived the "Crossing of the Equator" ceremony, a famous maritime tradition when polywogs were hazed by veterans, thereby initiating them into the Royal Order of Shellbacks. Seasoned enlisted men used this opportunity to dump garbage and whack with a fire hose rookie officers like Kerry because it was their first time south of the equator. "I am also directing the crossing of the equator ceremony," Kerry wrote home. "*Ouch!!!*"

Kerry's Pacific tour on the *Gridley* ended on a horrific and surreal low note. On June 5, 1968, as the guided-missile frigate was nearing its home port of Long Beach, California, both Thorne twins, Julia and David, would be dockside to greet him. They knew from his letters that as best he could from the middle of the Pacific Ocean, Kerry had been monitoring New York Senator Robert F. Kennedy's attempt to capture the Democratic nomination for president. Like an avid armchair baseball scorekeeper, he kept a chart of the various primary days, and used the ship's radio to ascertain the results of each. It was with glee that he found out his man Kennedy was racking up significant early victories: Indiana on May 7 and Nebraska on May 28. When Kennedy's campaign briefly stalled in Oregon, where he lost to the even more dovish Minnesota Senator Eugene McCarthy, Kerry was not discouraged. He knew that June 5 was also the day of California's Democratic primary. If Kennedy won that, and it looked likely, it was all over.

Instead of the festive welcome he anticipated, however, just as the *Gridley* approached shore, the news came over the radio that Robert F. Kennedy had been shot at his primary-victory celebration in Los Angeles. A Palestinian activist named Sirhan Bishara Sirhan, bitter over Kennedy's support for Israel, had pulled out a .22-caliber handgun and fatally wounded the senator as he passed through the kitchen of the Ambassador Hotel. Kennedy died on June 6. "It was just awful," Julia Thorne recalled. "We were in shock." Although over the years Kerry has spoken eloquently of the anguish he felt upon arriving in Long Beach, rushing to a TV set rather than a bar to celebrate his homecoming, he believes Kennedy's campaign speechwriter Jack Newfield had summed up his emotions at the time better than he could himself: "Things were not really getting better. . . . We shall not overcome."

With Julia Thorne, who was "stunned and slightly frightened," at his side, Kerry headed back to Massachusetts for a long leave, his dog-eared copy of RFK's political manifesto *To Seek a Newer World* close at hand. While on the *Gridley* he had lost his heroic friend Dick Pershing, and now he had lost his hope for 1968, Bobby Kennedy. All his future promised for sure was Swift boat training in San Diego and another, far more perilous tour of duty in Vietnam.

Chapter Five

Training Days at Coronado

At first the U.S. naval presence in Vietnam had been confined to a small branch of the Military Assistance and Advisory Group (MAAG). The Navy's efforts in Indochina grew gradually, in step with the top brass's understanding of the demands of guerrilla warfare in the jungles of Southeast Asia. Conventional approaches such as air power, amphibious forces, blockades, offshore gunships, and the onshore work of Navy Construction Battalions had to be reassessed, and overall strategy restructured to neutralize the extensive use of the country's inland waterways by the Viet Cong and North Vietnamese. "Not since the Mississippi flotilla was deployed to fight the Civil War battles of Vicksburg and Shiloh," *Newsweek* reported on July 3, 1967, "had the U.S. Army [*sic*] found use for an assault force designed specifically for river warfare. . . . But the war in Vietnam has seen the revival of many tactics and weapons of earlier days, and last week a U.S. river-borne assault force went on the attack in the Mekong Delta, its most important vehicle an unwieldy-looking craft that bears a striking resemblance to the ironclads of a century ago. The force was called 'River Assault Flotilla One' and its overall mission was to root out the Viet Cong from the river and swampland south of Saigon."

Given its druthers, the Navy virtually always opted for "blue water" operations rather than the messier, more direct "brown water" kind. Although the United States had conducted riverine naval operations in the War of 1812 and the Mexican-American War, in both brown water firefights had been limited. The situation in Vietnam circa 1968 called for at least a temporary change in that strategy. "The Naval presence is also being felt on the inland waterways, rivers, and canals in Vietnam," U.S. Admiral Horatio

Rivero boasted in his 1968 study *Riverine Warfare: The U.S. Navy's Operation on Inland Waters*. "On the theory that where water is, sailors will go, we have returned to river warfare, something we haven't thought much about since the Civil War." It was a high-risk decision for the Navy, Rivero continued, for, as Prussian King Frederick the Great had noted in the late eighteenth century, "the passage of great rivers in the presence of the enemy is one of the most delicate operations in war."

During the French Indochinese War (1946–1954), the colonialists tried to create an effective river force capable of severing the hold the Viet Minh (the nationalists fighting for independence) had on the Mekong Delta waterways. After the French departed, the South Vietnamese put what they had learned from their experiences early in their nation's conflict into developing River Assault Groups (RAGs). Many of these groups would use the same vessels the French had left behind. In 1962, U.S. naval advisors were assigned to the RAGs, marking the first American involvement in riverine warfare in Vietnam. Under American tutelage, the assault groups were structured to include various World War II amphibious craft and their personnel component was enlarged to some two hundred officers and men. In its early stages, the rudimentary, experimental effort had the capacity to carry a landing force five hundred strong.

In April 1966, the several naval commands in Vietnam were combined into a single entity—U.S. Naval Forces, Vietnam (NAVFORV)—which was responsible to the Military Assistance Command, Vietnam, under General William Westmoreland. In addition to the conventional forces used early in the war, several other programs were joined or created under the new command, including the Naval Advisory Group and the Coastal Surveillance Force.

With this change in administrative setup came a more concerted effort to check the guerrillas' freedom of movement in the Mekong Delta. At the same time, NAVFORV launched a new campaign to search out and destroy Viet Cong bases, putting the South Vietnamese Communist rebels on the defensive. To carry out the first part of this strategy, the U.S. Navy beefed up its River Patrol Force.

The River Patrol Force both spearheaded the Navy's inland operation and served as its mainstay. The effort revolved around the 31-foot, water-jet-propelled fiberglass craft the Pentagon called a patrol boat, river (PBR), manned by a crew of four and usually skippered by an enlisted man with the rank of chief petty officer or petty officer first class. With a mere eighteen-inch draft that allowed for maneuvering in shallow water as well as a heavy armament boasting three .50-caliber machine guns, the PBR permitted quick and powerful penetrations into areas that previously had been the province solely of the Viet Cong and their sampans. "The PBRs had much greater mobility than Swifts," recalled Louisiana native Garland Robinette, who served on one of the latter in Vietnam. "PBR[s] operated better in the canals."

The Navy's UH-1B helicopter gunships, affectionately nicknamed "Hueys," provided indispensable support to the PBRs. Whenever a river patrol boat came under fire, its crew could count on immediate aid from the choppers' big guns and 40-mm grenade launchers. Many veterans have claimed that the Viet Cong guerrillas plying the waterways feared the whup-whup-whup of the Huey's rotors more than any other sign of the U.S. presence in the Mekong Delta. "The helicopters gave us an unprecedented mobility," said retired Navy Captain Virgil Jackson, a Vidalia, Louisiana, native who served on aircraft carriers patrolling off the North Vietnamese coast in the late 1960s. "They allowed us to maneuver around the countryside and coastline at will, destroying enemy camps. And they were ideal for rescue missions. Without them we would have been like the French: bogged down."

The River Patrol Force added a whole new versatility to the U.S. Navy's operations on the rivers of Vietnam. Densely forested areas that had proved unreachable for years were soon under American surveillance around the clock. Thousands of sampans could now be checked for contraband before being allowed on their way. Troops onshore gained a powerful new source of support that made night ambushes of VC positions possible. All in all, the effort proved a qualified success in a war in which progress was hard to come by, much less define.

The success of the River Patrol Force spawned the later Mobile Riverine Force, a joint Army-Navy effort to conduct search-and-destroy missions throughout the Delta. The spin-off operation comprised small boats from the Navy's River Assault Flotilla, manned by troops from the Army's Ninth Infantry Division stationed at My Tho on the Tien Giang River. Using landing ship tanks and similar craft as mobile bases throughout the Delta's waterways, the Mobile Riverine Force could move around to deliver troops into many previously unreachable areas of Viet Cong territory. Long convoys of armored troop carriers (ATCs), bearing a platoon of infantrymen apiece, plied their way up small canals at a leisurely eight knots or less until they made contact with the guerrillas. At that point the soldiers would storm off the boats into battle, securing territory as they advanced into new regions.

Together, the River Patrol and Mobile Riverine Forces, with some support from the Naval Advisory Group, opened up areas of the murky Mekong Delta that had been completely under VC control. Slowly the Americans began to spread inland, securing territory and putting increasing pressure on the guerrillas' strongholds. Despite these successes, however, the Viet Cong managed to keep their supplies moving. In the wake of the brutal Tet Offensive, the American naval command determined to ratchet up to a far harsher strategy in the Delta. Thus was born the U.S. Navy's Operation Sealords.

The idea was to push harder and deeper into the Viet Cong's domain or, as the NAVFORV in Saigon put it, "to show the American flag and to prove to the enemy that we dared to take the fight to their backyard." It was an important, high-risk new venture, chest-thumping and all. "The Navy in particular spearheaded a drive in the Mekong Delta to isolate and destroy the weakened Communist forces," historian Edward J. Marolda wrote of the operation in his 1994 book *By Sea, Air, and Land: An Illustrated History of the U.S. Navy and the War in Southeast Asia*. "The Sealords program was a determined effort by U.S. Navy, South Vietnamese Navy and allied ground forces to cut enemy supply lines from Cambodia and disrupt operations at base areas deep in the delta." Operation Sealords, launched in late 1968, would depend largely on the Swift boat that had so captivated Ensign Kerry at first sight in Danang.

With a dual machine gun mount tacked on above and slightly behind the pilothouse and a mortar machine gun combination planted on the stern, the Swifts turned into what Kerry called "the bastardized descendants of the World War II PT boats." And Kerry knew something about PT boats. He had read John Hersey's *New Yorker* articles on their use in World War II, as well as journalist Robert J. Donovan's 1961 best-seller *PT-109*, about how a dashing twenty-six-year-old Navy lieutenant named John F. Kennedy had swum for his life and those of his men one night in 1943 after their 80-foot patrol boat was rammed by a Japanese destroyer near the Solomon Islands.

The experience marked the most heroic moment in JFK's young life, and it became part of the folklore of his public service.

Just as the shallow-draft wooden PT boat had appealed to John F. Kennedy, the shallow-draft aluminum PCF boat inspired John F. Kerry's dreams of adventure twenty-five years later. Although built a generation apart, both crafts were speedy and versatile. Either could afford the young officer in charge a chance to test his seafaring mettle, without too much supervision—to be of the U.S. Navy, but also apart from it.

The idea of high-speed Swift boats had grown out of a February 1965 study published by the Naval Advisory Group attached to the Military Assistance Command, Vietnam, titled "Naval Craft Requirements in a Counter Insurgency Environment." It recommended developing a new generation of reliable and sturdy shallow-draft patrol craft for immediate use in Vietnam. The new vessels would need to be self-sufficient enough to patrol 400 to 500 miles on one trip. They were to be equipped with high-resolution radar, reliable long-range communications equipment, and enough armament to mount a limited offense.

It fell upon a Navy commander named Cab Davis to find an American company that could quickly manufacture such a high-tech speedboat. Louisiana's Sewart Seacraft, which was already manufacturing water taxis used at offshore oil rigs in the Gulf of Mexico, seemed to fill the bill. The company agreed to take on the assignment. Davis obtained a rush approval of the blueprints and Sewart Seacraft built two prototype boats. The Naval Advisory Group's report had stressed that the boat's hull not be made of wood, so it could better withstand running aground in shallow waters, so Sewart Seacraft teamed with the Kaiser Company to pioneer a technique for attaching lightweight marine aluminum to the denser bronze propellers below the boats' waterline.

Once Sewart Seacraft's prototypes tested well at the Naval Amphibious Base in Coronado, an order was placed for fifty-four, followed by an additional fifty other boats. The watercraft were then officially named patrol craft fast (PCF). Designed for a crew of six—five enlisted men under one officer—each Swift carried a substantial arsenal. In addition to the twin .50-caliber machine guns mounted atop the pilothouse and another .50-caliber machine gun on top of the mortar at the stern, each PCF carried an M-79 grenade launcher, M-16 rifles, antipersonnel and concussion grenades, .38-

caliber revolvers, a riot gun, and any other weapon the boat's crew could beg, borrow, or steal. There was also an M-60 machine gun kept in a peak tank. Unlike the sturdier riverine assault patrol boats, the light little Swifts had no armor to speak of.

What they did have was power in the form of two weight-sensitive V-12 diesel engines capable of up to 2,800 rpm. Because of the weight of the ammunition stored on board, a PCF seldom got up to speeds above twenty-five knots. The passenger compartment alone had the stress resistance to hold as much as five thousand pounds of cargo. The design provided accommodations for five men—three forward under the pilothouse and two in the main cabin, apparently on the presumption that the remaining two crew members would be standing watch while the other four slept. A hot plate, icebox, and freezer afforded the ability to keep the boat out for several days.

After final test runs in the bayous of Louisiana, seven Swifts went to Coronado to be used for training and a half dozen were given outright to the Philippines. Starting in the fall of 1965, the rest of the order was shipped to Vietnam in groups of eight boats. Once in-country the PCFs were combined with Vietnamese patrol forces and American destroyers and escorts to carry out Operation Market Time.

"Sometimes it seemed that the North Vietnamese would order trawlers and junks to head south just to keep the U.S. Navy busy," recalled Lieutenant Commander Fritz Steiner, a 1957 graduate of the U.S. Naval Academy. "Market Time went on for a long, long time. But the impression I had was that we were spending too much time at sea. The VC went infiltrating in-country through the river system while we were patrolling the coast."

Begun under the leadership of Rear Admiral Norvell G. Ward, an able MACV naval-component commander, and fully implemented by Ward's starchy successor, Rear Admiral Kenneth L. Veth, Market Time was designed as a joint endeavor of the U.S. and South Vietnamese navies. "I decided to institute Market Time after South Vietnamese planes during the first two months of 1965 found two big trawlers unloading arms and ammunition along the coast, clear evidence that screening by a fleet of South Vietnamese junks was inadequate," General Westmoreland wrote in his 1976 memoir, *A Soldier Reports*. He went on to praise the efforts of the Swift boats involved in the coastal operation. "Whereas before 1965 the Viet Cong had received an estimated 70 percent of their supplies by maritime infiltration,"

Westmoreland boasted, "by the end of 1966 that had been reduced to a trickle of less than 10 percent." Later in his book, however, the Vietnam War's commanding U.S. general regretted that the Navy had not "earlier anticipated the requirements of riverine warfare."

After the PCFs were more frequently patrolling Vietnamese rivers, a commissioned Navy captain wrote a report analyzing the Swifts' durability. He recommended that the boats' crews should run the rivers in no more than twelve-hour shifts. The captain's assessment went largely ignored in practice. "Frequent crossing of the precipitated alluvial soil bars at the mouths of the rivers wore the brass speed props down, thereby decreasing speed until replaced by complaining repair and supply personnel, who with blue water mentality, blamed us for grounding our crafts," Elliott "Skip" Barker, one of Kerry's closest friends in Vietnam, later wrote in explanation of the difficulties which Swift boats faced. "Submerged fish nets, which Jacuzzi water jets would cut through or leap over, would foul our screws, bringing a V-12 diesel from full bore to a grinding halt."

The Swift boats operated out of five bases scattered along the coastline of South Vietnam. Farthest north was Coastal Division 12 in Danang. Next came Coastal Division 15 at the relatively quiet port of Qui Nhon. Midway down the South Vietnamese coast was the nicest base, Coastal Division 14, which shared an idyllic resort setting with Coastal Squadron 1, the Navy's main headquarters in an outlying section of the vast Cam Ranh Bay Naval and Air Force Base. Still farther south, marking the transition between the coastal hill country and the Mekong Delta, was Coastal Division 13 at the edge of the tiny town of Cat Lo. Only a few miles from the resort of Vung Tau at the mouth of the main shipping channel leading to Saigon, Coastal Division 13 offered the most exotic atmosphere for young Americans far from home. The southernmost base, Coastal Division 11, sat isolated on a floating barracks barge several hundred yards offshore from the town of An Thoi on Phu Quoc Island in the Gulf of Thailand. Of the five Swift boat bases, Coastal Divisions 11 and 13 provided the most support to Operation Sealords.

John Kerry would be training in San Diego for what he thought would be the purely coastal patrolling of Operation Market Time, whose objective was to establish a barrier along the coast of Vietnam to interdict the infiltration of insurgent supplies by sea. The scuttlebutt began to swirl at Coronado about the rough-and-ready captain who had recently taken over command of the Coastal Surveillance Force in South Vietnam. U.S. Navy policy on riverine warfare was about to change fast and drastically. Captain Roy "Latch" Hoffman, the new commander, was a proud Korean War veteran, a fierce believer in the serious use of force, and not the least bit inclined to sit in the headquarters at Cam Ranh Bay and watch passively. This was war, and that is where military reputations are made, and promotions won. "He was short, a bantam rooster who preferred the field to sitting behind a desk, especially if things were hot," Gregory L. Vistica wrote of Hoffman in *The Education of Lieutenant Kerrey*, a book about U.S. Navy SEAL Bob Kerrey, who went on to become a Democratic Senator from Nebraska. "But he was also a first-rate administrator, adept at tallying the numbers, or body counts, of Viet Cong his troops killed or wounded." Interviews with various Swift boat veterans turned up descriptions of Hoffman as "hotheaded," "bloodthirsty," and "egomaniacal." Not one, however, questioned his determination to win the Vietnam War.

In May 1968, while John Kerry was still seeing to the scrubbing of the *Gridley*, Hoffman was serving as Commander of Task Force 115 in South Vietnam. "I had command of the PCFs," Captain Hoffman proclaimed. "And I had all the Coast Guard Units, which included . . . the big gunners and . . . the smaller ones. And I had all inshore undersea warfare groups, including SEALS. I reported directly to Admiral Elmo Zumwalt, who had an unfathomable belief in young officers, perhaps sometimes to a fault. He also had a style of leadership which gave his combat commanders—like myself—full authority. You realized he had faith in you."

Accordingly, while in charge of Task Force 115, Hoffman demanded the same aggressive independence from his men. He considered Operation Market Time, despite its successes, a far too passive strategy for dealing with the enemy's supply lines. As long as there were Viet Cong anywhere near his jurisdiction, Hoffman wanted to go after them. And if there were no Viet Cong, he was not averse to expanding his jurisdiction until there were. He had little interest in procedural niceties so long as the job got done, and he would back up anybody who showed the due relentlessness in pursuing the Viet Cong. That was Hoffman's heart and soul in Operation Sealords: to destroy the enemy, with the more extreme dispatch the better. "I told the men in my Task Force, 'Either use your firepower or lose it,' " Hoffman declared. " 'I expect you to be aggressive; I expect you to use your head and do your job.' The rules of engagement in Vietnam were overly restrictive. We were being too conservative, not taking the fight to the enemy. But I changed that." Unbeknownst to John Kerry and his Coronado classmates, they would be applying what they had learned about skippering Swift boats in the bountiful blue waters of San Diego Bay to Hoffman's much more aggressive naval operation in Vietnam.

Although he was not required to report until August 21, 1968, the twenty-four-year-old Kerry arrived in San Diego a few days early to get settled in. He was very excited about the prospect of skippering his own boat along the coast of Vietnam. The monotony of the *Gridley* was over; the billet at Coronado was his ticket to action, and his way to join up with Freddie Smith and Danny Barbiero in the rice paddies and jungles of Vietnam.

Rather than live at the Naval Training Center (NTC), Kerry opted to rent an apartment in Mission Beach, so he could grab his surfboard and run out to catch the waves rolling in whenever the mood struck him, unencumbered by military-housing regulations. Whenever a longer block of free time presented itself Kerry would head to the auxiliary base at Brown Field and charter a T-34 or a Piper Comanche, usually to fly along the spectacular Baja coast of Mexico or to Palm Desert in California's nearby Coachella Valley. As he flew over San Diego, the white crosses atop Point Loma at the tip of Coronado would remind him of the body bags still being shipped home from Vietnam by the thousands. Everything else about San Diego smacked of youth and beauty and fun in the sun.

As a Navy town, freewheeling San Diego catered to the tastes of young servicemen. Broadway Avenue and the rest of the area around Fleet Landing were dotted with cheap-eats joints, tattoo parlors, and used-record stores. The area's main attraction for its young servicemen, of course, were the legions of camp-following women who sashayed their stuff around Oceanside's go-go clubs and La Jolla's carefree beaches. Nearly every day, the *San Diego Union-Tribune* ran stories about the Navy on its front page, truly newsworthy or not. On August 2, 1968, for example, the newspaper gave major play to an article on the announcement that Rear Admiral Edward E. Grimm would be replacing Rear Admiral David Lambert as head of the Navy's Pacific Fleet Training Command: *big* news in San Diego, but nowhere else. The Navy fueled the area's economy with myriad multimillion-dollar government contracts, and the entire region teemed with clean-cut bluejackets and white-clad officers, forming a steady stream of fresh faces saying good-bye to bucolic California for the pounding seas leading to the coastal plains of Vietnam.

John Kerry who had received a promotion to lieutenant (junior grade) returned to San Diego to attend classes at the Naval Amphibious Base on Coronado Island, a peninsula reachable from the city by a two-lane road or a ferry. Bracketing the city of Coronado between the North Island Naval Air Station to the north and the Naval Amphibious Base to the south, the 57,000 acres of U.S. Navy facilities on San Diego Bay formed a massive aerospace-industrial complex. Referred to as the Birthplace of Naval Aviation in honor of pilot Glen Curtiss's having first tested the hydroplane there, Coronado had since become home to some thirty commands, including the headquarters of the Naval Special Warfare Command responsible for America's elite maritime special operations forces, such as the Navy SEALS.

The classes were supposed to start on August 25, 1968, but a Navy mix-up delayed the beginning of classes until September 7. "Here I am with more time to think than I ever imagined," Kerry wrote David Thorne on September 4. "I've brought a whole slug of paperbacks and spend the whole day reading just to kill time." The books he had already finished included Margaret Mitchell's *Gone with the Wind*, Arthur Schlesinger Jr.'s *A Thousand Days*, and Dwight Eisenhower's *At Ease*. Luckily for Kerry, while waiting for classes to start, his sister Diana visited him. He took her to Tijuana to see a bullfight, which he recorded with great detail in the letter to Thorne.

At Coronado a team of chief petty officers spent almost two months teaching Kerry and five other aspiring Swift boat officers how to operate a PCF. They learned boat systems, maintenance, seamanship, weapons operations, boat language, and how to board and search a Vietnamese junk without getting their heads blown off. "Swifts were large at fifty-one feet, twenty-three to twenty-eight knots, twin screw, three and a half foot draft on step, thin skinned, an AK-47, .23-caliber round would pass through six sheets of aluminum, imbedding in the seventh, well armed, weight versus speed sensitive, high profile, very seaworthy boats," Skip Barker explained in a letter. "Essentially offshore oil rig supply boats, painted grey, with weapons and ammo lockers added."

Kerry's five classmates at Swift school were Skip Barker, Bob Crosby,

Chuck Mohn, Charlie Pfeiffer, and Clayton "Zeke" Zucker. "I don't have any great stories of Kerry at Coronado," Barker said decades later. "I recall him on the docks and on the boats, familiarizing himself with Swifts. Later in Vietnam, we truly bonded. John was erudite but not all-knowing, sophisticated but not foolish, social but not clingy, and genuine, even compassionate, sensitive to the entrustment of the lives and well-being of his crew to his care."

With his slicked-back hair and pugnacious open-mindedness, Barker resembled a young James Cagney, and he instinctively liked John Kerry. A self-described Southern "progressive," Barker was born September 10, 1943, in Mobile, Alabama, where his father worked for industrial giant Union Carbide. After moving to Birmingham, Barker graduated from the city's Ramsey High School and entered Auburn University, where he penned frequent, controversial articles for the *Auburn Plainsman* on such tinderbox topics as "The Failure of the Emancipation Proclamation: A Hundred Years' Perspective" and "Miscegenation: The Ultimate Solution to Racial Conflict?" Upon graduating from Auburn, Barker signed up for the Navy; like Kerry, he attended the Newport OCS, after which he was assigned to the destroyer *Richard S. Edwards*, where he served as combat information center officer, air intercept control officer, and gunfire liaison officer.

Before long Barker and Kerry became fast friends. "Kerry and I had some things in common," noted Barker, who after the war became both a cotton farmer and a defense lawyer in Alabama. "I recall John writing a lot," he said. "Every time I saw him lying on his bed, or maybe sitting at a table, he was writing. He and I also talked about politics—political issues and Vietnam. I had read Bernard Fall's journalism and we got into his view of the war some." (Fall's firsthand accounts of the French Indochinese War, *Street Without Joy* and *Hell in a Very Small Place*, offered critical analyses of earlier foreign involvement in Vietnam. Inspired first by Fall and then by the horrors he would soon see for himself in the Mekong Delta, in 1971 Barker would compose a paper at Samford University's Cumberland School of Law titled "The Innate Illegality of American Military Involvement in Vietnam.") At Coronado, Kerry and Barker enjoyed discussing politics, but in a couple of months, they would be in-country Navy officers in charge of their own Swift boats. Complaining about LBJ's post-Tet policy was hardly good preparation for instilling confidence in one's men.

With the success of Operation Market Time, the Viet Cong stopped using coastal areas to transport their ammunition and supplies, taking to the rivers and tributaries of the Mekong Delta instead. To answer this new threat, Operation Game Warden was formulated with PBRs as the main vessels involved. Its main objective was to interdict all contraband-running vessels using the Delta's *major* rivers. But once they encountered the U.S. and South Vietnamese patrol boats searching every vessel in the major rivers, the adaptable VC moved to the smaller rivers and canals of the Delta, pirate-style. The good news was that by the time John Kerry began training for Swift boat duty at Coronado, the U.S. Navy controlled the coastal waterways as well as the main rivers of South Vietnam. The bad news was that the many small tributaries running deep into the dense and swampy Mekong Delta remained in the hands of the Viet Cong. And the U.S. Navy wanted very much to change that.

In light of the success of Operations Market Time and Game Warden, Vice Admiral Elmo R. Zumwalt Jr., who was appointed commander of U.S. Naval Forces, Vietnam, on September 30, 1968, initiated Operation Sealords (for South East Asia Lake, Ocean, River, Delta Strategy), an extension of the Navy's riverine program. "By the time I arrived on the scene, the interdiction mission had pretty much been accomplished as far as the coast was concerned," Zumwalt wrote in his memoir *On Watch.** Zumwalt's new goal was to control the Mekong Delta.

The Mekong Delta was the most impenetrable part of South Vietnam.

**In all the vast literature published on the Vietnam War, Operation Sealords, even in* On Watch, *has been largely ignored. The likely reason is that the Mekong Delta remained under the jurisdiction of the South Vietnamese, making it a secondary priority for the U.S. command, and thus for the American media as well. As a result, many Americans back home had no idea how large a role the U.S. Navy was playing in Vietnam.*

A few books written after the Vietnam War do detail what the Swift boats did in the Mekong Delta, most notably Thomas Cutler's superb Brown Water, Black Berets: Coastal and Riverine Warfare in Vietnam. *A pair of first-person accounts—Russell H. S. Stolfi's "SEAL Raid in the Mekong Delta" and B. Anthony Snesko's "Swift Boat Rescue"—have appeared in the magazine* Vietnam. *The commanding officer of the* Greenwood, *Richard L. Schreadley, published the excellent* From the Rivers to the Sea: The United States Navy in Vietnam, *while Jimmy R. Bryant wrote a riveting account of the Navy's role in the Mekong Delta in* Man of the River: Memoir of a Brown Water Sailor in Vietnam, 1966–1969. River Patrol, *a documentary about the Swift boats, came out on videocassette in 1994. Still, that so little has been reported about Operation Sealords does a disservice to those who carried it out. Any comprehensive account of the Vietnam War cannot ignore the U.S. riverine war program on the waters of the Mekong Delta/Ca Mau Peninsula and the gallant efforts of some Americans, including John Kerry, there.*

Once the Mekong River itself crosses the border of Cambodia, it meanders through Vietnam before emptying into the South China Sea. Together with the Ca Mau Peninsula at the country's southernmost tip, the Delta region constituted about a quarter of South Vietnam's total area and accounted for more than one third of its population of 16 million. Since the start of the French Indochinese War in 1946, large sections of the Mekong Delta had fallen completely under the control of the Viet Cong, the Communist faction that had split off from the nationalist Viet Minh. If there was a "heart of darkness" in Vietnam, it was the Mekong Delta. The enemy came in many forms in that murky death zone: malaria-carrying mosquitoes, giant cockroaches, six-inch-long leeches, venomous water snakes, armies of black ants, and, of course, the treacherous, omnipresent Viet Cong. Temperatures routinely reached 100 degrees Fahrenheit, with saturating humidity that made just breathing a chore. The monsoon season, which lasted two thirds of the year, from April to November, featured inundating downpours that turned all high land into mud. Those lucky enough to escape malaria or dysentery grappled instead with body fungi and skin ulcers.

The entire region is crisscrossed by rivers, tributaries, and canals. The French Navy had found this web of waterways utterly impenetrable two decades earlier; by the time John Kerry was in training at Coronado, southern Vietnam's rivers and canals formed one of the most extensive transportation networks in the world, comprising more than four thousand miles of inland "water roads." This system provided the principal means of transportation by which the North Vietnamese Navy and the Viet Cong connected, the former protecting the South Vietnamese insurgents' free use of the Delta waterways to move their men and matériel. The task fell to the U.S. Navy to stop them. "The Delta's real roads are its waterways," declared historian Gordon L. Rottman in *The Vietnam Brown Water Navy: Riverine and Coastal Warfare, 1965–69*, "with sampans the main mode of transportation. Even villages were either floating, built on stilts, or with individual homes on earth mounds." Vietnam's sampans—wide, flat-bottomed, wooden skiffs often with mat-covered cabins for living quarters—were as common in the Mekong Delta as automobiles in the United States.

Besides receiving training in how to operate a Swift boat, which became second nature to the nautically minded Kerry, he also spent two weeks in a mandatory crash course in Vietnamese. Elegant native speakers with warm sepia skin and almond eyes, dressed in their country's traditional flowing silk *ao dai*, taught the young officers and selected enlisted men at Coronado's language school as many potentially lifesaving phrases as they could learn, and would certainly come in handy in Southeast Asia:

"Where is the enemy?"
"Go away quickly."
"Did the enemy blow up the bridge?"
"I need a doctor, quickly!"
"Stop. Or I'll shoot."

Kerry considered himself an above-average linguist, with some proficiency in German and French, and he studied hard to gain a working knowledge of Vietnamese. Although almost 50 million Southeast Asians spoke the language, few American troops ever mastered Vietnamese—including John Kerry, despite his game attempt. That said, he did do better than most of his classmates, all of whom were required to learn at least two hundred words. For example, one junior officer never managed to utter a single word. Only with enormous difficulty could he even mimic the teacher immediately after her painstaking, repeated enunciation of a phrase like "Are you smuggling machine guns?" But his inability to communicate one word of Vietnamese didn't faze him in the least. According to Kerry, his fellow officer would just burst out laughing: "Hell, if they don't answer me right when I point a fucking M-16 at their heads, I'll just fill 'em up with lead. Ain't no goddamn language problem then." It was Kerry's introduction to the challenge of winning Vietnamese hearts and minds.

As appalling as his remark sounded, that officer did get one thing right: the M-16 rifle really was the American serviceman's best friend in Vietnam. Yet it could also be a treacherous foe. Even though the Pentagon bought them from Colt Industries in droves (ordering nearly 840,000 of the rifles in June 1966, for instance, at the cost of $91.7 million), the M-16 had a reputation for frequent malfunctions. Powder residue had a way of causing spent casings to jam the rifle's chamber, which could prove fatal under a sudden attack. The problem became so glaring that in 1967 a House of Representatives subcommittee investigated the M-16's reliability. Its report stated that "the much-troubled M-16 rifle is basically an excellent weapon whose problems were largely caused by Army mismanagement."

New M-16 designs, including a model configured as a sniper rifle, soon appeared—as did a remedial lubrication kit to prevent jamming in the older models. What American troops liked about the M-16 was its lightness, so it was easy to carry above their heads as they slogged through the waist-deep mud of the jungles of Vietnam. By contrast, the Soviet-made AK-47, which served as the main weapon of the North Vietnamese, boasted fairly high quality, but its wooden stock rotted in the sopping humidity of Southeast Asia. The M-16 had the advantage here: a plastic stock that lasted in swamp warfare conditions.

While they were in San Diego, Kerry and his cohorts heard stories from in-country veterans about how their M-16s had saved their lives. They also listened to far more disturbing accounts of how other M-16s had earned nicknames like Black Widow, Widow Maker, and Mattel Rifle. But the tales of horror told by some recently returned Navy men had nothing to do with jammed M-16s, just with the hellish dangers lurking ahead in the Mekong Delta.

Every young officer at Coronado training for Swift boat duty had seared into his mind the story of how PCF-4 had been lost on Valentine's Day, 1966. Late that afternoon the boat had been quietly patrolling the coast of the Mekong Delta. "In the distance loomed three small mountains which the Americans called the Three Sisters, looking formidable in the surrounding alluvial plain," wrote Larry Wasikowski, a former Navy lieutenant and charter member of the Swift Boat Sailors Association. "The rumor was that a Viet Cong ammunition factory was hidden in the bowels of one of the peaks. It was not a friendly neighborhood."

Through his binoculars one of PCF-4's crewmen spotted a red, yellow, and blue flag flying from a pole on a bamboo raft floating a football field away. It had been a rather dull day, so the idea of snatching a Viet Cong flag for a souvenir appealed to Lieutenant Charles D. Lloyd, the officer in charge. As PCF-4 approached the raft, Lloyd took the precaution of ordering two grenades to be lobbed at it just to make sure the water wasn't mined. He believed he was even erring on the side of caution. The crew waited; they heard no secondary explosions. The VC flag looked like a risk-free trophy.

Lloyd eased PCF-4 up to the raft. A crewmate pulled out a knife and began cutting the flag rope. Wasikowski described the scene: "In the shadows of the mangrove trees on the nearby shore, a pair of eyes watched the great gray fish nibble at the bait. It was time to set the hook. Hands holding the wires of a crude but effective detonating circuit came together, allowing bare copper to touch bare copper in deadly union. An explosion erupted beneath PCF-4, ripping into her underbelly. The main deck buckled upward, crushing and trapping the coxswain against the overhead of the pilothouse; the gunner above was hurled out of his gun-tub and into the water. The Swift, her hull torn open and her frame traumatized, plunged beneath the frothing water almost at once and soothed her wounds in the cool mud of the bay's floor." The hull of PCF-4 had split in two.

Two Army aviators arrived on the scene of the sinking Swift, hoping to save its crew. Four of the men—David J. Boyle, Tommy E. Hill, Jack C. Rodriguez, and Dayton L. Rudisell—had joined the 58,000-plus names now listed on the V-shaped black granite Vietnam Veterans Memorial wall in Washington, D.C. The two survivors, Lieutenant Lloyd and Robert R. Johnson, were both badly wounded. Lloyd's shattered left leg was put in a make-do splint made from an M-14 rifle. Wincing in pain and grave with guilt, Lloyd was rushed to an operating room in Rach Gia. How could he have been so careless? Why couldn't he have lost his life instead of those of four of his men? In a voice quivering with the agony of remorse, he whispered to fellow Navy Lieutenant Gil Dunn, who was hovering over the operating table: "I don't hurt in my legs, Gil, I hurt right here." Lloyd's hand slapped at his heart.

The story of PCF-4—the first Swift boat lost in Vietnam—served as a focus of Kerry's training at Coronado. The lessons were obvious: one foolhardy act by a Swift boat lieutenant could kill or maim his entire crew. The Viet Cong, the instructors emphasized, did not play by the Geneva Conventions or the U.S. Naval Academy rules of conduct. In the murky realm of guerrilla warfare, a nifty flag could be a mine, a three-month-old baby a grenade, a smiling face a knife in the back. Vigilance was the primary quality in an effective Swift boat commander in a place where noticing a broken branch or a rousted bird or even too much silence could mean life or death. Only one thing was certain: nothing could be worse than the living hell Lieutenant Lloyd had doomed himself to for a souvenir. "Boy, did that story register with me," Kerry averred. "Because of it, when I got to Vietnam I was super-alert to not falling for the souvenir trick."

The instructors at Coronado issued an even more dire warning to Kerry and the other rookie lieutenants: at all costs, do not get captured by the enemy—the Geneva Conventions and the Uniform Code of Military Justice had little meaning in Vietnam. They spiced this frightening caveat with the tale of an Army warrant officer who had been taken prisoner. He was found hanging from a mango tree, skinned alive. "One man I met who had spent several tours in Vietnam told me how to make Vietnamese people cooperate during interrogations," Kerry remembered. "If the individual was a mother, simply take her child and throw it in the water with her watching. He said it was almost foolproof. 'Shit like that always works on gooks,' he said." Such ghoulish and bigoted talk made Kerry uneasy, but he learned just to shrug it off with a tough-guy nod; it was important for both his reputation and his own psyche not to be unnerved by warrior bravado. The same veteran also tipped Kerry to another "surefire" interrogation tactic: take a group of captured enemy Vietnamese up in a helicopter and then throw them out one by one until somebody coughed up some useful and detailed information. Kerry quoted the vet: "Fuckin' gooks, you ought to see their faces shit a brick as the first guy goes out." At first, Kerry didn't believe him, thinking it was just macho posturing or a tall tale to intimidate the uninitiated. Later, however, he was shown pictures, published in *Life* magazine, of a man being purposefully pushed out of a helicopter. "I was glad that I didn't know the name of the S.O.B. who had told me these so-called vital tips," Kerry avowed. "I would have hung him out to dry."

Years later, Kerry remarked that if there was one book he would like to write, it would be titled simply *1968*. As Thomas Houston Jr., an African American who served on PCFs that year, later summed it up: "First Tet, then [Martin Luther] King [Jr.], then [Robert F.] Kennedy, with people firing at me; man, no wonder we all got PTSD [post-traumatic stress disorder] and endless nightmares."

Without question, 1968 was a chaotic year. Every day's paper brought more tumultuous news. As an antiwar Democrat, Kerry interpreted even *good* news as bad news. For example, General William Westmoreland was fired, only to be replaced by the equally hawkish General Creighton

Abrams. And, of course, the news of Lyndon B. Johnson refusing to seek reelection had brought a cheer that Robert F. Kennedy would win the Democratic nomination for president, only for him to be gunned down in Los Angeles. Meanwhile, the United States and North Vietnam scheduled peace talks in Paris, and in Chicago, Mayor Richard Daley had ordered his police force to arrest protesters at the Democratic convention. Nothing seemed to make sense in 1968. Particularly burning a village in Vietnam to save the country.

Although John Kerry naturally grew more and more apprehensive that August, his 1968 was about to get much, much worse. The last week of August, as he waited for his PCF course to begin in Southern California, he spent hours glued to the television, riveted by the scenes of police violence and political intrigue at the Chicago convention. "I watched it intently," Kerry said. "I'll never forget the tear gas streams and screams, the chaos of people running around all of the hotels. Everything was in turmoil." He watched live coverage of the cerebral Tom Hayden of Students for a Democratic Society and the outrageous Abbie Hoffman of the Youth International Party denouncing the Vietnam War in revolutionary terms.

It was one more sign that America seemed to be unraveling into anarchy that year. On top of the deadly riots in Detroit, Newark, and other cities after the assassination of Martin Luther King, John Kerry's great fear, as he watched an estimated 8,000 to 10,000 antiwar protesters clash with some 12,000 Chicago police, 6,000 U.S. Army troops, 6,000 National Guardsmen, and 1,000 FBI agents, was that a bloody massacre of Antietam-like proportions was in the offing. The media dubbed the melee the Battle of Michigan Avenue. Fortunately, the violence was contained with no loss of life. But by the end of the week of the convention, city officials reported that more than 100 police and a like number of demonstrators had been injured, and 589 people had been arrested. "From the mags I've sent and am sending you should get some idea of the havoc that characterized the democratic convulsion," Kerry wrote David Thorne that September. "There's still an uproar here about police brutality and Mayor Daley and I suppose it will be hushed up."

What had happened inside the convention hall discouraged Kerry almost as much as what went on in the streets outside it. Vice President Hubert H. Humphrey had arrived in Chicago practically guaranteed the nomination, having secured about 150 more delegates than he needed to win. This bothered Kerry. He ardently supported Minnesota Senator Eugene McCarthy and was incensed that just a week before the convention Humphrey had appeared on CBS's *Face the Nation* embracing President Johnson's Southeast Asia policies. Resigned to the fact that Humphrey would be the nominee, Kerry did appreciate that Democratic Party officials had at least accorded Robert F. Kennedy the respect of showing a film tribute. He savored a few other highlights as well: the New York and California delegations singing "We Shall Overcome" in protest of the Johnson-Humphrey Vietnam "peace" plank; the fervent efforts of leading Democratic pols to persuade Massachusetts Senator Edward M. Kennedy to toss his hat into the ring; and civil rights activist Julian Bond's nomination for vice president, even though at age twenty-eight he was constitutionally ineligible. "It was terribly dramatic," Kerry remembered. "But one thing was certain: no matter who won the 1968 presidential election, I was headed for Vietnam. That was my reality."

Chapter Six

Trial by Desert

Before being shipped off to their Swift boats in Vietnam, John Kerry and his Coronado classmates first had to endure the U.S. Navy's grueling tests of their mettle in a Dante's inferno of worst-case scenarios. To help the young lieutenants and ensigns cope in the event they were captured, the Navy subjected them to six brutal days of physical, mental, and emotional training in what the military called survival, evasion, resistance, and escape (SERE). The program's primary mission was to teach naval aviators how to stay alive after being shot out of the sky by the enemy, but anyone serving in Vietnam would learn the myriad ways to boost one's chances of making it out of danger.

Neither Kerry, his Swift school classmates, nor any of the 121 other men who underwent SERE training that week in the fall of 1968 could have imagined what they were in for. In a letter to his girlfriend, Julia Thorne, Kerry described the experience as a "horrendous nightmare—more bizarre and surrealistic than any I have ever known."

On the eve of starting this survival school each man was issued a knife, a flashlight, and an old Army jacket. With these minimal "wilderness kits" in hand, the trainees piled into three huge vans—"cattle cars," Kerry called them—to be taken to Coronado's nearby North Island Naval Air Station. They spent the first of their SERE course's six days in a classroom there, listening to lectures on the basics for surviving in harsh conditions: "Code of Conduct," "Survival Psychology," "The Geneva Conventions," "Communist Indoctrination Methods," "POW Resistance and Escape," "Survival Medicine," and "Edible Foods." The class was then divided into six platoons, each led by three or four officers in charge according to seniority. Kerry's platoon comprised thirteen men.

The next phase of the SERE course ordered each platoon to put the classwork into practice. Kerry's group was sent to a nearby beach to get themselves food from the Pacific Ocean; because the men would be serving in coastal Vietnam, the Navy wanted to make sure they could feed themselves from the sea should their C-rations run out. The exercise had its uncomfortable aspects, as Kerry reported in a letter to Julia:

[We] passed bathers who stared somewhat open-eyed at this strange group of green-clad idiots strolling in formation past their surfboards and cabanas, each with a heavy coat on in the middle of the afternoon and each with a huge hunting knife attached to his side. Eventually we reached the far end of the beach, where we made a facsimile of a tent out of an old parachute and then started to try and find something for dinner. Using dead fish heads tied to lines we managed to pull some crabs up from the rocks out on a jetty and also to dig a few clams. One group caught a shark; another got a lobster with their bare hands—lucky bastards. By about nine that night we cooked some water and boiled our small catch in it and then, amidst groans of delight, devoured dinner for the night. In the end I tallied up one clam and a claw of a crab—not too bad for an evening meal on Survival.

Decades later, Skip Barker remembered that the hardest part of SERE training was feeling constantly starved for food. Suddenly, a full canteen was worth a fortune. "We were fed a lot of water, salt tablets, a few candy bars, and one ham and cheese sandwich a day," Barker wrote in an unpublished memoir. "We were quickly hungry. Survival training culminated in spending an afternoon and night on the beach with part of a parachute for shelter, your knife and canteen. We forged for sea life for food—pried open and ate barnacles, tried to catch minnow-sized yellow-fin, scavenged the beach for survival nature objects. We found some gnawed chicken bones, light-weight line, pieces of chicken wire, charred driftwood, and a blackened aluminum pot."

Catching seafood and sleeping on the sand in San Diego proved a picnic compared with what SERE had in store for them next. The trainees filed back onto the three big buses for the ninety-minute trip into the desert in northeastern San Diego County. The Navy had established a SERE training camp eight miles north of Warner Springs at an elevation of 3,200 feet that afforded a fine view of Palomar Mountain, and its famous observatory looming in the near distance. The facility consisted of an administrative building, a rustic staff barracks, two wells, a wastewater treatment plant, and a training compound. The rest was wilderness, less than a hundred miles from the heart of the largest desert in the United States.

"The Mojave is a big desert and a frightening one," John Steinbeck wrote in his 1962 memoir *Travels with Charley*. "It's as though nature tested a man for endurance and constancy to prove whether he was good enough to get to California." Or to command a Swift boat in Vietnam, apparently.

The trainees barely had time to take the topography in under the blinding white sun—before a team of tough, leather-skinned SERE survival experts began barking at them about staying alive in Vietnam under virtually any circumstances. Through their no-nonsense tutelage, Kerry and his colleagues learned how to build a fire in the rain, how to make a tent out of a parachute, how to squeeze water out of plants, and which bugs to eat to survive. Coping with thirst, hunger, and sleep deprivation formed the focus of the schooling. Ways to deal psychologically with known Viet Cong torture tactics such as being beaten with bamboo and having fingers broken or eyes gouged were also demonstrated. There was something in the program to terrify every trainee. "One thing that really did scare the living hell out of me was this cage at the end of the bungalow," Kerry confessed to Thorne. "In it were several huge, very ugly (phenomenally ugly as Grandma would say) rattle snakes. Apparently the place was crawling with them and we were warned to be particularly careful of them—a warning that I was not about to ignore for anything in the world. If there was anything that leaves me absolutely dry-mouthed and frightened it is the thought of meeting face to face with any poisonous snake—or even any snake. I just hate the thought of it. Well, here I was, walking around where the possibility of meeting one was greater than I have ever had before. Oh joy, oh bliss!"

A particularly harsh SERE exercise centered on letting the starving trainees loose in the desert for hours with the permission to eat anything they could find or capture. Because some eleven thousand hungry men had scoured the desert over the four years since SERE training began there, whatever natural food supply the arid land could offer had been depleted long ago. "Consequently, at about five-thirty in the evening, it was an extremely tired and pissed-off bunch of erstwhile Navy men who made their way back into camp empty-handed," Kerry groused to his girlfriend. "I think what they wanted to do was wear us down even further. It is hard to describe the heat and the thirst that one felt. I carried two canteens and within the space of the afternoon both of them were empty. Every step that I took I was watching for rattlers, and because I was the officer I felt [an] unfortunate necessity to lead the way. Eventually we got so tired that we just lay down under a tree and rested for about an hour—dreaming about the food that we wouldn't eat that night due to our industry and perseverance."

Just as all six of the platoons neared their breaking point, however, the SERE instructors showed up with a basket full of dead rabbits, plus a few wild pigeons. After a quick lesson in plucking and skinning, the ravenous trainees set to work chopping the game into 127 pieces, one for each man. "All this we threw into one pot and cooked in boiling water, thereby making what tasted like the most delicious and well-placed broth that I have ever had," Kerry enthused. "Man, did it hit the spot! By this time we were living somewhat like pigs, eating out of filthy hands—completely unwashed, bearded, smelly—a sight for the sorest of eyes."

Next came the hardest parts of SERE training: the so-called Navigation Hikes. For these, the instructors issued the young platoon officers different sets of geographic coordinates to get their men to specified rendezvous points within an allotted time. Three and a half decades later Kerry could still vividly recall the dried-out riverbeds, tall saguaro cacti, and desiccated arroyos he encountered wandering the California desert with a Sears, Roebuck compass. Worse, meals consisted of an orange or a quarter of a can of Spam. More acclimated to the breezes of Narragansett Bay, he cursed the choking desert heat, astonished that the human body could sweat so much. And he was one of SERE's successes: some men could not complete the course, done in by heat exhaustion and dehydration.

Unable to sleep easily on the ground, Kerry contemplated the stars those nights in the desert and felt grateful for reveille at four-thirty every morning. Bright and early one day all the SERE trainees were led to a riverbed and ordered to make it to a "freedom pole" two miles away without getting "captured" or "killed." The war game was an exercise in enemy evasion. "The real game had started and I ran like a scared bastard to get as far from the other people and as far down the course as possible," Kerry explained in his letter to Thorne. "Before too long I heard shots ring out on the course and we all knew that the aggressor was signaling his presence. I dove for a clump of sagebrush and breathlessly tried to gauge how far away the shots were. For a moment I was mad at myself for getting so out of breath and hot right at the start, but it quickly became apparent that I was correct in getting away from the mass. You can't evade with some ass near you making noise in the brush and moving when you know that movement is absurd." He strained to clamber up a steep ravine, crawling over cactus, his hands lacerated by thorns. He finally reached the top, where he hid in a clump of bushes. Away in the distance he heard voices shouting, "Get up, you Yankee pig!" and "Come out of there, imperialist!" in what sounded like a mix of Russian and German accents. "They spared no insult," Kerry put it.

Determined to evade the "enemy," he began slinking through the underbrush toward the freedom pole. His throat was parched, his lips chapped white. He drew close enough to the "detention camp" to hear the jeeps bringing the captured in for interrogation. "As I lay there thinking about what to do, one of the road patrol guards came sauntering unaware down the road," Kerry recalled. He dove into the brush, curled up in a ball, and tried not to breathe. "I froze . . . and the idiot walked right by me. I could have reached out from the bush and touched his leg. His uniform was a shock: it looked like a reincarnation of a German or Gestapo soldier. Dark gray pants with a red stripe, a dark gray shirt with red tabs on the collars and a star on that, and then the small Castro–de Gaulle type of hat also with a red star on it. For a moment I wondered if I was still in the United States."

Once the guard moved on, Kerry crept out from the brush. He dropped and rolled across the dirt road so as not to leave footprints, and entered a small stand of piñon trees. Moving slowly and stealthily, Kerry soon found himself just three hundred yards from the freedom pole. His chest heaved as he paused to survey the situation. The flag on the pole was red, signaling "danger." He would have to wait until a yellow flag before he could make his way carefully to the pole and his waiting reward of fresh fruit and water. "To my delight, within a few minutes the yellow flag was raised, and I snuck out quickly into the sagebrush and began crawling on my stomach to get to the reward that lay so close," he wrote. "By this time you have no idea how important a cup of water and a piece of fruit had become. It would be my first solid piece of anything since the small amount of crabs I had consumed. . . . My throat was so dry that air was sticking to it as I breathed and I could not have shouted or talked had I needed to."

Not far away to his right, Kerry spied another sweaty body scrambling desperately for the prize. "We joined up and began the last hundred yards' crawl together," Kerry wrote. "I was as amazed to look at him, as he must have been to see me. We were apparitions from some other world—our faces completely darkened with charcoal to keep reflection from sweat and body gleam away, our clothes ragged and filthy, our pockets filled with sagebrush stems for camouflage and a few extra leaves sticking out of our caps. Through the cover of charcoal was a week's growth of beard—two grubbier people have probably never met under more extreme and absurd circumstances."

Kerry was the first of the 127 trainees to reach the freedom pole, beating even his crawling partner to the cup of water and the orange. Within the next hour, twelve other men would likewise avoid capture. Feeling proud of himself, Kerry was suddenly taken aback when a "huge man" in his thirties walked over to the sailor seated next to him, grabbed him by his uniform collar, and started slapping him in the face. "Why don't you do as you are told, Yankee pig?" the burly goon kept shouting. As officer in charge, Kerry rose to the bait. "As senior man here I insist that you deal with me," Kerry intervened. That was all it took. Suddenly Kerry was being repeatedly "slapped fucking hard" in the face. Stunned and seeing stars, he demanded to speak to the officer in charge, only to be "belted" again, four or five times. He wound up facedown in the dirt with a rifle barrel in the back of his neck. The survival drill was still on full bore. Kerry was loaded into a truck and brought to a POW interrogation compound. In front were about a hundred "prisoners" who were digging pits with shovels in the scorching desert sun. A twelve-foot-tall barbed-wire fence surrounded the isolated compound. Kerry and the other new captives were instructed how to salute guards and pay proper homage to the so-called People's Army. "We were then made to strip down completely except for our underwear and to carry our clothes into what was called the pig pen—a small rectangular area with a barbed wire roof that was so low you had to get on your hands and knees to move through it," Kerry explained. "I caught my back on some barbs and ripped it open."

With people passing out in the heat, and fear draining them all, the moment everybody loathed most was hearing his POW number announced over the loudspeaker. When they called out 502, Kerry's assigned number, he tried to compose himself. He stared straight ahead as he passed a flagpole topped by the Soviet flag and headed to the interrogation chamber. On the floor were stacks of coffins and on the wall a picture of China's Chairman Mao Tse-Tung. The first torture drill involved forcing seven POWs into a box, crammed in one against another. After about thirty minutes claustrophobia set in. Then they were all shoved into solitary confinement cells half the size of a country outhouse. "I almost panicked," Kerry wrote his girlfriend of the ordeal. "My head was crammed down between my knees and I couldn't move it up. I have always had a fear of small places—of being shut in something and not being able to move. I have never thought that it might happen in such an awkward and frightening way."

Just in the nick of time, with Kerry at the end of his rope and pounding his fist, screaming for air, he was released. "I don't know how many pounds I sweated off in those minutes in the boxes but I was wet from head to toe," Kerry recounted, "and so thirsty again that I determined to steal water."

Things only got worse. A hood was placed over Kerry's head and he was ushered into a room for a private interview where the "big four" responses were name, rank, serial number, and, "I'm sorry, but my country will not allow me to tell you that." The verbal drilling started in earnest and Kerry refused to speak. After it was clear he wasn't going to cooperate, and once he had made the mistake of smirking again, he was punched in the face. "I have never in all my life been hit as hard as that and for a moment I just glared straight ahead and tried to pull the senses back into their proper places," he wrote. This went on for an hour with Kerry forced to do knee squats and push-ups. "I started to do the push-ups but would not count them out loud," Kerry noted. "We had been warned of this trick. They make you count them and when you say ten or twelve or whatever push-up you are on they record your voice. Later you hear a tape that asks the question: 'How many people's soldiers have you killed' and your voice answers on the tape 'ten' or 'twelve' or whatever number they chose to dub in. Finally, again, I collapsed exhausted."

While Kerry and most of the POWs managed to hold their own, a number of men broke down or joined enemy ranks within hours. "One poor sailor was brought before the entire company in tears to proclaim that he was a coward," Kerry wrote. That evening he studied ways to escape the POW camp but found none. With air siren drills and calisthenics at 3:00 A.M., it was difficult to maintain one's sanity. Sleep deprivation was part of the SERE drill. But at dawn the men were called out to the courtyard. "Up the flag pole went the Stars and Stripes in place of the Hammer and Sickle that had earlier been there and the guards all translated their bows into salutes as we knew them," Kerry wrote. "We stood there as relief and pride surged through every pore of the body."

A few weeks later, Julia Thorne came to stay with Kerry in San Diego. He gave up his Pacific Beach apartment, and they rented a room together at a coastal motel. She often journeyed back and forth to San Francisco, where her older brother, Lanny, a Marine, was also training for duty in Vietnam. On one occasion the three of them flew to Las Vegas on a whim. Julia Thorne smiled as she spoke of the jaunt with her brother and her boyfriend. "I had a blast. They were laughing so hard at me, because I had never seen anything like the Strip in my life. It was the most incredible thing." The trio played the slots at Caesars Palace and laughed at the bizarre acts at Circus-Circus. "They couldn't get me to stop pulling these levers," she said, laughing, years later. Not long after the Las Vegas trip, Lanny Thorne received his orders to Vietnam. His little sister saw him off in San Francisco. "I took him to the Greyhound Bus station," Julia said, "and I will remember for the rest of my life watching him walk off in his green fatigues [with] his gunnysack into the bus station and I never knew if I would see him again. And then I went back down to San Diego."

One unseasonably hot afternoon in October, John Kerry and Julia Thorne, just back from a quick trip to Las Vegas, strolled around the San Diego Zoo in the bright sun, holding hands and talking about the Vietnam War. "I just loved that zoo," Thorne smiled. "Although I didn't know it then, I was a budding naturalist, fascinated by seeing the animals in their habitats instead of cages." The zoo's latest acquisition, a pair of giant Komodo dragons, made her think of jungles, which made her sad knowing that in just a few weeks Kerry would be heading off to the Mekong Delta to risk his life for his country. When they got back to the parking lot, Thorne sat down on the hood of a random white Chevrolet sedan and proposed to Kerry. "I said,

'John, I think we should get married,' or something of the sort," she recalled decades later. "He was a little taken aback. I don't think he said anything. He did what I called the John Shuffle: he rubs his eye, his left eye, with his left forefinger knuckle, then he looks off and sort of wiggles his body. . . . He just gets shy and embarrassed. But he agreed. It was understood that when he got back from Vietnam we would get married."

The couple decided to keep their engagement secret, telling only their closest family and friends they had taken the leap. Kerry remarked that his main rationale for the secrecy was the death of Dick Pershing, who had been engaged to a woman named Shirley Gay. Kerry explained, "But Shirley and he were engaged publicly, and that was such agony. It made it tough on Shirley. I just didn't want Julia to have to cope with that kind of a public thing. We just decided we would not tell people publicly, just to reduce the pressure, reduce the intensity."

Because he would be leaving for Vietnam so soon, Kerry was granted some days off. He and his fiancée decided to head back up coastal Highway 1 to San Francisco to celebrate their engagement at the Mark Hopkins Hotel. "We went all out," Thorne remembered. "We rented a suite that had a wonderful view of the city, with perfect light shining in through our open windows. But then, in a moment I'll never forget, the atmosphere grew a bit dark." John called his parents with the news and they were ecstatic. Then, in her excitement over being engaged, tempered by the sadness that her fiancé would be leaving soon, she telephoned her father in New York. Her news did not go over well. "He was disappointed that I had gotten engaged," she explained. "So he pulled an 'imperial demand,' the kind only fathers can do, and made up some false emergency story so I would return to New York at once." Dutifully, she did, leaving Kerry to spend his last days stateside alone.

On the eve of Kerry's departure, Richard Nixon was on the campaign trail imploring Americans to believe in the Republicans' secret plan to achieve "peace with honor" in Vietnam. Nixon's appeal resonated with the patriotic citizens of San Diego, who saw destroyers anchored along 32nd Street, nuclear submarines bursting through the bay's waves, and flying aces roaring their F-4 Phantoms across the blue sky. At least in San Diego, victory in Southeast Asia still seemed possible.

Nonetheless, the antiwar movement—along with prejudice and suspicion toward the so-called radicals leading it—was active in San Diego. For instance, a local organization called the Military Order of World Wars publicly demanded that the University of California at San Diego fire philosophy professor Dr. Herbert Marcuse for his Marxist beliefs.

Paradoxically, Realtors and merchants in San Diego also discriminated against servicemen, in distrust of their rootless lifestyles. "I remember being turned away from several doors because I was in uniform," Kerry recalled. "Checks weren't cashed, because credit for men about to be shot at in Vietnam wasn't very good. Specials on just about everything for sale soaked the naïve servicemen of their already paltry paychecks. And worse still, a serviceman couldn't drive anywhere without getting stopped by the police for something: one tread of one tire over one half of a double line for one minute."

On balance, however, Kerry found his time in San Diego pleasant. He enjoyed tooling around the city on a Peugeot racing bicycle, often accompanied by Giles Whitcomb, soaking in the ocean breeze and Mexican ambience and quaffing tall, lime-twisted Coronas to help him unwind. Whitcomb had graduated from Harvard and was now a Naval Intelligence officer. "We practiced Vietnamese together," Kerry recalled. "We later met up in Saigon." At dusk, after each day of Swift boat school, Kerry would bike back to his apartment, taking a shortcut through the Marine Corps Recruit Depot, past the dusty fields where hundreds of recruits sweated through calisthenics and other drills. Above them on a platform was always poised a bulldog-faced drill sergeant, and on every fourth count the entire mass of molded young bodies would jump up at once, shrieking, "*Kill!*" Sometimes Kerry would see them practicing with bayonets, thrusting at imaginary enemies all around them. At the moment after the blades would have met flesh, they would drive in for the kill, all grunting together. "It sounded like a thousand men dying at the same moment," Kerry recalled. "Then they would return to their cadence and shouts of '*Kill*' and '*Hate*' and '*Gooks*' and other things."

On some days as Kerry cycled quietly by the fields he would see an honor guard practicing the regimented rituals of military funerals, precision-folding American flags into perfect tight triangles as bugles echoed with the haunting dirge of "Taps." One day the honor guard drills coincided with the enlistees' calisthenics, and the strains of "Taps" mingled with the hard, staccato shouts of "*Kill!*" across the Marine compound. The awful juxtaposition made an indelible impression on John Kerry, one so profound it pierced right through the cocky, gung-ho attitude he had assumed as armor against his emotions. He cried, reminded of Dick Pershing's senseless death. He thought of his friends already in Southeast Asia, Freddie Smith, Danny Barbiero, and Paul Nace, and wondered which would make it home. And he couldn't help but wonder as well how many of the young bodies doing push-ups in the dust would be consigned back to it in boxes by that honor guard across the recruit depot.

Other, more disturbing aspects of prewar life bothered him more and more. One day Kerry visited a lieutenant he knew who trained the Marine recruits. When he asked this fellow officer to join him for dinner, the lieutenant called a subordinate into his office to find out what was being served that evening. "Private!" Kerry's colleague barked. "What's for dinner?"

The enlisted man stood rigidly at attention and answered, "Sir, the private doesn't know, sir."

"What do you mean you don't know? What good are you if you don't know what's for dinner?"

"No good, sir."

"You like to eat, don't you?"

"Yes, sir, the private likes to eat."

"You're nothing but a worm," the lieutenant sneered back. "And don't smirk at me. Were you smirking at me, worm?"

"No, sir, the private wasn't, sir," the man replied.

"Find out what's for dinner, worm, and don't come back until you do. Hurry up, worm!" the officer concluded.

The enlisted man immediately disappeared, returning a few minutes later to reel off that day's dinner menu. Asked whether the food was good, he replied, "Sir, the private doesn't know if it's good," earning himself a fresh round of abuse before his superior finally let him go. Kerry winced at such a humiliation tactic—it seemed mean-spirited and juvenile.

As Kerry was leaving the Marine compound with the officer, another young recruit saluted him normally while carrying his rifle, a violation of regulations. The barrage of verbal abuse the recruit earned from the officer for this transgression sounded in Kerry's ears long after he turned the corner and headed for the beach. At that moment, Kerry grew determined to remain true to himself rather than the military system. He would follow orders, always, but he would also maintain an independence of mind and personal behavior.

Kerry completed his training at Coronado soon thereafter. It was late October. He went surfing off Windanesa Beach one last time, then jammed his uniforms, Bible, flight logbook, and personal belongings into a duffel bag and a backpack, and once again said good-bye to San Diego. He flew to San Francisco and spent a night at the Mark Hopkins, where he bumped into Yale University president Kingman Brewster at the bar. The next day he flew to McChord Air Force Base in Tacoma, Washington, and boarded a jet bound for Cam Ranh Bay in the Republic of South Vietnam. The plane stopped to refuel in Japan. Nearly two decades later, the time he spent wandering around the Japanese airport would come back to him as he watched the opening moments of director Oliver Stone's *Platoon.* In that Academy Award–winning 1986 Vietnam War movie, a baby-faced Army enlistee (Charlie Sheen) walks past a column of war-shattered veterans heading home from Vietnam. Sheen's still-innocent character, clean shaven and perfectly uniformed, reminded Kerry of himself that day in Japan; the stressed-out vets in their camouflage clothes reminded him of the ravaged men he saw in that airport, their hollow eyes staring straight ahead and still reflecting the abyss. Stone's actors exactly captured what Kerry called "the thousand-mile stare."

Also like Sheen's character in the film, Kerry heard the returning vets mumbling things at him like "You sucker" and "You motherfuckers don't know what you're in for." *Platoon* duplicated that queasy memory so precisely that he left the movie theater drained to the point that he had to sit for a while on a street curb to regain his composure in the clear Boston air. "Oliver Stone caught certain aspects of Vietnam brilliantly," Kerry explained. "That airport scene where the incoming were passing the outcoming men was exactly real and right. My welcome to Vietnam was straight out of *Platoon.*"

CHAPTER SEVEN

In-Country

"It was fairly cloudy over Vietnam as we came in, and breaking through the cover to descend on the bay was quite exciting," Lieutenant (j.g.) John Kerry wrote to his parents a few weeks after landing in Vietnam. "As we came in for a touchdown I could see a rainbow that ended splotch in the middle of the runway—significant either of the airport's value or more likely of the meaning it would have for me in a year's time (I am already counting the days). Once down we all suddenly realized that we were really there and stuck for one long year—twelve months—three hundred and sixty-five days and so on *ad nauseam*. Yeech! When the doors of the plane opened I was struck by the warmth—a sort of post-rain tropical mugginess that immediately clung to everything you had and to everything around you."

A tropical breeze blew across the concrete runway as a jet-lagged John Kerry disembarked at a Cam Ranh Bay airport clogged with lumbering camouflaged U.S. Air Force transport planes. His World Airways charter flight to Vietnam had been packed to capacity with everyone from soldiers, sailors, and airmen to civilians who were a mix of contractors and CIA agents. It seemed a surreal mélange of people to land in a war zone with. As he walked across the tarmac, Kerry could smell the scent of jungle foliage, and a strange sense came over him. It didn't feel real, being where he was, heading for what he was there for, but here he was in Khanh Hoa province; this was it. Other than a lone B-52 flying overhead, however, Kerry noted only one stark sign of the dangers lurking all around him: across the runway he could see a mortuary piled with dark green plastic body bags and silver metal coffins waiting to be shipped home.

"Vietnam was a swirling nightmare of warped anecdotes and bankrupt theories and operation orders when I first arrived in November 1968," Kerry wrote later. "Nothing made sense, least of all the fact that I was there. About the country I only knew what I read in the *New York Times*, or learned first-hand on the *Gridley*, or what I had been told while training in Coronado. About the real war—the soldiers' war—I knew nothing at all." He watched a platoon of South Vietnamese Rangers arrive, their slender frames straining under rucksacks almost as big as they were. "I was fascinated by my first view of the troops I had come to fight for and with," Kerry recalled. "There was a remarkable nonchalance in their movements, and the weapons they carried were treated with a reckless abandon."

As Kerry and the other new Swift boat officers were assembling in Southeast Asia, Richard Nixon defeated Hubert Humphrey in one of the closest presidential races in American history. Nixon's popular-vote mandate, such as it was, came from a scant margin of victory of 0.7 percent.* The entire nation, and particularly its young men in the armed services or of the draft age to join them, waited for their new president to unveil his campaign-promised "secret plan" for extricating the United States from Vietnam. When Nixon named a longtime GOP leader and former Wisconsin congressman, Melvin R. Laird, as his secretary of defense, and Henry Kissinger as national security advisor, it became clear what the so-called plan was all about. The straight-talking matter-of-fact Laird was a proponent of "Vietnamizing" the war, a policy that called for the United States to equip, train, and inspire the Army of the Republic of Vietnam (ARVN) to fight for their nation's freedom on their own, and then leave them to it. Laird's definition of a U.S. victory, simply put, amounted to pulling out and letting the South Vietnamese continue fighting the Communist North using American weapons. Hence, what Lyndon Johnson had initiated in desperation after the Tet Offensive and what Laird had first promulgated in Congress would be called the Nixon Doctrine:† allowing strategic allies to

---

**Because third-party candidate George Wallace's strength was concentrated in the South, the race was not nearly as close in the Electoral College. Nixon won 301 electoral votes to Humphrey's 191; Wallace captured five states and 46 electoral votes.*

†The Nixon Doctrine was not officially set forth until the President described it at a press conference on Guam on July 25, 1969. Nixon's most concise definition of his doctrine of "peace through partnership" appeared in his November 3, 1969, "silent majority" speech. A Nixon speechwriter accurately claims that his boss never stated he had a "secret plan" to end the war, that it is a myth.

be armed into regional powers by the United States in the interest of preserving international stability by containing the spread of communism.

The problem with Vietnamization was that its success depended on the motivation and the fighting ability of the ARVN, both of which were more than a little suspect by November 1968. "The South Vietnamese did not want the war to end," Clark Clifford, Johnson's last secretary of defense, explained, "not while they were protected by over five hundred thousand American troops and a golden flow of money." Ever since President Kennedy sent the first U.S. military advisors to Vietnam in 1963, complaints had poured back into the Pentagon from American soldiers and sailors throughout the war zone about the ARVN's lack of aggressive leadership, and particularly its utter inability to stymie any of North Vietnam's swift offensive strikes. Although the ARVN had unquestionably done a credible job of defending the important provincial cities of An Loc, Kontum, and Hue, it remained unable to mount and win decisive offensive battles. Part of the problem was the lack of coordination among the South Vietnamese forces, a deficit often attributed to personality clashes between various generals out in the field and those at headquarters in Saigon. Even more fundamentally, the ARVN had simply grown tired and war-weary, afraid of countryside attacks, and terrified of incurring more casualties. Exacerbating the fears of ARVN troops was its command's apparently correct conclusion that the United States intended to cut and run from Vietnam. With American extrication a foregone conclusion, ARVN troops naturally had even less interest in risking their lives for their country.

What seemed clear to John Kerry at the time was that Richard Nixon still dreamed of victory in Vietnam. After all, National Security Advisor Kissinger stated that the President refused "to believe that a fourth-rate power like Vietnam doesn't have a breaking point." Kerry found rather less disconcerting something he had recently read in a work by historians Will and Ariel Durant: "War is one of the constants of history, and has not diminished with civilization or democracy. In the last 3,421 years of recorded history only 268 have seen no war." Fiercely Democratic, if youthfully naïve, Kerry had hoped Robert F. Kennedy or Eugene McCarthy or even Hubert Humphrey would have been elected, stopped the Vietnam War, and given 1969 at least a better shot at joining those rare 268 years of peace on Earth. Instead, he had Richard Nixon as his commander in chief and the feeling that peace was a long way off. Kerry may have been a patriot entering a war zone, but politically he had become an even more uncommitted soldier. He was skeptical that arming and teaching the South Vietnamese how to fight would work. And he believed that there was a "credibility gap," as *New York Herald Tribune* reporter David Wise had claimed, between official U.S. government pronouncements on the war and actual policy and events.

Inside the Cam Ranh Bay airport terminal, Kerry scrutinized the faces of the South Vietnamese soldiers he saw there, and found in their expressions nothing but the blank stares of the uninspired. He also noticed that while all the native civilians appeared to be in a hurry, the ARVN soldiers ambled about listlessly, if in unison. So this was Vietnam. "It was strange to hear names like Vung Tau, Danang, Phan Rang, [and] Pleiku announced as destinations—places that till now were only words from newscasts and film and hell holes from stories that friends told you back at Coronado," Kerry wrote to his parents shortly after arrival. "We walked into the long bungalow, fought our way through a perfunctory customs and then changed all greenbacks into scrip. I thought back to PXs in Berlin and thought how strange it was to be playing with scrip again—particularly under these circumstances. The terminal had a USO booth where you could buy a heaping bowl of ice cream and I immediately did so. There were also a few places where one could buy reduced-fare tickets for flights around the U.S. and tax-free automobiles. A lot of good those would do me!"

As he waited outside the terminal for the truck that would take him to his post, off in the distance, Kerry spied a lush field high on a lovely mountain. He described it for the folks back home: "It had a clearing in the middle of it that was particularly green and close-shaven and with the sun of the late evening had that summery glow of a lawn freshly mown on which one loved to roll and come up smelling of cut grass and scratching all over." He remembered where he was and a "stupid vision," as he put it, came to him. He imagined being set upon by Viet Cong in black pajamas on that lovely mountain field, and he worried that his instinctive protests would betray his naïveté and fear: "Why do you want to kill me? I haven't done anything against you!" But then it occurred to him that just being an American armed with a rifle in Vietnam would give the VC all the justification they needed to kill him. "I saw myself waiting to be shot and then falling in slow motion, tumbling over and over again, to finally lie alone in the field with an eerie stillness muffling the echo of gunfire," Kerry wrote of the reverie in his personal journal. "The truck then arrived to transport us to the base."

Kerry and several other newcomers, including a few guys he had met at Coronado, crammed themselves and their gear into the vehicle and rumbled off down an eroded, rudimentary road. It was the first of many indications to him that the ancient rivers and French-built canals, and not the potholed roads, served as the country's main thoroughfares. The new arrivals' destination was the enormous Cam Ranh Bay Naval and Air Force Base on the central coast 185 miles northeast of Saigon. Little more than an airstrip and a naval office when the facility was inherited from the French, by mid-1965 Cam Ranh Bay had ballooned into the largest of the six seaport bases the United States erected in South Vietnam, the handiwork of Navy Seabees, Army and Marine engineers, and civilian contractors including Raymond International, Brown & Root, J.A. Jones, and Morrison-Knudsen. By the time Kerry arrived there late in 1968, the base boasted a ten-thousand-foot-long runway, ten berths capable of unloading deep-draft ships, ammunition and other warehouses as far as the eye could see, and accommodations for thousands of American personnel.

The Cam Ranh Bay airport sat at the northern end of the long peninsula covered by the base, with the Swift boats moored at the opposite end, on the sea. Stretching across the intervening miles along a once pristine, white-sand beach surrounded by mud hills, the vast base housed its manpower in rows upon rows of well-ventilated wooden barracks built on concrete foundations, facing water to the south. Amid these living quarters stood a post exchange (PX) that put a typical stateside Kmart to shame. The inventory was phenomenal, offering grocery, hardware, and pharmaceutical items to fill every imaginable need. Less material necessities than the PX provided could be found elsewhere: the base also maintained a sauna, a massage parlor, and clubs for both enlisted men and officers, complete with jukeboxes blasting the latest Motown, country, and R&B tunes. "When we finally got there after bumping crazily over the hills and some mountains it was too late for dinner, but we were shown to some small rooms in the wooden barracks that make home for all those stationed at this base," Kerry wrote to his parents. "The barracks are long structures with two stories and open slats that run horizontal to the ground, which let in the air. When it blows hard the sand (of which there is plenty) seeps through these slats unless one is lucky enough to put some plywood paneling on the inside."

Throughout the Vietnam War, Cam Ranh Bay served as the strategic nerve center for the U.S. Navy's Market Time anti-infiltration operations, which since 1965 had searched a half million vessels a year along the coast. The location was ideal. "Cam Ranh is considered the finest natural harbor in the world after Sydney, Australia," Neil Sheehan explained in *A Bright Shining Lie*, "but the region has always been sparsely populated, because the dark green of the rain-forested Annamites* touches the emerald green of the South China Sea there and what level land exists around the bay is either sand or sandy."

The large natural port at Cam Ranh was hardly a recent strategic discovery. In 1905, Russian Admiral Zinovi Rozhdestvenski had anchored his fleet in the Vietnamese bay before sailing to his navy's doom into a Japanese trap in the Tsushima Strait. This sparked the battle that ended the Russo-Japanese War. During World War II, the Japanese continued to control Cam Ranh until 1944, when U.S. Naval Task Force 38 bombed the port to smithereens. In 1968, as Kerry received orientation at Cam Ranh Bay, he marveled at how well organized everything seemed at the grandly rebuilt facility. His mind boggled at the number of petroleum tanks dotting the installation, not to mention its many amenities. "I wondered naïvely where the war was," he recalled. "It seemed dull. I had come to skipper a Swift—not fuel vehicles, sleep, shower, and shop."

At Coastal Division 14 on the far southern end of Cam Ranh Bay, Kerry presented his orders and immediately requested a transfer. If he stayed in Cam Ranh Bay, he would be insulated from the *real* Vietnam. He wanted his own Swift command, which could best be accomplished by being assigned to An Thoi or Cat Lo. It irked him to be assigned to the "fun in the sun and surf division"; he had not come to Vietnam for jukeboxes and cheap liquor. If he was going to avenge the death of Dick Pershing, he needed to see combat, or at least some sort of action beyond that offered by the diversions available at Cam Ranh. Like *Platoon* director Oliver Stone—who had served fifteen months in Vietnam with the 25th Infantry Division, receiving a Bronze Star for valor and the Purple Heart with a Bronze Oak Leaf Cluster denoting more than one award—Kerry felt inex-

---

**The Annamites were a people of Mongolian descent who lived in Vietnam.*

plicably drawn to combat. "Now I wanted to see another level, a deeper level, a darker side," Stone once explained about this morbid fascination with Vietnam. "What is war? Why do people kill each other? What is the lowest level I can descend to find the truth, where I can come back from it and say, 'I've seen it'?"

In the same spirit, Kerry asked the Coastal Squadron 1 administrative officer if he could go to Cat Lo or Danang instead—places where a brown water sailor could *feel* the war.

"No, you can't," the officer responded. "The names were drawn from a hat and it's decided. You're staying here."

"But what difference does it make if I change with somebody who doesn't want to go elsewhere?" Kerry asked.

"It doesn't make any difference," came the reply. "You can't do it because it's already decided."

The administrator did, however, add that Kerry could ask Squadron Commander Charles F. Horne III for permission to change assignments, and that the commander was usually impressed by a gung ho desire for combat. "So I went to see the commander and put my case to him," Kerry explained. "It was to no avail. I stayed in Cam Ranh Bay."

Commander Horne—who was also commodore of the five divisions comprising the Navy's dozens of Swift boats in Vietnam—would earn both a Bronze Star with Combat and the Legion of Merit with Combat for heroism in Vietnam. He made quite an impression on the young lieutenant (j.g.) who wanted out of cushy Cam Ranh Bay. A physical fitness buff, Horne had virtually been born a U.S. Navy man, on March 22, 1931, in San Diego, his father, Charles Horne Jr., serving under Admiral Richmond Kelly "Terrible" Turner, who helped formulate U.S. strategy in the Pacific during World War II, as head of the Navy's War Plans Division. The ten-year-old Charles Horne III got his first taste of war on December 7, 1941; his father was then stationed at Pearl Harbor. Young Charles never forgot seeing U.S. battleships on fire after the Japanese attacked. He won a commission to the Naval Academy at age seventeen, and after graduating in 1952 served in the Korean War as a deck officer on the U.S.S. *Valley Forge*. Like most career Navy officers, Horne moved from post to post and job to job, including running the Naval Officers Training Corps unit at the University of California–Berkeley and enduring the SERE program in Norfolk,

Virginia. In the autumn of 1968 Horne was ordered to Vietnam to help Captain Roy Hoffman make Operation Sealords a success.

Once in-country, Commander Horne took his mission to heart. On a wall of his Cam Ranh Bay office hung a captured North Vietnamese rocket, mounted like a stuffed trophy fish. It was Kerry's first glimpse of a phenomenon common among battalion and unit commanders in Vietnam; the paper pushers seemed to have a penchant for brandishing seized enemy weapons. Later in his tour Kerry would take note of the desk jockeys who flew into combat zones and bartered with the ground troops for war souvenirs, a particularly lucrative segment of the local free market. Kerry himself fell prey to such souvenir collecting, bringing back to America a captured VC flag, a broken Chinese rifle, and a used B-40 rocket.

Commander Horne spent much of his time in Vietnam on Marine helicopters, visiting his men. He was delighted to be in charge of five divisions, and to always have more than eighty Swift boats full of sailors and junior officers under his command. In the interests of morale, Horne made a point of addressing each new arrival from Coronado as "Skipper," which the young officers found very flattering. As none of them had yet been assigned a boat, however, they couldn't help wondering, "Skipper of *what*?" Still, the honorific did seem to get them to hang on Horne's every word as he expounded on the responsibilities and objectives of Operation Sealords, which was just getting under way. "By calling them all Skipper, I was trying to let them know it's awesome to have your own command," Horne reflected in 2003 from his home in Charleston, South Carolina. "And they were going into a pure combat situation up the rivers. The Viet Cong would be popping them all the time with B-40s, blasting at them in unrelenting fashion and then going back down."

According to historian and Swift boat veteran Larry Wasikowski, by mid-1966 the Navy had discovered that Swifts had an unexpected corrosion problem. After heavy use the brass shell cases would fall into the bilges. To cope with the defect, overhaul facilities were developed in Japan and the Philippines. By the time Kerry went in-country, makeshift repair stations had also been erected at Cam Ranh Bay and Qui Nhon. "At any given time there were sixty-five to seventy Swifts on patrol in Vietnam," Wasikowski recalled. "That meant they had an activated crew and were coming and going from shore bases."

At the time, Operation Sealords had many components besides Swifts: the U.S. Navy's Coastal Surveillance Force also operated two dozen Coast Guard cutters and 39 other vessels. The River Patrol Force ran over 250 patrol and minesweeping boats; the nearly 4,000-man Mobile Riverine Force comprised over 180 monitors, other armored craft, and transports. A Helicopter Attack (Light) Squadron provided air cover for these boats, while four or five 14-man SEAL platoons contributed reconnaissance, ground ambushes, and intelligence operations throughout the Mekong Delta. In addition, the South Vietnamese Naval Forces (SVNAV) had 655 ships, assault craft, and patrol boats, which were supposed to work in tandem with the Americans. As of Kerry's arrival, all of these vessels were ready for incursions into Viet Cong–controlled jungle and swamp hideaways. Just a week before he got to Vietnam, in fact, patrol boats, river, and riverine assault craft had opened up two canals, between the Gulf of Thailand and Rach Gia and the Bassac River at Long Xuyen. As Commodore Horne explained all this to his new skippers, Kerry grew even more eager for his own Swift boat, so avid was he to start searching and destroying and making Operation Sealords a success. "You might say I was conflicted," he confessed. "On the one hand, I wanted the Paris peace talks to end the war. On the other hand, I had trained to fight, and I wanted to."

After a week or so at Cam Ranh, Kerry began to settle into the rather comfortable routine of life at the enormous base. In Cam Ranh, each officer was assigned his own room, a true luxury when compared with the overcrowded dormitories of Coastal Division 13 in Cat Lo. What's more, the showers were only fifteen yards from his barracks, and fresh water was plentiful. On off days, after popping his requisite antimalarial pill, he had time to hit the beach and work on his tan. Most important, and unlike anywhere else in Vietnam, the mail arrived on time at Cam Ranh. Kerry was a prolific correspondent, regularly pounding out letters on a base typewriter to Julia and David Thorne, Freddie Smith, Danny Barbiero, and other friends, as well as his family back home. "I'd just use two fingers and peck away," Kerry said of his penchant for typing. "Whenever a free minute presented itself I tried to write." All of his college friends saved letters from him.

"In all honesty, despite the barren life that one is forced to lead while on a base, I am finding Vietnam fascinating," Kerry wrote on November 24 to his old friend George Butler, who was living in Washington, D.C., and about to start a stint with the VISTA program.

At last I can draw firsthand opinions and see for myself what is happening and what the Vietnamese people are really like. Thus far it is not very encouraging, but I cannot in fairness draw opinions in a finalized form. That will come at a later time. The coastline off which I am patrolling is both beautiful and alluring and I have to fight an inner desire to get off the Swift and walk along a beach or climb some hill that rises right out of the sea. It is a good chance to do some writing and there are quiet moments on patrol when I do just that. For the moment the seas are rough while we wait out the monsoon, but come next April or May I expect to see sunsets and panoramas that will rival cruising in the East. So far it has been quite lonely, but loneliness is good tonic for someone like me—especially as spoilt as I have been for the past few months. There is much more to write but I must save it for a later date as I hope to get to bed early tonight in preparation for tomorrow's sortie.

While at Cam Ranh, Kerry made little progress at finding out "what the Vietnamese people are really like." It was not for lack of trying, beginning each morning when a truckload of local workers crowded together like cattle, pulled into the base from a nearby town. He would scrutinize the faces that peered through the slats in the side of the truck and watch as the chattering women, pouting children, and wizened old men, all in straw coolie hats, poured out the back. He observed them as they were detailed to various jobs around the base, and as they went about their assigned tasks, from scrubbing the barracks to maintaining the roads. The "mama-sans," as the Americans called the local women who cleaned up after them, would make their way through the restrooms, without so much as a nod to delicacy, assiduously scouring toilets right next to ones men were urinating in. "I often looked for some expression that would tell me how these people felt, cleaning up behind us as we waged war in and on their countryside," Kerry later wrote. "But they never gave away thoughts. Perhaps that was why so few Americans could understand the Vietnamese."

The officers club at the Cam Ranh base was the best in Vietnam, and the center of the installation's social life, such as it was. Housed in a small Polynesian-style bungalow, the "O Club" boasted a snack bar complete with barbecue grill and popcorn machine, a small bowling alley, and a balcony that looked out over the gorgeous coastline. A machine even made ice cubes, just like in the bars back home, and a portable air-conditioning unit almost always stayed on full blast, neutralizing the tropical heat and humidity. Every night officers packed the place, getting drunk enough to dream themselves back to Fresno, Des Moines, New Haven, or wherever they could remember being happy. "I couldn't believe the set up," Kerry wrote to his parents. "It seems the military is great at perpetuating a favorite hobby of most of its members—drinking. Strange to know there was fighting not many miles away while one sat on a porch with flaming torches in the evening."

Before Kerry arrived in Vietnam, nurses from Saigon used to be flown in to Cam Ranh for wild parties at the O Club, joining the pair of alluringly pretty young Vietnamese women who served as its permanent waitresses. Stories of what went on at those shindigs bred more than a little resentment among the excluded. Finally, one of the enlisted men stationed at Cam Ranh grew so bitter that he couldn't share in the fun that he wrote about it to his mother, who promptly passed his complaint on to her congressman. That was the end of the O Club parties. "Occasionally, though, USO shows still came through and I was lucky enough to watch three Chinese girls bump and grind and tantalize in ways that I never knew existed," Kerry laughed, looking back. "To keep from getting to the club too early, one could always go to the outdoor movie. I remember watching *How to Succeed in Business Without Really Trying* and *The Green Berets* while a few yards away the waves rolled up on the beach, and still a little farther away someone was dying. The so-called war was very easy to take in Cam Ranh Bay."

Although panned by critics (and since renowned for its final scene showing the sun rising in the East), *The Green Berets* proved extremely popular with the men stationed at Cam Ranh. Based on a collection of short stories by Robin Moore, the 1968 release—directed by its star, John Wayne—characterized the ARVN as a garishly uniformed gaggle of lazy, two-faced cowards who "won't fight like men." The only ARVN soldiers treated approvingly in the movie were the torturing "interrogationists" shown hammering reeds and needles under the fingernails of captured Viet Cong. "Bleeding-heart do-gooders" were derided in gung-ho lines about the real-life enemy VC: "Funny thing about Victor Charlie: he thinks Americans are dickheads for coming over here and trying to drill water wells and build schools and orphanages. The only time he respects us is when we're killing him."

The Green Berets depicted in the movie are, as historian Garry Wills labeled them in *John Wayne's America: The Politics of Celebrity*, "tough realists—rough-and-ready American brawlers with an insatiable lust for wanton Vietnamese women. People who did not want to know about the actual Vietnam War could feel that the national unity and resolve of World War II might turn around this strange new conflict in the far-off jungles of the East." For this very reason Kerry disliked the film. His nature was too curious and reflective to let him be snookered by unrealistic prowar propaganda like John Wayne's dreadful Vietnam movie. He preferred the more skeptical perspective of in-country journalists such as Neil Sheehan and David Halberstam of the *New York Times*, Nick Turner of Reuters, *Newsweek*'s François Sully, and Stanley Karnow of *Time*. Like them, he wanted to know what was *really* going on—what the actual body count from an engagement had been, not what some self-serving Pentagon report claimed. Along with his friend from Coronado, Skip Barker, Kerry had learned much of what he knew about Southeast Asia from the reportage of French scholar Bernard Fall, who had been killed in the field in 1967 when his jeep ran over a land mine in a northern province of South Vietnam.

Unfortunately for John Kerry, some of his colleagues at Cam Ranh cottoned a lot closer to John Wayne than to Bernard Fall, at least rhetorically. It soon became obvious to him how easily a soldier uninterested in jungle fighting or democracy building à la Wayne could lounge away the war in the fifth-rate resort environs of the cushiest base in Vietnam, swapping stories about sexual conquests and combat derring-do in the comfort of the O Club as he counted the days to his next R and R in Hawaii or Japan or back home in the States. Those who did have the taste for warfare bragged about their death-defying exploits and how they had brought glory to their divisions. The most pumped-up told harrowing tales about what novelist Tim O'Brien would later call, in his novel *In the Lake of the Woods*, "the geography of evil: tunnels and bamboo thickets and mud huts and graves." Listening to this macho swagger about burning down villages and looting rice storages reminded Kerry of something Napoleon had once said: "Nothing will destroy an army more than . . . pillage."

As Kerry awaited assignment, he approached the war almost contemplatively: he liked to study what was happening around him, and why, and how he could make things better. He took a scholarly approach to the war, tape-recording the occasional worthwhile conversation and constantly scribbling impressions in his ever-present notebook like some latter-day Ernie Pyle. He read German-born American novelist Erich Maria Remarque's antiwar *All Quiet on the Western Front* to see what he could glean from its descriptions of World War I's gruesome trench warfare. And, of course, he went on several Swift boat patrols with other officers and their crews to ascertain just what he was supposed to be doing in Vietnam. Each time out he wished he had his *own* PCF to skipper.

Sometimes the boats would cross at Cam Ranh Bay and put in at the flyspeck island that served as headquarters for the SVNAV coastal patrol unit to pick up Vietnamese sailors to act as interpreters during the Swifts' board-and-search missions. Some of them proved good at their jobs as well as good company. More often than not, however, the South Vietnamese would beg off from the Swift missions on the grounds that the water was too rough during the monsoon season. To prove it, the allied sailors developed a tendency to get violently seasick just before patrol time. Virtually every seafarer has at one time or another suffered what Abigail Adams, the wife of President John Adams, once described as "that most disheartening, dispiriting malady." Faking seasickness to shirk duty was not only cowardly but a court-martial offense—and yet it was the apparent modus operandi of many members of SVNAV. Their faking caused considerable tension with and derision from their American counterparts. After all, as the U.S. Navy men saw it, they were getting buffeted around the seas by monsoons for the South Vietnamese, while their allies played sick, deserted in alarming numbers, and generally acted as though they didn't give a damn what happened to their country. "We could never count on having a Vietnamese interpreter with us," Lieutenant Chuck Mohn of PCF-78 recalled. "We did what we needed to do to survive."

As frustrated as Kerry grew with the behavior of so many South Vietnamese sailors, he felt even more disgusted by some of his own countrymen's bigotry toward their so-called allies, openly expressed in racial epithets like "gooks," "dinks," "slopes," "slants," "running dogs," and "zipperheads," the last a derogatory term for men who part their hair in the middle, as Vietnamese sailors sometimes did. One U.S. officer at Cam Ranh took particular pleasure in turning on the loudspeaker as his boat approached a pier and booming into the microphone, "Hurry up, you fucking zipperhead!" At the screeching American order, some uncomprehending South Vietnamese sailor would invariably go stumbling over the row of junks with what Kerry described as "the delicacy of an elephant on a tightrope."

He did encounter a few SVNAV interpreters who made sincere efforts to master English and to board and search apprehended vessels effectively. The majority, however, did neither. He recalled one Vietnamese who would climb aboard his assigned Swift at the Cam Ranh base pier, head directly below to the cabin, and slide into the skipper's bunk to sleep through most of the patrol. Finally, an American officer had enough. He went below, pulled out his .38 revolver, and put it to the snoring sailor's head to wake him up. The man sprang from the bunk, never to sleep on the job again, let alone ever take his eye off that officer. "There was no love lost between the U.S. and South Vietnamese Navies," Kerry explained. "At times it seemed *they* were at war."

Lieutenant Stephen Hayes, whose PCF-71 was once hit by a B-40 rocket, was among those tasked with trying to teach the Swift boat community how to implement the Vietnamization policy. Hayes's assignment fell under a "personal response" program the U.S. Navy had created to train its men to interact with their South Vietnamese counterparts in a less hostile way. "It was a cross-culture exchange," Hayes explained. "The goal was mutual respect." The new commander of U.S. Naval Forces, Vietnam, Vice Admiral Elmo Zumwalt Jr., considered this effort essential to making Vietnamization work. Unfortunately, the program did not pan out. Although some South Vietnamese sailors learned to operate the Swifts well enough, it proved impossible to make them want to. "I had mixed feelings about Vietnamization," Hayes reflected years later. "It seemed right that the primary responsibility for defeating the VC should fall on the South Vietnamese, so the basic premise of Vietnamization was sound. But it was also ridiculous, because the South Vietnamese didn't have the wherewithal to aggressively pursue the war. The main purpose for everybody involved was not democracy building but not getting killed—not putting your body in harm's way."

Don Droz, another Swift boat lieutenant, was less diplomatic in his assessment of SVNAV. It drove him crazy, in fact, that American lives were being lost for a nation that just didn't care. "While we are on the subject of the Vietnamese, however, let me just say that never have I been so disgusted with a group of people," he wrote to his wife on October 23, 1968. "The fact that we are here fighting their war for them becomes harder to stomach each day. We simply have to force the Vietnamese Navy to do anything—and then their performance is so half-assed and second-rate as to be a nuisance rather than a help. These people simply don't give a damn, Jud—so why should we? I'm to the point that I even cast a protest vote today on my absentee ballot. Count one write-in for Eugene McCarthy."

Don Droz, who became one of Kerry's closest friends in Vietnam, was a U.S. Naval Academy graduate with a penchant for fishing. Born September 29, 1943, in western Missouri while his father was off fighting in World War II, Droz had been a star at his high school in Rich Hill: winner of a national speech competition, valedictorian of his class, and state president of the Future Business Leaders of America. Although in combat he was fiercely competitive, Droz also had an easy-mannered way about him; friends and family came before his career. Missouri Congressman William Randall, a Democrat who served as a U.S. Army sergeant in the southwest Pacific and the Philippines during World War II, however, learned of his young constituent's scholastic successes and, recognizing Droz's potential, recommended him to Annapolis. Droz enrolled in 1962 and graduated four years later, along the way getting married.

The very sight of Don Droz's pudgy but swarthy baby face, with its Jimmy Durante schnozz and ever-present smirk, always made Kerry laugh—not at his looks, but in anticipation of the jokes and pranks and antics at which Droz had become a master in the interest of defusing tension. Such were his mirthmaking skills that he managed to get a laugh just by trying to grow a Fu Manchu mustache, around which a betting pool formed. He pulled off the feat, if with somewhat patchy results. Everybody appreciated Droz's openheartedness, Show-Me State directness, and Annapolis manners, but what made him so popular was his ability to laugh at the absurdity of duty as a Swift boat officer in Vietnam.

In September 1968, two months before Kerry arrived, the U.S. Navy transferred four Swift boats to SVNAV. A ceremony held at An Thoi, the U.S. Coast Guard base on Phu Quoc Island in the Gulf of Thailand, marked the transfer that signaled the start of the South Vietnamese Navy's eventual takeover of all of Operation Market Time. According to the November 1968 issue of the official U.S. Navy magazine *All Hands*, seventy-one South Vietnamese Navy officers had just successfully completed a Swift boat training course similar to Kerry's at Coronado. Zumwalt's predecessor as commander of U.S. Naval Forces, Vietnam, Rear Admiral Kenneth Veth, officially lauded SVNAV for its "professional competence" and "tremendous growth—not just in numbers, but more important, in efficiency and capability." Veth's generous words ushered Vietnamization into the Swift boat world. Actual Swift skippers like Don Droz knew better—that for the most part the South Vietnamese sailors remained unwilling to risk their lives, especially once they realized the Americans would soon be withdrawing their armed forces from Southeast Asia.

Lieutenant (j.g.) Kerry spent his first two weeks in-country on coastal patrols, cruising the boundless blue bays along Cam Ranh. These proved less than pleasant outings, however, not so much for fear of VC mines or mortars but because it was the tail end of the monsoon season, when the water turned its most turbulent. That November in the Gulf of Tonkin a weather menace so major it earned the moniker "Monsoon Mamie" was making its presence felt up and down South Vietnam's thousand-mile coast. Just ferrying local passengers the twenty-four miles from Cam Ranh to Nha Trang proved a five-Dramamine endeavor. "I felt like I was in a submarine riding a runaway missile—we would go smashing into some waves, rise completely out of the water with the screws screaming away and then come crashing down again to lose complete sight of what was ahead," Kerry wrote to his parents. "My arms were cramped from hanging on and several of the passengers were sick before long. I'm told by some of the OINC's [officers in charge] that after a few patrols like that they've come back pissing red and that several people have broken bones. Sounds like fun, no? However, when the monsoons are over and the water is calm there is supposed to be nothing as pleasant as a patrol—glassy water, etc. I can't wait 'til March or April when that starts—also it will mean only seven more months too!"

Kerry's initial schedule was intense: over ten straight days he would go out on five overnight patrols as substitute skipper, putting out of port one morning and back in the next, then out again the morning after that, and so on, sleep being pretty much out of the question. His job was to help secure designated areas by boarding and searching Vietnamese junks and sampans, to make sure they weren't carrying contraband. When the storms turned too fierce, his orders were to negotiate his little boat through the reefs to anchor in a harbor cove, where he and his men could only pray that the Viet Cong wouldn't ambush them. "I don't mind taking certain chances, and obviously there will be some, but I hate to just anchor and sit somewhere knowing that one presents a rather appetizing target," Kerry wrote home. After each ten days of this activity, the men would get four days off to rest, recuperate, and mend their boat.

Only on occasion did Kerry opt to regroup via alcohol at the O Club; more often he used his time off to study history and to read recent best-sellers such as Truman Capote's *In Cold Blood* and John Hersey's *Under the Eye of the Storm*. "If I continue this way—and I'm sure I will—I will run out of books before too long," he informed his folks. "I have a book for Business Boards, which I got to practice just in case, and also a couple of Barnes and Noble American History Summaries, which should keep my memory refreshed about things I want to remember. It was strange reading Hersey's book about a storm and then sitting here waiting for one to hit—it made his descriptions even more realistic than they were anyway." He also spent a good bit of time scouring the mail-order catalogues at the PX and mulling over how he could afford an engagement ring good enough for Julia Thorne in Bangkok. In the end he decided against that, admitting to his parents that it would probably be wiser to buy her a diamond of guaranteed quality back in Boston.

The rules of engagement in South Vietnam, from the U.S. Navy perspective, were quite simple. A curfew banned all water traffic on the rivers and up the coastline between 8:00 P.M. and 6:00 A.M. If a Vietnamese fisherman violated this basic rule, he could—and probably would—be shot at by U.S. forces. The greatest sin was for a Vietnamese vessel to be traveling on a waterway at night unlighted.

A sampan navigating in the shroud of darkness was assumed to be VC. It *would* be fired upon. "If the vessel was lighted, then the routine was to chase it down, search it carefully, and chase it home," Wade Sanders recalled in an unpublished memoir, "after you had taken the name and license number of the boat owner, to turn over to the Vietnamese Provincial Police."

Sometimes, to kill the boredom of patrol, Kerry would carefully study the Vietnamese fishermen he encountered in sampans and junks. To the Vietnamese, fishing was not just an occupation, it was a ritual interwoven with the religious and mythological aspects of the region. Certain families had laid claim to fishing grounds for over one thousand years. They believed that spirits and demons, both good and bad, determined a day's catch. He liked these fishermen—as did almost all the young Swift boat officers. But it was a mistake to overromanticize them. "We enforced the law, laws we didn't make, laws which were most often established by corrupt Vietnamese Provincial officials who extorted money from the fishing fleets," Sanders explained in his memoir. "I was not so naïve as to believe that there were never any Viet Cong hiding among these fishermen, but I am sure that our constant harassing, our frantic herding and bullying of these people made new enemies with each passing day."

After a few weeks in-country Kerry had grown bored of routine coastal patrols and impatient that he still had not been assigned his own boat. To shake things up he volunteered for a special mission on what the Navy called a skimmer but he knew as a Boston Whaler: a foam-filled-fiberglass boat that, according to its eponymous manufacturer, "would still float after taking 1,000 rounds of automatic weapons fire" and "could run when it was cut in half—as long as you've got the half with the engine." The plan worked, for good and ill. Kerry got his break from routine patrols, along with a piece of hot shrapnel in his right arm that earned him his first Purple Heart. "It was a half-assed action that hardly qualified as combat, but it was my first, and that made it very exciting," he demurred decades later. "Three of us, two enlisted men and myself, had stayed up all night in a Boston Whaler patrolling the shore off a Viet Cong–infested peninsula north of Cam Ranh, not very far from where Colonel [Robert] Rheault's Green Berets were supposed to have dumped the body of their suspected informer.* Most of the night had been spent being scared shitless by fishermen whom we would suddenly creep up on in the darkness. Once, one of the sailors was so startled by two men who

---

**Colonel Robert Rheault was commander of the U.S. Army Fifth Special Forces Group, from May to June 1969. He and six of his officers would be indicted for the June 1969 murder—or "termination with extreme prejudice," in military jargon—of a suspected Viet Cong double agent who had been working for the Special Forces. Although the case was dropped thanks to the CIA's refusal to turn over pertinent classified information, Rheault would resign from the Army in late 1969.*

surprised us as we came around a corner ten yards from the shore that he actually pulled the trigger on his machine gun. Fortunately for the two men, he had forgotten to switch off the safety. But had they been blown to bits and scattered in the water, the law said that it wouldn't have mattered. They were in a free fire zone as well as fishing in a no-fishing area after curfew, and I learned quickly that mistakes of that nature were always tallied as 'Viet Cong killed in action.' "

As it turned out, the two men really were just a pair of innocent fishermen who didn't know where one zone began and the other ended. Their papers were perfectly in order, if their night's fishing over. The fear was that they were VC. Allowing them to continue might have compromised the mission. For the next four hours Kerry's Boston Whaler, using paddles, brought boatloads of fishermen they found in sampans, all operating in a curfew zone, back to the Swift. It was tiring work. "We deposited them with the Swift boat that remained out in the deep water to give us cover," Kerry continued. "Then, very early in the morning, around 2:00 or 3:00, while it was still dark, we proceeded up the tiny inlet between the island and the peninsula to the point designated as our objective. The jungle closed in on us on both sides. It was scary as hell. You could hear yourself breathing. We were almost touching the shore. Suddenly, through the magnified moonlight of the infrared 'starlight scope,' I watched, mesmerized, as a group of sampans glided in toward the shore. We had been briefed that this was a favorite crossing area for VC trafficking contraband."

With its motor turned off, Kerry paddled the Boston Whaler out of the inlet into the beginning of the bay. Simultaneously the Vietnamese pulled their sampans up onto the beach and began to unload something; he couldn't tell what, so he decided to illuminate the proceedings with a flare. The entire sky seemed to explode into daylight. The men from the sampans bolted erect, stiff with shock for only an instant before they sprang for cover like a herd of panicked gazelles Kerry had once seen on TV's *Wild Kingdom*. "We opened fire," he went on. "The light from the flares started to fade, the air was full of explosions. My M-16 jammed, and as I bent down in the boat to grab another gun, a stinging piece of heat socked into my arm and just seemed to burn like hell. By this time one of the sailors had started the engine and we ran by the beach, strafing it. Then it was quiet.

"We stayed quiet and low because we did not want to illuminate ourselves at that point," Kerry explained. "In the dead of night, without any knowledge of what kind of force was there, we were not all about to go crawling on the beach to get our asses shot off. We were unprotected; we didn't have ammunition, we didn't have cover, we just weren't prepared for that. . . . So we first shot the sampans so that they were destroyed and whatever was in them was destroyed." Then their cover boat warned of a possible VC ambush in the small channel they had to exit through, and Kerry and company departed the area.

And that was that. They loaded their gear back onto the covering Swift boat and towed the Boston Whaler back to Cam Ranh. "I felt terribly seasoned after this minor skirmish, but since I couldn't put my finger on what we had really accomplished or on what had happened, it was difficult to feel satisfied," Kerry recalled. "I never saw where the piece of shrapnel had come from, and the vision of the men running like gazelles haunted me. It seemed stupid. My gunner didn't know where the people were when he first started firing. The M-16 bullets had kicked up the sand way to the right of them as he sprayed the beach, slowly walking the line of fire over to where the men had been leaping for cover. I had been shouting directions and trying to unjam my gun. The third crewman was locked in a personal struggle with the engine, trying to start it. I just shook my head and said, 'Jesus Christ.' It made me wonder if a year of training was worth anything." Nevertheless, the escapade introduced Kerry to combat with the VC and earned him a Purple Heart.

As generally understood, the Purple Heart is given to any U.S. citizen wounded in wartime service to the nation. Giving out Purple Hearts increased as the United States started sending Swifts up rivers. Sailors—no longer safe on aircraft carriers or battleships in the Gulf of Tonkin—were starting to bleed, a lot. Vice Admiral Zumwalt himself would pin the medal on John Kerry at An Thoi about six weeks after the doctor at the Cam Ranh base took the shrapnel out of the young officer's right arm. "He called me in New York to tell me he had been wounded," Julia Thorne remembered. "I was worried sick, scared to death that John or one of my brothers was going to die. He reassured me that he was okay."

Indeed, Kerry, the day after he was wounded, with a bandage on his arm, went on a regular Swift boat patrol with another officer. That excursion proved uncommonly dull. After they'd had enough of the monsoon-roiled waves tossing them all over the empty sea, they finally dropped anchor in a small cove and relaxed. It was well into the night before they saw something: the lights of a fishing junk moving through a curfew zone. The officer in charge seized the opportunity to show his young colleague how the U.S. Navy handled nationals who didn't know where they could and couldn't go about their business. "We went alongside a creaky, wooden junk that reeked of stale fish," Kerry wrote in his notebook. "The officer made a lot of threatening noises in his best Coronado Vietnamese and motioned the fisherman to go away: '*Di di, mau!*'—'Go faster!' "

The fisherman apparently did not pull his nets in fast enough. The lieutenant responded by pulling out a knife and cutting the man's nets off where they came over the gunwale into his junk. The hundreds of feet of heavy fishing nets quickly sank out of sight and beyond reach. Kerry watched the fisherman's face sink with the means of his livelihood. A horrible revulsion swept over him at both his senior officer and any policy that could permit such a free and senselessly cruel interpretation of patrol-boat duty. Yet, as a rookie who just barely knew what a Swift patrol really entailed, he knew there was nothing he could do in protest. A small boy who had been watching the whole episode from the bow of the junk moved quickly to pull up its anchor, and the junk glided away into the darkness.

At that, the lieutenant turned to Kerry and advised, "If you think they've been in a no-fish zone several times, just shoot a notch in the junk so you'll recognize it the next time you see it. Then, if they keep on doin' it, just shoot a hole in the boat below the waterline—that really gets 'em." This grotesque callousness toward the South Vietnamese infuriated Kerry. Ever since his first visit to Danang aboard the *Gridley*, he had pondered how the country's people and their supposed American allies could become true brothers in arms fighting side by side against the thuggery of Ho Chi Minh's communism. After the brutish behavior he had just witnessed, however, Kerry couldn't help but wonder what that boy on the junk would think of the United States from then on, if that was how America went about winning the "hearts and minds" of the people of Vietnam. It was a solemn lieutenant (j.g.) who realized he had just made an enemy, and the sheer dumbness of it would continue to gnaw at him for months to come. Kerry's only consolation came from the satisfaction of knowing that the bleak analysis of U.S. involvement in Vietnam he had first articulated at Yale University in 1966—and his conclusion that his country simply had no business there—remained sound.

The awful episode also reminded Kerry of a story his fallen friend Dick Pershing had told him from his own service in the 101st Airborne Division. According to Pershing, one evening a corporal in one of the platoons in his company had been ordered to set up a mortar position for perimeter protection through the night. Unfortunately, the corporal was stoned at the time and in his marijuana haze set the mortar up pointing in the wrong direction. When it was fired during the night, several innocent villagers and some livestock were killed. The next day the Army simply paid reparations in cash for the consequences, on a sliding scale topping out at fifty-two dollars for each dead adult within a Pentagon-determined productive age range, less for children and even less for animals, and nothing for, say an elderly grandmother. After that, the incident was officially forgotten, and the mortar rounds written off as the work of the Viet Cong. "It was all very convenient," Kerry still seethed decades later. "The Americans came out smelling like roses whereas the records tallied one more Viet Cong atrocity."

Shortly after the fishing-net episode, and less than a month into this second tour of duty in Vietnam, Kerry's division operations officer approached him and asked, "How would you like to go to An Thoi?" It was not really a question: Kerry had been away at the Air Force PX when the division had held a meeting to solicit volunteers, and by virtue of his absence he became one of them. He answered the offer with a grunt of confusion, for by then he had grown more comfortable in Cam Ranh, where his mail arrived regularly and he had the time to read and write. The stories he had begun to hear about life in Cat Lo and An Thoi were not appetizing and going into the danger zone was no longer so appealing, to put it mildly. But in response to his grunt of nonassent, the operations officer barked, "Good. You're going tonight. We're giving you the 44 boat, and you can pick up your charts and orders after you pack."

Kerry just stood there for a moment. Then he asked: "Do I have any choice?"

The operations officer shook his head. "You wanted to go a few weeks ago, so you just came to mind." He smiled. "Besides, there's no one else."

He was trapped, a victim to his own earlier enthusiasm. More disturbing was the seeming capriciousness of his selection for the dangerous duty; after all, he had not attended the meeting, and neither had the other man chosen to go to An Thoi, Don Droz. Stranger still, Kerry had not even been assigned a Swift boat of his own to patrol the rough waters of Cam Ranh Bay and yet here he was getting one on the spot to go run the rivers of the perilous southern Mekong Delta, where Viet Cong assassins lurked in camouflage behind the impenetrable jungle foliage. He would soon learn one of the main reasons he rated such dangerous assignments: he was single. In Vietnam, it turned out, an American officer's marriage could save his life, or at least lower the risk of being sent where he was likeliest to lose it.

Don Droz's wife sure came in handy when he heard the news that he had been chosen to take his Swift boat to An Thoi. Incredulous, Droz immediately began appealing to his close friend and fellow officer Tedd Peck, begging his bachelor buddy to take his place. "He pleaded with me that his wife was pregnant and they were going to have a baby in January," Peck recalled. " 'If I go down there I'll never see my kid,' he said. I told him that he had to follow orders, that there was no way I was going. I wasn't that generous or stupid. It was his bad luck." That night at the O Club, however, once Peck was full of booze, Droz continued his lobbying. "He told me he would do *anything* if I took his place." Peck laughed in retrospect. "I finally agreed I'd go south for three bottles of Johnnie Walker Black. Well . . . somehow he found them. So we shared a bottle, I packed the other two, and resigned myself to fate. It was not only the Droz factor, but my crew really wanted to go. It was a conspiracy."

Thus it was that the officer accompanying Kerry to An Thoi would be Edward "Tedd" Peck, a twenty-five-year-old fellow lieutenant (j.g.) from Syracuse, New York, who also had attended OCS in Newport and Swift boat school at Coronado. Barrel-chested and brimming with bravado, Peck had arrived in Vietnam that April. After a short stint in Danang he had been assigned PCF-57, complete with a makeshift crew who conducted coastal patrols and provided naval gun support to a detachment at the small staging outpost of Chu Lai—the Chu Lai Tigers, as they had dubbed themselves. "The Drunken Tigers would have been more appropriate," Peck could joke much later.

In his National Book Award–winning 1978 novel *Going After Cacciato*, Vietnam veteran Tim O'Brien characterized Chu Lai as little more than "rows of tiny huts" standing neatly on the sand, "connected by metal walkways, surrounded on three sides by wire, guarded at the rear by the sea." In short it was a combat outpost, yet not a bad place to be, as it had no division commander to spoil the men's fun. Some eight to ten Swift boats would be moored there at any given time, and on their off days the crews would pilot them into the bay for floating barbecues. Lieutenant (j.g.) Peck liked the setup well enough and even saw some action out of Chu Lai, but coastal patrolling was not the kind of warfare he had in mind. He had even less taste for the ambience at his next posting, the cushy, enormous base at Cam Ranh Bay. Sitting at an air-conditioned O Club bar downing Budweisers, chomping pretzels, and listening to Chuck Berry sing "Back in the U.S.A." on a jukebox ad nauseam was just not Peck's style. "I didn't care for Cam Ranh Bay at all," Peck sneered. "It was always full of brass and pompous bullshit."

On the afternoon of December 3, 1968, Peck and his core crew—Ken Golden, Michael Medeiros, Del Sandusky, and Gene Thorson—did manage to get into some fairly nasty action, and liked it. The following day Peck was trying to sleep the experience off when his men began banging on his door. It seemed that two Swift boats had been badly damaged by heavy fire deep in the Mekong Delta, and a pair of replacements was needed. Peck's crew, as bored silly as he felt in Cam Ranh Bay, was lobbying for one of the assignments. Peck wanted no part of it. "I told them to get the hell out of my room," he recalled decades later. "I was their boss. I didn't fraternize with these guys, memorizing their mothers' birthday and crap like that. There was no way I was leaving Cam Ranh Bay voluntarily to go up the rivers. That was a suicide mission."

But the pressure on Peck to lead his men to An Thoi increased. At the four o'clock officers' briefing at division headquarters on December 5, the topic turned to the Navy's need for volunteers to replace the two Swift boats that had been blown up in the Delta. All the PCF lieutenants were there except for Droz and Kerry. "Believe me," Peck declared, "nobody was so dumb [as] to volunteer. We were all cowards—smart cowards. Hell, if I had said yes, my nickname would have been Jose the Migrant Worker."

In the end, of course, Droz's fine whiskey and sob story worked their wiles, and soon Tedd Peck was paired with John Kerry for assignment to An Thoi. In addition to the perils inherent in going into the rivers, Peck did not look forward to facing the company of John Kerry. As one of his fitness reports pointed out, Peck had a "machine-gun mentality," completely at odds with Kerry's more reflective ways. Peck's own men on PCF-57—a crew that Kerry would soon inherit—all knew that their skipper was too quick to anger. And the one guy who got Peck's ire up the quickest was John Kerry, whom he found standoffish and condescending. "I didn't like anything about him," Peck proclaimed. "Nothing." For his part, Kerry liked Peck, and decades later recalled none of this supposed animosity between them.

Whatever their differences, Kerry and Peck now had no choice but to head into combat on the rivers together. For his part, a piqued and unenthusiastic Kerry packed his duffel bag, ate a last half-decent dinner, and wrote a few letters to family and friends with the news of his immediate change of address. As loath as he was to go, however, something in him thrilled just a little at the prospect of heading into the unknown. By the time he had to leave Cam Ranh, an almost perverse excitement crept into his demeanor. It was in that curiously conflicted mood that Kerry took the wheel of PCF-44 and pulled away from the pier with her new crew, pointing her bow toward who knew what lay ahead on Phu Quoc Island in the Gulf of Thailand and beyond. Finally John Kerry had the helm of his own patrol boat and was piloting it off to fight for freedom, just as the men he admired from World War II and Korea had done a generation before. He knew his heart and mind were in the right place.

CHAPTER EIGHT

PCF-44

As Lieutenants (j.g.) John Kerry and Tedd Peck pulled away from the pier at Cam Ranh just after midnight on December 6, 1968, a few of their in-country friends—including an elatedly grateful Don Droz—fired flares into the ink-black sky in good-luck tribute to the brave crews of Patrol Craft Fast-44 and -57. The sight reminded Kerry of the Boston Whaler night. Except this time the figures silhouetted on the shore did not scatter like gazelles but stood waving and shouting "go get 'em" and "good wishes." Then the flares faded away, and with them the comforts of Cam Ranh Bay. The Swift boats were on their way to whatever fate awaited them in the wild and dangerous river system of An Thoi, where the enemy hid in a secret sanctuary. The two 50-foot patrol boats held course about a mile off the coast as they made their way south toward the Mekong Delta, an area approximately 14,000 square miles. Peck stopped his PCF-57—which had a flashlight taped to the .50-caliber machine gun to enable night vision—to briefly talk with the men on the other Swift. After the crews said their good-byes, Kerry took his PCF-44 five miles out, where gigantic waves made navigation hazardous and sleeping near impossible. "It was my first night as a skipper," Kerry recalled. "I was especially tuned in."

First the crew had to clear enough bunk space to lie down. The boat's main cabin looked liked a room in a frat house, strewn with personal items, from Kerry's pillowcase full of books and two guitars to a football, a poncho, electronic gear, and several transistor radios. The crew had not had time to unpack and stow their gear; they didn't even know what they may have forgotten. Fortunately, the rest of PCF-44 merely appeared to be in the same shape as its cabin. The vessel was just a year old but already looked battle-scarred. Dozens of bullets had pierced its aluminum sides, making the craft's very seaworthiness suspect to the casual eye. Swift boat veterans knew better: bullet holes were not necessarily problematic. Commander George Elliott, who saw a lot of damaged PCFs from his operational post in An Thoi, believed the genius of the Swifts' construction lay in their thin aluminum skins, which could absorb enemy fire. "Aluminum was good because rocket-propelled grenades—which the VC were attacking our boats with—are called a HEAT (High Explosive Anti-Tank) round," Elliott explained in 2003. "That means they don't have any metal in the round itself but the shrapnel from the high explosive takes the metal from the target and spreads it around. But if the round hits this aluminum there's not even enough shrapnel to explode. So you'll see pictures of about 2.5- to 3-inch holes in these Swift boats with little other damage, because unless the round hit the engine block there was not enough solid metal to cause damage. It'd just go through the boat."

A PCF's biggest maintenance problem came after rough weather beat it to pieces, often separating the frame from the aluminum hull. "Besides the frame problem, the Swifts caused little real problems," Elliott said. "The engines in particular were loud but reliable." He certainly had it right about the noise. Any officer who spent time commanding a PCF invariably complained about trying to shout orders over the din the Swifts made, a roar akin to that of a Harley-Davidson climbing a hill. There was nothing stealthy about the boats. Even their pilothouse's small windshield wipers made an irritating squeak. "The engines were loud, twin five-hundred-horsepower General Motors engines," recalled Stephen Hayes, who commanded a PCF. "They made a real steady and consistent groaning noise. But when we were in a firefight—look out. With two big machine guns blasting over our heads, the noise was deafening. The end result is that I, like many others, experienced a residual hearing loss."

When PCF-44 pulled out of Cam Ranh Bay, its skipper barely knew his crew. In fact, Kerry met two of them for the first time only minutes before the boat set out for An Thoi. One of the newcomers would serve as his lead petty officer, the enlisted man who acted as second in command: the extremely capable Radarman Second Class James R. Wasser of Kankakee, Illinois. "I immediately liked Kerry," Wasser declared years later. "He was in total control, and willing to be aggressive. He wanted to take the fight to the enemy."

Wasser was already a seasoned Mekong Delta veteran by the time he joined PCF-44, and had only a few months to go in his yearlong Vietnam tour. He had trained at Coronado until early in 1968. Usually broke and always up for a dare, he had made extra cash in San Diego through a form of performance art at the enlisted men's club there, the walls of which were infested by hungry cockroaches. Wasser could usually find somebody to give him a dollar for every one he ate. "I would pound the wall and out the cockroaches would run," he laughed. "Then I popped them with my fist, not enough to squash them. I would put a little salt on them and swallow them whole. Beers were a dollar and it took me a lot to get drunk, so needless to say I did a lot of cockroach eating back then." This skill could have come in handy had Wasser taken his SERE training in the California desert like Kerry, but after he completed his Coronado training he went to Navy survival school on Whidbey Island in Washington State's Puget Sound, where it was too cold even for cockroaches. "Two damn inches of snow when I was there," Wasser grumbled decades later. "I was always numb, my ass constantly freezing. But I did learn how to escape from the enemy."

Wasser was then sent to South Vietnam, in March 1968, and assigned to a succession of Swift boats. He earned a reputation for fearlessness under fire, an asset born of his fiercely passionate loathing of the Viet Cong, and Communists in general. The pair of American flags tattooed on the back of his right shoulder attested to his political leanings, which tended toward the view that the antiwar protesters, conscientious objectors, and rich kids who finagled cushy stateside assignments to fulfill their military service obligations were all "chickenshit punks." His opinion of the draft dodgers who fled to Canada or Europe is unprintable. Never one to mince words, Wasser simply could not comprehend how any true Americans, even pacifists, could not want to help their country in wartime. "I still hate Jane Fonda's guts," he proclaimed thirty-three years after he returned from South Vietnam and thirty-one years after the actress's controversial visit to North Vietnam. "I wouldn't give her oxygen if she were dying."

With his prematurely thinning hair, toothy grin, and round face dominated by big, black-rimmed glasses, Wasser looked to Kerry more like a Southside Chicago bagman than a full-bore hawk petty officer in the U.S.

Navy. One thing was certain: Wasser sure loved his guns. His motto in Vietnam held that his two best friends were Mr. .50 and Mr. .50. "Where I grew up in Illinois," he explained, "if it flew, we shot it." Staying in peak physical condition had also become a near obsession for Wasser. On every PCF he served, he would use two .50-caliber ammo cans filled with nine yards of cartridges each on both ends of a broomstick to make a weight to work out with. "Pumping iron was important to me," he stated. "I wanted to be physically ready for the enemy."

Wasser had volunteered to go to An Thoi on PCF-44 out of sheer boredom after spending the previous month at Cam Ranh Bay. He preferred combat to inactivity, and the crashing of bombs to a quiet naval base. Yet beneath his gung ho hawkishness stood an astute student of the war. At Coronado, Wasser had learned to speak Vietnamese exceptionally well and to understand its nuances enough to communicate the essentials of U.S. board-and-search missions much better than most U.S. soldiers in-country. "They spoke with three different dialects: Hanoi, Hue, and Saigon," Wasser explained. "The key for telling the Hanoi dialect, thereby making them our enemy, was the way they pronounced their *s*'s." Wasser had picked up on what a Vietnamese barmaid character in Robert Olen Butler's 1993 Pulitzer Prize–winning short-story collection *A Good Scent from a Strange Mountain* described as the key to learning her language: "We used tones to make our words. The sound you say is important, but just as important is what your voice does, if it goes up or down or stays the same or it curls around or it comes from your throat, very tight. These all change the meaning of the word, sometimes very much, and if you say one tone and I hear a certain word, there is no reason for me to think that you mean some other tone and some other word."

Although proud of his linguistic talents, serving as a board-and-search interpreter proved too passive a job to satisfy Wasser's patriotic taste for combat. His goal was to get back onto a Delta river and see some more action before he went home. "Give me just one good firefight and I'll be happy," he used to say. The die-hard conservative Democrat petty officer thus got along fine with his liberal Democrat superior Kerry. "You've got to understand Kerry was an extremely aggressive officer, and so was I," Wasser avowed. "I like that he took the fight to the enemy, that he was tough and gutsy—not afraid to spill blood for his country. Some people were suspicious of him because he would talk to himself into a tape recorder. And I didn't like his later antiwar attitude. But I'm talking straight: he always put his men's welfare first, and was tough, tough, tough. He was a great leader. Plus he served his time in-country and fought for the right to have his opinion."

Before his sudden transfer to PCF-44, Wasser served on the Swift boat that had covered for the Boston Whaler Kerry was on the night he was wounded. While Kerry and crew had been strafing the shore, Wasser had been on the Swift boat interrogating the Vietnamese fishermen the patrol kept bringing back. When one old man loudly vented his frustration at the slow pace at which the Americans were processing detainees, Wasser replied that at midnight he was going to kill him. The fisherman believed the threat—thanks to the petty officer's expression as well as the fact that every so often, Wasser would come up to him, look pointedly at his watch, and then glare at the old man menacingly. Naturally, the fisherman grew more and more agitated. Finally, near midnight, he began to cry. Precisely at midnight, Wasser started to laugh. Somehow he managed to get the old man laughing with him. "Everyone, I was told, thought it was very funny," Kerry recalled, "and at the time, considering all the crap that Wasser and men like him had to put up with, I suppose it was."

This rather cruel little game reminded Kerry of a similar incident he found far less amusing. It had occurred just a few days after his arrival in Cam Ranh, when one of the Swift boats inspecting a junk had turned up a man whose documents apparently were not in order. The crew detained him despite the man's passionate insistence that his identification papers were legitimate. After they brought him aboard the Swift he continued to jabber excitedly, flapping his arms about in an even more agitated frenzy. To keep him from causing trouble, rather than have him ride on the open fantail, as was customary with detainees, the PCF crew shut him up in the tiny, watertight forward compartment, where he was battered about all the way back to the base through the monsoon-tossed seas. At the base he was then made to wait for several hours until the area intelligence officer arrived at the division headquarters to collect him for further interrogation.

While the fisherman was kept waiting, the commander of the division and several other officers, in on the shakedown, walked up to him and asked if he was VC. Each time the man shook his head vigorously, his eyes bulging with fear. The officers would then laugh and run their index fingers across their throats to indicate what would happen to him if he was. The old man's face contorted through several levels of fear and confusion as he remained squatting on the floor, pathetic and alone, through the ordeal. Kerry found out a day or so later that the man was, in fact, a legal fisherman whose papers had been misread by the individual who had boarded and searched his junk. The division acted as though it were a minor mistake—an inconvenience—of wartime.

The other volunteer new to Kerry on PCF-44 was Boatswain's Mate Third Class Drew E. Whitlow, who knew a little bit about every aspect of Swift boat duty, from navigation to engine maintenance and first aid. Whitlow had grown up in southeastern Oklahoma, and remained a shy Great Plains country boy through and through. After joining the U.S. Naval Reserves in 1964, he took courses at Eastern Oklahoma A&M, followed by a stint at Oklahoma State University. In late 1967 his mother, who had herself served in the Navy during World War II, urged her son to go to Vietnam to fight for his country. "'Duty and Country,' she used to say," Whitlow remembered. "She knew we had to defeat communism."

On January 1, 1968, he volunteered for duty in Vietnam and was quickly accepted. He went through training at Coronado's Swift boat school and found himself in-country on June 1. The first boat he served on, mainly on coastal patrols, was PCF-28. Although he proved a good sailor, he suffered from a few unfortunate tendencies, worst among them his habit of falling asleep at the helm while the boat was under way. But Whitlow could be utterly relied upon to slap ropes around the holds, tie knots, point his gun in the enemies' direction, and man the radio. A character not unlike the lovable Ensign Chuck Parker played by Tim Conway on the TV sitcom *McHale's Navy*, the always baseball-capped Drew Whitlow was tough in the clinches and a loyal friend.

All the men on PCF-44, and John Kerry in particular, learned to respect Whitlow. In his folksy Okie twang he called the Swifts "water boats," the U.S. Navy "Uncle Sam's Service," and peppered a "daggone it" through nearly every utterance. Standing over six feet tall but weighing only 145 pounds, Whitlow looked like he would get blown overboard by the first fresh tropical wind that came up. He kept himself on deck with a wiry strength and the gravitational pull of his cowboy boots. Whitlow stayed anchored psychologically as well, thanks in part to a good-luck charm his grandfather, a Choctaw Indian, had given him to bring to Vietnam: a rabbit's foot, which the boatswain's mate rubbed three times to ward off every impending danger. Indeed, the charm apparently worked like one: even though he participated in many skirmishes on the death-dealing Delta rivers, somehow Whitlow managed to emerge physically unscathed every time, never suffering so much as a skinned knee or a bloody lip. He liked to joke that the Pall Malls he chain-smoked in Vietnam threatened his life a lot more than any Viet Cong AK-47. "My goal in Vietnam," Whitlow's colleagues remembered him deadpanning to their howls of laughter, "is to protect my heinie." Yet when his tour of duty ended on June 1, 1969, instead of quitting the Navy, as he had planned, he decided to make it his career. Whitlow would put in a total of twenty-six years before retiring from the service in June 1990.

The "wild man" of PCF-44 was Gunner's Mate Stephen M. Gardner. By all accounts Gardner was a fearless hothead and was nearing the end of his tour; he simply didn't care too much what the upper brass thought of him. What he did care about was having fun, and for Gardner there was little better fun to be had in Vietnam than grabbing the big guns and blasting away at the enemy. Like James Wasser, he joined PCF-44 itching for the chance to unload its six-hundred-rounds-a-minute M-60s on the VC one more time before he went home.

The Swift boat's engineman, Petty Officer Third Class William M. Zaladonis, proved the hardest crewman for Kerry to get to know, if perhaps the most competent and focused on his job. "He didn't care what happened, when, or whether anything ever happened at all," Kerry recalled. "Underspoken and understated, he saw to it that the engines ran, that the oil was at the right level, and that he got to eat on time." They both shared a love of Harley-Davidson motorcycles. Sequestered in the Swift's stiflingly hot engine room, Zaladonis would pass the time listening to doo-wop music by groups like the Five Satins and the Moonglows. When really bored, he would peel the gray paint off an 81-mm motor-recoil basket with his forefinger, chip by chip. He became good friends with Wasser. "He was nuts," Zaladonis laughed years later, reminiscing about the radarman. "He was crazy, always looking for a good time, a real sarcastic wiseass."

The final member of PCF-44's crew was another boatswain's mate: twenty-one-year-old Stephen W. Hatch, an independent-minded would-be pirate from Altoona, Pennsylvania. Short, stocky, and in tip-top shape, Hatch, who was born July 17, 1947, grew up in a military household: his father, a produce wholesaler, had been at Guam and Midway during World War II. After a short spell in the National Guard, he enlisted in the Navy and was sent to Danang for onshore duty. He was ordered to drive trucks to the dock to pick up merchandise for the PX. Just over a year later, in February 1967, he was assigned to a Swift boat as boatswain's mate. Fond of rock 'n' roll classics like "Johnny B. Goode" and "Tutti-Frutti," he once got a tattoo in a Hong Kong parlor of a devil riding a black panther with "The Wanderer," the title of a hit Dion song, written underneath the image. He was proud of his other three tattoos: a cross-and-rose with "Mom and Dad," a Navy ship with a rose, and a busty sailor girl with a mop and broom. "The mop girl had clothes on her," Hatch clarified. "My dad told me if I ever got a naked girl tattoo he'd kill me."

Hatch, who was slated to leave Vietnam in January 1969, delighted in reminding his crewmates that he had only thirty more days left in his tour of duty, and then "it was heigh-ho off on the fuckin' freedom bird and back to the States." Hatch was hardly alone in this sentiment; virtually every member of the U.S. military in Vietnam dreamed about the big chartered Pan Am and Northwest Orient commercial jets that streaked across the sky above them, on the way to Honolulu and home. Watching the shiny planes climbing away from Vietnam while one was slogging through the jungle muck on patrol exerted a powerful pull on men on the cusp of combat, who wondered if they would ever see their own families again. Soldiers still stuck in-country for a while resented "short-timers" like Hatch, who made things worse by singing Simon and Garfunkel's 1965 anthem "Homeward Bound." The military's official one-year-tour policy for U.S. soldiers in Vietnam also spawned a more serious problem than the annoying habits of those close to going home: many short-timers evinced a considerable reluctance to keep fighting, regarding their real mission as avoiding the dangers of combat in their last few weeks. "Hatch never let us forget it," Kerry remembered of the boatswain's mate's impending departure. "For someone who had just arrived in-country, it was the last thing that I needed to think about."

Short-timer or not, however, Hatch joined Wasser in assuring Kerry that with roughnecks like them on board he had nothing to worry about; they were riverine old pros, hard-bitten veteran sailors up to whatever artillery came their way. "Since I was a neophyte, I said nothing and hoped that they were right," Kerry admitted in his journal. "They seemed like men who could get the job done!" Whatever the source of his crew's confidence, the blue-collar men of PCF-44 were ready for the debate champ of Yale to lead them into action. All five of them—unlike their skipper—believed in the stated goals of the U.S. military engagement in Vietnam, and that they were bringing sweet freedom to the farmers of the Mekong Delta through the barrels of their M-60s.

It's not hyperbole to say the men of PCF-44 fast became a band of brothers. On Swifts, the division between officer and his enlisted men got broken down. There was no private place for an officer to hide. Back at Cam Ranh Bay the officers had their own club and living quarters. On PCF-44 rank meant something only when under fire. On a personal level the playing field was even. Shortcomings, faults, and phobias could not be camouflaged by a bar or patch. The old World War II slogan "We Are All in This Together" was literally true.

Thus it was that together this eclectic crew pointed the bow of PCF-44 in the direction of the Delta and continued down the coast of South Vietnam. "Being a simple country boy, I at first worried about Kerry," Drew Whitlow admitted decades after the fact. "I remember thinking here was a northern type who was going to get us into trouble. But he was the best. On other patrols you never knew where you were going. With Kerry you knew exactly what that day's goal was. We didn't just cruise around. He was anything but wishy-washy."

The first night aboard his own Swift boat passed slowly for John. Sleep proved out of reach: his nerves were jangled, his muscles couldn't relax, and every time he started to doze off, it seemed a particularly large wave would pick the Swift up from the rear and send it surfing down into the trough ahead. In the bunks down below, it felt as if the boat would somersault right over. What's more, every time the onboard radio crackled to staticky life Kerry couldn't help but spring up to monitor what was going on. "Hearing that radio static was eerie," Kerry remembered of the sound that underscored his role on the waterways of Southeast Asia decades later. "If I hear that static today on a radio I sink back in time. The sound triggers a great flood of smells, and I can almost feel the air, the warmth, the humidity clawing at you [when] that static on the radio would come on and flash a message. I was on alert, [to] my constant alarm bell—my connecting clock to this strange other world. I'd be sitting there in placid beauty, watching little Vietnamese kids play on both sides of the bank, and the radio would crack on and bang! That was the world of reality, that the military was talking to you and you had a job to do."

At dawn, to the delight of PCF-44's crew, on the horizon shone the lights of Cap St. Jacques marking the end of the Vung Tau peninsula and the entrance to Cat Lo, where Coastal Division 13 was based. Kerry's buddy Wade Sanders had preceded him to Vietnam and was a skipper of a Swift boat, PCF-98, with Coastal Division 13. "Cat Lo was a small Vietnamese fishing village," Sanders recalled. "It was smack down in the middle of town. It was a dusty little base the U.S. Navy shared with the Coast Guard." Directly outside the base was a little string of whorehouses and bars, one of which had a sign reading "Massages and Drinks, Cheap." Kerry's boat navigated its way slowly into the Vung Tau Bay at the mouth of the Soi Rap River. This was the main shipping route to Saigon. A few miles from the opening of the river, the Swift's crew spied huge cargo vessels under the flags of India, Japan, Brazil, France, Greece, and seemingly the rest of the United Nations jockeying for position in the harbor. "Hundreds of ships always anchored in the bay ready to unload the machinery of war," Sanders recalled. "Ammo, tanks, trucks—you name it." Each vessel groaned with a cargo of war matériel and waited for clearance to churn it up the channel to Saigon. "We passed them and the patrol boats assigned to guard the entrance to the harbor," Kerry recorded in his notebook. "I didn't know what the light code identifying us as friendly forces was for that day. There was a moment of panic and then Wasser said, 'Hell, switch on any combination—they don't know what they are either.' He was right. We glided through the harbor with no communication."

Once inside the inner harbor, Kerry spoke briefly with the Swift boat lieutenant assigned to patrol the area. After they swapped a few Coronado stories, he offered Kerry some tips on how best to navigate through the narrow channel leading to Coastal Division 13's headquarters. Kerry and his crew maneuvered PCF-44 up the small estuary to the base, where they refueled and ate breakfast. This quick logistical stop was all the shore time they were permitted; their orders made it quite clear that they were not to remain moored anywhere overnight unless it was a true emergency. The crew grumbled, Wasser the loudest, at having to pass up the pleasures of Vung Tau, including the Grand Hotel. (Later in this tour of duty Kerry would repair there on a day off with Wade Sanders, enjoying the French onion soup and Cuban cigars. "John and I loved that hotel," Sanders recalled. "You could envision French soldiers dancing under huge ceiling fans only a generation ago. Now it was us.") It didn't seem fair that PCF-44 had to pass it by, but their job was to get to An Thoi and its dangers as fast as possible.

Kerry had a chance to talk to Sanders, who looked tired and embittered, about Operation Sealords. Lieutenant Sanders had good reason to feel bitter. On September 7, PCF-98's 81-mm mortar had gotten jammed after a phosphorescent illumination round misfired, leaving the fin and propellant section stuck in the barrel. Sanders himself was ordered by his division commander, Lieutenant Dan Knight (then visiting the LST), to go forward to the Swift pilothouse and turn over his boat to his coskipper, Robert Emory. It was there that Sanders heard the mortar discharge and ran back to find his second-class boatswain, John McDermott of Pittsburg, Kansas, convulsing on the wet deck of the Swift, with half his head blown away. The round had detonated during disarming procedures at the precise moment that McDermott looked down the barrel. Earlier McDermott had told Sanders that he wanted the fin section for a souvenir. Sanders had told him to stay away until the round was clear. For reasons that Sanders does not know, and that have tortured him with nightmares ever since, McDermott, who was always an upbeat, positive crew member, did not heed him and it cost the sailor his life. Kerry recalled how that day Sanders "spoke emotionally of holding his crewman's head in his arms, knowing that there was nothing that could be done to save him." Even decades after McDermott's death, Sanders grew emotional talking about him. "I love him," he said. "He was my brother. The hardest thing I've ever done was sitting down and writing his wife, and kids, that Mac wasn't coming home."

Sanders felt obligated to tell Kerry the rotten truth he had learned in the Delta. "You know, John," he whispered ominously, "what we're doing is really crazy. I went on a raid the other day with no air cover—nothing—and we were just plain lucky to get out with our skins. It's not what we thought it was going to be at Coronado."

Kerry, still pretty much a rookie at this Swift boat business, did not fully grasp the magnitude of what his friend was trying to tell him: that there was little glory to be found in the Mekong Delta, just fear, misery, and senseless death. Appropriate to its unlucky numerical designation, Coastal Division 13 plied waters that were considered the Bermuda Triangle of Southeast Asia: what went in often never came out. While Sanders's warning rattled Kerry, and his account of McDermott's gory death made the newcomer queasy, his words could not pierce the armor of optimism Kerry had forged to get himself through Vietnam. Only a fool would have been eager to get shot at, but he still believed that until he had been in a more heated firefight than the one on the Boston Whaler, he would not know what real combat was all about. He felt he just couldn't talk or write home about the war with authority unless he truly experienced it first.

Sanders, who would become a deputy assistant secretary of the Navy during the Clinton administration, had to leave to go on patrol, but he had more he wanted to say to Kerry. The two men agreed to call each other on an arranged radio frequency after PCF-44 had finished refueling and departed the Coastal Division 13 base. "From that moment onward, we talked by radio quite a bit," Sanders reminisced decades later. "We were using unauthorized call signs, and using unauthorized frequencies."

Of course, the crew of every PCF in Vietnam used its radio as a source of amusement as well as an aid in survival. Games would be played with other PCFs over the apparatus, such as by coding rendezvous coordinates in such terms as "Babe Ruth's home run record minus two or Carl Yastrzemski's batting average plus nine." Like doing crossword puzzles, it was a good way to kill time. In even duller moments Swift crews would radio Army personnel on the shore just to call them "mud eaters" and be ragged as "anchor clankers" back. Coming up with cool radio call names provided another diversion for the officers. Willie Auxier was from West Virginia and thus became "Appalachia," while Doug Armstrong's feat of delivering a baby on a sampan earned him the moniker "Big Daddy." Teetotaler Pete Dodson had gone to Harvard, so he naturally got labeled "Harvard Square," and Jim Colombo's taste for cheap wine made him "Red Mountain." Friendly Gary Blinn was "Sweet Pea," scrawny Robert Bolger "Twig Echo." James C. Carolan's size-14 feet rated the handle "Clodhopper J," while Stephen Hayes's curly hair made him "Steel Wool." PCF-57's Tedd Peck was known as "Chubby Checker" for sharing the King of the Twist's October 3 birthday as well as a knack for the dance. As for the lieutenant (j.g.) on PCF-44, Kerry's call sign was "Rock Jaw," commemorating his prominent long face.

PCF-44 already needed a few quick repairs, which were completed in the Vung Tau harbor only after Kerry filled out work-request forms for each procedure, in quintuplicate. As soon as the military-bureaucracy nonsense was taken care of, the boat was off again back down the river, past the ramshackle huts that stuck high up out of the mud on the banks, past the rickety old junks piled to the gunwale with nets manned by rail-thin local fishermen, past the once carefree resort of Vung Tau, and out through the shallows into the deep water six miles off the coast. From there, navigation proved precarious. "It would get shallow and you could run aground," Whitlow recalled. "If you didn't play the tide right you'd get stuck in the mud for half a day. No amount of pushing would get you out."

As Kerry would soon learn, Swifts made wonderful coastal patrol boats but they often got stuck in rivers. On the right, as the PCF headed south, sprawled the Mekong Delta, so flat for so many miles that its landmass barely registered on radar. Kerry felt overwhelmed by its vastness, awed by this watery prairie that seemed so utterly unsuited for war. Only occasionally could a few reassuring green tufts of mangrove be seen rising from the horizon. Between the shore and the course they were following some six miles out, the water depth dropped to six inches in places, and seldom reached more than four feet anywhere except in a few scattered channels the current sliced out in ever-changing patterns. Each time they passed the mouth of one of the big rivers feeding out of the Mekong—the Co Chien, the Bassac, or the My Tho—the water would grow rough and rock the Swift boat uncomfortably through the muddy waves, testing its seaworthiness. "It got so choppy and rough it would jar your kidneys," Whitlow said. "The water just popped us around until it hurt."

Fortunately for the crew of the PCF-44, ever since John Kerry had been a young boy sailing off Cape Cod, he was fascinated by maps—not the glove-compartment-road-atlas kind, but the detailed navigation charts that noted every brook and eddy. "Because I loved sailing," Kerry noted in an interview, "maps were my ticket to adventure." The French had been the first of Vietnam's many invaders to thoroughly map the Mekong Delta, and

Cau Mau Peninsula, and for the most part the U.S. military navigation charts the Swift skippers used simply duplicated their careful surveying. Kerry thus enjoyed an advantage over most other officers in the brown water navy: he could read the old but still useful French maps. "But he still got lost on occasion," Sanders recalled. "We used to tease him about getting lost."

As the day wore on and Kerry acquainted himself with the old maps, PCF-44 headed down the coast to the point of Ca Mau Peninsula at the southernmost tip of Vietnam. Around midnight, just four miles from the point where they would cross from the South China Sea into the Gulf of Thailand, they rendezvoused with Peck's PCF-57 and a landing ship tank (LST) and refueled. Huge fenders were lowered over the sides of the landing ship so the Swifts could tie up alongside. "The LSTs were the mother ships to PBRs and PCFs," Commander Charles Horne explained years later. "They could travel up the main rivers of the Mekong Delta fifty to eighty miles. The Swift boats would pull up to them to refuel and spend the night. Sometimes the men just stopped by to get a decent dinner. It was like a destroyer tender. In Vietnam the LST was a really top-notch makeshift tender." With their maintenance complete, PCFs 44 and 57 continued south, sterns seaward, firmly pinioned to their large friend. "I looked up about twenty feet to the deck of the ship and saw a ladder come tumbling down on us with an invitation to come aboard," Kerry recalled. "We quickly accepted."

The LST's captain interrupted a late-night showing of a Jerry Lewis movie to welcome Kerry and Peck with coffee and conversation. Kerry wanted neither but acceded to the latter, which meant listening to the captain boast about a recent operation his ship had conducted with some Swifts. He seemed at pains to clarify a *Stars and Stripes* news story that had criticized his vessel for missing some of its targets during the preraid bombardment. Suddenly Kerry realized that the captain was speaking of the raid that had taken out the men and boats he himself and Peck were being sent to An Thoi to replace. Curious, Kerry pressed him for details. The captain related more about what eventually would become known as the Bo De massacre. What astonished Kerry most was that the LST officer appeared to be less concerned with the damage to the Swifts and their crews than he was with the press coverage.

Nevertheless, his chilling story intrigued Kerry, who recognized it as a cautionary tale. The operation began with five Swift boats being detailed to penetrate the Bo De River on a Sealords raid. They were to enter the Bo De from the South China Sea, run through it to the Cua Lon, transit that, and then exit out its mouth into the Gulf of Thailand. In keeping with the basic concept of Operation Sealords, throughout the entire transit the Swifts' crews were to take "targets of opportunity"—sampans, people, and the like—under fire to show that "the VC didn't own the river." The U.S. Navy wanted to make the point that they had the guts to go right into Victor Charlie's backyard.

The five Swifts had shoved off from the captain's LST as his ship began bombarding the shore to cover their entrance into the Bo De. The *Stars and Stripes* reporter covering the operation criticized the LST because her shells had for the most part fallen short, landing harmlessly in the South China Sea. As the mostly useless bombardment ended, the boats started for the entrance of the river with the anticipation that their scheduled air support would arrive. After all, the entire mission had been predicated on it.

But in what became a not infrequent Sealords foul-up, the air support never showed. This failure did the mission in. The Officer in Tactical Command was Bob Brant—"Friar Tuck" to friends, in recognition of his girth. A native of Michigan's upper peninsula, Brant was involved in numerous firefights, eventually getting shot in the neck; he survived. "Friar Tuck was one of the original river rats who secretly explored the mouths of most of the rivers in the Delta," Kerry noted. "Before coming to An Thoi he had seen one of his men killed while toying around the entrance to the Ganh Hau." Just before the incident, Brant had complained about the lack of air support to Area Commander A. L. Lonsdale. Undaunted, Lonsdale radioed back and ordered the boats to go in anyway. They did. Within five minutes of entering the Bo De River, the VC started firing from both banks and sliced the Swifts to ribbons. Seventeen men were wounded in the fiasco; one officer's leg was shot off. It seemed a miracle that no one was killed.

An ABC News correspondent riding on one of the Swifts tape-recorded some of the action that awful night. "Helicopters had become passé," Brant recalled. "At that time it was in vogue for reporters to ride in Swift boats." Brant shared a copy of the tape, which Kerry borrowed and had copied. Although the reporter's tape recorder got shot, capturing only part of the action, it provides the most vivid real-time documentation of an Operation Sealords raid known to exist. The hectic sounds of the action come through even in this partial transcript:

> Hit the starboard side. . . . It looks like we're going to fire on that starboard side there in another moment. . . . They're firing on the starboard side now. . . . I see the tracers hitting the trees and there seems to be a bunker there. . . . Apparently they received some fire from that area. . . . There doesn't seem to be any more coming from there. . . . It's a funny thing: once we got inside the mouth of this river, which we are about one half mile [up] at the present time, the boats seem to have slowed up somewhat. . . . Maybe it's just my imagination. . . . It seems like we were going a lot faster when we were out in the South China Sea, but it seems like we've slowed up considerably and it's a very uncomfortable feeling here, being between two banks and in enemy territory. . . . I see a very well-fortified bunker on the starboard side. . . . We have not received any fire from it. . . . It looks like . . . I can't see . . . I can count one, two, three, four, five bunkers right here on the starboard side as we're passing. . . . The skipper has not called for fire yet. . . . We are receiving fire now. We are receiving fire now; we are receiving . . . oh my God! Oh my God! All the guns have opened up at the same time. . . . I don't know what's going on. . . . I don't know what they're firing at, but everybody is firing at the same time. . . . He's keeping his head straightforward and he's firing. . . . There's a shot fired on the . . . they're firing the mortars now . . . they're firing the mortars. . . . There seems to be fire all the way around now. . . . Who? His gunner's mate is hit . . . one of the gunners has been hit. . . . I don't know on board which ship . . . or which boat but one of the gunners is bleeding badly.
>
> 38-32, this is 31; my gunner's mate got a bad leg. . . . It's all tore up. . . . My forward gun tub is out of commission; over.
>
> This is 38; roger. . . . The gunners have opened up again. . . . 31, your gun tub is out of commission. . . . The gunner seems to have been killed. . . .
>
> 31, this is 38: abort the mission; head out . . . we're going to

abort the mission . . . we're going to abort the mission . . . we're gonna head out. . . . 31, this is 38: do you read me? Over.

This is 31, negative, say again, over.

This is 38: abort mission, head out. . . . We are heading out now. . . . We are going to turn around and head out. . . . We are going to turn around and head out. . . . Romeo 8, this is Abbey November 38: we're going out; over. . . . The .50 calibers have opened up something . . . on the port side now as we're going down . . . following the leader here. . . . I'm going to need a medevac immediately . . . one man needs a medevac right away. . . . Abort the mission. . . . Abort the mission . . . get the hell out of here . . . apparently all boats abort the mission. . . . The order's just been given for all boats to abort the mission. . . . Hit the starboard side . . . hit the starboard side. . . . They're just keeping their heads down. . . . Everybody's gonna get outta here. . . . It seems like we walked into a trap. . . . They're firing the mortars point-blank at each bank. . . . The .50s are firing like crazy . . . port side. . . . Keep firing . . . keep firing, the captain says . . . the skipper says. . . . All boats abort the mission . . . all boats are to abort the mission. . . . Shoot your way out. . . . Looks like we're walking into a trap. . . . Hit that port side . . . that port side. . . . Get the hell out and shoot your way out. . . . All units, this is Abbey November 38: our boat's been hit . . . max speed 1200 rpms; over.

82 is hit amidships, 82 is hit amidships. . . . We're covering ya. . . . Everybody's to fire their way out. . . . Everybody's . . . the boat in front of us seems hit . . . it's smoking . . . our boat number 38 . . . our . . . our . . . I think he's been hit. . . . We haven't taken any hits yet. . . . I don't know. . . . Our aft gun's just let go . . . they're just letting go now. . . . There seems to be a great deal of fire from every side . . . no matter where we go there seems to be fire. . . . There's no escaping . . . they're all over the place. . . . Our engines have picked up some speed . . . I don't know how much, but the .50s are keeping those heads down. . . . One or two of the boats have been hit . . . one has been hit amidships. . . . The boat is literally, literally filling up with cartridges . . . big ones all over the place. . . . The skipper has got his gun out the left-hand side and

> he's firing automatic M-16 on the port side. . . . This is one hell of a mess here. . . . Let's keep those .50s firing . . . better keep those guns goin'. . . . Are we hit? We're hit . . . I thought we were hit, but we weren't . . . there are rounds coming over this boat. . . . I can hear them . . . the .50s are firing like crazy now . . . everybody's firing . . . everybody's firing . . . we're hit . . . we're hit. . . .

Listening to the tape play brought home to John Kerry the grim reality of what he was about to face. First he had heard Wade Sanders detail the grisly death of his crew member and now this. Kerry brooded anew about why he had been sent down to An Thoi: to kill Viet Cong deep in so-called Indian Country. The days of Market Time coastal patrols had been replaced by Sealords river raids. He wondered about the fellow officer who got his leg blown off in the Bo De massacre—according to the LST officer, just six weeks after he had arrived in-country. He also remembered Dick Pershing's death and the story he had heard at Coronado about the sinking of PCF-4. Kerry felt more opposed to the war than ever—yet, oddly, equally more compelled by patriotism to fight it. "But I guess that until you've been in it, you still want to try, because when we arrived in An Thoi, we requested that we be sent on a mission," Kerry revealed many years later, still perplexed about why he and his men had so wanted to see action. "Some strange drive made us ask. It wouldn't be many days before we'd ask to get the hell out of there."

Had there been no Vietnam War in the 1960s and 1970s, some smart developer might well have found An Thoi the ideal spot to build a luxurious resort. Nature had graced the southern tip of Phu Quoc Island in the Gulf of Thailand with sparkling water, sandy beaches, balmy temperatures, and a spectacular waterfall. As it was, An Thoi remained a small fishing village on this island, its main claim to fame that it manufactured *nuc mam*, the "champagne" of Vietnamese fish sauce. Of course, as the U.S. military presence in Vietnam grew, generals, admirals, and other brass learned of An Thoi's loveliness and began flying into the tiny airfield to swim in the turquoise waters of its bay. But to the U.S. servicemen who fought in the area and had some awareness of such ghastly guerrilla tactics as booby traps, underwater mines, and sniper nests, all An Thoi signified was danger. This tropical paradise teemed with Viet Cong. For starters, a prison camp located just outside the village held more than 17,000 hard-core Communist rebels. At this remote U.S. Navy base they were interrogated, fed, and watched over twenty-four hours a day. Unfortunately for the American troops stationed there, another 500 or so free VC were hiding out in Phu Quoc Island's densely forested mountains; from here, they boated back and forth from their Cambodian bases to help the prisoners escape. These guerrillas also infiltrated the island's villages to collect taxes from many of its 12,000 inhabitants.

To the men on the PCFs on the Ca Mau Peninsula, beautiful but deadly An Thoi symbolized all the uncertainties inherent in Operation Sealords. Decades later, at a Swift boat reunion in Norfolk, Virginia, a score of Vietnam veterans were asked to say the first thing that popped into their minds when they heard "An Thoi." They came up with remarkably similar, graphic responses: "rats' nest," "godforsaken pit," and "horrible outpost" are among the printable. As a U.S. senator, John Kerry tried to sound a little less colorful, but his answer amounted to the same foul memory of the place. "It meant the boats which had been hit and the men who had been wounded," Kerry intoned. "It meant the Bo De massacre and all it evoked from our portentous visions of the first blood spilled in an effort that promised to spill a great deal."

That An Thoi built such a singularly menacing reputation may have been a bit unfair to its magnificence, given that other Navy combat units were just as involved in Operation Sealords, and the men of Coastal Division 13 at Cat Lo certainly took their share of chances and casualties. In the early stages of Sealords, however, An Thoi, home of Coastal Division 11, indeed suffered the brunt of the VC guerrillas' resistance. With each new casualty, calls for replacements would go out to the other divisions. As the messages started to come in more and more frequently, the potential replacements began to wonder if the task on Phu Quoc Island was simply to serve as cannon fodder. James Wasser summed up the ethos of the place: "You knew when you were at An Thoi that you were going to get into some heavy shit. The countryside may have been beautiful, but no matter where you went from there you were going to get your ass shot off."

When PCF-44 arrived at the floating barracks anchored off An Thoi late on December 7, 1968, a pall had fallen over the installation. One of its enlisted crewmen, Steve R. Luke of Provo, Utah, had been killed in an ambush just forty-eight hours earlier. Gregarious, fun-loving, and fierce in his love of country, the young Steve McQueen lookalike had been shot in the Bo De River by an AK-47 round. Operation Sealords was still new enough that the people involved had not yet had time to feel their way through the first raw emotions that come with combat and the death of young colleagues for ill-defined purposes. As a result, the servicemen of Coastal Division 11 carried themselves with a contrived outward bravado that might as well have been an unwritten regulation. For the same reason, the Navy tried to make An Thoi as pleasant as possible for those stationed there. The mess hall offered pretty good sandwiches as well as an honor-system beer and soda dispenser, and the recreation-room tape deck played soothing popular hits like Otis Redding's "(Sittin' on) The Dock of the Bay" while the enlisted men shot pool and threw craps. Yet underlying the niceties seethed an inchoate tension, which couldn't be escaped or ignored or erased even with a whole bottle of Jack Daniel's every night.

The day after Kerry got to An Thoi, Swift boat Commodore Horne himself flew down from Cam Ranh Bay to lend his presence to a memorial service for Steve Luke. Virtually all the men and officers of Coastal Division 11 filed into the enlisted mess hall and sat down quietly. The base chaplain spoke first, mouthing all the usual platitudes. The eyes of Luke's comrades in arms wandered in search of something more than the impersonality of it all. "Luke was married only nine days before he went overseas," James Wasser remembered. "We had been together in boat training at Coronado and at Cat Lo Base patrolling. His death broke my heart. I had known him real well."

After the chaplain, Commodore Horne got up to speak. He began by thanking the dead man for his noble actions, and then assured the assemblage that he stood 100 percent behind the Swift boat program and what it was trying to accomplish in the Ca Mau Peninsula and the Mekong Delta. "We'll do everything we can to see that Swifts have all the advantages possible for their missions," Commodore Horne told the men. They would need every advantage they could get to make it out of Vietnam's rivers alive. "I just want you to know that I'm behind you all the way," the commodore emphasized.

There was nothing really wrong with what Horne said, and his very effort in coming to An Thoi to say it indicated his sincere desire to let his men know he cared how they felt about the loss of their comrade. But to Kerry's ear, Horne's rah-rah rhetoric about Luke's dying for the American way struck a sour note; it seemed inappropriate for a memorial service, and in some way demeaning to the memory of the man to laud him not as an individual but for his willingness to have served as a cog in the great U.S. military machine. A few words about who Steve Luke was and why his specific loss mattered would have been more ennobling; a simple moment of silence or a short, sincere prayer would have been in better taste. As it was, no one left the memorial service feeling uplifted. In fact, the mood at Coastal Division 11 sank from numb to morose.

Yet Kerry respected Horne, and while he had not been impressed by the commodore's remarks at the memorial service, the pep talk did further pique his curiosity about Operation Sealords. The endeavor had become practically the only activity on An Thoi that Navy Swift boats and Coast Guard cutters took part in, and Kerry wanted to know why. Commander Horne had talked about Sealords as if it were the end-all and be-all of the entire American naval effort in Vietnam. "I've come down here today to make sure you men understand that we're behind you," Horne had stressed. Kerry and most of his cohorts would never have put in for duty on the rivers if there had been any doubt about that. What Kerry could not understand was why there was suddenly such a fuss about the Sealords program. The extraordinary emphasis placed on the splendor of what the Swifts were doing made not only Horne's rhetoric itself but the entire enterprise sound contrived, as if it were all for propaganda purposes.

The U.S. servicemen detailed to river and coastal patrol duty in Vietnam, had long been boarding and searching sampans, detaining suspects, and providing artillery-fire support on the fringes of the war. Now, it struck Kerry, at last they seemed to be making *Navy Times* headlines and taking home medals. He began to fear that careerism in the Navy's upper ranks had something to do with it—after all, the more headlines and medals a commander's men earned, the greater his chance of making admiral. So what if some lives were lost in the process? "Lives are always lost in war"—that's what the bulk of Americans sitting safely in their living rooms watching the TV news would have then said. They would say it over and over to keep from thinking too much about how the growing casualty lists were getting as common on the news as the college football scores every Saturday night. The more often they said it, the more the "fortunes of war" would become accepted as an excuse for the increasing loss of life. And the more accepted the excuse became among the public back home, the easier it would get for the military minds in Vietnam to justify losing still more lives. In his journal Kerry called this "The Law of Increasing Excusable Casualties."

But what troubled him even more, and spawned his first real doubts about Operation Sealords, was the growing tendency among America's military leaders, including Commodore Horne, to measure every mission's success in terms of enemy bodies counted, huts burned, sampans sunk, and medals won. There was something downright sinister in the cold and calculating way Johnson administration officials like Secretary of Defense Robert S. McNamara and National Security Advisor Walt Rostow turned to the language of the dismal science to discuss statistics like "kill ratios." Worse, by now this risk-assessment approach to warfare had trickled down through the ranks.

How different things had become since World War II Supreme Allied Commander Dwight D. Eisenhower confided to his wife one night in April 1944, just weeks before the D-Day invasion: "How I wish this cruel business of war could be completed quickly. Entirely aside from the longing to return to you (and stay there), it is a terribly sad business to total up the casualties every day—even in an air war—and realize how many youngsters are gone forever. A man must develop a veneer of callousness that lets him consider such things dispassionately, but he can never escape recognition of the fact that back home the news brings anguish and suffering to families all over the country." Kerry couldn't help but wonder if men like Lyndon Johnson and Richard Nixon also saw the faces that went with the dead bodies, and shared Eisenhower's compassion over their loss. He doubted it, and that doubt began to spread to the whole point of the Vietnam War.

As Kerry reflected on Horne's effusive praise for the Swift crews' gallant efforts, his first meeting with the commodore at Cam Ranh Bay came to mind. Horne had then talked about Operation Sealords, waxing enthusiastically about the latest improvements made to the PCF design and betraying a special pride in the M-60 machine gun placed on the bow of each Swift boat. He had fairly glowed when he told his young officers how the sailor manning the gun would stand in the forward watertight compartment with only the upper half of his body exposed, the rest remaining shielded by the quarter-inch aluminum hull. Horne neglected to mention that that hull might not be able to stop a BB at a thousand yards. Kerry also recalled how, using a plastic model of a Swift boat that he moved around the conference table like a kid playing the old Mattel board game Battleship, the commodore had demonstrated the various ways a PCF could demolish huts and sampans.

"Gentlemen," he had declared then, "this is a real opportunity for you to test your mettle. We only want the best men in this operation and we think we've got them now. Two of the men in An Thoi have already been recommended for the Silver Star. So you see, there's quite a lot of action. We're real proud of what those boys have done down there."

After Steve Luke's memorial service, Kerry met one of the men who had been put in for the Silver Star, Lieutenant (j.g.) Mike Bernique—Ol' Bernique, the commodore had nicknamed him. It was obvious how brightly the dazzle of his decoration was intended to reflect on and inspire the entire Swift squadron. Born on December 26, 1943, in Fall River, Massachusetts, Bernique had spent his growing-up years in Germany, France, and Washington, D.C. With his burly build, thick neck, and red balding head, Bernique, a University of Notre Dame graduate, was tough as nails. At Horne's orientation briefing in Cam Ranh Bay, the commodore had boasted how Mike Bernique's exploits had even gotten a river named after him. With equal relish Horne had told Kerry and others how the incident that ended in "Bernique's Creek" had almost gotten its honoree court-martialed. It was a wild take indeed, one that began with the other of the commodore's "boys" who had been put in for the Silver Star, Lieutenant Mike Brown.

Once Kerry arrived in An Thoi, he finally learned why the commodore was so high on Bernique and Brown. As it turned out, Horne's version had not revealed just how much both episodes had to do with the official participation of Swift boats in Operation Sealords. In September and October of 1968, enemy infiltration through the barrier set up by the U.S. Navy's Coastal Surveillance Force had been almost completely curtailed. As a result, the patrols had become so boring that the Swift boat crews began to look for any diversion they could find from the board-and-search routine. For those boats stationed near the Mekong Delta, a fine source of entertainment presented itself: making hell-for-leather dashes up the Viet Cong–infested rivers through "free fire zones"* with guns blazing, mostly for kicks and to have something dramatic to write home about. These excursions up the Delta rivers and in the Ca Mau Peninsula became a regular game among a small group of boats from Cat Lo and An Thoi, respectively, and made a fine way to break the tedium.

Ironically, given that Operation Sealords would soon force generally unwilling Swift crews to do the same thing these early cowboys devised for fun, the voluntary shoot-'em-up runs through Viet Cong territory were conducted against the strict orders of division commanders, one of whom promised to court-martial anyone proven to have gone up a Delta river. In one such instance a commander had gotten in the face of a skipper friend of Kerry and snarled, "I know you're going in those rivers—I know it. And I just want you to know that if anything ever happens to your boat or your men, or if I ever catch you in a river, I'll have your stripes and have you before a green table so fast you won't know what happened to you. Is that understood?" Under Sealords, the same officer was awarded two Bronze Stars, a Navy Commendation Medal, and a Purple Heart for doing exactly the same thing in the rivers.

Then, shortly before Operation Sealords commenced, the command of the Coastal Surveillance Force changed hands, and policies began to change as well. The new commander, hawkish Captain Roy Hoffman was ecstatic about Sealords. He knew that military reputations were made in wartime, and he was determined to make his in Vietnam. What's more, he had a genuine taste for the more unsavory aspects of warfare, and truly wanted to smoke the Viet Cong out of their tunnels, burn their jungle outposts, and annihilate them once and for all. Decades later, many Swift boat veterans under Hoffman's command would compare him with the rough-hewn colonel in the movie *Apocalypse Now* who boasted that he "loved the smell of napalm in the morning." In short, Captain Hoffman sought to convince his Swift boat skippers to do whatever it took to notch splashy victories in the Mekong Delta and thereby get him promoted.

**In 1965 the U.S. Department of Defense officially forbade the term "free fire zone" on the grounds that it suggested a policy of allowing indiscriminate killing. Former free fire zones thus were redubbed "specific strike zones." But nobody in Vietnam ever used the more gentle term. They were forever known as free fire zones.*

Kerry would never forget how ardently Captain Hoffman lauded the exploits of one "enterprising officer" from the Danang Swift division. The officer had surprised some thirty Vietnamese who were fishing in round, floating baskets just off the shore of a peninsula in an area that was, unfortunately for them, a free fire zone. Hoffman considered it ideal military thinking that the Swift skipper had shown the presence of mind to sneak his boat in between the baskets and the shore, cutting the fishermen off from escape and then opening fire on them. All the baskets were sunk, and so were the fishermen. "Fantastic," Hoffman reportedly proclaimed upon hearing the news. Kerry himself would later hear Hoffman praise such "industriousness" at a remarkable meeting in Saigon. Clearly, the Navy had undergone a sea change. Not only were cowboy antics on the rivers of Vietnam no longer frowned upon, they were rewarded with medals. "I was in-country after this happened," Wade Sanders later wrote. "When the OINCs heard about this, and heard that the butcher was being considered for a Silver Star, a howl or outcry was heard all the way to Cam Ranh Bay and Captain Hoffman reversed the consideration of a medal. We all thought the son of a bitch should have been court-martialed."

It was in this gung ho, new ethos that Lieutenant Mike Brown made his bold move. "No doubt he had no idea his action would transform the quick dashes into the rivers from clandestine games to public policy," Kerry wrote. "But they did just that." It happened when Mike Bernique was assigned to gather intelligence at the mouth of the Cua Lon River, the largest waterway in the Ca Mau Peninsula at the southern end of the Mekong Delta. The Cua Lon region was reported to be the lair of several thousand Viet Cong, although the U.S. military really couldn't say for sure, as no American had been there in years. For this reason the river was, of course, designated a free fire zone.

After entering the mouth of the Cua Lon on the Gulf of Thailand side of the peninsula, Mike Brown decided to run his Swift boat the entire length of the river. When he and his crew reached the end of the Cua Lon, they turned into the Bo De River, and exited from it into the South China Sea—but not without incident. Near the river's mouth, the Swift came under fire, and Lieutenant Brown and several of his men sustained minor wounds.

At first Brown and Bernique attempted to keep the full details of their excursion secret, but word naturally seeped out. When it reached Captain Roy Hoffman, Lieutenant Brown was recommended for the Silver Star. The concept of river incursions into VC territory had clearly found a sympathetic advocate in the commanding officer at Cam Ranh Bay.

Ten days later, on October 14, Bernique made his own dash up a small river, the Rach Giang Thanh, near the Cambodian border at the northern end of the Gulf of Thailand. With the help of U.S. Army troops riding along on his Swift, Bernique overran a Viet Cong tax station that had been extorting funds from the local citizenry. He and his cohorts killed several of the VC tax collectors and captured their weapons. Unfortunately, because the action had taken place so close to Cambodia and several of the dead had fallen on the other side of the border—which at the time the U.S. Navy took great pains to observe—the incident was not so easily lauded, or forgotten. "A furor erupted over Bernique's violation of the ban on the Rach Giang Thanh," Thomas J. Cutler wrote in *Brown Water, Black Berets*, "and it appeared that he was facing a court-martial for his actions."

The next day Bernique was flown to Saigon, where the naval command debated whether he should be court-martialed or decorated. The panel decided on the latter, and Lieutenant (j.g.) Mike Bernique joined Brown in being recommended for the Silver Star. Years later, Kerry could still remember how Commodore Horne constantly declared, "We couldn't do anything at first because the Cambodians said that when Bernique's Swift fired, it killed several Cambodian citizens on the other side of the border. But when that was cleared up we put him in for a Silver Star." Horne's thinking reminded Kerry of a scene in Joseph Heller's *Catch-22* describing pilot Yossarian's decoration for dropping his bombs into the water, killing only fish. To Kerry, it seemed that a like absurdity had gotten into the brown water navy in Vietnam, where breaking orders and international law had become a Swift lieutenant's surest way to a serious decoration. Killing civilians apparently was no longer a disqualifying factor, provided one killed enough enemy to offset it.

Kerry recognized, of course, that Bernique could not be faulted for the official aftermath of his actions. "What he had done took a lot of guts, in the context of war guts," Kerry maintained. "It was the military charade which followed that disgusted everyone—including Mike. When he came back from Saigon, though, he brought with him the official beginnings of Operation

Sealords. A few months later his medal followed him." The deliberations on whether Bernique should be punished or rewarded for his derring-do on the Cambodian border typified the top-level confusion in the U.S. military as to how to prosecute the war in Vietnam. "Everything was right and wrong at the same time," Kerry summarized the problem in his journal. "In war things are like that, or so we were told. It all depends on who and what is involved. The U.S. military was always right, though. We quickly learned that. It was right because the men running the United States couldn't bear to have it any other way."

As soon as he arrived in-country, Kerry began to recognize the way the hierarchy worked. At Cam Ranh, for instance, he learned of a VIP delegation's visit to the grand base a few days before he got there himself. The Johnson administration had invited some twenty-five corporate chieftains and other distinguished American citizens to Vietnam, and it wanted, Kerry was told, its important guests to get an honest picture of what was going on in Southeast Asia. To see that they did, among other demonstrations, a Swift boat excursion was arranged so the likes of NBC chairman Robert Sarnoff and Pennsylvania Governor William Scranton could get an up-close look at the small-boat war. When the delegation boarded the boats the VIPs found themselves treated to a real show: a fast ride across Cam Ranh Bay, followed by a hasty mortar firing for which the rather elderly visitors got to don flak jackets and play soldier, and then another fast ride back to the pier with the Swifts flashily crisscrossing one another's wakes to commands radioed to them by the military equivalent of a choreographer. The entire performance, which the crews had practiced the day before, took forty-five minutes, and from that the VIPs supposedly learned all about the Navy's Swift boat operations.

Unfortunately, such alleged fact-finding tours were hardly uncommon, and some much worse, considering what came out of them. On one occasion General Earle Wheeler, then chairman of the Joint Chiefs of Staff, traveled through South Vietnam's provinces to talk with U.S. military advisors. One after the other, the advisors boldly told the general how badly the American effort was faring. Finally, Wheeler found an Army major duplicitous enough to detail all kinds of good things about Vietnamization. General Wheeler was heard to remark, "I'm sure glad I finally found someone who knows what he's talking about."

Only a few weeks later, Kerry, on a PCF-44 patrol, observed four troop battalions from the Ninth Infantry Division at Dong Tam and five Mobile Riverine Force squadrons staging an assault for the benefit of Secretary of Defense Melvin Laird. "To facilitate putting on a good show, an area was picked out for the landing where the chance of guerrilla contact was minimal," Kerry sneered. "Nothing was to mess up the show for the secretary of defense."

Unfortunately, a reconnaissance company did meet resistance in the proposed "assault" zone. A few men were hit by booby traps and the company came under heavy fire several times. This actual guerrilla presence forced the exercise's stagers to move the prospective battlefield at the last minute to another "safe" area miles away, near the South China Sea. Codes, radio frequencies, and virtually every other operational detail had to be changed, with enormous effort and at considerable expense to the American taxpayer. The next day Melvin Laird flew over the staged landing in a helicopter. "It was the only 'action' he saw in the entire Mekong Delta," Kerry said. "No doubt he could report to the President, and the President could relay to America, that the war in South Vietnam was winding down smoothly and that Vietnamization was a blooming reality."

During Kerry's first days in An Thoi, he talked with his men a great deal about the war in general and Vietnamization in particular. He would, in fact, become known for sparking dialogues about current events with nearly everyone he spent time with. Kerry liked to talk about more immediate concerns as well, and he learned a great deal about the area's rivers before he ever navigated one by asking questions about sandbars, poisonous snakes, booby traps, and everything else he thought he should know to stay alive.

"You want to go in a river?" Kerry would be asked with a wry smile. "Christ, you can take my place anytime." Every variation on this conversation ended with the veteran chuckling, shaking his head at Kerry, and walking away.

A few rare individuals, such as Mike Bernique, did like the rivers, though. Bernique was as close to being the perfect soldier as you could get—the kind of first-rate officer who considered river eel and skinned paddy rat delicacies. Although even he would later grow tired of serving as cannon fodder under Operation Sealords, during its first weeks in late 1968

Bernique could not have been happier. "If the opportunity had presented itself he would have beached his boat and chased after Ho Chi Minh himself," Kerry stated. "Mike still believed in what the U.S. was doing in South Vietnam." To Bernique, the Sealords river raids were merely a logical extension of the American military's overall Vietnam policy. "Goddamnit, John," he once asked his fellow Swift officer, "don't you see how if we were to leave here the whole of Southeast Asia would fall?"

Kerry replied, "No, I don't, and even if that did happen—so what? Nobody has been able to tell me why we have to do what we're doing here. Nobody can really explain in terms that make sense why it is [that] the United States has to be the one to lose its men and spend its money for a supposed threat, which few people can define. Besides, I don't see how what we're doing is helping the Vietnamese anyway."

"But it's not the Vietnamese that we're really trying to help anyway," Bernique shot back. "It's ourselves—by beating the Communist advance. You have to draw the line somewhere. This is where we have to make the stand."

This strong-minded pair liked to argue about why the United States had to make a stand in South Vietnam, about the reality of the Communist threat, about Americans' paranoia regarding communism. Kerry, for example, considered the domino theory—that if one country in Southeast Asia turned Communist, other nations in the region would follow—ludicrous. Bernique, on the other hand, insisted that if South Vietnam went Communist it would just be a matter of time before Cambodia, Laos, Thailand, and the Philippines would follow suit. The two men's discussions would go on until the arguments were exhausted on both sides. No ground was ever gained by either, and in the meantime Operation Sealords went on.

Before long Kerry would discover that Bernique had an uncanny knack for getting himself shot at. Any Swift boat skipper who went into a river with him invariably soon heard Bernique's voice coming over the radio saying something like, "This is 50—we're taking some AK[-47] over here. I'll just sit here and suppress it for a while," followed a few moments later by: "We're still takin' some AK over here—do you want me to move or stay where I am?" Given his brazenness, it hardly seems possible that the whole time Bernique spent in An Thoi his boat was never hit once. It was always shot at but never struck. That kind of luck is the stuff of folklore, and before long tales were told about Bernique "the Unscathed." For the most part, Kerry saw Bernique as the ideal Swift boat skipper: a skillful navigator and a natural leader with a fearless demeanor. They formed a mutual admiration society. "I would go up a river with that man any time," Bernique enthused to the *Boston Globe* in 2003 about Kerry. "He was a great American fighting man."

Kerry's days in An Thoi seemed to go by quickly. It was easy to keep busy; only a few hours after he got there, for example, PCF-44 was sent out on patrol north of Phu Quoc Island along the Cambodian border. The area proved so quiet that his crew swam and relaxed the entire time. During the night, however, a message came in from the U.S. Army base midway down the island, requesting the Swift boat's fire support on a harassment-and-interdiction (H&I) mission. From Kerry's perspective H&I was not only a farce but an immoral way to go after the enemy. "It required us to shoot indiscriminately at a target area we couldn't see," he explained. "The purpose was simply to harass any possible movement by guerrilla troops. One never knew what was being hit. If someone other than the guerrillas happened to be in the target area—too bad. They also were harassed and interdicted. Such was the war in Vietnam."

Almost as disturbing as H&I was the near impossibility of communicating between an Army camp and a Navy boat. Neither service ever knew which code was being used for radio transmissions; because each used separate codes, messages between the two forces generally caused nothing but confusion. A typical exchange went something like this:

"Headhunter: this is Apple Blossom 44. What code are you using?"

"Aah . . . this is Headhunter. I'm using the OSP-2—over."

"The whaaaa?"

"The OSP-2."

"Aah, roger Headhunter—what date do you have on that?"

"Wait one . . . 44, this is Headhunter. I can't find the date . . . are you sure you know what you're asking for? . . . oh . . . the 21 . . . I say again, the 21."

"Aah, roger, this is 44. I don't have the 21 date. Interrogative—you have another date?"

"44, this is Headhunter—roger . . . aah, what date do you have?"

"This is 44—I have. . . ."

So it would go, late into the night on an H&I mission, until the shells finally went whistling off through the darkness, toward wherever a dubious code aimed them.

On another day during Kerry's brief stint in An Thoi, his PCF-44 patrolled near the U Minh Forest, a notorious VC-infested expanse into which five hundred French paratroopers had been dropped in 1952, never to be heard from again. Somehow the Swift boat managed to run over the fishing nets hanging off several junks full of children and old people. "The nets became tangled in our propellers," Kerry recalled. "We had to put a man in the water to cut us free and then, showering the boss fisherman with profuse excuses for the damage done to his nets, we proceeded to the next junk. The value of the phrase 'excuse me' became very clear. We were always saying it for one reason or another. *'Xin loi, xin loi omg'*—'Excuse me, mister'—the most necessary word in the lexicon of the American fighting man."

At the end of another patrol, Kerry and his crew found an island to visit, where they managed to offset what they had done to the fishermen's nets. Wasser, Gardner, and Hatch went ashore in the life raft. They befriended the island's villagers with the gift of a first-aid kit and some candy bars, thereby proving in the same way GIs had in World War II that guns were not as good as butter at winning over people's hearts and minds. When the trio headed back to the beach to return to their Swift boat, they were stunned to see that the life raft had been towed out to sea by a fisherman. Panic-stricken, Wasser and Hatch started waving frantically from the shore like some latter-day Robinson Crusoe and his man Friday. Only Gardner kept his cool, perhaps because he realized the South Vietnamese fisherman had taken the raft to fill it with a gift of coconuts for the Americans who had given them candy. Once that was sorted out, the trio got back in the raft and paddled out. As the life raft neared the Swift, Wasser capsized it several times, coconuts and all. Kerry, Whitlow, and Zaladonis howled with laughter but appreciated the booty, which all six began cracking open and devouring.

After this idyllic patrol, they returned to the base at An Thoi to discover that they were being transferred to Cat Lo—immediately. "You must be joking," Kerry protested to the operations officer. "We just got here. The men are tired. Why send us?" But sent they were. After only six days in An Thoi,

Kerry and his crew set sail for Cat Lo to reinforce Coastal Division 13. "I tried to fight the change—not because we wanted to stay in An Thoi and be shot at, but because we didn't want to have to move and resettle once again," Kerry noted. "Our mail was already lost, and the trip back against the monsoon seas promised to be nothing but a bitch. It was just that."

Everything went wrong on the way to Cat Lo. The radar on the boat traveling with them didn't work, and once a huge wave crashed against PCF-44's radar scanner, theirs didn't either. Then everything went to hell. The lights shorted out. The pilothouse leaked. The radio went silent. The force of the boat dropping from one wave to the next made it impossible to move around. Kerry and his men lay on the bunks or the deck, trying to alleviate the battering from the collisions between the Swift boat and the sea. As Kerry described the scene: "With each thud the man in the driver's seat just hung on and grunted. Following a particularly resounding crash a cacophony of four-letter words filtered up from the main cabin. Food and clothing were strewn all through the boat."

An early-morning inspection of the two compartments revealed that both were full of water. The Swift therefore arranged an emergency rendezvous with a Coast Guard cutter patrolling the waters they were traveling through. The cutter lent PCF-44 a pump to supplement its own, which couldn't get the water out fast enough. Once pumped dry, the Swift resumed course to Cat Lo.

The officer in charge of the Swift accompanying PCF-44 was being transferred because he had seen too much of An Thoi. The perils of Operation Sealords had gotten to him. Each time he had departed for the southern patrol area, an assignment that generally called for mounting a series of raids before returning to base, he had formed the habit of taking one of his fellow officers aside to show him his locker, pointing out particular items in the event he didn't return and his belongings had to be sent back to the States. Often the poor man would just sit in the An Thoi officers' lounge staring at a wall, and smoking Camel after Camel, his hands shaking so hard he could barely hold the cigarettes. Kerry remarked that he couldn't have been a nicer guy; he just had too much imagination and did too much thinking.

Combat zones are rarely conducive to thinking and imagining. For every James Webb—who served in the Fifth Marine Regiment in Vietnam winning the Navy Cross, the Silver Star, two Bronze Stars, and two Purple Hearts, before eventually writing a best-selling novel about his experiences, *Fields of Fire*—there were a thousand U.S. soldiers unable to reflect on their combat experiences. They followed orders plain and simple, leaving the politics and poetry of war to the Webbs and Kerrys. Among the most useful qualities in combat, in fact, is the ability to forget—not permanently but functionally. John Kerry differed from most American soldiers in Vietnam in this regard; he didn't want to forget a thing. In fact, he wanted to learn all he could about his experience and memorize it all. He insisted on thinking and adamantly refused to numb his mind. Yet, in his voluminous "Vietnam War Notes," Kerry recorded Ernest Hemingway's opinion that "danger only exists at the moment of danger. To live properly in war, the individual eliminates all such things as potential danger. Then a thing is bad only when it is bad. It is neither bad before nor after."

Kerry used to contemplate Hemingway's view at great length. In the back of his mind simmered the notion that one day he would write about all that he had seen and done in Vietnam. He had an idea how such things were done; he had studied the World War II dispatches of artful reporters such as Bill Mauldin and John Steinbeck and devoured the World War I poetry of Wilfred Owen and Siegfried Sassoon. He dreamed of writing an autobiographical novel about his Swift boat experiences along the lines of James Jones's 1951 *From Here to Eternity* or Norman Mailer's 1948 *The Naked and the Dead*, but with a dose of the absurdity found in Heller's *Catch-22* and Kurt Vonnegut's 1969 *Slaughterhouse Five*. Once stationed in Vietnam, he kept detailed journals. He wrote long letters to his family and friends. He tape-recorded the stories of fellow Swift boat officers and crewmen in the Mekong Delta. Sometimes he wrote poems. "There were no pretensions that I could be either a good soldier or a good writer," Kerry revealed more than three decades after he left Vietnam. "In fact, there were many questions whether I wanted to be either. What I did want was to keep the imagination alive—keep thinking. For a while I tried both but gave up in favor of both, staying alive and sane. Besides, what I had seen and heard of the war made it seem absurd to want to be a 'good soldier.' The South Vietnamese certainly didn't care. In fact one of them once defined to me his view of being a soldier. 'You come here for a year,' he said. 'You can afford to charge a few bunkers. The times you may be killed are not as often as for me. You go to Hong Kong or Bangkok for your rest. Me—I am here for maybe thirty years as a soldier. I have to fight all that time. Why should I chase the VC? He doesn't bother me and I doesn't bother him. It's okay like that. When he come after me, I fight—okay?' "

Why, then, care about being a "good soldier" in an unwinnable war fought to help a people who didn't give a damn themselves? That was the question that gnawed at the enlisted men and junior officers of the "second wave" of American soldiers stationed in Vietnam late in 1968. Kerry believed it was obvious to most "second wave" U.S. troops in Southeast Asia that at best they were fighting for a standoff, via a slow, steady attrition, for a long-term gain that never didn't seem in doubt. "Being a good soldier often involved being a dead man, and being a dead man for a long-term, doubtful goal, which the majority of Americans thought was a mistake, didn't appeal to many of my friends," he explained. "The command would keep telling us, 'But there's light at the end of the tunnel.' We'd tell them, 'Yes. And it's a VC with a lantern and a hand grenade.' "

Thoughts like these filled John Kerry's restless brain during the miserable trip to Cat Lo, which felt as though it would never end. But thirty-five hours after PCF-44 departed from An Thoi, its crew spotted the lights of Vung Tau Peninsula. A few hours after that, they were asleep—ashore, in beds that didn't move, between sheets that felt like home. But these comforts would prove short-lived.

Chapter Nine

Up the Rivers

It was not an Operation Sealords raid that first brought Lieutenant John Kerry into a Mekong Delta river, though it might as well have been. PCF-44's first patrol out of Cat Lo—the headquarters of Coastal Division 13 under Commander George Strulie—was scheduled to last four days. Kerry didn't know it then, but those days would afford him a sweeping view of the riverine war that would dispel any illusions he may have had that much constructive could come out of Sealords. Individually, the events of those four days seemed insignificant compared with many incidents later in Kerry's second tour of duty. Taken as a whole, however, the stint proved instrumental in giving Kerry and his five-man crew some valuable perspectives on the war. Perhaps most remarkable was how quickly Kerry's earlier curiosity about the Mekong Delta's rivers was satisfied. "Within a day the tedium of the patrol got to each of us," Kerry wrote in his ever-present notebook. "As a result, the events which did give the insight to the war were viewed with a much colder eye than if I had been caught up with feelings of importance for what I was doing."

On the bright side, the four-day cruise proved that Lieutenant (j.g.) Kerry—who had just turned twenty-five, on December 11, 1968—was a fine leader of his men. In fact, in the short time since PCF-44 had left Cam Ranh, his crew had come together into a cohesive unit. Radarman James Wasser, Engineman William Zaladonis, Gunner's Mate Stephen Gardner, and Boatswain's Mates Drew Whitlow and Stephen Hatch took pride in PCF-44 and easily accepted young Kerry as their officer in charge. They also felt comfortable joking around with him and had howled with laughter when Kerry broke out a baseball cap he had bought at the Cam Ranh Bay

PX with "Boss" printed across the front. They had also gotten the point. The five working-class enlisted men bonded quickly with the Yale graduate who led them. Class distinctions meant nothing on a patrol craft fast in Vietnam. "In the darkness and solitude of night, or parked in a cove before a mission, or in the beauty of a crimson dawn before entering the Bay Hap, the My Tho, the Bo De, or any other mangrove-cluttered river," U.S. Senator Kerry recalled at a Swift boat dedication ceremony in 1995: "We shared our fears and, no matter what our differences, we were bound together on an extraordinary journey, the memory of which will last forever." They pulled away from the pier at Cat Lo with spirits high, feeling satisfied with the way things were going for them. They had no lust for battle, but they also were not afraid. Kerry wrote in his notebook, "A cocky feeling of invincibility accompanied us up the Long Tau shipping channel because we hadn't been shot at yet, and Americans at war who haven't been shot at are allowed to be cocky." In the coming weeks, he would call PCF-44 his "sanctuary" and "cloister."

In truth, PCF-44's crew were too busy loading the machine guns and laying their M-16 rifles out on the deck to worry too much about their vulnerability. They were well aware, of course, that in the event of an ambush their reactions would have to be instantaneous, as even a few seconds' delay could mean death. The only way to ensure their survival was to know exactly where everything on the boat was at all times. The instant reacting would come of its own accord if and when the time came for it. Besides, they weren't going to have to kill anybody, because they weren't going to be shot at, because they were invincible, like GI Joe. "We thought so then," Kerry marveled in retrospect. "We were full of confidence and hubris." They also took comfort from the fact that the U.S. military enjoyed a spectacular technological advantage over the enemy, including such powerful assets as helicopter gunships and satellite-based surveillance. Superior weaponry may not have kept France from being routed out of Vietnam, but the American troops could reflect that in the entire time the French were in Southeast Asia they had deployed only seven helicopters; the United States had brought more than four thousand. Virtually every U.S. soldier in-country would find reason to consider these helicopters not only a security blanket but his best friend.

* * *

John Kerry's war journals about his command of PCF-44 may provide the best documentation of the activities of any Swift boat in Vietnam. Some thirty-five hundred American sailors served on Swifts during the Vietnam War, but what each of them did cannot be ascertained, as no deck logs were kept on the boats. Only commissioned U.S. Navy vessels—ships with names, such as battleships and destroyers—keep books. Or at least only those books are kept by the U.S. Navy, which discarded the reports PCF officers in charge were required to file with their division commanders after every river mission in Vietnam. The longest of John Kerry's reports ran six pages, but it was distilled with the rest by his commander into a single annual summary of about forty pages of boiled-down, detail-free, bureaucratic ass-covering. The absence of a thorough official paper trail on the Swifts helps explain why so little has been written on the riverine war in Vietnam. The details of what particular boats did and how individual sailors performed come largely from oral-history recollections and travel orders. Fortunately for the posterity of PCF-44, Lieutenant (j.g.) Kerry kept his own records, from which much in the accounts here has been drawn.

PCF-44's inaugural four-day riverine mission began when the boat entered the Long Tau River at its mouth, across the bay from Cat Lo. From there the river ran through the bleak mudflats of the Rung Sat Special Zone, an area dubbed the "forest of assassins" that had been set up with extra defenses because of its proximity to Saigon and the Viet Cong's eagerness to disrupt the cargo ships delivering war supplies to the capital. Thoroughly defoliated by thousands of tons of American herbicides, the Rung Sat Special Zone looked like a post-atomic wasteland. The U.S. military command in Vietnam believed that the easiest way to eliminate the dangers of the country's jungles was to eliminate the jungles themselves. With the greenery gone, would-be enemy ambushers would have no natural cover to spring from, and therefore wouldn't try to—or so the theory went. Aerial defoliant-spraying missions thus became commonplace between January 1965 and April 1970. Agent Orange—a mixture of the potent weed killers 2,4-D and 2,4,5-T—was the newest poison of choice. Manufactured by the giant U.S. chemical firm E.I. du Pont de Nemours & Company, the reddish brown, dioxin-spiked herbicide was shipped to Vietnam in orange-striped barrels. No one yet knew, of course, that Agent Orange would turn out to be deadly to fauna as well as flora, causing serious human illnesses, including a variety of cancers.

The crew of PCF-44 found out on their first day of Mekong Delta river patrol together that at least some Viet Cong thumbed their noses at what the Americans' thousands of tons of herbicides had accomplished. No sooner had Kerry and his men turned off the Long Tau to assume their patrol station on the Soi Rap River than would-be VC assassins used the cover of the Rung Sat mud to fire some rockets at a freighter, hitting it twice. "We listened over the radio while the headquarters at Nha Be detailed two PBRs to move into the area of the attack," Kerry recorded. "Shortly afterwards, we were asked to move from our position to cover the PBRs while they searched small estuaries of the Long Tau. We raced upriver at full speed with the feeling that we were fighting the entire war in those few moments. The glory was short-lived. Nothing suspicious could be found."

Nobody on the freighter was hurt, as Kerry learned was almost always the case on the cargo ships. The Merchant Marines who manned the freighters in fact had quite a good thing going. Every time they were shot at they earned extra, easy-to-come-by "hostile-fire pay," just for staying down below in the safety of a huge steel hull. When added to their regular wages—sometimes for doing practically nothing for weeks in the safety of the Vung Tau harbor while they waited clearance to sail to Saigon—this bonus hostile-fire pay made the Merchant Marines some of the biggest money-earners in the war zone. "Meanwhile, a Marine lieutenant was being paid far less to have his head blown off somewhere in the mountains of Vietnam," Kerry noted in his journal. "It was an irksome comparison if you believed having your head blown off was worth any sum."

Shortly after PCF-44 returned to its assigned area from the attack on the freighter, an American air strike commenced just a few hundred yards from the boat. There had been no warning—nothing. "I had been eating a peanut butter and jelly sandwich when the scream of a jet some few hundred feet above sent me scrambling for cover," Kerry remembered. "I ducked and then looked: kaboom! A billowing cloud of flame and black smoke exploded into the sky."

The roars of more jets turned to shrieks as the planes dive-bombed down from several thousand feet. "They dropped in graceful, silent arcs, silent until right above," Kerry recalled. "Then they changed from small specks in the distance into large, menacing bombers. They would drop their load and drive almost straight upward, gaining altitude as quickly as possible and rolling out from a nearly vertical position at the maximum point of climb."

From his vantage point, Kerry had a fine view of the "black eggs" the jets hurled from their wings. As each pilot pulled back on the stick, his bombs would fly forward, propelled by the jet's momentum and aimed by the angle it was climbing at the instant the projectiles were released. The napalm or high-explosive payloads would fall on their targets an "impersonal" few hundred yards ahead of their launch points; by the time the bombs hit, the planes would be too high into their climbs for the pilots to see precisely where they landed.

Kerry radioed the tactical operations center and politely asked if an air strike had been scheduled in PCF-44's area, and if so, why he had not been notified. His boss replied, "Yes, there is one. Disregard it." And so, despite the jets screaming overhead and dropping their ordnance so close by, Kerry and his crew disregarded it. "I never found out whether my life had been saved by something I didn't know about or whether it had almost been taken by something I did," Kerry averred. "There was a quality about these jets that I just didn't trust. Not too many months before I had arrived in Vietnam, one of the Swift boats had been shot right out of the water by an Air Force jet mistakenly identifying the Swift as a North Vietnamese PT boat. The jet fired one rocket into the Swift, and only the skipper and one crewman survived. The skipper is now disabled for life. I consequently always felt nervous when some jet jockey made a low pass for identification purposes or for pleasure." Later in his second tour of duty, he would make a point of investigating how many U.S. soldiers had been killed by friendly fire since World War II.*

Kerry's first night on Delta river patrol found PCF-44 supporting a Provincial Reconnaissance unit (PRU), one of more than two hundred such eighteen-man squads in a CIA-funded program aimed at destroying the Viet Cong via assassination, kidnapping, and sabotage carried out by Vietnamese mercenaries. Paid four times as much as ARVN privates, plus hefty bounties per kill or captured prisoner or weapon, many PRU members were hardened South Vietnamese criminals given the choice between life imprisonment and joining up; the rest were North Vietnamese and Viet

---

**Estimates of the number of U.S. casualties accidentally caused by friendly forces during the Vietnam War range from 2 percent to as high as 20 percent.*

Cong defectors recruited for their willingness to do anything for a price. The commander of the PRU that Kerry's Swift had been assigned to support boasted that five of his men had been decorated for bravery by Ho Chi Minh himself. Now, for money, they were shooting their former comrades in arms, local peasants, or anyone who broke curfew or trespassed.

This policy of recruiting a mercenary army of proven traitors and vicious criminals instilled deep doubts in Kerry about the eventual success of the United States in Vietnam. The outright encouragement of enemy defections to the ARVN caused great resentment among many South Vietnamese army regulars, who were envious of the money lavished on the defectors by the U.S. government. Why should the regulars fight if enemy deserters came so far ahead? What's more, the regulars never knew if the defectors could be trusted—in fact, many defectors later returned to the Viet Cong. If "Vietnamization" meant stacking the ARVN with enemy sympathizers, it hardly seemed likely that the South Vietnamese would ever be able to sustain, much less win, the war on their own.

Kerry wondered about all this as he and his crew helped the mostly ex-VC of the PRU land in a supposed guerrilla stronghold on his Swift's opening night in the riverine war.

After the PRU had disappeared into the dense mangrove, Kerry beached PCF-44 on the mud a few hundred yards downstream from where the unit had landed. "There we sat silently, waiting to help if called upon," he wrote later. "Hours passed slowly by. Then, late at night, a red flare shot into the sky from the PRU's position. It meant 'Emergency Extraction'—get the PRUs out as fast as possible." Two PBR craft anchored close by also sprang to action. Before Kerry and crew could get their Swift off the mud, "the PBRs had disappeared up a small estuary." The skipper of one was on the radio shouting to headquarters, "Emergency Extraction requested—moving in now—Emergency Extraction requested—moving in to coordinates XZ 0310."

Kerry had scant idea what they were supposed to do. "The disorganization was incredible," he declared. "We had never worked with the PBRs before. The Operation Order given to us that morning contained no contingencies for Swift boats. No one had ever expected that the Swifts would be included in an operation like the one we were participating in." During a planning session prior to the mission, all the PBR skipper had said on the subject at hand was: "If there's an Emergency Extraction we just try to get in there and get them out—okay?"

"Where do we go?" Kerry had asked.

"Up to where they are. Give them fire support if they need it, and get 'em out." Apparently it was that simple.

With that caution in mind, Kerry headed his Swift up the estuary in the wake of the PBRs. He and his crew could hear their engines, but it was so dark they couldn't see where the boats were. Suddenly Wasser yelled, "Over there, Mr. K! Over there!" as he pointed to an even smaller stream that disappeared around a bend.

Then Kerry heard another voice shout, "Hey, skipper! I saw someone move in that hut over there—should I open fire?" It was Gardner, yelling down from the gun tub.

"Where?" Kerry shouted back. Before Gardner could tell him, he looked out the other door of the pilothouse and saw that his Swift was only ten feet off the bank, and that ten yards in from the river's edge stood a long thatch hut with a light shining from it. At that moment shots came from the vicinity of the PBRs. There were a few more M-16 bursts and then the .50-calibers started firing in earnest. Some of the tracer bullets whizzed right over the Swift crew's heads.

Kerry pointed PCF-44 up the small stream after the PBRs. A fishing stake spanned the waterway bank to bank, opening in the middle just wide enough for the slender PBRs to get through but not enough for the Swift to follow. "More tracers came our way," Kerry recalled. "For a moment I hesitated and then I said, 'The hell with it.' We smashed right through the wooden pylons. They broke on contact with the bow. Still, we couldn't see the PBRs. The shooting was sporadic by now. The stream had narrowed so that we barely had room to turn around. I wondered if we should continue when the decision was made for me. We ran aground."

"A few more shots were fired," Kerry continued. "I couldn't tell from where or in whose direction. I ran out of the pilothouse and took the controls at the aft helm, where it was easier to see. Slamming the gears first into reverse and then into forward, we managed somehow to come unstuck. I turned the boat around, hoping that the PBRs wouldn't come tearing around the bend and crash into us, creating a gallery of sitting ducks for whatever had prompted the shooting in the first place. We were lucky. We managed to get out into the larger estuary, where we waited for the PBRs. I cursed the size of the Swifts. Wasser said, 'Shit. I was hoping we could've gotten up there and seen something.'"

Kerry's account of that night's events continues:

> *Eventually the PBRs appeared. They had a sampan in tow and were moving very slowly, confident that the shooting was over for the evening. Hatch nursed the Swift alongside the PBR. I jumped aboard to talk with the Chief Petty Officer in charge.*
>
> *"What happened?"*
>
> *"The PRUs were patrolling through the area when they came on a hut with two people in it. Man and a woman. PRUs went in and found the woman writing a letter to her VC boyfriend. So they took 'em into custody. As they were comin' back they spotted a sampan with four people in it. They took 'em under fire and that's it." It seemed like an every day occurrence to him.*
>
> *"Were the people killed?" I ventured timidly.*
>
> *"Hell, yes. PRUs don't miss when they shoot."*
>
> *"But the people in the sampan didn't fire or anything?" Just shooting them seemed incredible.*
>
> *The Chief talked on. "Doesn't matter. They shouldn'a been there. Besides, one of the PRUs says they had guns but that the sampan tipped over and the guns were lost in the water."*

Then, Kerry wrote, he looked over at the young woman they had detained, "who was squatting in the rear of the PBR." He went on: "She was defiant. She sat very calmly, watching the movements of the men who had just blown four of her countrymen to bits. She glared at me. I wondered about her boyfriend who was fighting us somewhere else. The PBR crew said that the men in the sampan got what they had coming to them, but I felt a certain sense of guilt, shame, sorrow, remorse—something inexplicable about the way they were shot and about the predicament of the girl. I wanted to touch her and tell her that it was going to be all right, but I didn't really know that it would be. Besides, she wouldn't have accepted my gesture with anything but scorn. I looked away and did nothing at all, which was really all I could do. I hated all of us for the situation, which stripped people of their self-respect."

Kerry returned to his boat, and as PCF-44 moved back out to the Soi Rap River, he turned around and saw the PRU mercenaries talking animatedly, no doubt discussing the lucrative killing they had just done. Kerry described the scene: "One of them mimicked the expression and the position that one of the dead had assumed at the instant he had become one of the dead. It had been easy. No shot had been fired at them. Besides, the dead didn't matter at all. They were now just four more casualties of the Vietnam War. The United States would now pay each of the PRUs X number of dollars for the people they had killed. And the total body count was now four higher than before."

The next day, from the air base at Nha Be, Kerry went up in a helicopter to quickly inspect a bunker emplacement his PCF-44 had been assigned to destroy from a river. The reconnaissance flight turned into a medical evacuation. The chopper was ordered to land at a tiny base post near the mouth of the Long Tau River, where a skeletal South Vietnamese soldier was ushered into the seat next to Kerry's. The man had been wounded in the face and his entire head was swathed in gauze soaked a dark red. "I didn't know what had hit him," Kerry stated. "I tried not to imagine. He kept feeling the bandage with his hands, his head lolling around in an uncontrollable manner. Occasionally his head would sidle against my shoulder. I found myself feeling a strangely perverted sense of nausea." It disconcerted Kerry that he was forced to witness this gruesome apparition right before going on a dangerous mission. He was supposed to be on reconnaissance, not medevac duty. "My feelings were irrational, I know, but I couldn't change them," he admitted. "I felt stupidly that by getting hit the Vietnamese had gypped me of a peaceful ride over the Delta, and had reminded me that we were not at all as invincible as we had felt the day before."

The experience may have been a portent; PCF-44's mission against the bunker proved a failure. One of the two helicopters covering the Swift boat suffered a malfunction so both returned to their base, leaving Kerry and his crew alone on the river. They were in no position to carry out the mission without the helicopters; the tactical operations center had read the tide table wrong, sending the Swift in at low tide. As a result, the crew couldn't even see over the banks of the river, let alone spy their target.

On another occasion, PCF-44 was pressed into service as a water taxi to ferry some South Vietnamese troops to their assigned position. With them was a U.S. Marine captain who served as their advisor. The captain was nearing the end of his tour of duty, so the ever curious Kerry decided to ask him to assess his experience.

"What's it been like?" Kerry queried, fishing to learn more about the war than what he had seen himself.

"Kinda varied," the captain replied. "There's no way to sum it up really."

"How would you describe the war?" Kerry pressed. "Do you think we've accomplished anything?"

"I dunno," the Marine demurred. "Been pretty bad, I guess. We had a job to do. We did it as best as possible, I guess."

He clearly didn't really want to talk, at least not to a stranger, so Kerry stopped trying to draw him out. For several long moments the two officers stood silently in the Swift's pilothouse, the Marine staring absently down the river, his brow furrowed as he squinted against the tropical sun. Suddenly he blurted out, "I can't really say it was worthwhile. I mean, I can't see what we've gotten done. We've torn up a lot of villages and killed a lot of people that probably shouldn't have been killed. Then we've lost a lot of good men, too. I dunno. It's hard to say. I sure as hell know that we can't win over here. . . . Nothing to win, anyway. You run through a fucking village cleaning out the VC and then you come back a few weeks later and they're all in the same place again. You walk over booby traps, booby trap after fucking booby trap, and there's nothing you can do about them. Just keep going and hit some more. I dunno. I'll just be glad to get out of here and forget."

"Were you always down in this area?" Kerry asked him.

"No. I was up around Danang for a while," the captain answered. "Then they shifted me down here to take on this advisory bit. Not really a normal way to come about it, but I did. . . . Man, that was a scene up there. We used to sit around on some mountaintop waiting for weather to lift with battalions of North Vietnamese regulars closing in on us. That was hairy. You felt alone out there, just sitting on a hill waiting for the gooks to sneak up and shoot your head off. That was a hell of a setup. But we got out of it—lucky, I guess."

"What do you do down here?" Kerry ventured.

"I've been helping these guys to set up a perimeter defense for their village," the Marine replied, pointing to the ARVN soldiers sitting about Kerry's boat. "But it's harder than hell, because no one wants to sit out on the perimeter and man a gun. They all insist on coming into the village at night because they feel safe. They're really chickenshit."

Soon after that, they arrived at the designated village, where PCF-44 deposited the Marine and his entourage. "I didn't envy him having to stay overnight, but it seemed as though he had seen a lot worse," Kerry concluded.

Indeed, virtually every U.S. Marine in Vietnam had seen a lot worse. "Our greatest accomplishment was holding Khe Sanh in 1968," Marine Sergeant Carter Kirk of South Carolina recalled. "That was Johnson's Dien Bien Phu. The Marine heroism there was awesome." It was nearly impossible to describe the hell most of them went through, charging up steep hills to be taken, then afterward carrying away their countless dead buddies, only to find the North Vietnamese reentrenched back in the same place a few days later. The American public had no idea how often that happened. Nor would the folks back home hear how many of their soldiers had been killed by their own country's artillery and bombers. When they did learn of these friendly-fire incidents "The Law of Increasing Excusable Casualties" always seemed to go into effect, nearly everybody appearing to prefer to forget as quickly as possible just how horrible the facts really were.

One of John Kerry's best friends in the U.S. Navy would be killed by friendly fire, and he has no intention of forgetting about it. The man's name was Bob Crosby and he had been an only child. Although he had lived not far from Kerry, in South Hamilton, Massachusetts, the two never met until the U.S. Navy brought them together. They had trained in the same Swift boat class at Coronado and then traveled to Vietnam on the same flight. Lieutenant Crosby would be stationed in An Thoi for the majority of his tour. He hated everything that went on there, as well as the war in general, but he served faultlessly. "He had been up the Song Ong Doc on a mission when they were hit by recoilless," Kerry recalled. "A piece of shrapnel tore open the throat of one of his men and Bob immediately rushed back to the outpost at the village of Song Ong Doc where they waited for a medevac to arrive. While waiting, the Vietnamese gathered around the boats and snickered and laughed while Bob cradled the head of his dying crewman in his lap. He looked up at them several times, imploring them for some decency and privacy, but the Vietnamese only laughed and snickered and pointed all the more. Bob's boy died in his arms. He used to tell me how impossible it was for him to reconcile going back up a river to try and help those people. I remember how he would shake his head and tighten his mouth when he spoke of it, but he was too soft-spoken to say all that he thought."

Finally, having done his share of the grisliest duty, Crosby had been transferred to the staff of Coastal Division 12 in Danang, as an instructor. On September 24, 1969, when he had only six weeks left in-country, he was brought on a work barge to the Danang pier to indoctrinate the new crew of PCF-13. As Crosby stepped onto PCF-13—whose officer in charge was Elmo "Bud" Zumwalt III, son of the commander of U.S. Naval Forces, Vietnam—a .50-caliber machine gun, with one round carelessly left in the chamber, accidentally discharged. It hit Crosby in the groin at point-blank range. The kidney machine in Danang was broken so he was flown to the Third Field Hospital in Saigon, but by the time he got there it was too late. Crosby died two days later. "It was a bitter pill for everyone in the division to swallow," Kerry recalled, "but no amount of second-guessing was going to bring Bob back."

John Kerry's disillusionment with Vietnam in general and the Navy's riverine war in particular began long before the senseless death of his friend Bob Crosby. For instance, Kerry's notes from November and December 1968 make it clear that while he loved his crew, he was in some ways disappointed by PCF-44 as a vessel, especially with its lack of mobility. And their engines made so much noise that enemy guerrillas always had ample warning of a PCF's presence. Depending which way the wind was blowing, they could be heard as far as three miles away, giving would-be ambushers an even greater advantage than they already enjoyed. "For anyone wanting to smuggle contraband, we actually made the task easy," Kerry noted. "All they had to do was hide in the mangrove or in a small canal until we had passed around a corner. Then they could dart across the river in a matter of seconds. It was no wonder we rarely caught anyone. And we tried everything—sneaking back on our tracks, drifting down with the current, engines silent—everything. You just couldn't win."

At night the problem became even worse. If Kerry called for a searchlight to inspect from a distance, it only made the boat a better target. After a while he resorted to beaching the craft and then shutting down both engines as well as the generator. In silence, he and his crew would lie in ambush for hours, waiting for someone to cross whatever river they were on, which no one ever tried in their presence. Sitting in the mud this way increased the chances that the VC would sneak up on them from the shore of course, but luckily that never happened either. Instead, some of the men on PCF-44 came to enjoy its nighttime patrols best, both for the cool tropical breezes after the sun went down and for the seafood they could catch in the dark, quiet hours. "We had plenty of delicious squid dinners," Wasser remembered fondly. "I would cook them up on a hot plate." Undaunted by the enemy, he would sometimes shine the searchlight into the clear black water in search of night-feeding octopuses; when he spotted one, he would toss a grenade into the river and, boom!—dinner. "Darkness was my friend," Wasser explained. "From my way of looking at it, it was harder for the VC to find us when the sun went down."

Worse than any of the drawbacks caused by the Swift boat's design, however, were the pitfalls attendant upon board-and-search duties. No PCF could be expected to search every junk and sampan that passed by, of course, but by picking ones to inspect at random, the Swift crews invariably missed so many as to make the ratio between stopped and ignored vessels ludicrously unbalanced. The crux of the problem was that some junks were so large or so loaded that it took hours to search them properly; it proved nearly impossible for the Americans to squeeze past everything the boats carried to get into the lower parts of their holds. At one point on PCF-44's first river patrol, Wasser climbed aboard a junk piled from gunwale to bilge keel with sand. He and the rest of the crew started to dig toward the bottom hoping to find a cache of Red Chinese AK-47 rifles destined for the Viet Cong. After about an hour of pawing through the sand, they gave up. In the meantime, twenty or thirty other junks sailed right on by. "You could never get them all," Wasser said. "If you let one go by, which we did often, that could be the one with the contraband on it."

To maximize the number of Vietnamese vessels they inspected, Kerry turned to anchoring in the middle of a river, and hailing the various craft as they came by. If a junk or sampan tried to continue past, pretending not to notice them, one of his crew would fire an M-16 round across its bow. The tactic invariably worked. "One couldn't help but think what it would be like to be cruising down the Los Angeles Freeway or the Connecticut Turnpike, and have a Mexican or a Canadian who was helping the U.S. government search automobiles fire a shot across the front of your car to make you stop," Kerry recorded in his notebook.

Inspecting the locals' identification papers amounted to an even bigger farce. The ID cards they were handed never looked anything like those they had seen in training. "None of us were fluent enough in Vietnamese to read them anyway," Kerry griped. "We used to pretend that we knew what we were doing, though, grunting and nodding officiously, hoping the Vietnamese in question might betray himself. They usually looked at us and laughed."

The Navy provided its Swift officers with blacklists of suspected Viet Cong, but the lists never included the proper accent marks, so—since Vietnamese is a diacritical language—each name on the list could belong to a hundred or more people in any given area. One day Kerry and his crew took an entire ferry into custody because its helmsman's name was on the blacklist and nobody else on the ferry knew how to steer. Wasser stayed on board and held a gun to the suspicious man's head as he steered the ferry back to the U.S. base at Nha Be. PCF-44 followed behind, its crew gratefully eating rice offered to them by a Vietnamese woman who had been taking the ferry. At Nha Be the helmsman's papers were sorted out; he proved innocent of everything but the accent marks in his name, as was usually the case.

It took just one day of board-and-search river patrol for Kerry and his men to regard it as a joke. "And it kept us laughing," he exclaimed. "Once, while we were anchored with about thirty junks around us, another sampan approached our boat too fast, sending people, chickens, paper, everything, spilling into the water. It was one of the only times I ever saw Vietnamese yell at each other." Another time, Wasser was searching a sampan for war matériel when a disgruntled chicken on board bit him so hard he fell over backward into a cargo hold full of vegetables. Its comic relief was welcome, of course, but the sad fact remained that board-and-search duty was an utter waste of time. "That chicken got me pissed off," Wasser recalled. "I wanted to shoot its head off with my 12-gauge. I always kept that shotgun with me when I did searches. I wish I would have bitten the bastard's head off." During Kerry's entire stint in Vietnam he never found a single piece of contraband on a junk or sampan, unless one counts a U.S. military-issue anchor he confiscated from a Vietnamese barge.

Frequently just as ineffective and perhaps even more absurd were the psychological operations (PsyOps) the United States mounted in Vietnam. Concocted in 1954 by World War II Air Force officer turned advertising executive turned CIA agent Edward G. Lansdale, the initial U.S. PsyOps in Vietnam aimed at persuading the country's Catholics to move from North to South and back the Saigon government in the interests of their freedom to worship. By 1965 the inevitable bureaucracy had grown up around PsyOps into the Joint U.S. Public Affairs Office. Set up by the U.S. Information Agency (USIA) in Saigon, it deluged the entire country with anti-Communist propaganda leaflets, and air- and ship-borne pre-recorded loudspeaker appeals. "The PsyOps tapes we were ordered to play on Swifts were ludicrous," Wade Sanders recalled. "One was called 'The Ghost of the U-Minh Forest'! The tape was supposed to be of VC ghosts who had been killed by the U.S. or ARVN in that mysterious region. It was like a bad Halloween tape. A Vietnamese voice instructing them to surrender was intermixed with moans and groans. We thought that the VC who heard it must have had a hard time muffling their laughter. So we ditched the tape and blasted out James Brown and the Stones instead."

The problem was that the North Vietnamese did this sort of thing much better, having a clue about their own culture, and not only swayed many South Vietnamese into the VC but put doubts about the war in U.S. troops via their own version of World War II Japan's notorious Tokyo Rose: Trinh Thi Ngo, the "Hanoi Hannah" whose English-language broadcasts over Radio Hanoi started spewing anti-American propaganda in 1965. Evidence of the relative effectiveness of the two sides' hearts-and-minds efforts presented itself to John Kerry in the form of sailors' utter lack of interest in the riverine war. The United States had given the South Vietnamese Naval Forces (SVNAV) several new Swifts, but Kerry's crew rarely saw them on patrol. The South Vietnamese sailors always seemed to be painting them gray, claiming the paint had to dry out before they could patrol. On the rare occasions that an SVNAV Swift did pass PCF-44 on a river, all the former's crew did was wave as they went by; none ever went on any raids if they could help it. "Generally they were anchored, and everyone aboard was asleep,"

Kerry recalled. "We would pass about two feet from their boat at full speed." Then his crew would watch and laugh as the noise from their engines and the waves from their wake roused their South Vietnamese counterparts, however briefly. "One of their crew always poked his head out of the hatch to see what was happening, and then he disappeared from view," Kerry added. "When they weren't asleep they were usually fishing."

The mild antagonism between the American and South Vietnamese sailors assigned to the Mekong Delta built to a more serious head in An Thoi. Several Vietnamese Swifts had finally been persuaded to go out on a Cua Lon River run, below Rach Gia, but got ambushed. In the heat of the action one of their American advisors—a gunner's mate petty officer—was blown overboard. The boats' crews refused to stop in or even near the area to look for the lost man. In fact, they fled all the way out of the river without even considering it, and decided not to go back to search for him because, as one of their officers put it, "We don't want to be ambushed again." It didn't seem to matter to them that the advisor might still be alive; going back for him was out of the question. Instead they put a call in to the American base, which immediately sent three U.S. Swift boats out on a search at three o'clock in the morning, including Lieutenant Skip Barker's PCF-31.

"It was a particularly dark night, either no moon or overcast," Barker recalled. "The Decca 202 radars clearly painted the riverbanks, but the banks were so uniform that distance traveled could only be estimated by dead reckoning. . . . Eighteen pairs of eyes searched from bank to bank while moving slowly upriver. We detected nothing." For two miles they searched, back and forth, popping flares, retracing every inch of the river where their fellow sailor had gotten lost. The three Swifts had no luck. When they ran out of flares, they gave up their search and began a quiet trip back to An Thoi. They never found the American. "It was hard to care about working with the South Vietnamese after that," Kerry stated. "They refused to help us find our man."

By contrast, the crew of PCF-44 proved ready and willing to proffer emergency aid to the South Vietnamese. On the typically exhausting last night of a four-day patrol, Kerry received a call to pick up a Vietnamese woman in urgent need of immediate transportation to the U.S. Navy base. PCF-44 rushed to rendezvous with another Swift, which transferred her to

Kerry's boat on the Soi Rap River. The woman was suffering an extrauterine pregnancy and appeared near death. She was clearly in excruciating pain, clutching the hands of another woman who came aboard with her all the way to Nha Be as she lay on the deck of the cabin.

Kerry had radioed ahead to the base, where the officer on watch said they would be expecting them. But when PCF-44 arrived they were greeted with nothing but bureaucratic consternation. "You can't bring her here," Kerry was told. "She'll have to go to the Vietnamese pier."

"The Vietnamese pier!" he shouted, dumbfounded. "I don't believe it. Where the hell is that?"

"A few hundred yards further down the river," came the reply, accompanied by a finger pointed into the darkness.

Kerry piloted PCF-44 to the Vietnamese pier as fast as the boat would go. As they drew near, the Swift hit several stumps submerged in the murky water, causing terrifying delays in a situation where human life hung in the balance. The Swift's crew breathed a collective sigh at the sight of a team of waiting doctors who were hovering over the woman the instant Kerry's men passed her stretcher to the pier. The last Kerry saw of her, she was being borne around the corner of a building by a group of soldiers. "We never found out if she lived," Kerry recorded in his notebook. "For the first time, though, some part of the patrol had been worthwhile."

When PCF-44's patrol ended, a day later than expected due to a scheduling snafu, Kerry went to the tactical operations center at Nha Be in quest of information on just how the U.S. Swift boats' missions were devised. The center was shared by American and South Vietnamese naval personnel, and it quickly became clear that the Americans couldn't make a move without first checking with their SVNAV counterparts. The officer on watch clued Kerry in to the dilemma: the South Vietnamese officers didn't seem to know what they were doing or what was happening and didn't seem to care to find out. All they cared about was maintaining their authority. "Whenever you want to get a plan moving," the officer told Kerry, "you explain to them the details of your idea. That takes a while to get through to them. Then you wait for the Viet to say, 'Why don't we do this?'—which is exactly what you proposed to begin with. If they don't suggest it, nothing happens. If they do—which is generally the case—you congratulate them on their good thinking and move on to the next problem." Incredulous, Kerry could only marvel at the Catch-22 absurdity of the way his life and the lives of his men and all the others doing their duty for their country were being run.

Kerry asked the tactical operations man if there were any waterway restrictions on traffic to Saigon. Every night on patrol he gazed longingly upon the bright lights of the big city just seven miles away, as seductive as the City of Oz or Las Vegas glittering beyond the horizon. Saigon had boutiques, bars, black market deals, nightclubs, brothels, and fifty-six thousand registered prostitutes (plus who knows how many freelancers). There were some river-travel restrictions, the officer informed him, but then he confided that there were also a few loopholes, too. On the spot, Kerry decided to exploit those. The key was to have a plausible excuse at the ready should his boat be stopped on its way into the city. "We knew that while we were cruising up and down the river at night, someone was sitting at the bar of the Majestic Hotel in the center of the city, drinking, probably with a girl at his side," Kerry explained. "It seemed wrong. We were jealous at any rate and wanted to share it. So instead of turning right as we left Nha Be after being relieved, we went left, up the Long Tau, and into the heart of the city. We didn't have permission from the division or anyone else, but we felt that we deserved an irresponsible, personal moment, so we did it anyway." Boys will be boys after all, and they just couldn't resist—though the mere attempt to taste the storied capital proved excitement enough as Kerry described in his journal:

> *The world of Saigon, which had been silhouetted nightly by flares lighting the perimeter defenses, suddenly became a real world of freighters tied up to the docks, of countless barges housing refugees, of billboard advertisements, of cars and buses weaving in and out of traffic, of large government buildings dominating the waterfront, hundreds of shacks raised above the mud banks on wooden poles, dilapidated Korean Navy LSTs docked in front of the Vietnamese Naval Headquarters, of water taxis skirting across the harbor, and of one U.S. navy Swift boat parading gallantly through the middle of the city where she had no business being. We didn't stay long. We didn't even go ashore. I chickened out. But for a few minutes Vietnam had shown us a kind of reality we hadn't seen. We vowed we would return.*

But the boyish grins on the crew of PCF-44 faded fast upon their return to base. Kerry's journal continued: "When we reached Cat Lo we were greeted with the news that one of the boats had been hit badly on a raid." It happened on December 17, 1968, as PCF-51 was out on a PsyOps mission, transiting the Bang Cung River (known as Rocket Alley). The officer in charge, Lieutenant (j.g.) Robert Emory, had had to be medevaced. One of his men, John Ramon Hartkemeyer, of Hamilton, Ohio, had been killed.

Born on August 24, 1944, Emory had grown up in Evanston, Illinois, just blocks from the campus of Northwestern University. All-State and talented in mathematics at Evanston High School, he was accepted into the Naval Academy and, in a literature course there, discovered the Greek and Roman classics, which would become his passion. Upon graduating from Annapolis in 1966, Emory was ordered to the U.S.S. *Sylvania*, stationed in Naples, Italy. All things considered, it was a cushy assignment. But Emory's world was shaken on June 8, 1967, when the intelligence ship *Liberty* was hit by an Israeli torpedo in the Mediterranean Sea, apparently but still disputably by accident.

As communications officer on the *Sylvania*, Emory was the first to hear of the *Liberty* disaster, in which thirty-four lives were lost. The *Sylvania* immediately rushed to the *Liberty*'s side, arriving the next morning with first aid and logistical support. "The bombing of the *Liberty* was my initial introduction to combat," Emory stated. "It made me think that perhaps I should volunteer for Vietnam." A second motive arose soon thereafter, sealing the deal. Emory's predecessor as communications officer aboard the *Sylvania*, Jan Gilbertson, had sent him an 8-mm film (of himself motoring around the blue coastal waters of South Vietnam in a PBR, along with a letter describing his first firefight). "Gilbertson had gotten the ribbon on his beret snipped while I was on easy duty in Naples," Emory recalled. "I decided that I too wanted action. So I wrote my detailer requesting an assignment to Vietnam. He wrote me back saying, 'Go to Swift School at Coronado.'"

Emory first went to Vietnam in July 1968 and stayed in-country his full three hundred and sixty-five days. He was assigned to Coastal Division 13 and his first job on the southern coast of Vietnam was working off the LST *Page County* while sharing command of PCF-98 with Wade Sanders, John Kerry's friend from Long Beach. Early in his tour of duty Emory witnessed the horrible mortar accident on PCF-98 that claimed a sailor's life and which Wade Sanders had already told John Kerry about. The death of Boatswain John McDermott remained seared into Emory's memory decades later.

Not long after that gruesome accident Emory was assigned command of PCF-51. It was late October 1968, Operation Sealords was in full swing, and Emory was spoiling for his first firefight. His Swift was in fine shape, despite having taken some mortar fire itself, but it would not stay that way long. Disaster soon happened. "I was on a twenty-four-hour patrol run between the Ham Long and Cua Dia Rivers," Emory related. "The canal I was in was too small to even turn my boat around." Accompanying him on the mission was a Swift commanded by Lieutenant D. C. Current, who just a few months earlier had been shot in the head and hospitalized in Japan; he was now back in action, if rather more cautious than before. "Together our boats came out of the dense jungle and into a wider part of the canal," Emory recalled. "There were suddenly rice paddies and humans farming."

Part of their mission, as usual, was to play the PsyOps propaganda tapes over the loudspeaker whenever they saw Vietnamese within earshot along the riverbanks. So Emory ordered Boatswain's Mate John Hartkemeyer to rush below and put on one of the anti-Communist USIA tapes. "As he went underneath, I was standing at the portside door," Emory said, describing the scene decades later from his home in Vienna, Virginia. "Suddenly a rocket-propelled grenade was fired from forward of the boat at the pilothouse. The grenade actually made contact with the boat just behind my legs." Rocket-propelled grenades explode on contact rather than in the air; the VC sniper's obvious intent was to blow up the metal on PCF-51 into its own shrapnel, and succeeded. Emory's body was instantly riddled with metal fragments. "It looked like somebody had taken an extremely coarse pepper grinder and ground pepper up and down the backs of my legs," he remembered. "I was covered from ankle to buttocks with shrapnel."

The rocket slammed into PCF-51 just behind the portside doors of the pilothouse, skidding diagonally across the boat and exiting at the right starboard corner of the main cabin—which was where Hartkemeyer was standing below playing the PsyOps tape. In searing pain from the molten shards burning into his flesh, Emory managed to grab the radio and shout, "We are under fire—under fire!" By this time both of the Swift's gunners were blasting away on their M-60s in the direction the grenade had come from.

Meanwhile, the boat's engines were failing. Luckily, Current came to their aid. Wounded but still in charge, Emory ordered Boatswain's Mate Third Class Steve Bredenko to drop his gun and go check on Hartkemeyer. A long minute passed before Bredenko emerged at the top of the stairwell, ashen. "I can't help him," he reported to his injured skipper. "His head is gone." The two men just stared at each other.

Anxious to get both his and Emory's Swifts out of harm's way, Lieutenant Current gave his colleague a couple shots of morphine and ordered him medevaced out of the canal by helicopter. "By the time I got to the field hospital I was morphine-inebriated," Emory remembered. "A nurse hovered over me on the operating table and all I remember saying is 'Ahhh . . . a woman.'" After a successful operation, Emory—who would be awarded a Purple Heart and a Bronze Star—was sent to the hospital at Binh Thuy for three weeks of recuperation. Upon his release, he was taken off Swift boat patrol and made maintenance officer at Cat Lo. While still in the hospital Emory wrote a sincere condolence note to Hartkemeyer's family in Ohio. He never received an answer. "The parents refused to accept that it was their boy John," he explained. "They refused to believe, because the corpse was headless." As for Bredenko, although the nightmarish sight of his decapitated crewmate never left him, he had the satisfaction of having killed seven Viet Cong with PCF-51's M-60s, including a regional commander—or so the Navy claimed.

After he heard the awful details of the incident, John Kerry went to see PCF-51, which had been towed back to the base at Cat Lo. "A huge hole ran through the pilothouse and into the main cabin," he wrote in his notebook. "While I stared at the boat several men were on board, cleaning up. One of them turned to me with a look of disgust as he picked some hair and teeth out of the ceiling. The pleasure of Saigon and the feeling of invincibility disappeared. It was all very real again. I figured we were next."*

*****_PCF-51 ended up suffering more fatalities than any other Swift boat in Vietnam. During the brutal months from December 1968 to May 1969, PCF-51 lost five crewmen including John Ramon Hartkemeyer; Quartermaster Third Class Thomas Eugene Holloway on April 12, 1969; Engineer Third Class Dewey R. Decker on April 15, 1969; Gunner's Mate Third Class Richard W. Stindl on May 15, 1969; and Boatswain's Mate Third Class Robert A. Thompson on May 19, 1969._

Chapter Ten

Death in the Delta

As 1968 drew to a close, American troop strength in Vietnam reached a new high of 540,000. Over the course of that year, combat claimed the lives of 14,592 U.S. servicemen and -women, as well as nearly 28,000 South Vietnamese military personnel. John Kerry and his crew had a great deal of company in wondering whether they might be next to die. "You'd have to be a fool not to fear for your life every time we took to a river," PCF-44's Drew Whitlow recalled. "On every patrol—particularly at the start—we thought about death. I kept my little Gideons Bible on hand *all* times. God had taught me that I wasn't going to live forever. Still, my stomach had butterflies. But we tried to bat fear away by keeping alert and praying and working together. We all knew that you're most likely to get a bullet when we weren't paying attention."

Christmas Eve, 1968, turned out to be memorable for the men of PCF-44 though not in the jingle-bells sense folks were enjoying back home. The only concession to the holiday spirit was that morning's rare breakfast of scrambled eggs, after which the crew headed their Swift north up the Co Chien River to its junction with the My Tho only miles from the Cambodian border. Because they were only an hour away from that neighboring country, Kerry began reading up on Cambodia's history in a book he had borrowed from the floating barracks in An Thoi. He learned how the great Khmer Empire, which lasted from the seventh century to the fifteenth, grew to reign over southern Vietnam, Laos, Thailand, and Burma. Its magnificent capital was Angkor Wat, until the empire began to crumble under Vietnamese invasions beginning in the thirteenth century. He learned of Cambodia's post-Khmer history of invasions, by nations from old Siam to modern France, and how they had splintered its people. He even read about a 1959 Pentagon study titled "Psychological Observations: Cambodia," which, as William Shawcross later noted in his book *Sideshow: Kissinger, Nixon and the Destruction of Cambodia*, painted the nation's inhabitants as "indifferent farmers, incapable traders, uninspired fishermen, [and] unreliable laborers." The Pentagon report went on to state that Cambodians "cannot be counted on to act in any positive way for the benefit of U.S. aims and policies."

Lieutenant Kerry kept that last point in mind as PCF-44 patrolled the watery borderline between Cambodia and Vietnam. After all, Norodom Sihanouk himself—installed as king of Cambodia in 1941 during French rule, and after it, in 1955, as president—complained to the White House about Mike Bernique's cowboy antics on his country's rivers. Cambodia's history made Sihanouk nervous about any breach of its territorial integrity. He insisted upon remaining neutral, desperate not to anger either the North Vietnamese or the United States. President Sihanouk just wanted Cambodia to be left alone, pleading that "We are a country caught between the hammer and the anvil, a country that would like to remain the last haven for peace in Southeast Asia." Raids through his rivers like Bernique's were not conducive to preserving Cambodia's official neutrality.

Kerry was keyed up that Christmas Eve by PCF-44's position; he had never been that far up the rivers before, in the upper reaches of the Mekong Delta where the brown-water navy earned its stars and bars. He and his crew were *deep* in enemy territory. "The Swift seemed to sense the spirit of exploration as she kicked a symmetrical wave out behind us, breaking an even spray to both sides of the bow," he wrote in his notebook. "It was early morning, not yet light. Ours was the only movement on the river, patrolling near the Cambodian line. We moved along quickly, trusting the radar sweep, which illuminated islands, junks, and sandbars protruding from the water."

A PBR swept by them on one side. The Swift looked very large and well protected next to the small-hulled vessel. In fact, the PBR presented less of a target, as it was painted green to blend in with the elephant grass and mangrove of the surrounding jungle. It was also vastly more maneuverable than the PCF, capable of turning on a dime at full speed. "We chatted briefly with her crew," Kerry recalled. "The patrol officer warned us of a sandbar just ahead, and I congratulated myself for stopping to accept advice about the regions we had never navigated. Had we continued on our course prior to passing the PBR we would have run straight into the bar, which was obscured under a foot or so of water."

Kerry recorded his thoughts that day in a philosophical, second-person, present-tense voice:

> *Everything around you is quiet and the only a [hum] breaking an otherwise still Southeast Asian morning is the now-high whirl of your engines. All across the river, in splotches of green, are pieces of mangrove that have eroded away from the banks and which are now plying a drifting and uncertain route with the tidal current that sweeps through the [river]. It makes you think of the story of the wooden seagull that followed the air currents of the world and that saw the movements of all the world's people below its graceful and motionless wingspan. You wish that you could be transformed into that itinerant nothingness that lets you watch the world pass by with all its gross trimmings but which demands nothing of you. To be free so that you can comment or not comment as you see fit and then just hop on a breeze and be blown restlessly to some new horizon with new hope and strength.*

PCF-44 arrived much earlier than expected at the River Patrol Force LST stationed at Long Binh. The crew refueled quickly and stole a hasty morning snack of rice and tea. From the Task Force 116 barracks next to the LST, Kerry watched, intrigued, as the Cobra gunships and troop-carrying helicopters of the Air Cavalry flew in and out of the nearby airfield.

In a later notebook entry, Kerry would assess the notorious derring-do of the American chopper pilots in Vietnam. "The Cobra gunships, thin rocket-carrying helicopters, were a phenomenon of the war," he wrote. "They were instrumental in helping countless numbers of Americans out of predicaments, out of ambushes and battles, but they were driven by jockeys with a reputation for being stark raving mad. The pilots were mostly 18- and 19-year-old Army Warrant Officers who received their commissions with their training to fly the gunships, directly after they finished high school. They had guts, the kind of reckless guts that kills a lot of men and makes heroes out of the survivors. Their performance was always in the best traditions of war." Yet Kerry found himself troubled by the realization that some of these wild boys could kill innocent people from a distance and not be bothered by it. He was no pacifist—far from it—and understood that "accidental atrocities" were a part of war. But it disturbed him that even a few trigger-happy American yahoos considered killing Vietnamese civilians a sport. As he confided to his journal:

> *It seemed incredible to me that we could so freely give them rockets and machine guns to fire with life-and-death—usually death—determination over people traveling in the rivers. Later in my tour, I was to work with two Cobra gunships that were to cover us on an operation. We specifically asked the pilots to remain close to us, restraining their trigger fingers so that if we were ambushed they would have plenty of ammunition. During the mission one of the pilots sighted a sampan and without asking permission, radioed to say that he was going to check it out. The next thing we knew he'd blown it out of the water because it was in a Free Fire Zone. Later, having used much of his ammunition, he requested permission to leave in order to rearm, an operation that left us uncovered for more than forty-five minutes in an area where cover was essential.*

But Kerry wasn't focused on the chopper jocks that morning of December 24, 1968. By the time his Swift was finished refueling at Long Binh, it was fully daylight. Once he and his crew got under way, barges, junks, and sampans of every shape and size moved with them up the river. The barges burst with the bright colors of freshly washed clothes draped across them to dry in the sun. Women did their domestic chores atop the roofs of the cabins while their children played on the decks. The barges glided with a silent grace that underlined the simplicity of everything around them. The crew of PCF-44 couldn't help but feel a certain peace, though they knew it couldn't last. Soon they came upon the small canal on the left that paralleled the main river and led to their destination: Sa Dec, a town that had been ravaged during the Tet Offensive. They turned into the canal and entered an even lovelier and more peaceful panorama.

Emerald-green rice paddies stretched away from them on both sides of the placid waterway. The morning sun turned the landscape brilliant with rich colors. Huge, muscular black water buffalo rubbed their backs lazily against the trunks of the trees that grew from the canal. Kerry felt criminal disturbing this serenity with his boat's chugging engines and the ripples from its wake. He slowed the motor to its minimum revolution and guided the Swift quietly through the mist. "It was so serene," Radarman James Wasser remembered. "It took your breath away."

Kerry brought the boat to a full stop for a moment to let a small boy in a sampan herd a string of water buffalo through the canal in front of them. The child sidled his craft up to the rear of the last animal in line, raised his arm, and brought the switch in his hand down hard on the beast's hindquarters. The buffalo grunted and moved ever so slowly ahead. "I'll never forget that little boy herding these seven to ten buffalo," Wasser said. "He was so young, yet he was their master."

A few moments later the ambience changed as PCF-44 arrived in Sa Dec, an active little town situated right on the canal. Women hawking and haggling over various wares scurried back and forth through an open-air marketplace lined with shanties ladling out regional specialties like *pho*, a hearty beef noodle soup. South Vietnamese flags fluttered from trees, houses, and poles, lending the village the air of jousting tournament day in Arthurian England. The entire life of the town seemed to be jammed into a few hundred yards along the banks of the canal. "Make no mistake about it," Wasser exclaimed, "Sa Dec was a hot zone!"

PCF-44's crew headed to the middle of town, to the tactical operations center that ran the local river patrol unit. Kerry described the scene in his notes: "As we walked through the marketplace, children ran around us yelling, 'Number One—you Number One—okay, okay,' all of which we thought voiced great approval for your presence. Then they showed us why they approved by stretching their arms toward us, waiting for a handout. They were like all the children who watched us from junks and hovels as we went by in the Swift. Their tiny faces stared at us, wide-eyed and wondering, sad and bewildered, knowing death but not knowing why, knowing life but living it half dead."

Kerry and his men left Sa Dec as soon as they completed a routine intelligence investigation of village leaders. To get back into the main river from the canal, they had to negotiate their way under a rickety drawbridge, which boats could clear only at low tide. The moving part of the bridge drew up at an angle, suspended from only one end, which meant PCF-44 would have to make it under with its radar apparatus just inches from certain destruction—and the investigations and court-martials that would ensue in that event. But they didn't have time to wait for the right tide, and decided out of cockiness to give it a go anyway. A crowd gathered to watch the Americans smash up their boat. They cheered in admiration when the Swift somehow managed to squeeze under the drawbridge unscathed.

Bursting with self-congratulations, Kerry and his men sailed farther up the river. At a bend just as they were approaching the Cambodian border, two PBRs met the Swift. Kerry's crew stopped to talk to the men about their plans for the night's patrol as well as about the three-day Christmas truce scheduled to go into effect in just a few minutes, at six o'clock that evening. Kerry's notebook detailed what happened next: "Suddenly there is an explosion and a mortar lands on the bank near all three boats."

"Gimme the binoculars!" Kerry yelled to Stephen Hatch. The boatswain's mate passed them up. Kerry scanned the bank but saw nothing.

"Where the hell did that come from?" Wasser asked.

Kerry radioed headquarters that someone had fired a mortar round at them. He inquired whether any friendly troops were operating in the area. "Negative" came the response. The mortar fire must have been one of the Viet Cong's never-ending sniper attacks.

Kerry was still on the radio when another round crashed into the water only fifteen yards away. All three boats sprang into action. "Scramble, goddamnit," shouted several voices at once. "Get the hell out of here!" The Swift got up to full speed before the two PBRs could even untie their lines. While Kerry radioed back to headquarters that they were still receiving fire, the trio made a sharp right and sped up a river behind the area the mortar rounds had come from. Two men came running down the bank, yelling and waving creaky old rifles. They motioned to the boats to come in to the shore. They complied, gunning their engines and charging forward, not caring whether it was sand or rock or mud in front of them.

Wasser moved up to the bow and in his best Vietnamese asked the men what they were yelling about. The pair told him that a squad of VC had attacked their village, wounding one man, and then had moved down toward PCF-44's position. Kerry glanced at his watch and noted that the Christmas truce had begun—supposedly—three minutes earlier. "We were in real trouble," Wasser recalled. "We were the most-inland unit in the whole damn Navy."

From his post in the gun tub, Stephen Gardner called out, "Hey, skipper! I see a couple of sampans down there—back toward the river." Kerry backed the Swift away from the shore and moved slowly toward the inlet the sampans had disappeared into as two villagers watched them from the bank. PCF-44 couldn't squeeze into the stream after them, so they waited nervously at its mouth for the PBRs, which they would cover as the slimmer boats headed into the inlet.

In his journal later that night, Kerry wrote, "You look around and hear your own breathing, smell the heat in the air, see your men now in flak jackets and battle helmets and ready for what might come. The PBRs arrive and then again, from the bushes, we start to take sniper fire, small arms weapons that can kill but which you feel is a ploy by Charlie to bring the fire in." Kerry ordered his men to hold their fire; he was not going to let the VC lure his boat into killing any innocent villagers onshore, much less the men on the PBRs. "Charlie had .51s, we had .50s," Wasser later griped, speaking of the Viet Cong's relative firepower. "Charlie had an 82[mm] mortar, we had the 81. Those bastards—Charlie was always a millimeter ahead of us. It ticked me off."

What happened next ticked Lieutenant Kerry off too, as he expressed in his journal:

> *Suddenly, in a flash, that is a moment of hell and blindness, the reeds erupt and bullets walk out across the water at your boat and those around you. Then screaming flashes of tracer, greenish-yellow,* and deadly, come at you with a terrifying suddenness that catches all by surprise and you watch for a moment as red streaks move at you in a three-dimensional kaleidoscope and with equal suddenness your hand jams down on the throttles and the boat leaps out of the water. From PBR and Swift a cacophony of explosions as they answer with anger, shame, and surprise the wall of fire that met them. Quickly, too quickly, you are past the ambush point and you wheel your boats around to run back and out into the main river. From somewhere reason calls you and you grab the loudspeaker and yell to your men to hold your fire until right on top of*

**U.S. rounds used red tracers; VC and/or North Vietnamese tracers flashed greenish.*

the spot and then there is thunder again and no hearing and only red streaks tearing towards the land. You are in the river and away and you slow your boat. The PBRs are with you and you stop to catch your breath.

In the heat of the moment, all of the men of PCF-44 had been a bit less philosophical, their remarks along the lines of "Jesus fucking Christ! Did you see that?" and "Hot damn—that was somethin' else!" They were elated. They had come out of their first major firefight unscathed. Kerry now knew exactly what Winston Churchill had meant back in 1898 when he wrote of his combat experiences with Britain's Malakand Field Force the year before: "Nothing in life is so exhilarating as to be shot at without result."

The same did not hold true for the best of those who *did* the shooting *with* result. Still giddy, one of the Swift's enlisted men asked, "Hey, skipper—did you see that one guy flip over when the .50-caliber hit him? I think he was one of those farmers, or somethin'."

"I didn't see him," Kerry answered. "Where was he?"

"Standing on the bank during the first pass. I think the PBR in front of us hit 'im, 'cause when we got up to him he was lyin' flat in the field."

All the while his crewmates were celebrating their survival. Wasser remained silent. He was in no mood for crowing, as he realized that *he* had killed the old man who had been leading a water buffalo in the line of fire. "Christmas has never been the same," Wasser confessed thirty-five years after the fact. "Two ARVNs also got badly shot up. I'm haunted by that old man's face. He was just doing his daily farming, hurting nobody. He got hit in the chest with an M-60 machine-gun round. It may have been Christmas Eve, but I was real somber after that. We did nothing wrong. But still, to see the old man blown away sticks with you." Some U.S. servicemen tried to assuage guilt over the occasional accidental killing of an old Vietnamese man or woman the same way nineteenth-century U.S. cavalrymen dealt with the deaths of aged Indians at their hands on the Great Plains: they convinced themselves that Native Americans weren't really people. In 1960s Southeast Asia, reference would be made to an old Chinese game played with forty tiles representing different animals. The thirty-fifth tile was the *ba muoui lam*—an "old goat." The U.S. military's tacit position held that

American soldiers needn't worry if a thirty-five went down by mistake—but Wasser did.

Kerry had not seen the old man get hit. Yet when Wasser told him about the civilian casualty, he certainly felt bad. PCF-44 had only its crew on board, and no troops to send ashore to ascertain whether the old man was indeed dead. "It was crazy," Kerry recalled. "We had perhaps immobilized several VC and knocked out their machine guns, and we couldn't even find out. [It was] possible we hadn't done a thing. But for all the shots exchanged, we would never know what had happened. We did find out that two friendly troops who had moved into the area without our knowing it had been wounded—by whose fire, again we didn't know. The PBRs went in to the bank of the river to medevac a half-dozen wounded. We sat out in the river covering them. It was dark by this time."

After the medevac left, the Swift set about rearming the PBRs. As they bobbed on the river, passing ammunition and supplies back and forth, twenty rounds of sniper fire suddenly came their way from the other bank, opposite from the site of the last ambush. The rounds fell short of the three boats, however, so nobody gave a damn. "Charlie was always taking potshots at us," Wasser recalled. "It was part of the game."

This holiday was not proving very happy for the Delta river patrol crews. On their way back to Sa Dec to make their reports, "the night darkness is broken by tracers flying up and out of a Vietnamese outpost that is celebrating Christmas," Kerry recorded in his journal. "The bullets pass dangerously near your boat and you think of the stupidity of this whole thing and the ridiculous waste of being shot at by your own allies and so angry you jump on the radio and ask who the hell is shooting at you and inform your seniors that they had better [get it] squared away before you return fire. Apologies are quick to come but they mean nothing amidst all the chaos and waste."

The PBRs had initially intended to drop the rest of the wounded men off at the ARVN outpost, but because of the celebratory shooting they couldn't get close to it. One of the skippers called headquarters and barked, "Please have those goddamn gooks stop shooting so that we can complete the medevac." The plea was to no avail. Finally, the chief petty officer on one of the PBRs flat refused to try to get in to the outpost's pier.

So the three vessels continued on to Sa Dec. When a few more bullets whizzed their way, Kerry sent up a flare to identify their boats as friendly. The South Vietnamese shot at it. PCF-44's entire crew asked if they could fire back, but Kerry said no. Once again he got on the radio to tell headquarters what was happening: "Look, the Vietnamese are shooting at us. It seems that they don't really want to stop. Do you think that it would be beyond reason to call them and ask them to stop? Thank you; out." Again the watch officer apologized but the shooting continued until the boats gave up and got out of range of the outpost.

By the time they reached Sa Dec, the air had cooled off considerably. "The night for once is comforting," Kerry wrote, "and you take a Coke and some peanut butter and jelly and go up on the roof of the cabin with your tape recorder and sit for a while, quietly watching flares float silently through the sky and flashes announce disquieting intent somewhere in the distance."

Silhouetted by the lights shining from the junks parked along the canal, Kerry could see empty fishnets swaying from their teak poles in the gentle breeze. He realized that the nets would not be filled that Christmas Eve, or any other night as long as the war ground on; nighttime meant curfew for the South Vietnamese, by a law—American law. It was their country, but the United States imposed and enforced the edict that at night anything that moved could and would be shot, and damn the consequences. After all, one could never know if a movement in the dark was innocent or hostile. All a PCF officer could know was that somebody, for some reason, was breaking curfew.

Inspired by these reveries, Kerry wrote in his notebook: "You call down to one of your men and ask him to draft a message to the Admiral in Command of all Naval Forces in Vietnam and also to the commander of market time. It says, 'Merry Christmas from the most inland market time unit.'" He meant to be clever and to point out to his superiors the incongruity in a U.S. Navy Coastal Surveillance Force boat crew spending their holiday on a river canal not far from the Cambodian border northwest of Saigon. The message had to be sent in code, of course, so as not to compromise PCF-44's position. "You hope that they'll court-martial you or something," Kerry continued, "because that would make sense," or at least more sense than everything else that had happened that day. "We were getting close to Cambodia," Wasser explained later. "We were out there all alone in the darkness."

For the first time in weeks, Kerry dreamed of home, of his father reading the *New York Times* in Groton, of his mother festooning a fresh spruce with ornaments while his siblings tore open gifts and laughed. "No snow, no sleighs, no fat jolly Santa Claus with frosted lips and frozen hands and an outstretched arm that begs for the little money that people have at this time of year," Kerry wrote home that night. "Indeed, there is no familiarity with the date." Yet a gentler tone still managed to creep into his journal. "But the night soothes everything," he penned, "and the people and things that are close to you dart through the mind and bring the only warmth and peace that there is. Visions of sugarplums really do dance through your head and you think of stockings and snow and roast chestnuts and fires with birch logs and all that is good and warm and real. It's Christmas Eve."

On the morning of December 29, Lieutenant John Kerry and his crew nearly lost one of their own in a small canal just off the Co Chien River. They had been probing the waterway with another Swift boat on a minor Sealords raid, and on their way back out had come under fire. "We went into a dangerous area that had numerous hooches and sampans," James Wasser recalled. "The enemy was thick. Once we got in the canal we took a lot of small-arms fire, followed by mortar. Our adrenaline was racing; we went right back at them with all the firepower we could muster. That's when Gardner got hit."

Just before the shooting started, Gunner's Mate Stephen Gardner had shouted: "There's somebody running over there. . . . He's got a gun . . . on the port side—on the port side!" PCF-44's crew had been firing at thatched huts on their way out of the canal, and the reports of their own guns had muffled those of the shots being fired at them. Suddenly, Gardner shrieked, "I'm hit!" and stopped firing for a moment. Before Kerry could ask his condition, Gardner shouted from his post: "I'll be okay," and went back to firing his two .50s. After the Swift exited the canal, Gardner came down from his gun tub. Sure enough, blood trickled down his arm from where the bullet had pierced it just above the elbow. His crewmates patched him up and then started for the primitive base at Dong Tam, which had the best medical facilities in the area—which wasn't saying much. "In truth, Dong Tam was a hole in the wall, but it was all we had," Wasser explained. "It wasn't much of a village, but it's where we got ammo, mended wounds, and took fuel."

Gardner was the first man under Kerry's command to get shot. Given how narrow the canal was, however, Kerry and his crew still felt extraordinarily lucky. The round that hit the gunner's mate came from an AK-47, the Chinese-made automatic rifle that dispensed "little pieces of copper-jacketed death," as Philip Caputo put it in his 1977 Vietnam memoir, *A Rumor of War*. "Any remnants of the stupid feeling of invincibility which had been with us in the Long Tau completely disappeared with the sight of his small, fleshy wound," Kerry recorded. "None of our earlier actions, including my shrapnel episode outside of Cam Ranh Bay, had the same sobering effect as having been shot at within the confines of that small canal and then hearing the yell—'I'm hit'—and seeing blood." Kerry averred, "We had learned something. We learned that pieces of metal flew at one indiscriminately. One time they could land in flesh and one could laugh. The next time it would be the brain or the eye. We were to learn that lesson several times over."

No matter how mentally prepared a soldier may feel for combat, no boot camp can ready a human being psychologically for the gut-wrenching sight of a healthy young body torn apart by violence choking its last breath and dying. As Sergeant Thomas Oathout of the 173rd Airborne Brigade summed up in a letter from Vietnam to his mother back in Delaware: "To shoot and kill somebody, turn your head and walk away, isn't hard, it's watching him die that's hard, harder than you could imagine, and even harder when it's one of your own men."

While John Kerry had heard about the death of many men in Vietnam, including his dear friend Dick Pershing as well as others he had known in the States, he had not seen it for himself firsthand—that is, until the day he rushed Stephen Gardner to the U.S. Army's Third Surgical Division. Upon their arrival at the makeshift hospital at Dong Tam they came upon a handful of seriously wounded young South Vietnamese soldiers. What Kerry—with Wasser at his side—experienced while waiting for Gardner to be treated would numb some part of him forever, as he recorded in his war diary:

VIETNAM: 29 DECEMBER 1968

He was Vietnamese. I didn't know his name. Nobody in the tent did, I think. He was completely nude and his bony, minute body was stretched out on the brown plastic mat covering the operating table. Figures in green pushed in and out through the two doors which marked the pre-operating section of the Third Surgical Division, U.S. Army. An eerie, fluorescent light shone down on his face. His chest moved up and down without rhythm and with very little strength, sucking breath in slow gasps. My eyes darted between the operating table and a huge plastic tube for air-conditioning which ran across the overhead. It dominated all the other sceptic [sic] trimmings of the emergency ward.

I watched while a young medic worked to prepare a fourth pint of blood for transfusion. With a pump like those used to take blood pressure, he would squeeze the blood through a plastic tube and into the half-dead Vietnamese. Now and then the Vietnamese's feet would twitch and his arm would try to move up towards his head, movements which were strangely disconnected from the rest of his body and from normality.

I will call him Nguyen. He was a Tiger Scout, a forerunner for one of the platoon of infantrymen at Dong Tam, the Ninth Infantry headquarters. Someone whispered to me that he had been hit by a booby trap. Someone else said it was gunfire. No one knew, obviously. From where I was I could see his neck bleeding. His head was arched back and his eyes, only half open and dazed, were searching for something. There was nothing close here for this man—his was a moment of complete loneliness, I thought. No one to hold on to. No one to talk to, because he could not speak English and we could not speak Vietnamese, and how, anyway, does one bridge the gap at a moment like this?

His left hand was wrapped in gauze. The gauze had turned almost completely red. A pool of blood had gathered on the table below the green army stretcher on which he lay. Everywhere there was blood pouring out of him. Even the transparent, plastic splints around both legs assumed a red tint. I felt weak. My stomach began to twist and sweat poured all over me. I sat down on the floor because I thought I was going to be sick.

Suddenly Nguyen's right arm moved straight out, grasping towards the door. He grunted desperately. A doctor quickly took his

> *pulse and his blood pressure. His toes, sticking out from [the] plastic splints, twinkled back and forth. He tried to raise his head and look—perhaps ask something—perhaps a last twinge of fight—and then he was quiet. His right hand, still reaching, came down slowly onto his chest, and his other arm, bandaged and absent, lolled over the side of the stretcher. Nguyen was gone. No words. No cry.*
>
> *It seemed absurd—a man dying alone in his own country. I wanted to cry but I thought that I couldn't let myself and so tears just welled up in my eyelids. Now I wonder why I didn't and I'm sorry.*

Nguyen's corpse was carried out of the room. A nurse came in and patted large wads of gauze into the pools of blood darkening on the table. When the fluffy white wads turned to sodden red sponges, she tossed them into a rubber container and readied the table for the next man. A medic then escorted Gardner from the room. Having witnessed part of Nguyen's struggle to live had made PCF-44's gunner's mate queasy as well. He felt relieved to get out of there and to another room to have his arm x-rayed.

Three and a half decades later, when asked about watching Nguyen die while Gardner received treatment, Kerry turned somber. "It hardens you a little bit," he replied. "It reduces the mystery a little bit. You have to learn to sort of compartmentalize it and shove it away—put it in a different place so it doesn't affect you. But it always does affect you, and you're always confronting this ominous question of 'one moment here and [the next] moment gone.' But it almost conditions you to it: this is where we are and this is what's happening. You just get through it. Think of the plight of the people that got through the Holocaust; think of the people that survived Dresden or Stalingrad or D-Day—you go through it, and the human psyche has extraordinary ways of helping you weather those things. Suddenly, events like Nguyen's death became part of the experience of being among the living, and I think you're a little less scared of being among the dead."

At Dong Tam, after Gardner's X rays were taken, he and Kerry were directed across the helicopter pad to a building that housed an enormous outpatient clinic. While a surgeon cut into Gardner's arm to remove the metal fragment, Kerry walked around and watched the streams of wounded coming in. The lines were as long as those they had all waited in for physical examinations at boot camp. Here, however, instead of unscarred young recruits, the patients all had blood-splotched bandages wrapped around their heads, legs, arms, chests, or stomachs, making the scene as grim as the goriest pages of *The Red Badge of Courage*. "At the entrance to the building were boots and combat packs, which were covered with mud, still wet," Kerry recorded in his notes. "Some of the mud was mixed with red streaks of blood. Only minutes earlier those packs and boots had been stalking Viet Cong in a rice paddy or along a path a few miles away. Now the men who had worn them were waiting in the building to have someone dig around in their muscle to pull out pieces of grenades and bullets."

While Kerry was waiting, Nguyen's remains were carried in. Kerry watched in silence as the medics tugged and pulled at the corpse to get it into one of the dark green rubber body bags used to transport the dead from battlefield to morgue. The body, like all the bodies Kerry would see in Vietnam, had taken on a surprisingly cumbersome waxy quality; it rolled "uncontrollably" as the medics tried to put it in the body bag. Then they zipped it up and Nguyen was gone, driven away to be buried in some forlorn rice paddy along a riverbank somewhere in the land he had died for.

A depressed Kerry went outside and waited near the helicopter pad while the doctor finished with his gunner's mate. A chopper landed and a triage unit scurried about assessing the wounded—two of them screaming for more morphine—then taking them in order of severity to the building Kerry had just exited. The worst off were borne on stretchers to the building where Nguyen had died. Somewhere on the base Kerry heard artillery starting to fire. A steady series of explosions shook the hospital buildings and the ground he stood on. Eventually he had enough of the Mobile Army Scenic Hospital (MASH) scene, and returned to the room where Gardner was being treated. "When the doctor finished sewing Gardner's arm, we left for the Swift," Kerry recalled. "Gardner was all right. No heavy lifting for a few days, but otherwise totally all right. He kept the piece of metal for a souvenir."

On the way to the hospital at Dong Tam that morning, PCF-44 had passed some twenty vessels of the Army-Navy Mobile Riverine Force. The combined effort presented an impressive armada. Some of the Navy's Task Force 117 craft were so heavily armored their hulls could stop a B-40 rocket fired at close quarters. Others carried platoons of troops from the Ninth Army Division's Riverine Forces. If ambushed, together they could conduct an assault on the offending riverbank and secure territory as they advanced. "We were astonished to learn they were going into the very canal we'd been ambushed in," Kerry remembered. "It didn't make sense. Two one-quarter-inch-aluminum-hulled boats get shot at in a canal and have a man wounded, only to see an armada of troop-carrying assault craft go in the same canal a few hours later. The men on one of the assault craft told us they would never consider going into that canal without at least ten or twelve of their armored vessels for an escort. They told us we were crazy to have gone in there."

Luckily, the consequences had not been too dire, and the Dong Tam doctors had done a good job on Gardner's arm. It almost made up for the disappointment Kerry's crew had suffered at Dong Tam. Then, on the way to their patrol station, they had stopped for fuel at one of the LST stationed in the middle of the My Tho River a few hundred yards from Dong Tam. The officer who refueled PCF-44 mentioned that most of his colleagues were at the Dong Tam base attending a "Bob Hope Follies" USO show. The legendary comedian was on a USO East Pacific tour, performing for U.S. troops in Japan, Korea, Taiwan, Thailand, Okinawa, Guam, the Philippines, and Vietnam. His troupe included actress Ann-Margret, former Los Angeles Rams football star Rosey Grier, and swing band leader Les Brown. "I'd planned to spend Christmas in the States, but I can't stand violence," Hope joked at the beginning of his monologue. "Besides, it was the perfect time to come to Vietnam—the war has moved to Paris." The Bob Hope extravaganza had been going on for about an hour, the officer told the crew of PCF-44, so if they hurried they might be able to see the finale, which promised to include some gorgeous female dancers known as the Gold Diggers. "We needed no second invitation." Kerry smiled, looking back. "Ann-Margret looks good in most circumstances, but after a few patrols she looks like everything you've ever wanted. With visions of her and the rest of the troupe leading us on, we headed toward the base. Unaware of the docking facilities, we went straight down a canal that seemed to parallel the base, thinking that perhaps we would come right out on the location of the show itself. We were even dreaming of showing Ann-Margret around the Swift."

After they had gone a mile or so, the Dong Tam airfield disappeared and jungle engulfed the canal on both sides. The waterway suddenly took on a very hostile aspect. Kerry decided, with little discussion, to turn around and head back. PCF-44 headed at full speed out of the canal, which could not have been more than sixty feet wide. When the Swift finally made it back to the My Tho River, the crew confronted the heartbreaking sight of a huge Navy landing craft ferrying the troops back to the LSTs. The USO show was over. "The visions of Ann-Margret and Miss America and all the other titillating personalities who would have made us feel so at home hung around us for a while," Kerry noted, "until we saw three Chinook helicopters take off from the field and presumed that our dreams had gone with them."

One of the men on the landing craft showed Kerry the entrance to the docking facilities, which were hidden around a small lip of land. When PCF-44 finally got there, the crew learned that their trip down the canal had taken them through "Route 66"—one of the most notorious Viet Cong–infested waterways in the vicinity. "We had stupidly cruised through it, without manning the guns, searching naïvely for Bob Hope," Kerry confessed to his notebook. "It simply hadn't occurred to us that Bob Hope's presence did not change the status of the war. The VC still existed. The Freedom Foundation should have presented us with an award for perseverance in the search for the American dream." The Swift's quest for Bob Hope and his attendant beauties betrayed its crew's innocence, the same kind of all-American dumbness Norman Mailer had captured so well in his 1967 analysis *Why We Are in Vietnam,* which ends with his young protagonist gushing, "We're off to see the wizard in Vietnam. . . . Vietnam, hot damn."

Later, their pursuit of Bob Hope would make Kerry think of the wounded men he had seen in the hospital at Dong Tam while waiting for Gardner to get patched up. The sight had reminded him of the thousands more lying in hospitals throughout South Vietnam, Japan, and the United States. Many of those damaged youths would never be able to walk down a street or run down a field or make love to a woman again. "Bob Hope and Ann-Margret had visited many of them," Kerry reflected. "The men were told that everything was going to be all right. 'Hi,' Ann-Margret would say. 'My name's Ann-Margret. How are you?' And before some kid with his eyes bandaged so he couldn't see answered, they'd say, 'It's nice to see you,' and swinging their hips and other things go on to the next bed—Ann-Margret, Miss America, or Miss Universe, and other live pneumatic dolls. Their visits made everything all right."

Then Kerry thought about the wounded U.S. soldiers and sailors bedridden in South Vietnam, who knew that a four-star general or a vice admiral would visit them in the hospital, because in war generals and admirals are supposed to do that. "It bridges the gap between the illusions of those who direct the war and the disillusionment of those who fight it, and it makes the general or admiral human," Kerry wrote. "You think how nice it is for him to take the time to come down and see you when really you should ask what the hell you're doing there in the first place." Kerry went on in his journal to imagine just such an encounter:

" 'How do you feel, son?'

" 'Fine, sir,' answers an 18-year-old with his leg blown off at the hip.

" 'Good. We'll have you home soon and you'll be better than ever.' "

Then Kerry speculated how the wounded teenager would wonder what home in Ohio, Florida, or Oregon was going to be like without his legs or arms or face. "Lying in a hospital bed, he can't even imagine yet what effort it will take to get the veterans' hospital to help him to be 'better than ever,' " Kerry continued bitterly. "But it doesn't really matter about his leg anyway. He lost it for his country, fighting for freedom, and that's worth a leg or two, even in Southeast Asia. So all the 18-year-olds are told. And they give them Purple Hearts."

Not long after Gardner's injury, PCF-44 was back out on patrol, gliding up and down the Co Chien River on one engine to preserve gas. It was during this patrol that the crew devised their own sport: the Swift boat duck hunt. Huge flocks of ducks often landed on the lower reaches of the river near the South China Sea, and PCF-44 would drive toward them at full speed while one of the crew stood on the bow with a .12-gauge riot gun. Every time, before the Swift could get within a hundred yards of them, the flock would rise off the water and fly away, prompting laughter and shouts of encouragement from the crew: "Fly, baby, fly—just a little closer, a little closer, damn it . . . over there, over there . . . let's go get 'em!" Then blam! blam! and the chase continued. Not once did the mighty hunters manage to shoot a single duck in flight. "Sometimes I went into the gun tub on top, to get up high and shoot at them," Wasser remembered. "But they were smarter than us."

They did better with sitting ducks. One evening PCF-44 was patrolling close to shore when the crew spied a Vietnamese teenager on a sampan driving a flock of ducks along the muddy bank of the river. They brought the boat near to the bank and motioned to the boy to paddle over to them. He did, whereupon Gardner pointed to the ducks waddling along the bank and then to his M-16 by way of asking if he could shoot one. The teenager seemed reluctant until the Americans pulled out some fresh fruit and C-rations to trade. At that point the boy grinned broadly, motioning first to the ducks and then to Gardner's rifle, nodding his approval.

Blam! went the M-16, sending one of the ducks into the muck. The others waddled forward as fast as their webbed feet would take them. Gardner quickly stripped to his underwear and pranced off through the mud to collect his trophy. "With the white of his jockey shorts accenting the white of his untanned legs churning against the dark, black mud, Gardner proved far more entertaining than Ann-Margret and her cohorts could ever have been," Kerry noted. "Even the ducks stopped en masse and turned to watch him pick up their comrade and slosh his way back to the boat. The Vietnamese looked at us as though we were crazy, and was still staring as we disappeared around a corner to prepare a duck dinner." Wasser took photos of the comical episode, and years later critiqued that night's meal. "We had a little makeshift grill." He laughed. "But the duck sucked. It wasn't tasty, but we ate it. You have to do what you have to do."

Later that night, as the crew sat on the deck eating the bland-tasting duck, Kerry surveyed the water and noticed an ominous stillness all around them. No sampans were moving on the river. No birds were flying overhead. Everything was perfectly silent, with not so much as a ripple disturbing the water. "Even the evening sun seemed to be tainted, shining halfheartedly," Kerry recalled. "I could sense a sound moving toward us. I don't know why I felt it, but I did. At first it was a silence that one could hear, then a hollow whooshing, a vague displacement of air. Then it was a cacophony of explosions, like thunder rolling through hills—one quick explosion after the other, thundering on top of each other, cracking and echoing for miles. Just as suddenly, everything was quiet again. Not a motion."

"B-52 strike," Kerry said, turning to Wasser. Thunder itself couldn't make a terrifying sound like that. Then a huge black cloud rose above the trees several miles away. Practically in unison all six of the men on PCF-44 vocalized their fear, crying out things like "Jesus Christ," "Holy shit," and

"Mother of God." All Kerry could think about was something he had once heard regarding the impact of America's air power in Vietnam: that after every B-52 strike, twenty or thirty local citizens would turn themselves in as enemy sympathizers to the nearest U.S. or ARVN troops or Saigon government authority in hopes of receiving shelter from future bombardments. It was this same terror of the unscheduled U.S. bombing raids that brought refugees to Saigon by the thousands over the course of the war, turning that once beautiful city of 500,000 into a teeming slum of 3.5 million displaced Vietnamese. Between March 2, 1965, and October 31, 1968, when President Johnson "ended" it, the Operation Rolling Thunder bombing campaign flew more than 350,000 missions over North Vietnam, dropping 655,000 tons of bombs. In 1968 alone, American bombers made 172,000 sorties over North Vietnam and another 136,000 over Laos. "When the B-52s flew past, we could feel the water shake," Wasser recalled. "We hoped they had their shit together and knew what they were trying to hit. There was a lot of friendly fire going around."

As he sat in his Swift boat on the river listening to the strike, Kerry could easily understand why the American bombs generated such terror. He and his crew stayed quiet for a while, feeling the silence build up again until the thunder pounded their way once more. He was glad PCF-44 was far enough away that the B-52 bombs couldn't hurt them. "The second time around it shook the Swift boat, and the water appeared to tremble with the rest of us," Kerry recalled. "I thanked God that I was safe where I was and not underneath those thousands of tons of bombs, all dropping in the same terrifying moment in the same godforsaken place. One never even heard the planes—just the displacement of the air by the myriad bombs, and the bedlam that followed."

That night, at one point Kerry awoke with a start, thinking that something was wrong. He swung down out of his rack, in his haste making a loud clank when his feet landed on the metal deck. In his slumber, for some reason, the Swift's main cabin had suddenly become unsafe. The instant his bare feet hit the cold metal, however, the cobwebs cleared away and he realized sheepishly that his concern grew only from the conglomeration of the Swift's usual nighttime noises, the fitfulness of the little sleep one could get on a wee-hours patrol, and the psychological aftereffects of having heard the B-52 strike. It was the fourth time that night he had been startled awake and out of his bunk by his own mind. Each time he had found his boat riding along smoothly. One of his four starts from sleep had some justification, however: the usual late-night patrol noises had included concussions from an artillery barrage so close that Kerry had dashed up to the pilothouse and grabbed the throttles, only to find himself laughing with the man on watch at his overreaction.

Lack of sleep proved one of the biggest problems for the crew of PCF-44. They all longed for an entire night of uninterrupted sleep, but on a moving Swift boat in the middle of a war zone all they could get was an uneasy doze. Eventually, however, Kerry learned to trust his men on watch enough not to jump out of his rack every time he woke up. "There is the constant temptation just to let go and relax and sleep all night—trusting to the nth degree the young men who man your boat and who make up your watch sections," the conscientious young lieutenant (j.g.) wrote to his parents. "Eventually you have to succumb and leave your life and that of the boat in their hands—but always, as last night reminds you with a dry taste in your mouth and with eyelids that cascade down over dirty cheekbones, the sleep is light and restless." And all the radio had to do was crackle "priority" and Kerry would be wide-awake, listening intently for information that could spell life or death for his crew. He and his men all, of course, looked forward to each time they could return to port and wash the grime and tedium of river patrol duty off in a long, cool shower. But a day later, or two at the most, they would be back in harm's way again, cruising the Mekong Delta in search of enemy guerrillas. "You're always getting up, boarding and searching," James Wasser remembered. "The adrenaline never left our veins. You would jump from boredom to sheer terror in two seconds. And remember: we never got the first shot. They always shot at us. That sort of thing keeps a man on edge." Yet for all their efforts, the Swift boat crew never could see that anything was being accomplished.

Wasser, as PCF-44's second in command, was responsible for putting the Swift's siren on and searching the detained sampans—always a risky proposition. Sometimes sampans that were not even carrying contraband panicked and ran at the sight of an American patrol craft. "One time we chased a couple of boats up a canal," he recalled. "As we neared them they abandoned ship and fled into the jungle. When I entered, their rice was still hot. I still have the rice bowl and three smaller eating bowls, plus a large pewter serving spoon which said 'France.' I took them as souvenirs." In addition to leading the board-and-search operations, it was Wasser's responsibility to stand by his young skipper in the event Lieutenant Kerry committed a blunder. "Sometimes he got disoriented and misread the navigational maps," Wasser allowed. "It was easy to do. Once, we hit a sandbar and couldn't get loose. We didn't call it in because we didn't want to get John in trouble. We just sat around for hours, waiting for high tide. It eventually came, and off we went."

Indeed, Kerry inspired both respect and protectiveness in his men. What's more, they just plain liked him. Even at the age of twenty-five, and in the potentially awkward position of commanding men both older and more experienced at the tasks at hand than he was, John Kerry displayed the crucial combination of qualities that make for a good leader. He made his authority known clearly but unassumingly; he took a loose, commonsense approach to military rules and regulations; he put his men's safety far above the pursuit of medals or promotions; and, perhaps most important, he showed his crew not only compassion but also that he knew how to have fun, and considered both essential.

Chapter Eleven

Braving the Bo De River

The new year of 1969 started out well for the crew of PCF-44. They were momentarily out of the Mekong Delta river system, patrolling the open sea off the coast of Vung Tau from their Cat Lo base. The lovely port city some fifty miles southeast of Saigon served both as a U.S. Army-Navy Mobile Riverine Force base and, better yet, as the main R and R center for American troops in South Vietnam. Called Cap St. Jacques in the country's French colonial era, Vung Tau had much to recommend it to any U.S. serviceman, and particularly to John Kerry. As he had written to his parents on December 21, 1968, Vung Tau "is an old resort town, greatly influenced by the French and perhaps the main pleasure dome to which the bigwigs of Saigon would retreat for a weekend of sand, sun and fun. It is also about the only town in all of Vietnam that allows troops liberty and which is still open in the military sense of the word."

Still marveling at their luck, the men of PCF-44 had been enjoying this coveted, comparatively safe assignment to cruise the resort's aquamarine coastal waters for about six hours when an "immediate action" message came over the radio. It said, "Return to Cat Lo Without Delay," whereupon six hearts sank. Every member of the Swift boat's crew knew that the ominous order meant one of only two things: either they would be leaving immediately for an Operation Sealords raid or they were being transferred out of Cat Lo. "I don't know why we were so sure that transfer was in the wind, but the day before, the skipper had mentioned that two boats were going to have to go to An Thoi and replace the most recent casualties," Kerry recalled. "He had designated myself and someone else to go, but at the last minute the order had been rescinded. I knew it wouldn't last."

By the time PCF-44 made it back to Cat Lo, its crew was exhausted from slapping away mosquitoes all day, and because it was three o'clock in the morning. No one at the base seemed to be awake, but Kerry was in no mood to wait to hear his fate, so he entered the officers' barracks and roused Lieutenant Skip Barker, his friend from Coronado. "He confirmed that it was An Thoi and [his] was the other boat going," Kerry remembered. "We didn't say any more. I went to bed pissed, and he rolled over and went back to sleep. There was no way to release the frustration that I felt; everything just welled up inside of me. In those instants, I hated the military for its impersonality. I felt like a Ping-Pong ball, cracked and relentlessly smashed from one side of the table to the other. I even pictured myself being killed in An Thoi without knowing anyone, and the idea of dying alone annoyed me more than the idea of just dying."

Kerry had learned to accept the deaths of others, be they dear friends like Dick Pershing or complete strangers like the Vietnamese soldier he called Nguyen. Contemplating his own was another matter. He was only twenty-five, for god's sake, and he had big plans for the future: he had beautiful Julia Thorne to marry, wonderful children to conceive and raise with her, and a career in law or politics to forge. What's more, this wasn't the Great War of 1917 or the Good War of 1941, but a murkily-motivated military action he never had been sure needed to be waged at all. Kerry had once wondered if it was different for Dick Pershing; he had an epic ancestral heritage to live up to, and die for. And he had hoped it was different for Nguyen, too; it was his native land he gave his life for. But even their deaths seemed so senseless and unnecessary.

After a few hours of tossing and turning these thoughts over, John Kerry accepted that he had to go back to An Thoi, and perhaps even die there, because he had chosen to be a U.S. Navy officer, and as such had sworn to preserve, protect, and defend however he was ordered to, and the one thing the best, if not necessarily the brightest, always do is keep their word. But he still didn't have to like where it got him. "It seemed so stupid, though, when I thought back to my first moments in Vietnam, when I had asked for An Thoi," Kerry reflected in retrospect. "If they had sent me in the first place, there wouldn't be anything to get mad at. But the incessant changing—this is what annoyed me more than anything, and the crew. They were furious. No sooner did they arrive in one place than they had to pack up and say good-bye and send off forwarding addresses and hope that their Christmas mail would catch up to them someday. I wished that the skipper could have seen their faces each time I had to tell them that we were moving again."

PCF-44 got moving again shortly before noon one day early in January 1969. Skip Barker—whose one month's seniority over Kerry made his PCF-31 the lead boat—had set out twenty minutes earlier in order to stop by the Army supply pier to pick up some butter pecan ice cream for the trip. He and his crew met up with PCF-44 in the Cat Lo bay, but the two Swift boats did not get very far. The seas around the mouths of the Cua Tiey and Cua Dai rivers were inordinately rough, so the pair of PCFs stopped, turned their sterns seaward, and radioed back to base that unless directed otherwise they were returning to port. "It's the first time I ever saw sailors voluntarily putting on life preservers," Barker remembered. "It was just awful out there. I made the sensible decision to turn back." No one on either Swift disagreed with him, especially Kerry. Returning to Cat Lo would mean a chance to go into Vung Tau for one last evening on the town. The Swift crews arrived back at the pier hoping that perhaps their orders to An Thoi would once again be rescinded, but they were only postponed until the monsoon season died down.

Nevertheless feeling like a teenager on a snow day, Kerry immediately went to the base transportation headquarters with Barker and managed to secure an oversize Chevrolet pickup. While there they ran into Commander George Strulie, who told the young officers they had done the right thing in turning back from the monsoon-tossed seas. Thus vindicated, Barker and Kerry drove to Vung Tau and savored the sights and sounds of a warm tropical night, joyful just to be alive. "A strong evening light shone mysteriously through [the] trees along the road lending a serenity to the environment that was not deserving," Kerry wrote in his journal. "On all sides of [us] were scores of Vietnamese peasants on bicycles making their way home after a day at the market or in the fields."

One fact of the war that fascinated Kerry was how Viet Cong riding fifteen-dollar bicycles always seemed able to not only outsmart but outmaneuver the Americans with their six-million-dollar aircraft and other high-priced, high-tech weaponry. He marveled as the intrepid Vietnamese managed to carry as much as five hundred pounds of supplies up hills, and at how their French-made Peugeot bikes were often so loaded that the malnourished porters had to walk rather than ride them down the country's pitted dirt paths and mountain trails. Stories also abounded of bomb-bearing VC bicycles wheeled into Saigon hotels or Da Nang nightclubs and then detonated. "Season after season, regardless of the weather, their bicycle supply system continued to roll," wrote Lieutenant Colonel William Scheck in a 1998 *Vietnam* magazine article. "In the end the protracted hostilities showed that the best armed and best equipped army in the world could not choke off the clandestine supply system."

The bicycle brigades reminded Kerry that Vietnam truly was a guerrilla war, and that the guerrillas were frequently outfoxing the United States. The patient North Vietnamese Davids constantly made fools of the American Goliaths. Even when the rain pounded down so hard the Swift boats had to stay in port, Kerry would see hundreds of drenched Vietnamese pushing their bicycles forward, refusing to be thwarted by the elements. Yet many of the very same men would feign seasickness when the U.S. Navy asked for their help on board-and-search missions. No matter how many histories, Graham Greene novels, or Bernard Fall dispatches he read, Kerry realized he would never fully comprehend the enigmatic people of the land he had come to aid.

Kerry accomplished his main objective on his day off: he had reached Julia Thorne by telephone. "His calls were always reassuring," Julia recalled. "I lived for them." But the reunion moment was short-lived. Before he knew it, Cat Lo and the relative freedoms of Vung Tau were behind him. The Swift boats were on their way to An Thoi. "Skip led the first part and I went below to read and think," he recorded. "Before long, though, Skip stopped, and we turned our sterns to the seas as we had before and debated whether or not to turn back once again. The seas were extremely rough and we were experiencing difficulty in maintaining control." The PCFs' design gave them a tendency to surf down large waves and plow into the troughs of the next ones, creating a constant danger of broaching to broadside. Barker decided to radio in once again that unless otherwise directed, they were turning back. It wasn't long before they received the encoded reply: "Proceed to ten-fathom curve and continue on assigned mission." Kerry and Barker were livid. "Those assholes, I thought," Kerry recalled. "What did they know about what it was like out here? I hoped that a huge wave would bowl us over, and then I could go raging back to headquarters and dramatically cuss at and then KO some poor slob watch officer who had not the faintest idea what it was like to bob around in a fifty-foot aluminum boat in a rough sea."

And so the two Swifts continued on. Kerry went below deck on PCF-44 for one of what Robert Penn Warren called the "great sleeps" of his character Willie Stark, a fictionalized version of Louisiana's infamous political "Kingfish" Huey Long. "I remembered how the hero of *All the King's Men* would go to sleep for hours at a time to pass over the rougher spots of life," Kerry explained. "For some reason this made a lot of sense at the moment. But with the boat crashing against the concrete water, sleep was far away."

Instead he kept stewing and feeling frustrated by the loss of Cat Lo's familiarity. He railed internally at how the news of his transfer had "come like a thunderhead that suddenly appears to drown out a summer sky—borne from nowhere and carrying all the danger of a tropical storm." Days later, he confided his feelings to his diary: "I felt like a small child who hides in attics on rainy days and creates his dreams, lulled by the pitter-patter of raindrops on the attic roof. I wished that I was that child with his toy trains, lost in the winsome, boyish pleasure that his loneliness brought him. I wished that I could walk out of that attic to the warmth that waited on the floors below rather than into the hail of bullets that waited in an ambush somewhere in the Mekong Delta."

Indeed, while Kerry waxed melancholy in private, his PCF-44 followed Barker's PCF-31 past the now familiar landmarks of the Bo De River, the island of Hon Khoai, and the Ca Mau Peninsula, Vietnam's southern cape. They were well into VC country, where the jungle grew so dense it took a machete to bushwhack through the underbrush just to see beyond one's reach.

Morning brought the two Swift boats into the clear turquoise waters of the Gulf of Thailand, and before long to the floating barracks of Coastal Division 11 at An Thoi. There Kerry immediately sought out Division Commander George Elliott, who lent a sympathetic ear to his young officer's efforts to get his boat and crew locked into one place. But the commander stressed that he had no say in the distribution of forces. He could only promise that sometime in the future, when an opportunity arose, the crew of PCF-44 would be rotated out of An Thoi. Kerry relayed this vague assertion to his men, who all put on their best faces and feigned optimism for the sake of one another's morale. "My main concern was to get the job done without loss of life," Elliott maintained in a 2003 interview. "I never played favorites."

That night in January 1969 proved a bad one for Kerry. Sleeping fitfully in the topmost of the three-tier racks in the barracks, in his dreams he was still far offshore, crashing PCF-44 against the surging waves of the South China Sea. He called out loudly in his nightmare-riven sleep, "Back her down, boats, back her down!" A few hours later he snapped upright in his rack so fast he hit his head on a pipe, mercifully knocking himself back to sleep. It was neither the first nor the last time Kerry would suffer nightmares; in the wee hours of one patrol his men were awakened by the sound of their skipper crawling around on his hands and knees on his bunk, and when one of them asked what he was doing Kerry answered, "Looking for the throttles, but I can't find them. Don't worry, I'll get oriented." On another occasion at An Thoi, Kerry got into his rack, dozed off, and minutes later started yelling, "Slow down!" and then leaped from the top bunk to the deck below. The man in the lower rack howled with laughter at the spectacle of his nocturnal visions. "I sprained my foot and gashed my hand and noticed nothing until the next morning," Kerry recalled of the incident. "Somewhere over in a corner, someone else was counting down in his sleep: 'Ten, nine, eight, seven, six, five. . . . ' A den of nuts, I concluded. We all laughed about it the next day." As Skip Barker summed up Kerry's nighttime antics: "John was quite a sleepwalker."

Every evening the officers at An Thoi assembled for a briefing in the division office atop the barracks. Those returning from raids or patrols would detail what had taken place, after which the operations officer would read off everyone's assignments for the next few days. The meetings tended to be raucous and good-humored, with even the handing out of assignments taken lightly. But the mood soured quickly for the unlucky Swift officer informed that his boat was being sent up the Bo De River. The expansive, muddy Bo De, which spanned nearly five hundred yards across at its mouth, flowed from the Gulf of Thailand across the Ca Mau Peninsula and emptied into the South China Sea. The river ran deep enough for Chinese-made submarines to get up it, and had earned its reputation as a graveyard. On November 24, 1968, for example, *Stars and Stripes* had published a story on the recent vicious Viet Cong ambushes on PCFs 31, 38, 72, 82, and 93. The Swifts had been fired at from massive bunkers stoutly reinforced with teak logs. "The Bo De had the same macabre fascination that any forbidden place had," Kerry's Coronado colleague Wade Sanders recalled. "We often just cruised past the mouth of it and quaked. You could see the first bend. That's where people got killed. . . . You could feel the suction. . . . The river's reputation as a killer diffused the air around you. There was this overwhelming desire to get the hell out of there. The feeling was like when you're a small child and get spooked in the darkness. You start sprinting to the light in the distance, but you're worried that what's chasing you will get you first."

After the briefing that brought the unwelcome assignment for PCF-44, Kerry went down to his boat and told the crew that they were going on a Sealords raid, designated number 210, up the Bo De River with five other Swifts, and should thus make sure they had enough ammunition on board. "We joked for a while, and then I watched a half-witted English comedy with Julie Christie in a minor role, and then I went to bed"—Kerry laughed in retrospect—"again in the top rack, much against everybody's advice."

In the brown-water navy, preparations for a mission almost always proceeded slowly, but for Sealords raid 210 they were made in a hurry. Shortly after noon on January 8, 1969, the six Swift officers selected for the task were called into shore for a briefing with the lieutenant commander who was Captain Roy Hoffman's chief of staff and had flown down from Cam Ranh Bay to lead the operation. Kerry arrived late for the briefing, but quickly determined that he "could easily have missed the whole thing and still have missed nothing at all." When he did get there the lieutenant commander was officiously spouting gung-ho military boilerplate: "Well, I want you to get a maximum gunfire-damage assessment, and I'll leave it up to you officers in charge as to what you use and when. . . . We'll go up the river here," he blustered as his finger pointed to a blue squiggle on a map, "and I'll split you up there, two boats going up each canal. . . . I'll leave it up to the OINCs as to how you destroy—"

At which one of those officers in nominal command interrupted: "But, sir, if we go up there simultaneously, won't we be shooting at each other?"

"Well," the lieutenant commander replied with the indifference of inexperience, "I'll hold the middle group back until the outside ones get finished, and that should take care of that." Aye, aye, sir and anchors aweigh, Kerry thought.

A few hours later the six Swift boats left An Thoi for the treacherous Bo De River. It was a beautiful afternoon, and Kerry sat on PCF-44's cabin roof in the tropical splendor finishing F. Scott Fitzgerald's first novel *This Side of Paradise*, in which the protagonist spends a two-year, four-page "interlude" from his utter self-absorption desultorily fighting in World War I after graduating from Princeton. As the story ends, however, the main character grows up a little when he discovers that an old gent he had been haranguing about the merits of socialism is the father of a classmate killed in the war, and rues his own egotism. "I thought jealously of the fact that Fitzgerald was only twenty-three when he wrote it, and the last chapter stayed with me for some time," Kerry recorded of his impressions. "My days as a senior at college came back to me, and I remembered how Vietnam and death had been something that happened in books, automobile accidents, hospitals, old-age homes, and television programs."

As he mused, PCF-44 continued south, past the luscious-looking islands that had seemed so alluring on his first visit to An Thoi. But this was strictly a business trip, and with the lieutenant commander's boat out in front Kerry couldn't exactly deviate from the prescribed course to check the islands out. By late afternoon, as the cerulean sky faded to slate, the string of Swifts entered the shallow muddy waters at the tip of the Ca Mau Peninsula. Off in the distance rose the island of Hon Khoai—or Pouli Obi, as the French had called it—in the lee of which the little task force would lie quietly until the LST assigned to support them arrived.

As PCF-44 crept slowly into a cove that protected a ten-hut village as well as one of the South Vietnamese Naval Forces' Coastal Groups, dark descended and each of the six boats withdrew into its own cove for a brief respite. Lights glimmered from their cabins as the crews cooked their dinners, which on Kerry's Swift were eaten while the men went over the mission chart, marking in the designated code names for aircraft and navigational reference points.

Later that evening some of the Swifts' crews opted to mingle with their cohorts on the other boats, and Kerry struck up a conversation with a pair of PCF veterans who had already seen action on the Bo De River: Radarman Dorsey E. Lee and a crewmate of his Kerry remembered later only as "Deuce." Their skipper was Lieutenant (j.g.) Zeke Zucker, who had been in Kerry's six-man class at Swift school in Coronado. Both Lee and Deuce had been in-country for more than a year, and had been shot at more times than either could count. Kerry was eager to hear about their experiences on the Bo De, and Lee in particular proved willing to recount them.

Like Kerry, Lee—who hailed from Jennings, Louisiana—had arrived in Vietnam in 1968, but he would stay in-country for sixteen months. Assigned initially to Cat Lo, he became notorious for hanging out in the brothels outside the base gates, where he picked up valuable pointers for his later career as proprietor of the only licensed adult bookstore in Louisiana's Calcasieu Parish. While in Vietnam Lee also developed a sideline that made him quite popular in the Mekong Delta: cooking. His specialties included fried shrimp, homemade potato chips, and pineapple upside-down cakes. "Bologna sandwiches only go so far," he later joked. "Some of those boys on the boat would have starved without me."

One afternoon not long after he had arrived in-country, Lee was told to run an errand to a U.S. morgue in Saigon. There, to his horror, he had seen human bodies piled up in plastic bags. Shortly after that he had been sent up the Bo De River, an equally wrenching experience his new acquaintance John Kerry was itching to learn more about. That night, in the cove of Hon Khoai, Dorsey Lee and his pal Deuce agreed to sit down with Kerry and let him interview them about their experiences under fire on the deadly waterway. Kerry tape-recorded the conversation, excerpts of which follow:

John Kerry: You guys have both been up the Bo De before . . . right?
Dorsey Lee: *Beaucoup* times.
Kerry: How many times did Cat Lo make the Bo De raid?
Lee: Once . . . twice.
Kerry: And then An Thoi took it over?
Lee: Well, Cat Lo and An Thoi ran it the first time.
Kerry: Well, when was the raid . . . you know, when I first came down here . . . the boats had just been smashed up.
Deuce: Right.

Lee: That was the third raid that didn't make it—not counting Mike Brown's run.
Kerry: You made it on the first one?
Lee: Yeah, we had three boats from An Thoi and two from Cat Lo, or something. . . . Anyway, we started, and we met in the middle and the An Thoi boats didn't have any problem, but the Cat Lo boats took recoilless up there in that one corner there [*pointing on chart to the first turn the Swifts would make the next day*]—the main corner in the left turn.
Kerry: And ever since then it's gotten tougher and tougher each time?
Lee: No, the second time we took some small stuff. . . . It was nothing.
Kerry: What about that Song Ong Doc that we went up the other day? That's supposed to be the second worst river in Vietnam.
Lee: Yeah, but since Deuce and I are on the crew, you won't get those rockets fired at you.
Kerry: That's good. . . . I'm glad I've got you guys now.
Lee [*to Deuce*]: How many times [have] we been up the Bo De?
Deuce: Oh, shit, all kinds of times.
Kerry: You mean on unofficial runs—just going up it?
Lee: Well, we weren't allowed to go up, you see, for a long time.
Kerry: So you just went up there?
Lee: Yeah. We'd sneak up in there, but if they found out, your OINC got a letter of reprimand and everybody got into trouble.
Kerry: You mean that you once got court-martialed for going in there, and now they send us in? How many times would you say it was probed?
Lee: Oh, a hell of a lot.
Deuce: In fact, that one boat . . . went up there at midnight until six o'clock in the morning. . . .
Kerry: That first bend is where you get hit, isn't it?
Lee: We got hit there. After we got past that first bend I could've sworn we never came under fire again, but Mr. Bernique was with us and he swore we did. . . . The way we did it, you'd go up there

until you'd get fired at and then you'd fire back, and then you'd haul the balls out. We'd try to go past that first bend.
Deuce: No, we got .50-caliber fire—I saw the [PCF-]93 in there.
Lee: Well, all right, but I never knew we [were] under fire after that.
Deuce: You can't tell, though, 'cause the boats following your Swift are still shooting. [*Kerry expresses surprise that all the bunkers had not yet been knocked out, and asks why.*]
Lee: There've only been two whole runs all the way through. And the first time they went in and met halfway.
Kerry: Yeah, but which part of the river are the targets in?
Lee: Right after you get around the bend . . . starting there [*pointing to the chart*] and all the way down.
Deuce: There must be eight million huts and . . . just *beaucoup* fucking sampans.
Kerry: Maybe that's what we're going in for—them?
Deuce: There's a hell of a lot of it . . . oh, shit . . . we saw five VC flags when [we were] in there.
Kerry: Did you see those signs saying "Kill Americans"?
Deuce: Yeah, that's right, on the corner where the canal is [*pointing to a small tributary on the chart*]. You'll see them.
Lee: We shoot at them but you can't do anything to them. . . . One of the first crews that came out said that they saw . . . maybe five hundred to a thousand sampans in there—right on the banks you can shoot at.
Deuce: They told us to run through there, and they said to take targets of opportunity under fire but don't slow down—run through. You see, our whole mission was to probe the canals in the Bo De. Run through the canals and probe 'em. But we went in there on the wrong tide, and we probed two, so then we just, as we [were] going in, we [were] shootin' .50s at sampans, and just shit like that.
Lee: There will be more targets than you can destroy, going by them. If you're going five hundred rpm, you can't do it. They're on both sides, and if you have one guy shooting one side and one guy the other, you still won't get 'em all.

Kerry: That place had got to be full of VC?
Deuce: We weren't finding [any] sign of people.
Kerry: But where did they all go?
Lee: Inland.
Deuce: They hide in bunkers.
Lee: There's one stretch . . . it must be three miles long, and there must be bunkers every fifty yards.
Kerry: Well, this B-52 strike might solve that?
Lee: That won't solve it. . . . They've had them in there before.
Deuce: But we weren't planning on finding these people. Like I said, when we went up this one canal by ourselves on the 98 boat—Mr. Johnson's crew—while the other four boats were waiting for us, and there was that one sampan that had moved from one side of the creek to the other side, and there was a hut there that still had a fire burning and—
Lee: You could see where the food had been cooking and the clothes are still dripping wet—
Deuce: And shit like that—we went in and came back out, but where we went in there [were] a few footprints that were leading from a sampan up to the banks. . . . We went in and we weren't going to take anything under fire until we came out. . . . We came out, and oh boy, there [were] all kinds of footprints where they'd been taking shit out of the sampans and going back in the bushes. We got the sampan, though.
Lee: Yeah—they'd recamouflaged all of their stuff.
Deuce: Yeah. You could see the camouflage—you know how they tear the bushes off, and the bushes hadn't had the time to get wilted to any degree and, you know, you could see where it had been broken off—and they were all fresh bushes. That's how we could tell where the camouflaged sampans were . . . they were all the fresh bushes, and all the rest [were] kind of dead-looking.

There [were] a couple things . . . as we were going in you could see a sampan real easy, and then we'd come back out and somebody would have camouflaged it over. . . . They've got two signs there. They've got one pointing up that long canal that said "Saigon" or something or other; then they had the other one—"Kill

Americans." We didn't know what it was. We just had one of the guys copy it down real nice and easy-like, you know, and got all the pronunciation marks, everything, and took it back and was asking those gooks and nobody would tell him, so he went over there to this one . . . gook officer we [had been] working with pretty close.

Kerry: What did the whole sign say? . . . It said, "Come Join the—"

Deuce: "Come Join the Viet Cong and Kill Americans."

Kerry: Is that right?

Lee: It'll still be there, I'm sure.

Deuce: What's the deal tomorrow? Are we supposed to go all the way through and go out the Cua Lon?

Kerry: I know nothing about the mission yet.

Deuce: If they tell us we're going to go in the Bo De and come out the Bo De, I'm not going to like it.

Kerry: I don't know. We'll find out tomorrow.

Deuce: If we run all the way through . . . we'll have a good time—you'll expend all your ammo.

Deuce: Yeah, but those people are living there right now.

Zaladonis: Man, I bet [you those] people up the Bo De right now are sitting there talking and saying, "Hey, day after tomorrow [a] bunch of Swift boats are coming up. . . ."

Deuce: I'll tell you what's piss-poor: those boats coming up. They stop at Pouli Obi and out of all these Vietnamese forces, there's VC in every one of them and they've got radios and they'll call them up and say, "Oh man, you better be ready, 'cause there's six boats sitting right here. . . ." They figure that there's at least ten percent of these junkies that are either hard-core VC informers or VC sympathizers. They keep checking the records, and checking the records, and never find anything.

Lee: That doesn't include the PRUs—just the regular fuckers.

At that point Kerry heard the engines of the other boats start up, and figured it was time to move out to the rendezvous with the LST. "I thought about what the men had said, though—about the VC knowing that we were coming," he stated. "Several of us had talked about it before, and it seemed fairly obvious that they must have some channel of communication to be as prepared as they were on the occasion. Each time a raid had been held on the Bo De or the [other] rivers in that area, the boats had grouped at the island of Pouli Obi beforehand. And yet people proceeded as though the enemy had no brains or foresight whatsoever. Perhaps we should have let Deuce or Lee run the war and then there might have been some result."

Prominently penned in bold in John Kerry's "War Notes" about Operation Sealords raid 210 appear the words "Lambs to the Slaughter." Next to the phrase, however, he noted for posterity that in fact he felt optimistic, in the same sense that a young Winston Churchill did when he fought with Britain's Malakand Field Force in what is now northern Pakistan: "Bullets—to a philosopher, my dear Mama—are not worth considering. Besides, I am so conceited I do not believe the gods would create so potent a being as myself for so prosaic an ending. Anyway, it does not matter. . . ." Looking at this quote in his journal decades later, Kerry chuckled. "I remembered the words that Winston Churchill had written his mother before doing battle in 1897," he explained, "and I took solace in the fact that others had gone before and it couldn't be that difficult." Like many GIs in Vietnam, Kerry, for both family consumption and posterity, would send home tape cassettes describing his daily life in the brown water Navy. "We were always concerned about John," his sister Diana recalled. "Besides writing us these amazing letters he constantly sent tapes. They meant so much to us. We would get copies made and circulate them between us."

Just past three o'clock that morning, after some floundering around in the dark to find one another, the six U.S. Swift boats led by the lieutenant commander left their support LST and set course for the mouth of the Bo De. It was time to think about the matter at hand, and the best and stealthiest way to enter was a small, untried canal above the Bo De and move down onto the big river from there, the idea being to surprise any VC lying in wait. "We were all a little uncertain how we could surprise anyone with the noise of the boats, but questions were not ours to ask," Kerry noted in his journal.

In fact the operation would involve two rivers and four canals. The Bo De itself would not be touched until the end of the raid, at which time the boats would turn the big Delta river and make a rush for the exit down it to the open sea. At that moment, fixed-wing aircraft were scheduled to swoop down and strike at any enemy fortifications on the river's banks. Up until then, while the Swifts were inland, they would enter waterways and strike targets through and on the canals and smaller rivers surrounding the Bo De.

The six vessels had about three hours to travel north, so Kerry went below and lay in his bunk to think amid the constant drone of the Swifts' engines. With death on his mind, the self-proclaimed "uncommitted soldier" could simply not fully fathom the reckless nature of the mission he was on. They would be taking six 50-foot aluminum boats up tiny rivers into the heart of Viet Cong territory, waterways that in some places extended only just wide enough to allow a Swift a twisting turnaround and in others not even that, and they would sit there on those narrow streams making tempting targets for ambush. Because the PCF crews wouldn't be able to see through the tall, thick jungles on either side, it would be impossible to aim at anything on the banks with any semblance of accuracy. "Nothing I had ever heard of seemed as tactically stupid," Kerry recalled. "Certainly a great many men at the Battle of Midway had done things much braver, but then they were the attacked and it was not the kind of risk they were asked to take every day. Finally, I thought it silly to look for parallels, and I slept until the driver woke me up when we were about ten minutes from the mouth of the river." PCF-44's crew still had to break out their small arms, lay out the ammunition, and put everything in the boat in pinpoint order. Once the shooting started there would be no time to be asking where some gun was; it had to be right where it was supposed to be and it had to work. "I pulled on two pairs of flak pants," Kerry noted, "to protect the family jewels, if nothing else—and we laughed."

The six Swifts fell into line, PCF-6 taking the lead followed by PCF-38, PCF-3, PCF-44, and PCF-50, with PCF-71 bringing up the rear. Suddenly, PCF-6 darted off at top speed, but the boat maintained radio silence so nobody on the other Swifts could find out why. Whatever the reason, PCF-6 soon returned to formation, just as Kerry saw the air-cover team swoop toward them over the trees. Just as suddenly, the string of Swifts found themselves ten yards from the entrance to a canal and going in.

The canal was twenty yards wide at most. Mud banks rose two to three feet high on either side, making the boats look boxed in. "At the very mouth the foliage was sparse, but within a hundred yards they were surrounded by a thick jungle of trees that rose well above them." Kerry described the setting. "It was like moving down a canyon of green limbs with a small brown mud line at the level where the canal walls met water. Perhaps, when going slowly enough, they could see ten or fifteen feet back into the green. The boats slowed down in order to negotiate a tight turn—so tight that the rudder alone would not get one around it. The engines had to be worked against each other, outboard forward and inboard back, in order to twist around the pivot point."

Three of the Swifts had made it around the bend when the lead boat—PCF-6—opened fire on the left bank. The second and third boats, PCF-38 and PCF-3, followed suit, and the canal thundered with high-caliber automatic weapons. The remaining three boats put a few short bursts of cover fire into the brush, just in case any VC were there and thinking ambush. "Nothing was heard on the radio," Kerry related. "A stupid situation. There was everyone firing, and no one even knew if anybody had taken fire or if there was a specific target. Then suddenly it was quiet again. Still no explanation. With the noise, there was no doubt in any mind that Charlie would be setting up in the Bo De. The exit was going to be costly and excruciatingly difficult."

Nevertheless, the six Swifts continued slowly down the canal. PCF-6 moved out onto a slightly larger river and started around a bend to the right, then stopped. At that instant the river shuddered with a resounding explosion. First PCF-6 and then every boat began to pound the banks with everything they had. Trees along the water toppled as the .50-calibers crashed through them. M-79 grenades exploded short of their intent as they clipped the branches and brush only twenty yards from where they were lobbed. The deafening explosions reverberated down in the jungle canyon without pause.

"I thought for a moment, 'Jesus Christ, you're really in it now,'" Kerry remembered, "and then I just shot and shot and shot, at nothing in particular, just trying to fill up the brush and hit anything that might have been shooting at me."

The lead boat began to turn around and the rest followed, still with no word from PCF-6 on the radio. Once reversed, the leader's throttles rammed full forward, and soon all six boats moved out at full speed in what Kerry called "their twenty-yard bathtub, firing cover shots on the sides with our wakes slamming up over the banks."

At that point Kerry yelled, "Cease fire!" to his men, and for most of the trip out they shot only an occasional cover round. When they finally spied the exit, Kerry fired his last M-79 to clear the grenade launcher's bore. "We tore out into the South China Sea at full speed, in confusion, until we were clear of the shore and out of mortar distance," Kerry recorded in his diary. "The boats slowly collected and we grouped together to learn what had happened."

PCF-6 had been hit by recoilless small-arms fire, and Kerry could see a hole through the pilothouse door. He also saw Frank Gilbert, the lead boat's officer in charge, come out with a battle dressing on his arm, trying to stanch the blood trickling down his hand. It turned out that his had been the only boat hit, and Gilbert the only man wounded on the raid. "We had been remarkably lucky once again," Kerry recalled. "I sat on the boat and quietly savored the moment of rebirth. For an instant, worries are gone. There is almost always a feeling of great calm after a firefight. The relief alone that one is out lends a tremendous sense of accomplishment when in fact nothing has been accomplished. There'll be more moments of terror and you'll worry again, but for the moment, the life around you and in you is yours."

PCF-44 bobbed around in the chop of the South China Sea while some of the other boats mortared the area of the ambush and the support aircraft emptied their ordnance into the VC bunkers they had discovered from above. Then PCF-6 darted off in the direction of the LST, again without a word.

"The lack of communication at this point was ridiculous," Kerry reported. "No one even knew who was in charge. I felt that the mission had been run in an abortive manner from the start, and there wasn't too much doubt in my mind that the friendly lieutenant commander from Cam Ranh Bay had dropped the bubble while we were in the river. Be that as it may, they had run into a barricade of poles sticking into the water, each about a foot apart and stretching for about two hundred yards down the river. Charlie had been ready for us in every aspect and obviously wanted to keep us out of the area we had been heading for. I suspect that we had run into the guards for that entrance and they had summoned the guts to take us under fire. Perhaps they didn't need to summon guts. It had long been my theory that they dig a deep hole to hide in and put a lid over it. As we approach, they take aim and put a few hasty rounds out whereupon they plunk down the lid having done their damage and sit safely underground while our barrage went merrily over them. After all, they knew that we had never stopped in the history of the raids. When hit, the boats would just shoot back and get the hell out—usually with one or two wounded or dead on board. We never found out what damage had been inflicted. Some of the men who had been in the 'massacre' on the Bo De reported that they saw guns shooting at them from small peepholes in the banks of the river."

After PCF-6 took off there was even more confusion. Nobody on the other boats had any idea who was supposed to do what. Finally, the senior officer on the LST who also served as operations officer for the Swifts that worked off it took charge. He ordered PCF-44 up another entrance to try to get onto the Bo De and execute the mission that they had originally been sent on.

This next canal, however, was not even as wide as the first one they had gone down. The foliage was thicker too, and the combination made it impossible to go faster than a few knots. "Suicide, I thought," Kerry remembered, marveling. "But luckily no one shot us. When we finally turned around, it was because the passage had become too thick and in order to get the Swift turned we had to run it into the bank in front of us and then back down onto the one astern. I swore as branches rubbed against the windshield and swarms of red ants poured on board. What in the name of God did we hope to accomplish up there besides getting our asses shot off?"

PCF-38 and PCF-50 were waiting for Kerry's boat when it got out of the canal. A few moments later PCF-6 arrived back on the scene and started toward another entrance. One of his crew turned to Kerry and said, "Jesus Christ—how many fuckin' times do we put it on the line today?" Kerry didn't have an answer. Apparently there was someone watching for them to pass from the sea into the mouth of the next river up the coast, so they raced back to the LST, where a meal was waiting for the Swift crews. After lunch Kerry went back to the LST's officers quarters, put his feet up on the lone desk, leaned back, and in a well-pleased-with-himself manner, proceeded to listen to the tape he had recorded while on the rivers. "For a moment, I was angry at not having brought my camera with me, because the width of the river we had been on and the sight of six Swifts negotiating the turns should have been recorded for posterity," he recalled. "However, I had given in to an earlier mood that determined it would make more sense to worry about my boat and shooting back than standing amidst whizzing bullets and rockets taking pictures."

As Kerry was lounging in the quiet, air-conditioned quarters, Lieutenant (j.g.) Will Imbrie, a Rutgers University graduate and the operations officer for those now on the LST, came bursting in and announced that the Swifts would have to get under way and proceed to the island of Hon Khoai while the LST rendezvoused with another ship. "We would join her again just off the island at eight-thirty that night," Kerry recorded in his notes. "For a moment I was annoyed because it meant getting under way again and everyone on the boat was tired. We had been going [in] overdrive since we left An Thoi, and everyone had been hoping to use the afternoon to catch up on much-needed sleep. But then Will told me that we could leave the boats in the small harbor and go up on top of the mountain to visit the Navy radar site and lighthouse. This idea gained quick approval. The islands in the area we were working in were all, without exception, beautiful and enticing, and I knew that the one-hour journey it would take to get there would be rewarded by a quiet lagoon and an evening at anchor with a cold beer—a fitting reward after a day at sea."

It was about three in the afternoon when Kerry and company left the LST and headed the fifteen or so miles south to the island. Hon Khoai lay about nine miles off the tip of the Ca Mau peninsula. "On the map it looks like a squiggly cloud or like Casper, the Friendly Ghost," Kerry wrote of the island.

> *From the sea it rises steeply to its height of some 1,100 feet, with heavy green jungle draping the sides from the rocks by the water to the pinnacle. There were three hills, really, the second in line being the tallest, and between the first two humps, on the western side, was a lagoon with the tiny village that was home for about seventy-five Vietnamese, fifteen dogs, four pigs, two goats, one cow, a Vietnamese Coastal Group, four American advisors, and their pet monkey—who, quite incidentally, smelled, and bit all newcomers. I was in no hurry to get there and so I kept the boat throttled back slightly—both to keep from hurrying and also to give the engines a rest. I thought with pride and gratitude about the engines.*

> *PCF-44 had been from Cam Ranh to Cat Lo to An Thoi to Cat Lo to An Thoi and on patrols that easily tallied up enough miles to complete that journey several times more, and they had not yet had a serious engine failure that required running on one engine for a period of time—or worse, that required a tow. The engines had become a part of the whole character of the boat, and we didn't want to annoy or betray them by pushing too hard.*

And so the Swift lolled through the sea, its crew enjoying the warmth of the evening, the glint off the water, and the peace that enveloped the scene. "I watched entranced as the island grew larger and larger in front of us," Kerry remembered. As his men sprawled about the boat relaxing, Kerry stayed alone with his thoughts in the pilothouse, giving in to the hypnotic effects of the setting and the swaying of the sea. Before long they reached Hon Khoai. "I pulled the boat alongside [PCF-]50, the one on which Mike Bernique had been operating for the day," Kerry wrote. "The 71 boat with Will [Imbrie] had developed engine trouble and we had passed him coming into the lagoon. Originally, Will had been planning to go up on top of the mountain, and he had called the people there to alert them to our arrival. By luck it happened that we had arrived at the moment that a huge, two-bladed Army cargo helo was making the monthly replenishment run to the island. Thanks to Will's call and the 'misfortune' that kept him on his boat in the harbor, Mike and I were free to capitalize, and we began in good fashion. The officer atop Hon Khoai had detailed the helo to come down and pick us up at the village. We were grateful, because it turned an hour's fatiguing hike into a five-minute jaunt. The huge, ungainly bird managed somehow to settle down on the rocky border between the water and the village, one wheel precariously poised at a different angle and level from the others. It seemed that at any moment it would go toppling into the sea. That it didn't was credit to the extremely skillful piloting of the army major in command of the aircraft."

After the chopper landed, the men all stood around on the rocks and chatted for a while. The major photographed the assembly for posterity as some of the villagers loaded goods into the helicopter. Kerry wasn't paying much attention to what was being put on board, although he did notice that many of the boxes were labeled Johnnie Walker, Ballantine, or Schlitz, and he wondered just what kind of operation they were running at this deserted naval base. As they all loitered on the rocks, Mike Bernique was holding forth on the day's operation to an attentive audience, as much because of his clear voice as for the hawkish flamboyance of what he was saying. Once the provisions were aboard, Kerry reported, "We walked into the belly of the helo after fairly rudely refusing some petty officer who desperately wanted us to take an ensign and first-class petty officer out to the Coast Guard cutter *Ingham*. Somehow the urgency for our use as a water taxi was met with an inward hostility after the events of the day, and I reasoned that our rest was earned and that I didn't have what it took to get the crew under way again. Besides, I could taste a beer and the enjoyment of the helo ride and wasn't about to freely give up either. Both Mike and I told the guy to commandeer one of the junks and send the Vietnamese navy on this important mission of mercy. Goddamnit, it was their war—let them do some work for once."

The whine of the chopper's rotors grated on Kerry's ears but exhilarated his spirit. How great it would be, he imagined, to have access to a biplane or a T-34 trainer to go screaming over the patchwork of waterways that made up the Mekong Delta in that "aero-inebriated fashion" that acted as such a tonic upon his mind and body. It occurred to him how much more sense it would make for him to be killed, if that were to be his fate in Vietnam, doing an Immelmann or a barrel roll or even a simple loop in a plane like that, rather than being shot by a sniper while stuck in the middle of a river on a loud, clumsy aluminum boat.

As the chopper rose above the ground and the Swifts below grew smaller and smaller, Kerry yelled out the open side and gave his crew the finger, at which they hooted and waved. "We circled briefly and went down on the other side of the island where we again picked up some cargo, this time from a hook that they lowered from the midsection," Kerry recorded. "And then, like the elevator in the Empire State Building, we shot skyward to the small plateau on the crest of the mountain."

Landing proved just as difficult this time as it had down below on the island, and the first attempt had to be aborted. "On the second pass, we let down with a small bump on terra firma and Mike and I got up and walked out of the tail of the bird," Kerry wrote. "A boy with a huge smile greeted Mike with a friendly 'Hello, Mr. Bernique, nice to see you again.'" By this time, of course, Bernique had become quite a legend, thanks to the daring Delta river raid that had won him the Silver Star. His boisterous demeanor made him even more unforgettable in person. As the mismatched pair walked the few feet from the small clearing where the helicopter had landed to the lighthouse at the very top of the mountain, they found themselves awed by the beauty of the remote spot. Below them on the windward side of the island, whitecaps broke across the rolling waves that smashed upon the rocks. "On the horizon was a wispy mist that reduced one's vision but nevertheless we could see tiny junks bobbing with the swells as the fishermen that manned them and worked the nets sought a small amount of protein to sustain themselves and their families," Kerry wrote in his journal. "On the leeward side it was calm and we could see across to the end of the Ca Mau peninsula that was the southernmost point of Vietnam. Below us was the village that they had just left behind with their Swifts at rest in the cove, like toy models in a bathtub."

Directly before them stood a French villa built of stone that Kerry later learned had been carried up the mountain on the backs of coolies in 1926, when the French had installed the lighthouse. A stone patio graced the rear of the villa and led to the gray tower of the lighthouse itself. Atop the structure, outside the glass that protected and magnified the light, extended a balustrade that afforded an even more inspiring view. A stone walkway surrounded the entire villa, and below it were the sides of the mountain and the jungle that engulfed them. Kerry and Bernique went into the living room, where they found a refrigerator bearing very cold and tasty beer. They immediately opened a couple and for a while just relaxed in the cool room, glancing through magazines, including an old *Playboy* and an issue of *Time* that to Kerry's surprise and delight informed him that Apollo 8 astronauts William Anders, Frank Borman, and Jim Lovell had been chosen 1968 "Men of the Year." A young Vietnamese woman with an engaging smile came into the room to set the table. Bernique turned to Kerry and said, "*Belle cuisses*"—French for "beautiful thighs"—at which he smiled in agreement and they both laughed.

Apparently, the eight-man U.S. Navy detachment billeted in this small paradise had a Vietnamese family living with them, the "Mama-san" of which did all their cooking and housekeeping. "A difficult existence to say the least," Kerry joked. "They talked briefly about the war and the idiocy of it all. They came to the miserable conclusion that perhaps it was a panacea for other human failings. The most interesting discussion centered around the strange behavior that war brought out in men—the way it changed them and forced them to try and prove things that were really inconsequential and stupid."

Unfortunately, their fine dinner had to be cut short because it was getting dark and there was no helicopter to take them back down the mountain. As it was, Bernique and Kerry had to follow the path through the jungle precisely, as they had no flashlights and there wouldn't be time to backtrack before it was pitch-black out. Their fortune-favored hosts kindly fixed up a backpack for Kerry to cart off his shiny new toys in, along with an invitation to return the next day to share the detachment's Sunday steaks. Bernique and Kerry, tipsy from the beer, set off from the Shangri-la and began plodding their way down the mountain.

"The path was well worn but it was steep and our legs were accustomed to sea and not hidden roots and stones," Kerry wrote. "In a slight state of inebriation then, after a morning dash with Charlie in a small river, after a transit through a calm evening sea, after an arrival at a Pacific Island lagoon and a breathtaking helicopter ride to a small oasis on Mount Olympus, and now an expedition through thick jungle down a steep mountain slope in the dark, I wondered if the day were real." Bernique turned to him and in his husky hero's growl remarked: "We'll have something to tell 'em over a drink in New York, won't we, John, m'boy?" Kerry recorded in his journal that all he could do was agree.

Chapter Twelve

Taking Command of PCF-94

The good time perched atop Hon Khoai was a rare reprieve from the war for the men of Coastal Division 11. Swift boat duty in Vietnam had become a dangerous gambit for its young officers. Gone were the coastal patrols—it was now mainly riverine warfare. "When I signed up for the swift boats, they had very little to do with war," Kerry said in a little-noticed contribution to a book of Vietnam reminiscences published in 1986. "They were engaged in coastal patrolling and that's what I thought I was going to be doing." Now the somewhat easy billet had turned deadly. Under Admiral Elmo Zumwalt's command, Swift boats were aggressively engaging the enemy. Zumwalt, who died in 2000, calculated in his autobiography *On Watch* that sailors in Operation Sealords had a 75 percent chance of being killed or wounded while on duty in Vietnam.

No neophyte naval officer in his right mind could have been comfortable with those odds, and neither were the PCF veterans. "Our divisions were on the brink of mutiny," PCF-98's Lieutenant Wade Sanders declared of the mood in the junior officers' ranks. "The people that were managing Sealords were sending us up the same river on the same day at the same time, week after week. They might as well have handed out our schedules to the VC. We were also all pissed off that we were being ordered to kill livestock and burn hooches even if the enemy wasn't anywhere around. There was just a lot of seething discontent when we arrived in Saigon."

Indeed, the Swift skippers would get quite a chance to air their views about Operation Sealords at a very high level. At 5:30 in the morning of January 22, 1969, a select group of Swift boat and other naval officers from Coastal Division 11 were awakened by the steward aboard an auxiliary personnel lighter (APL) barracks craft, and immediately swarmed the tiny restroom's three sinks to get cleaned up for a big day. From the moment they left the APL at 6:15, their schedules ran, for once, with true military precision. They were met at the An Thoi pier by several jeeps and a half-ton pickup and shepherded from there to An Thoi International Airport, which consisted of a single landing strip topped by a metal runway built by the Seabees. There the lieutenants climbed aboard an old DC-3 cargo plane and into the inward-facing canvas seats lining its cabin. A few minutes later they were airborne and heading for Saigon, to a remarkable and highly unusual meeting with not only the commander of U.S. Naval Forces, Vietnam, Admiral Elmo Zumwalt, but also the overall commander of the U.S. Military Assistance Command, Vietnam (MACV), Army General Creighton Abrams.

According to Lieutenant Rich McCann, a New Jersey native, the entire "Saigon summit" was sold to the junior officers as a "steak and beer" getaway. "It turned out to be hamburgers and Kool-Aid," McCann later laughed, "but, hey, we got a day off." Over the past months a special bond had been created between Admiral Zumwalt and the men of Coastal Divisions 11 and 13. "While we didn't like getting shot at," McCann recalled "we had great admiration for Zumwalt himself." On a number of occasions Zumwalt just landed on the beach at An Thoi in a helicopter and headed to the barracks to pat his young officers on the back and urge them to keep fighting hard. "He had a genuine affinity for the Swift boat officers," McCann recalled. "He treated us extremely well. So while many of us may have had beefs about Sealords, we didn't have any complaints about Zumwalt as our superior officer."

Upon the young officers' arrival at Tan Son Nhut Air Base, MACV's headquarters on the outskirts of Saigon, "a couple of photographers recorded our visit for posterity," Kerry reported in his notebook. Indeed, standing in the middle in the group shot, with Mike Bernique on his right, Kerry did look vaguely like a lankier young John F. Kennedy, in his Navy khakis and thick-thatched officer's haircut. At the airport they also "were met by somebody's representative and whisked in a bus through the streets of Saigon to the residence of Admiral Zumwalt," Kerry wrote. "The bus had screens across all the windows to prevent a terrorist from riding up on a motorcycle and tossing a grenade into the passengers' midst, and the result was that

I felt like an animal going to a circus or a prisoner being carted to court from the jailhouse. But it was for my protection, and so I settled down and watched the sights of civilization go by, for Saigon was Vietnamese civilization as we had not seen it in months."

From the road Kerry saw refugees milling about everywhere, most wretchedly in the squalid Cholon district of the capital, which made a place like the vice-ridden, impoverished Olongapo City in the Philippines look pretty decent by comparison. Yet other parts of the metropolis seemed downright inviting. "I could have driven and looked all day but we weren't there for pleasure and very quickly we were at the Admiral's residence and were ushered into the backyard where the members of Coastal Division 13 from Cat Lo were waiting," Kerry's journal entry continued. "I went up and saw my old division mates and again missed being in Cat Lo. Wade [Sanders] was there and Ked Fairbank, and there were smiles on all sides. I hadn't seen them since before my hasty departure from Vung Tau several weeks earlier. We didn't have very long to chat though before we were ushered into the admiral's residence and escorted to the living room, where they had chairs set up for all of us. We each selected one and then stood in front of them waiting for some signal to sit down. Before it came we heard 'Attention on Deck' and in walked Vice Admiral Zumwalt and General Abrams himself."

This was a big deal on both counts. To a history-minded junior officer like John Kerry, Elmo R. Zumwalt Jr. of Virginia was already a naval legend at age forty-eight. Known for his commonsense balance and steadiness, Zumwalt had graduated seventh in his class (1942) at the Naval Academy in a shortened wartime program. Zumwalt then served on the U.S.S. *Robinson*, a destroyer that saw action in the Battle of Leyte Gulf. He received the Bronze Star with Combat V for helping thwart a Japanese torpedo attack. Again he showed valor during the Korean War when, serving as the U.S.S. *Wisconsin*'s navigator he successfully brought the battleship through a maze of mined and restricted waters. By the time he took control of the U.S. Navy's Vietnam operations in September 1968, he was already the stuff of legend throughout the Navy. Fiercely intelligent and hardworking as well as exceptionally open-minded for a military man, Zumwalt would become the youngest four-star admiral in American history. Unorthodox in approach, Zumwalt wrote poetry in his leisure time and was more interested in military results than martinet minutiae. "Zumwalt truly adored the young officers who served in Vietnam," former Secretary of the Navy Paul Nitze recalled. "He never forgot what it had been like for him in the Leyte Gulf during World War II and then the Communist-infested waters around Korea. He knew that the backbone of the navy was its junior officers."

On January 22, 1969, Admiral Zumwalt, a perpetual smile in place, had to face a score of young officers in awe of him—and some of them were angry. Besides Operation Sealords, some of the men had questions about why he had ordered the aerial spraying of Agent Orange throughout the Mekong Delta.* They realized its purpose was to destroy the vegetation along the riverbanks, which the VC were using as sniper dens, but they worried that it might also be harmful to them. "The admiral took the stand and introduced General Abrams after we had been given permission to sit down and we listened with open ears for whatever special word the 'man' himself could give us," Kerry wrote in his notes, regarding the reckless nature of Operation Sealords and the promise of the recently renewed Paris peace talks. "For several weeks there had been conjecture that a major change in strategy was due and that we were going to be cut in on it here," he continued. "But no. The general, in his portly manner, talked to us about the conduct of the war and told us how what we were doing was terribly important to the war effort. He congratulated us and expressed his admiration and then exhorted us to carry on and continue in our present vein. The talk lasted about twenty minutes and about the only impressive thing was the fact that we had been flown to Saigon to be exhorted by a four-star general and the commanding officer himself. I turned with a questioning look to the officer next to me when it was over and we wondered together without saying anything what it was that we were meant to garner from this exhibition."

Kerry was hardly the only skeptic at the so-called Saigon summit. "They tried to pump us up," recalled Lieutenant (j.g.) Bill Schmadine of PCF-5, who had been skippering Swifts in-country since June 1968. "They didn't ask for our advice. But after months of being together, seldom mentioned in *Stars and Stripes* or patted on the back, it felt good to at least be recognized

---

**When Admiral Zumwalt's son served in the brown water Navy, he was exposed to Agent Orange. After Elmo III died of lymphoma in 1988, Admiral Zumwalt began lobbying politicians to appropriate necessary funds to treat the thousands of Vietnam veterans suffering from Agent Orange–related diseases. His efforts proved successful in 1996, when President Bill Clinton signed an order that made such ailments eligible for medical benefits from the Department of Veterans Affairs.*

by Zumwalt and Abrams." According to Lieutenant Larry Thurlow of PCF-53, the entire powwow occurred because a few lieutenants "whined" about being shot at regularly. "The whole thing smacked of a high school pep-talk," Thurlow recalled. "They made us convene to pat our backs and tell us to stop second-guessing." Reflecting back on the event decades later, Skip Barker recalled the snowballing atmosphere of dissent which led to the meeting in the first place. "Our division commanders had grown tired of hearing their officers express concern about our mission," Barker explained. "Collectively, at least most of us felt that running rivers, showing the flag, and shooting things up was too imprecise."

What Kerry, Schmadine, Thurlow, Barker, and the other junior officers could not have known was that MACV-commanding General Abrams—who had served as tank commander under "Old Blood and Guts" General George S. Patton in World War II—himself harbored deep-seated worries about the timetable for Vietnamization. It troubled Abrams, and Zumwalt for that matter, that the deadlines were continually being moved up as the U.S. government became more and more anxious to get out of Vietnam. What's more, there were no plans to maintain a permanent U.S. military presence in South Vietnam, as had been established in West Germany and South Korea after America's last two wars. In fact, when military historian Lewis Sorley interviewed Abrams about U.S. policy in Vietnam in the wake of the Tet Offensive, the general revealed views nearly identical to those of John Kerry. "We started out in 1968," Abrams told Sorley as early as March 1969. "We were going to get these people by 1974 where they could whip the hell out of the VC—the VC. Then they changed the goal to lick the VC and the NVA—in South Vietnam. Then they compressed it. They've compressed it about three times, or four times—acceleration."

As it was, the January 1969 meeting in Saigon did not leave the contingent of Swift officers either less peeved at or very impressed by Abrams; they were, however, glad to be getting Army helicopter support for their river runs. "Once the general had left we were shown over to the main part of the Naval Forces, Vietnam Headquarters and there we were given an intelligence briefing," Kerry recorded. "Again, we met with nothing that we didn't know and I suspect that any member of the officer corps in the division could have stood up and given a better presentation with no preparation at all. At the end of it, the admiral came back in and made a few more remarks—to the effect that the general had been very impressed with the cut [of the jib] of the men he had spoken to. The admiral [said he had] replied that he felt that in that room was a future Chief of Naval Operations, which seemed strange to me because almost everyone that I knew was planning to get out. Perhaps the admiral was referring to himself?"

Upon completing his remarks Zumwalt opened the floor to questions, and they came hard and fast. One of the officers from Kerry's division asked about the accusations Lieutenant Colonel William Corson had made in his recent book *The Betrayal*. Corson, an expert on Southeast Asia, had been sent to Vietnam to command a tank battalion in 1966, during which assignment he had learned how to win the confidence of the Vietnamese people. In a *New York Times* interview the following year, Corson offered what would become his primary thesis in *The Betrayal*: that the United States would lose the war if it continued to prop up what he considered the utterly corrupt Saigon regime. "The peasant sees that we are supporting a local government structure he knows to be corrupt," Corson noted, "so he assumes that we are either stupid or we are implicated. And he decides we are not stupid." Then the retired Marine counterinsurgency expert, who had served in three wars during his twenty-four-year military career, went on to detail the chronic dishonesty among local South Vietnamese leaders.

Zumwalt somewhat obliquely claimed that he had not read Corson's book. "First off, you've got to consider the source," Zumwalt snapped. Then according to Thurlow, the admiral shot a glare that said, "'You better not have a follow-up question, buddy.'" Lieutenant (j.g.) James Galvin of PCF-22 asked a question about an article he had read in the *Saturday Evening Post* about VC switching labels on materials in port cities like Danang and Saigon. Galvin, a native of Somerville, Massachusetts, and graduate of the UNSA class of 1966, was not complaining. He merely wanted to know what Zumwalt thought of the article, written by one of the authors of *The Ugly American*. Galvin, years later, laughing about the incident, admitted he was shot down. "I had my hat handed to me by Zumwalt."

Kerry decided to open another matter of concern, as he recounted in his journal:

> *I asked how, if our job was ostensibly interdiction of the movement of supplies, they could justify offensive actions such as we had been*

> *sent on—attempts to draw the enemy into ambush and then destroy his ambush capability. He said that the purpose was to show the American flag—an answer that seemed very strange to me when I considered that it was the Vietnamese flag that we were supposed to be fighting for. Why didn't we show their flag, or better yet let them run up the rivers and show their own flag? Many friends of mine in the Marines told me about their operational orders and necessity for artillery wherever they go. The admiral went on to say that he knew Navy men found it hard to go out and find the enemy, but that the Army did it all the time and that we should get used to it. I wanted to point out that the Army was equipped and trained differently than us and that they had some form of support beyond that which we had, but then I thought better of it and silence was the better part of virtue.*

After some parrying, the admiral was finally rescued by his trusty aide Captain Roy Hoffman, who stood up and made a few remarks about the unavoidability of innocent people being killed in Southeast Asia. Zumwalt declared this normal—fortunes of war, as it were, and to be expected. Then he went on to laud the actions of a PCF boat driver in Danang who had stumbled upon a cabal of VC off the coast of the Batangan Peninsula and killed them all. This was the aggressive kind of officer Zumwalt felt his riverine war needed. "Hoffman started quoting Winston Churchill, telling us Coastal Division 11 was doing the most important work in the U.S. Navy," Thurlow recalled. "We all looked at each other and thought, 'What is this crap?' "

In coming months, life would only get tougher for the PCF lieutenants. Jim Galvin, for example, would earn three Purple Hearts and two Bronze Stars with Combat V for getting shot at with pronounced regularity. Although he would go on to serve thirty years in the Navy, becoming a captain, he marveled about the unfairness of Operation Sealords in a singular regard. "We were only earning $55 a month for hostile-fire pay," he recalled. "That was the same amount I would later earn for working calmly on destroyers off the coast. It wasn't fair." With dreadful pay and the odds of getting shot at hovering around 75 percent, it was hard to fathom that all of these junior officers had *volunteered* for Swift boats. "I left this whole

Saigon operation just a little bit sicker and a little bit more depressed than when I came in," Kerry recorded in his journal. "Even the intelligence officer there told us he thought that what we were doing was a mistake, but he couldn't control his boss and that was that. The unfortunate thing about advisors is that they tend to tell the advised what they want or need to hear—particularly if they want to move up."

The bus ride back to the Tan San Nhut airport was bittersweet for Kerry, who had no idea when he might see motorbikes, girls, and cafés again. The quick brush with civilization made him ponder just what civilization meant to him anymore. All the accustomed, genteel people, places, and things he cared about—Julia Thorne, Dick Pershing, his parents, Yale, sports, Cape Cod, Groton House, and the rest—seemed somehow lost to him forever. For the first time he wondered how, *if* he survived Vietnam, he would ever be able to fit back into society again.

Back in An Thoi, Swift officers had to learn not to grow too attached to particular boats or crews because they were often reassigned to other available PCFs as soon as those they were used to had gone in for repairs. That is precisely what happened to Tedd Peck, whose PCF-57 went in for an overhaul the same day as the Saigon summit, after which he was assigned to replace Lieutenant Tom Heritage on PCF-94. "I was ticked off, because the M-60s on the 57 boat had never fired tracers; they were in mint condition," Peck remembered. "I was leery of the 94." And he wouldn't be on it for long. A week after he got the boat, on January 29, PCF-94 was tied up to U.S.S. *Terrell County*, an LST aboard which Peck and his crew were all sleeping soundly. At five o'clock, the morning still blanketed in predawn darkness, the 94's new skipper received orders to go out on a river raid. "I screamed, 'Goddamnit!'" Peck recalled, "and told my men we had to go."

Paul Connolly, a desk commander at Cat Lo, had ordered six Swifts, one with him aboard, up the Cua Lon River. Wanting to see some action for himself, once in motion Commander Connolly ordered two of the PCFs to go up one of the first canals they encountered. Boom! Boom! Boom! came the echoes from up the canals, some from explosions that left one of the Swifts badly damaged. "I thought this was going to be a short day," Peck reminisced in a 2003 interview from his home in Arizona. "It was obvious to me that Connolly, a desk officer, would want to get the hell out of the river system. And certainly there would be no more canals to explore."

But Peck's instincts proved wrong. As the six Swifts proceeded up the Cua Lon, Connolly again ordered a pair to head up yet another canal to look for signs of Viet Cong. Lieutenant Bob Hildreth's PCF-72 and Lieutenant Tedd Peck's PCF-94 got the call. Hildreth cleverly seized the backseat, immediately radioing: "Ninety-four boat, you go first." Peck had seniority over Hildreth, but pride kept him from dickering over the protocol. "Aye, aye" he replied, only a little snidely, and started up the dangerous canal. About two hundred yards into the waterway, Peck claimed he had an out-of-body experience, actually feeling his spirit leave his corporeal form. "I was resigned to the fact that I was going to die," Peck explained. "I was looking down on my boat watching my men prepare for a firefight."

And what a crew they were. The crackerjack outfit that manned PCF-94 began with Leading Petty Officer Del Sandusky, a burly five-foot-nine, 220-pound natural pugilist who wore his black horn-rimmed glasses even in his sleep and served as the no-nonsense soul of PCF-94. Born December 30, 1943, in Streaton, Illinois, Sandusky grew up mostly in Illinois, Florida, and California; his father worked for Shell Oil and the Santa Fe Railroad and moved the family around constantly. After graduating from high school in Chicago, Del Sandusky went to Navy boot camp in Coronado. Eager to serve abroad, he got his orders to Vietnam in early 1965, just as President Johnson was escalating U.S. troop involvement in Southeast Asia. Sandusky quickly rose to the post of division staff coordinator in Danang, but when Operation Market Time got under way, he was pulled from his desk job and assigned to PCF-57. "That's when I began seeing real river action," Sandusky recalled. "By the time I was assigned to PCF-94, along with Gene Thorson and Michael Medeiros, we were in the thick of it all."

The Swift's engineman, Eugene "Gene" Thorson, nicknamed "Thor," was born on September 21, 1945, in Fort Dodge, Iowa. His father, a farmer who grew soybeans and corn, had been exempt from U.S. military service in World War II because he was helping to feed the troops. In 1965, the less agriculturally inclined young Gene Thorson signed up at the Navy recruitment office in Ames, Iowa. He was sent for training in San Diego and Charleston, South Carolina, specializing in repairing combustion engines. After two yearlong stints, in the Mediterranean and off Puerto Rico on a minesweeper, Thorson was sent to Vietnam in May 1968. Before long, word of his mechanical talents got out, and Thorson became coveted as a Swift repairman, a man who could fix engines even after they'd been damaged by exploding underwater mines. His crusty-bumpkin demeanor and ferocity in battle inspired Thorson's friend and crewmate Michael Medeiros to dub him Popeye the Sailor.

Medeiros, PCF-94's twenty-one-year-old boatswain's mate, was born on January 14, 1948, in Oakland, California, where his father worked in a mill that made prefabricated doors and windows. In 1965, while a junior in high school in San Leandro, California, Medeiros joined the Naval Reserve. Infatuated with the study of electronics, he believed the Navy offered him the best career opportunities in the field. After he studied at Chabot Junior College, his first Navy assignment was to a diesel submarine based in Hawaii. From there he was sent to Vietnam. On May 20, 1968, he started a yearlong tour of duty. Smart, conscientious, and organized, Medeiros made friends easily while serving on a number of Swift boats, including PCFs-57 and -94. Only five foot six and 130 pounds, Medeiros earned a reputation for acting as chaperon when his buddies went looking for trouble. Patriotic to the core, he would remain a Navy man, staying in the Reserves until 1983. "My father had been in the Navy in World War II," Medeiros explained. "So I learned to love things nautical. I joined the Sea Scouts as a boy and was hooked. I had no antiwar sentiments at all while I was in Vietnam."

The gunner's mate on PCF-94's crack crew was David M. Alston, born on April 3, 1947, in Rock Hill, South Carolina. Growing up African American in the deep South during the Jim Crow era had not been easy on Alston, who had suffered many encounters with the most visceral bigotry the times had to offer. Anxious to escape the South, he moved to Chicago and joined the Navy. On his first assignment, aboard the destroyer escort U.S.S. *Newell*, Alston was disappointed to discover a fair amount of racism in the Navy too. "On board ships," he noted later, "it was always white and black."

While on the *Newell* in San Diego, Alston often helped refuel the Swift boats used for training at Coronado and liked the idea of being on one. In addition, he averred, "I wanted to become more a part of the Vietnam War. So I volunteered for Swift School, and got accepted." Every one of his classmates in the Swift program turned out to be white. On the first day of classes Alston overheard one of them, Mike Manno, point him out, saying, "Looky there—we've got a black!" More determined than ever to succeed, Alston made it through Swift training with flying colors, then was sent to Warner Springs, like Kerry, for SERE training in the harsh California desert. One afternoon the already depleted trainees were forced to march miles across the arid landscape in scorching heat, and Manno ran out of water. None of his fellow Caucasians would share their canteens with him.

"Only I stepped to the plate," Alston remembered. "He was forced to drink out of my—a black man's—canteen." And from that moment on, the two became friends. "We were tight after that," Alston continued. "That racism b.s. was gone."

Upon arriving in Vietnam, Alston was assigned to PCF-21 and then to Tedd Peck's PCF-57, on which he was wounded during a January ambush. He had taken two bullets and had to be medevaced to a Coast Guard cutter. As he was being moved onto the operating table, a small King James Bible fell out of Alston's back pocket. "The doctor handed it to me and said, 'You should always keep this. It saved your life,'" Alston related. "And you know, it did. God must have heard my prayers when I was being shot at, for I had so many rockets flying by me I should've been dead."

Rounding out Peck's five-man crew on PCF-94 was Radarman Thomas M. Belodeau, whose shy, demure demeanor masked a fighting instinct that had already made him a decorated seaman. On July 5, 1968, Belodeau had been serving on PCF-27 when he spied a Viet Cong suspect running from a riverbank and went after him. As enemy fire exploded all around, Belodeau had gone in and pulled the suspected VC from the water for interrogation, earning a Bronze Star with the Combat V device for his bravery. "I cannot adequately convey or describe to you the measure of this man at war—screaming up a river in the dead of night, no moon, fifty yards from Cambodia, literally bouncing off the riverbank, waiting for a mine to go off or a rocket to explode," Kerry would later marvel at Belodeau. "And always, always dependable—always there for the rest of the crew."

As Lieutenant Tedd Peck and his brave crew ventured up the canal off the Cua Lon early in the morning of January 29, all six sets of eyes carefully scanned the thick jungle on both sides of the waterway in hopes *not* to see any Viet Cong. Although the men of PCF-94 all had on flak vests and pants as well as their helmets, the tremendous roar of the Swift's twin engines made them feel awfully exposed, having to concede to the enemy the element of surprise. And the VC indeed took advantage of it; suddenly there was a booming explosion that literally lifted PCF-94 off the water. Peck was standing in the pilothouse doorway with an M-16 at the ready to start strafing the jungle. Just as he pulled the trigger he was hit by two machine-gun bullets, one in his arm and the other in his chest. The Swift's only chance to survive the attack depended on helmsman Del Sandusky's getting them the hell out of there. If the pilothouse got blown up, the entire crew was as good as dead. "It was the most intense moment of my life," Sandusky claimed, looking back. "We were under fire by the VC on both banks; their rifle fire was coming at us from all directions. With Peck down they knew I was in charge. The radio was gone; the radar was gone. I shouted on the loudspeaker, which *was* working, where Thorson and Belodeau should aim fire."

Although bleeding profusely, Peck managed to continue firing back, covering Sandusky as he hightailed PCF-94 out of the canal at full speed. At that point a third bullet hit Peck, smashing into his ankle so hard that it broke his leg. The pilothouse had now caught fire and, as he clutched onto the M-79 grenade launcher, Peck realized he couldn't move. Meanwhile, Sandusky recalled, "I was in zombie mode. I just kept saying, 'Oh, shit,' 'Oh, shit,' a hundred times." David Alston, who had been grazed in the head by a bullet and took another in the arm, was fighting away like a madman when Tommy Belodeau rushed into the pilothouse with a fire extinguisher just in the nick of time. With so much ammunition on board, had the fire spread, PCF-94 and her crew would have been blown to smithereens. "Only God saved us," Alston recalled. "It was a miracle. We should have all been gone."

As he struggled to snatch his breath, Peck remembered, he suddenly smelled fresh air, which meant Sandusky had pulled them out of the canal back onto the Cua Lon River. They were safe. The entire skirmish had lasted only about ten minutes, though it had felt like several days in hell. "I didn't know how bad I was hit until I saw Bob Hildreth looking over me," Peck recalled. "I couldn't talk; my vocal cords were tight as steel. I was going in and out of shock. They were trying to give me morphine, but I waved them off. In case we were hit again I wanted to be able to swim."

A Coast Guard cutter arrived on the scene, and its medics put Peck in a stretcher and ordered him taken to a nearby high weather endurance craft (WHEC). A medic there immediately began pumping Peck with IVs and irrigating his wounds with saline solution. Although in terrible pain and only half conscious, Peck remained as feisty as ever. When a few men on the WHEC snapped photographs of him lying there wounded, he excoriated them as "ghouls" and glory seekers who wanted to send home proof that they were stationed near enemy territory. "Don't take my picture, you assholes," Peck kept screaming while flipping them the bird with his one good arm. Once a helicopter arrived, Peck suddenly found himself spinning around in the sky dangling from a rope, the deck of the WHEC getting smaller and smaller below. The doctors at the MASH unit quickly decided that Peck was going home. The first two bullets to his upper body had gone right through him, but they had to take care removing the third one, which had lodged in his ankle. The next morning, Peck was medevaced to a Saigon hospital.

It was John Kerry who inherited Tedd Peck's boat, including PCF-94's first-rate crew. All five men of PCF-94 still loved their country and remained proud to serve it in the Navy. But even before the horrific attack they barely survived on January 29, the crack Swift crew had begun to come around to the view veteran Randolph S. Forrester would sum up in the October 2002 issue of *Vietnam* magazine: "Any government can make mistakes." After seeing Lieutenant Peck so badly wounded, PCF-94's brave complement had new reason to wonder whether Uncle Sam was sacrificing its youth for a lost cause that could easily take them next. "We were in shock for a few days," Sandusky explained of the aftermath of the doomed mission. "We eventually got PCF-94 to a Coast Guard cutter. I got a welder to fill the gaping hole. But even when the boat got fixed, I wasn't. I still have bad dreams about the ambush. I can't watch war movies. Period. I still struggle with PTSD [post-traumatic stress disorder]."

Before Kerry was assigned to take over PCF-94, Commander Elliott asked Del Sandusky what he thought of Kerry. Sandusky had some reservations. "I was leery at first," he recalled. Years later, Sandusky explained that after the Peck ambush, the crew was suspicious of any new officer. "But Kerry turned out to be excellent," Sandusky admitted. "Any doubts I harbored were dead wrong. He had real courage and ended up saving my life."

Upon meeting his new officer in charge on January 30, Gene Thorson felt PCF-94 was in good hands with this young lieutenant. "From the start Kerry was a leader," Thorson recalled. "He kept you on your toes. He motivated you; kept you alert, told you you were the best. He used to say, 'We're gonna go get the VC, so be ready.' Our weeks with Kerry, however, were scary. We were deep, deep, deep into VC territory. We were a traveling bull's-eye, basically. If you couldn't hit us, you had a real bad pair of eyes."

New Englander Tommy Belodeau felt an immediate kinship with his new lieutenant based on simple regional pride. "The crew didn't have to prove themselves to me," Kerry explained in retrospect. "I had to earn my spurs with them. When the chief petty officer, Del Sandusky—known as Sky—finally gave me the seal of enlisted man's approval, Tommy was the first to enthusiastically say: 'I told you so, Sky—he's from Massachusetts!' "

Spending weeks together on a small boat could create friction among six disparate individuals, especially over small things like taste in music. This never became an issue on PCF-94 under John Kerry, who shared his crew's preference for rock 'n' roll on cassette tapes over the offerings of Armed Forces Radio, which they generally listened to only while fixing up the boat or tuning its engines. Kerry's favorite group was the Doors, and he would blast "Light My Fire" and "Love Me Two Times" while patrolling the Delta rivers, finding empowerment in Jim Morrison's powerful baritone to send the VC fleeing. The psychedelic Los Angeles band had taken its name from British author Aldous Huxley's *The Doors of Perception*, a 1954 account of his experiments with hallucinogens. Kerry, however, never touched drugs in Vietnam. "I like the lyrical intensity of The Doors," he explained. "Morrison was a poet, in my opinion. The drug part didn't interest me."

Indeed, every man he served with maintained that Kerry never took so much as a puff of marijuana, which he, in fact, banned from his boats even on trips to Saigon, where packaged joints were sold under cigarettish brand names like Craven A and Park Lane. It troubled Kerry how easily available opium was throughout Southeast Asia, and its popularity among U.S. servicemen there downright sickened him. The statistics alone were disturbing: in early 1969, when he took command of PCF-94, the mountainous Golden Triangle region of Burma, Laos, and Thailand was producing more than a thousand tons of raw opium annually; the area's largest refining laboratory turned a hundred kilograms of it a day into heroin aimed for sale to

American troops in South Vietnam, at two or three dollars a dose. Although Kerry claimed never to have seen anyone use the narcotic, a 1974 U.S. Office for Drug Abuse Prevention study would report that a staggeringly high percentage of U.S. servicemen surveyed admitted having "commonly used" heroin in Vietnam.

In addition to the Doors, other favorite PCF-94 tapes included Bob Dylan's *Blonde on Blonde*, Van Morrison's *Astral Weeks*, and the Grateful Dead's *Anthem of the Sun*. When the skipper was really in a mood for musical intensity, his men would agree to blast some Bach or Mozart, shrouding the Swift in a strange symphonic magnificence hanging heavy in the jungle air. In any case, the men on Kerry's boats all maintained that they played music on patrol to alleviate their own boredom, rather than to frighten the Viet Cong, as in Francis Ford Coppola's movie *Apocalypse Now*. Yet Kerry—and all the men of PCF-94—reject the Coppola version of rock 'n' roll Swift boats zooming up and down the Mekong Delta rivers like psychedelic sailors. "Not content with rock background music and smoke bombs and unending river banks littered with bodies and burnt helicopter shells, Coppola reinforces his purely hallucinatory image of Vietnam by showing us some final moments of the river trip through the eyes of an acid-popping member of the patrol boat crew," Kerry wrote in a *Boston Herald American* review of the film's premiere. "And then, adding insult to injury, Coppola, by introducing Dennis Hopper as a demented journalist, brings us the Easy Rider of the Cambodian Jungle. The vision with which Coppola began the movie seems to have eluded him. Somewhere, just like the country in its involvement in Southeast Asia, Coppola got lost."*

The simple fact was that the VC were not that easily frightened, much less that easy to recognize. The battlefield was everywhere in Vietnam, the enemy sometimes a barefoot child carrying a bomb in a satchel. As a result, for the most part the rule on a Swift boat was "better safe than sorry." Every Asian was seen as a potential sniper. If a noise came from the thick mangrove on a riverbank, it was deemed wiser to spray the entire area with machine-gun fire than to make a closer investigation. And if in doing so one accidentally killed a civilian, it was better to keep it to oneself. The unspoken

---

**The one fairly realistic Hollywood portrayal of special forces in the Mekong Delta has been* Go Tell the Spartans, *a 1978 movie based on Daniel Ford's* Incident at Muc Wa *(1967).*

standard operating procedure in Vietnam was to concentrate solely on one's own life.

One of the most horrific moments of Kerry's tenure in Vietnam occurred late one afternoon toward the end of that winter, after PCF-94 had headed out from the LST at Song Ong Doc and down the coast to the mouth of Square Bay. Accompanied by Lieutenant Rich McCann's PCF-24, they anchored near the entrance to the Cua Lon River, shut down their engines, and slept until fifteen minutes before their night patrol began at ten thirty. "Rich led the way into the river and got lost near the entrance," Kerry recorded. "Then he ran aground and we began a comedy of trying to get back into the channel."

The night was pitch-black, neither Swift's search or boarding lights were working properly, and both boats kept getting stuck on the bottom of the shallow channel. "Many minutes of silent patrolling had gone by when one of the men yelled, 'Sampan off the port bow,' " Kerry wrote. "Everybody froze and we slowed the engines quickly. But the sampan was already by us and wasn't stopping. It was past curfew and nothing was allowed in the river. I told the gunner to fire a few warning shots and in the confusion all the guns opened up. We moved in on the sampan, and taking one of the battle lanterns off the bulkhead shone it on the silhouette of the craft that was now dead in the water."

Technically, the two PCFs had done nothing wrong. The sampan, operating past curfew, was undeniably in a free fire zone; what's more, there had been more than a few incidences of sampans trying to get close enough to U.S. Navy vessels to toss bombs into their pilothouses. But knowing they were following official Navy policy didn't make it any easier to deal with what the crews saw next. "The light revealed a woman standing in the stern of the sampan with a child of perhaps two years or less in her arms," Kerry wrote. "Neither [was] harmed. We asked her where the men from the stern were, as one of the gunners was sure that he had seen someone moving back there. She gesticulated wildly and I could see traces of blood on the engine mounting. It was obvious that they had been blown overboard. Then someone said that there was a body up front and we moved in closer to see the limbs of a small child limp on the stacks of rice. She had already covered it, and when one of the men asked me if I wanted it uncovered I said no, realizing that the face would stay with me for the rest of my life and that it was better not to know whether there was a smile or a grimace or whether it was a girl or a boy."

Almost every American who served in Vietnam witnessed innocents getting killed. For many veterans, including John Kerry, it was these civilian casualties that would haunt their consciences the most. That late winter night in 1969, as the ill-fated sampan sank into the Cua Lon River, Kerry prayed that the memory would "not stay with me too vividly." But he knew it would, and in that sickening moment he despised the Vietnam War with a vengeance. As hard as all the men on both boats tried to console themselves that what happened had been an accident and that the VC were to blame because they often used women and children as decoys and shields, it was no use. Their feelings of guilt were beyond assuaging, for it was impossible to rectify a mistake that resulted in death, or to rationalize it in their own minds. "The child was still dead," Kerry wrote, "and we had done it."

Many sociologists have written about ships as "total environments, floating human laboratories in which even the tightest-lipped men will reveal their deepest secrets and inner truths to one another." It's a psychological theme that Herman Melville explored in *Moby Dick*, and Stephen Crane in his celebrated short story 'The Open Boat." The confessional phenomenon seems to hold true for shorter-term seafaring vessels as well, including in Vietnam. Yet on the Swift boats that plied the Mekong Delta, one truth of war was tacitly banned from so much as being mentioned: American atrocities. It was all right to criticize—as General Bruce Palmer Jr. would in his 1984 book *The 25-Year War: America's Military Role in Vietnam*—the U.S. free-fire-zone policy, on the grounds that its reliance on random military might resulted from poor intelligence gathering on the ground. One could even get away with speaking of the slaughter of as many as four hundred unarmed Vietnamese civilians by a U.S. infantry company at My Lai on March 16, 1968. But within the close confines of a Swift boat, no one ever used the word "atrocities." The potential for them made the term too psychologically loaded under the circumstances.

In retrospect, the result of the American media's lingering focus on civilian deaths committed in Vietnam by U.S. soldiers, particularly the My Lai incident and the Thanh Phong incident, is that all the brave humanitarian deeds of the U.S. soldiers and sailors have been overshadowed. Even some of the images Americans associate with their nation's shameful behavior in Southeast Asia, in fact, portray nothing of the kind. For example, Associated Press photographer Huynh Cong Ut's 1973 Pulitzer Prize–winning shot of napalm-burned nine-year-old Kim Phuc, running naked and screaming down a village road has become an emblem of the terror the United States wrought upon the region. What remains rarely noted is that this frightened girl was fleeing in anguish from North Vietnam's bombing of her village of Trang Bang. The United States actually had nothing to do with that incident. And while Bob Kerrey came under fire in 2001, soon after he left the U.S. Senate, from the *New York Times*, *The Nation*, and CBS's *60 Minutes* for his role in a Navy SEAL unit's alleged joy-torching of the village of Thanh Phong, no such attention has been paid to the other side of the story of what American servicemen did in South Vietnam, such as the rescue of forty-two civilian detainees from starvation by Lieutenant John Kerry and his crew.

Just before Kerry took over PCF-94, he was ordered to take a Swift on a "show the flag" mission on a tributary of the Duong Keo River with a few other boats. PCF-44 had gone in for repairs, so his assigned vessel and its crew were entirely new to Kerry. Lieutenant Skip Barker, whose PCF-31 went on the mission, provided the most detailed account of what ensued after the Swifts got on the Duong Keo. Every hundred or so yards the boats moved forward, he remembered, the tributary grew narrower. Then they saw signs of habitation in the form of crudely built hooches on the left bank. "We kept going until the stream and its banks' vegetation narrowed to less than fifty feet and decided to turn around," Barker recalled. "We drove our bows into the vegetation and mud banks, twisted, and finally started back down. As we approached the clearing on our right, one of the gunners reported seeing someone run into a dirt mound back near the tree line. We used five-hundred-watt speakers to get anyone within hearing distance to come out [and] give up. No one appeared."

The officers could see earthen mounds built near the tree line, situated some seventy to a hundred yards back from the riverbank the Swifts' bows were now touching. The crews kept calling to whoever was back there to surrender, still to no avail. It was at that point that Kerry, clutching his M-16 and clad in a green T-shirt, camo pants, flak vest, helmet, and boondocker boots, jumped off his boat onto the bank. "It was a truly dangerous, ballsy move," Barker marveled. "[He] must have been asleep or absent during the

Swift-school session at Coronado when we were taught about *punji* stakes, booby traps, and [the] other terra-firma horrors we were privileged to avoid by being in the Navy—[that is,] on water."

As a rule, Swift officers in charge were not supposed to leave their boats to go ashore after the enemy, but Kerry chose to disregard that rule. To just sit in the river, he decided, was tantamount to suicide. Another crewman quickly joined him and as they crept up to the suspicious mounds, dozens of Vietnamese women, children, and old men popped up with their hands in the air. Keeping his M-16 trained on this assemblage, Kerry directed them all to the riverbank. "A search of the mounds, now known to be shelters, produced some NVA flags, pictures of Ho Chi Minh, and U.S./V.N. PsyOps packages," Barker recalled. "The people were scared, thin, weak, probably diseased, and except for the children, crying. The children were curious and touching these strange, armed foreigners."

Concerned about their detainees' obviously shaky health and well-being, Kerry reported their location to the nearest LST and requested instructions as to where to take the civilians. He did not like the encrypted radio response he got: tactical command had no interest in his forty-two noncombatant detainees. All they wanted to know was the number of enemy combatants killed in action (KIAs). "Negative replies evoked more inquiries as to KIAs," Barker related. "A decision was quickly made by Kerry: [the] Swift boat officers were ordered to load the people up and take them to the LST."

Instead of burning the hooches, Kerry was set on saving the half-starved civilians' lives, and the other men on the mission quickly got into the spirit of what he was trying to do. As the column of Swifts proceeded down to the Duong Keo, out to sea and the LST, the various boats' crew members rigged the stern decks with ponchos to shield the villagers from the sun. "Water, bug juice, bread, and canned fruit [were] shared," Barker described the scene. "Tears dried; hopeful smiles began to appear. Awe and fear reappeared as we came alongside the [LST]. The people were too weak or small to scale the cargo nets; davits with canvas bags were used to off-load them."

Kerry caught hell from his superior officers for bringing in the forty-two civilians. Nevertheless, he recalled, "I insisted that they get proper medical attention and were properly fed. For an afternoon, it felt good to really be helping the Vietnamese instead of destroying their villages. These people brought up the best in all of us." After they received treatment for dehydration, lice, and malnutrition, and were interrogated, the civilians were loaded on the Swifts and taken to Son On Doc City. "That one felt good," Kerry summed up the effort. "It's always good to help."

The men on the Swift boats found ever less satisfaction in the tasks they were ordered to perform. In fact, it became a black-comic routine among the crews to lampoon the logic behind their Operation Sealords river runs, each being deemed even more absurd than the last. The apotheosis of this lunacy may have come with the fittingly named Operation U-Haul, a pet project of the Swifts' administrative commander, Charles Horne, who was always itching to come down to An Thoi and run a river raid. Kerry liked Horne, but considered his Operation U-Haul pretty much a suicide mission. "This operation was so fantastic that it almost had to come from Al Capp or Charlie Brown to be for real," Kerry reflected. "For weeks there had been talk of setting up a SEAL outpost at Cai Nuoc. They had already almost been rubbed out trying to get the SEALs up there to look at the village, and as a result of the VC reception they had met on that ferry run, the mighty SEALs were now very loath to put a foot on a Swift boat. Our following was decreasing, because the Army had already practically refused. Now we were losing followers from within our own fold."

Throughout his "War Notes" Kerry evidenced nothing but respect and admiration for the Navy's elite Sea, Air, Land Commando unit; he'd witnessed SEALs attacking the Viet Cong from river patrol craft, from underwater as frogmen, from helicopters, and by high-altitude parachute drops. When they first arrived in 1966, only a few SEALs operated in Vietnam; by the first two months of 1969 there were more than a hundred in-country, most of them attached to Task Force 116, the Navy's River Patrol Force in the Mekong Delta. The VC called them the "men of green faces," and feared the smart, tough SEAL teams more than any other American troops. "'Watch out for the men with green faces' was a phrase that troubled the sleep of many a VC, we found out later," recalled Senior Electronics Chief Leonard Waugh. "And that fear was something we wanted out there."

Kerry's primary SEAL-related job as a Swift boat skipper was to pick a designated team up, take them upriver, drop them off at their destinations to conduct their daring overland incursions, and then pick them back up when the commandos were finished. "They wanted to be inserted after dark in an area near the Cambodian border, where they would wait in ambush for VC squads or sweep through a village and try and find someone willing to pass information on to them," Kerry recorded in his notes.

He and his Swift boat cohorts appreciated the SEALs above all else for their facility at detecting and defusing underwater mines. Whenever a PCF had to enter a canal, for instance, it certainly eased the crew's fears to have a SEAL diving into the water ahead of them to make sure the course was clear. The commandos were also founts of useful information about the enemy. As historian Kevin Dockery wrote in his *Navy SEALs: A History of the Early Years*: "The SEALs in Vietnam were in a unique position to take immediate advantage of any intelligence that came their way. Using their own intelligence networks, as well as information brought to them from other sources, the SEALs could quickly put together as complete a picture as possible of a given target. And with their experience, organization, and support, the SEALs could react to a sudden target faster than any other U.S. military organization."

Neither Kerry nor the other Swift officers and crewmen objected to the basic goal of Operation U-Haul, which was to use the PCFs to support the SEAL operations in the Delta. What troubled them was how they were ordered to do so, as the plan sounded insanely dangerous. On a clear February 1969 day, Commander Horne arrived in An Thoi decked out in faded greens and a brand-new jungle-camo hat. He was eager to go on a key operation to the Ca Mau Peninsula's An Xuyen Province. Over the past few weeks, he explained, some of the finest military minds in Cam Ranh Bay had been put to figuring out the best way to bring fuel up the Bay Hap River to supply the special, very small gasoline-powered boats the SEALs used for many of their incursions into VC territory. "The SEAL boats had Ford V8 gasoline-inceptor engines," noted PCF-31's Skip Barker, "just like those highway patrol cars. The SEALs burned up their gasoline like crazy."

With all the logistical capabilities of the massive U.S. military forces in Vietnam, the geniuses at Cam Ranh had concluded that the best way to get this highly flammable fuel up the Bay Hap—a waterway notoriously infested with VC snipers, who had claimed many American lives and boats on it—would be for Swift boats to tow huge rubber bladders full of gas five hundred yards off their sterns all the way up the river to the village where the SEALs

:hard Kerry in front of his favorite airplane. (Courtesy of Julia Thorne)

Above: John Kerry and Julia Thorne moments before a flight. (Courtesy of Julia Thorne)

Left: John Kerry at Yale University, where he headed the political union. (Courtesy of Julia Thorne)

As head of the Yale Political Union, Kerr got to meet with man dignitaries. Here he i with Senator Thomas R. Kuchel of Californ in 1965. (Courtesy of John F. Kerry)

Kerry with President John F. Kennedy at the America's Cup races in 1962. (Courtesy of the JFK Library)

rry with close friend David Thorne in front of the U.S.S. Gridley *in Long Beach, California. ourtesy of Julia Thorne)*

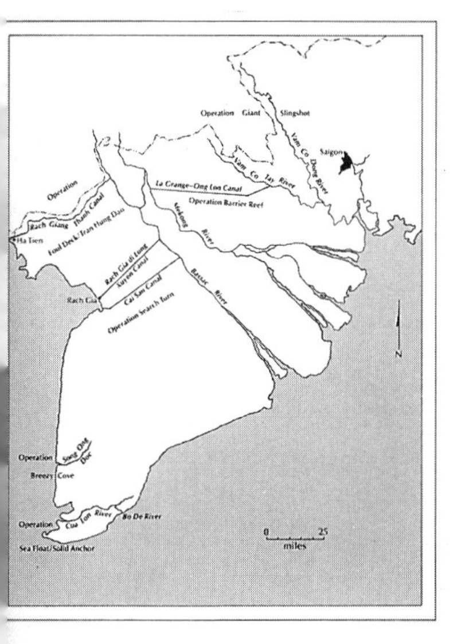

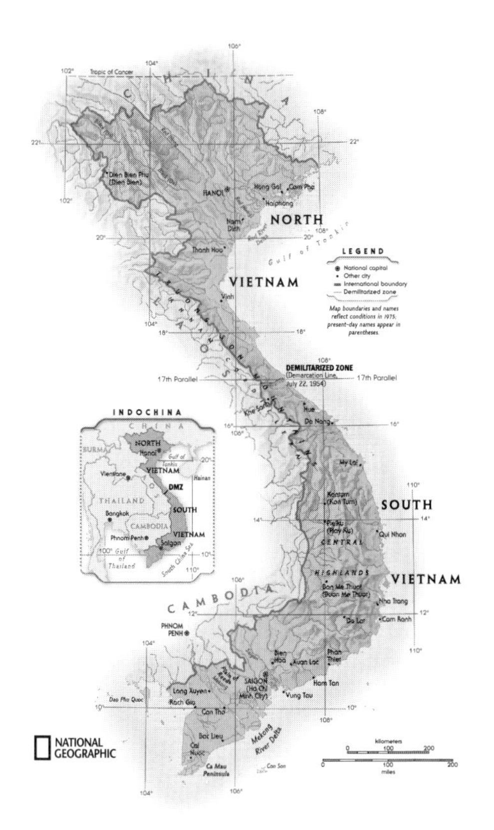

ove: A map of the Mekong Delta and ι Mau Peninsula in South Vietnam at the ıe of Operation Sealords. (Courtesy of ıval Institute Press)
ght: Vietnam circa 1968. (Courtesy of ıtional Geographic Books)

Kerry (top row, fifth from the right) *at the U.S. Naval School Command in California. (Courtesy of John F. Kerry)*

River traffic in Vietnam. (Courtesy of Bill Zaladonis)

Kerry showing South Vietnamese how to use a camera. (Courtesy of Bill Zaladonis)

Beaching of a Swift boat. (Courtesy of Bill Zaladonis)

Helicopter photograph of Kerry's PCF-44. (Courtesy of the U.S. Naval Historical Center)

Pulling out a dead body. (Courtesy of Bill Zaladonis)

A captured Viet Cong guerrilla fighter. (Courtesy of Bill Zaladonis)

Wading through the mud. (Courtesy of Bill Zaladonis)

bove: Crew of PCF-94: (rom left to right) *Gene horson, David Alston, om Belodeau, Del andusky, and Kerry.* (*Courtesy of Michael Iedeiros*)

ight: Lieutenant (j.g.) erry aboard PCF-94. (*Courtesy of Michael Iedeiros*)

Above: U.S. Navy SEALs aboard PCF-94 preparing to be dropped off in the Ca Mau Peninsula. (Courtesy of Michael Medeiros)

Top: A mine-sweep operation. (Courtesy of Michael Medeiros)

ove: Two Regional pular Force bounty nters. (Courtesy of ichael Medeiros)

ght: A Swift boat on astal patrol. (Courtesy John F. Kerry)

Swift boat convoy heading up the Bo De River. (Courtesy of Stephen D. Hayes)

Oftentimes Swifts were ordered to patrol small canals in order to find Viet Cong strongholds. (Courtesy of Stephen D. Hayes)

Above: Kerry receives the Bronze Star for combat valor. (Courtesy of John F. Kerry)

Right: Kerry and his fiancée, Julia Thorne, in 1969. They would get married less than a year later. (Courtesy of Julia Thorne)

Admiral Walter Schleck Jr. greeting Richard Kerry at his son's wedding in May 1970. (Courtesy of Richard Kerry)

erry writing a VVAW speech hile traveling in 1970. Courtesy of George Butler)

Kerry playing a game of "roof ball" in Cape Cod, Massachusetts.(Courtesy of Julia Thorne)

Kerry organizing a VVAW rally in Washington, D.C. The protest was known as Dewey Canyon III. (Courtesy of George Butler)

Above: Kerry's historic testimony before the Senate Foreign Relations Committee on April 22, 1971. (Courtesy of AP)

Right: Devastated by the emotion of the Vietnam veterans throwing away their combat medals and ribbons, Kerry, sitting with Julia Thorne on the far right, breaks down crying. (Courtesy of George Butler)

Kerry turns from the podium after addressing thousands gathered in front of the U.S. Capitol for an antiwar rally on April 24, 1971. (Courtesy of AP)

Above: Armistead Maupin with President Nixon at the White House. (Courtesy of Armistead Maupin)

Right: Kerry, leading the VVAW, walks with his hands behind his head as the police escort veterans and supporters from Lexington Green on May 31, 1971. It was his first—and only—arrest. The demonstration Kerry was participating in was known as Operation POW. (Courtesy of AP)

Above: Kerry with ex-Beatle John Lennon at a protest rally in New York's Bryant Park during the summer of 1971. (Courtesy of AP)

Left: John Kerry with his daughter Vanessa at a New Hampshire barn. (Courtesy of George Butler)

d Kennedy campaigns with Kerry in Lowell, Massachusetts. Kerry was seeking to win a U.S. ngressional race in the Fifth District. (Courtesy of George Butler)

A meticulous Kerry working on a speech in October 1972. (Courtesy of AP)

Above: Kerry rubbing the name of his friend Dick Pershing at the Vietnam Memorial in Washington, D.C., on March 26, 1992. The ceremony commemorat the memorial's groundbreaki (Courtesy of AP)

Left: Kerry walking with his close friend John McCain down a U.S. Senate corrido Together they fought to reso the issue of POW/MIAs and eventually reestablished U.S Vietnam diplomatic relation (Courtesy of AP)

kept their gas guzzlers. "And so someone had designed a special halter that would help the rubber balls stay buoyant and follow in a straight line behind the boat," Kerry reported. "Since the boats had to make the run up the river anyway with the balls behind them, someone else figured that they might as well carry other goods, and accordingly, they were loaded down with planks of lumber and with prefabricated, precut building materials."

According to both Kerry and Barker, the briefing beforehand took account of every imaginable detail. The Swift boat officers selected for Operation U-Haul had been herded into the wardroom of the LST where Commander Horne was to make his presentation. There he spread out several large sheets of brown wrapping paper, on each of which appeared a diagram of part of the mission. Neatly hand-drawn at the top of one sheet was a PCF-type hull, and on it were the carefully lettered words "Swift boat." A straight line extended from the artist's rendering of the stern, ending at a circle with "gas" written in the middle. Next to the line, midway between the diagrammed stern and the circle, was a box marked "miles." This was the distance at which the gas-filled bladders were to be towed by the Swifts.

Commander Horne then explained how important Operation U-Haul was to U.S. strategy in the Mekong Delta. Particularly important, of course, was that Horne's immediate boss, Captain Hoffman, would be watching the mission closely, which the commander clearly believed made all the preparation worthwhile. "We had to keep at the enemy," Hoffman recalled. "And fuel is an essential part of waging war. Getting gasoline to the SEALs wasn't fun but a reality."

So it was that the crew of PCF-94 were rousted from their bunks early on the appointed February morning to make a solo advance run up the Bay Hap River to Cai Nuoc. There, they picked up a company of regional South Vietnamese Popular Forces troops, then started back down the river to deposit the local soldiers at various strategic spots along the banks in hope that later that day they would cover the Swifts' slow progress back up the Bay Hap with the gasoline bladders. "I don't think that [Commander Horne] was quite aware how slow three knots would be when traveling through a hostile zone," Kerry remarked. "We spent about three hours getting into Cai Nuoc, unloading the troops and waiting for them to settle into their proper positions, and then finally returned to the [LST] where the loading of two other Swifts with lumber had already commenced."

Both of those Swifts' main cabins had been crammed so full of wood and other supplies that it became impossible to pass through them. The gunwales were likewise piled high with long planks that at least protected the cabin windows. A third swift was towing gas. Kerry's own PCF was anointed the mission's command boat, so on it embarked Commander Horne himself, who planned to run the operation as the officer in tactical command (OTC). For the past day Horne had kept telling Kerry that although he would act as the overall OTC, he would let Kerry run his own boat. Once aboard PCF-94, he kept his word and immediately told Kerry to take over. Lumber bulging from the Swift's sides, at Horne's command, PCF-94 made a tour of the other three boats also slowly making their way toward the coast.

Every two minutes, it seemed, Horne would have Kerry radio one of the other Swifts to ask how the bladders were holding up and whether the boats were going as fast as they could. Each time they got the same answer, but Horne kept sending Kerry back to the radio to ask again every few minutes anyway. The Swifts simply could not move any faster than five knots, because the bladders would start diving like submarines, putting them at risk of breaking against the bottom. As it was, the Swifts had to drag the sloshing rubber balls over mud on the flats to get into the river, slowing their speed down to a glacial one knot or so. "It was scary as hell," recalled Barker, whose PCF-31 trailed right behind Kerry's lead boat. "Just imagine—one sniper round in one of those bladders, and the river would have been on fire. And in all likelihood the Swift boats would have been blown up in a chain reaction."

A jittery Horne began asking Kerry more questions as the boat inched toward the mouth of the Bay Hap. Did he think they would get fired at? What had his experiences been on that river? It seemed so odd—just a month earlier it had been Kerry asking all the questions. Now he was the closest thing to a riverine warfare expert around. "But Horne was smart," Kerry recalled. "He asked a lot of good questions. He at least had the courage to get into the river with us."

After Kerry quickly rattled off his answers, Horne reaffirmed his desire that Kerry give the orders unless something drastic happened. Kerry had to laugh inside. He knew from experience that if anything drastic should happen there would be so much booming noise and wild confusion no one would hear any orders even if he managed to give some. As they chugged closer and closer to the river entrance, Kerry remembered a jittery Horne, who kept turning to Kerry and saying, "Well, Johnny, how do we look?" An empathic Kerry recognized the high-strung nervousness in Horne's voice—he sounded like Kerry had just two months earlier.

At long last the Swifts got onto the Bay Hap River. They proceeded up it uneventfully for a couple of miles, all the way at a painstaking four knots. Somehow they managed to negotiate all the bladders through the fish stakes, and nothing exploded. About a mile from Cai Nuoc, however, they took a few sniper rounds. PCF-94 returned fire for a moment, until Kerry realized they were too close to the friendly Popular Forces troops they had put on the riverbanks that morning. He ordered a cease-fire. Horne asked Kerry what had happened. Kerry replied that they had been shot at, and the commander broke into a broad grin and said, "Really?"

When they eventually arrived at the village, they couldn't find anybody willing to help unload all the lumber. Instead Vietnamese children clamored around the Swifties begging for handouts. For a fleeting moment they thought about enlisting the children to help carry the lumber. But they didn't. They knew that the children would disappear into their houses with planks, never to be seen again. So Horne supervised the Swift crews as they toted all the wooden planks into the village. A few children nevertheless managed to swipe wood. That arduous chore completed, the Swifties returned to their boats, and they set off to pick up the South Vietnamese troops who had been lining the banks of the river to cover them.

Finally the Swifts headed back to the LST, the crew of PCF-94 having spent some thirteen hours at GQ over the course of the day. Upon their arrival, one of the boats was called back to An Thoi, and Commander Horne was eager to leave on it. He ordered Kerry to write up the report on the mission, emphasizing that he should specifically detail how the gasoline bladders had reached Cai Nuoc safely. More important even than that, Horne wanted Kerry to make sure to mention his code name—"Rusty Jones"—in his report to make sure "people up there" would know he had been present on Operation U-Haul. According to Kerry, Horne wanted to prove that "the administrative commanders cared enough to get off their duffs and come down where the action was." The junior officer did exactly as the commander wished.

The longer he was in the bug-ridden Delta river system, the more Kerry's "War Notes" began to reflect his increasing distrust of the war's high command. He grew more and more uncomfortable with the notion that an American life was worth so much more than a Vietnamese life. "Most of the men found that 'it was really easy to kill a gook,'" he opined sadly at one point in his journal. "The popular view was that somehow 'gooks' just didn't have very much personality—they were ignorant 'slopeheads,' just peasants with no feelings and no hopes. I don't think this was true among most of the officers, and this made me wonder how much of it was feigned among the enlisted so that they would look good in the eyes of their more chauvinistic comrades." In fact, Kerry never heard any such callous barroom bravado from his own men, but the opposite: the entire crew of PCF-94 were as clearly shaken by the loss of the child on the sampan sunk in the Cua Lon as PCF-44's James Wasser had been after accidentally hitting the old man with the water buffalo a few weeks earlier.

"I know that most of my friends felt absolutely absurd going up a river holding a loaded weapon that was supposed to be used against someone who had never really done anything to you and on whose land you were now trespassing," Kerry revealed. "I had always thought that to kill, hate was necessary, and I certainly didn't hate these people." In truth, he added, scanning the shore for suspicious movements to shoot at made him "feel like the biggest ass in the world." The twenty-five-year-old had explored similar uneasy feelings in a letter he had written to his parents back on December 21, 1968. Describing the pleasant scene of American soldiers and their Vietnamese girlfriends strolling down the streets of Vung Tau one sunny afternoon, he reflected on the crucial difference between occupiers and liberators of war-torn places. "I asked myself what it would be like to be occupied by foreign troops—to have to bend to the desires of a people who could not be sensitive to the things that really counted in one's country," Kerry wrote. He had been reflecting on Germany's occupation of France during World War II, he added, when "a thought came to me that I didn't like—I felt more like the German than the doughboy who came over to make the world safe for democracy and who rightfully had a star in his eye."

Less than three months later, experience had brought him to another melancholy observation. "It was when one of your men got hit or you got hit yourself that you felt most absurd—that was when everything had to have a meaning in order for it all to be worthwhile, and inevitably Vietnam just didn't have any meaning. It didn't meet the test," he wrote in his diary. "When a good friend was hit and perhaps about to die, you'd ask if it was worth just his life alone—let alone all the others or your own." Kerry never could convince himself that it was.

"But the ease with which a man could be brought to kill another man, this always amazed me," he went on. Even more troubling to him was the official imprimatur the U.S. military accorded this coldheartedness. To illustrate his point, he referred to the messages that would come in from the brass at Cam Ranh, praising the Swifts' gunners whenever they had killed a few VC and ending, "Good Hunting." Kerry confided his distaste to his diary: "Good Hunting? Good Christ—you'd think we were going out after deer or something—but here we were being patted on the back and receiving hopes that the next time we went out on a patrol we would find some more people to kill. How cheap life became."

But sometimes risk taking actually saved lives in Vietnam. On February 26, PCF-94 was sent on a mission to establish a listening post not far from Square Bay. They were part of a two-Swift convoy. Suddenly their radar detected a sampan in a curfew zone, moving in zigzag fashion from left to right. "It was a free fire zone and we could have shot at it," Michael Medeiros recalled. "But Kerry didn't want us to." Instead the two Swifts chased the sampan with a searchlight. Briefly, their boat took VC fire—nothing too much. When they caught up to it, the sampan was motoring with nobody apparently onboard. The VC had jumped into the river to escape. They forced the sampan to beach. Then, in an extremely dangerous move, Kerry ordered it to be boarded and searched. "We found rice and demolition wires," Medeiros explained. "But no people."

An all-out search was on. Both PCFs beamed their searchlights into the muddy, marshlike terrain, to no avail. No human was in sight. Just as they were about to give up, they spotted two VC frozen onshore, refusing to move, hanging on fish stakes. They looked like muddy mannequins. Kerry ordered the bow of PCF-94 to approach them—a truly dangerous proposition. "We knew they were bad guys because we had been shot at," Kerry recalled. "So we weren't going to just ignore them and leave. We didn't want to shoot them, because that was against the rules." Because he was the tallest and the mud wouldn't rise to his neck, Kerry jumped off PCF-94 to apprehend these men.

Medeiros followed him into the mud, ordering the frozen men not to move. His crewmates had their weapons aimed directly at the two VC. One false move and they would have been blasted away. As they approached the men, somebody on PCF-94 yelled, "Skipper, there is a snake over there coming at you!" Ever since childhood, Kerry had been petrified of snakes. Many that he encountered in Vietnam were by far worse—one bite and you had about a minute to live. "Shoot the motherfucker," Kerry yelled. Boom, boom—Sandusky obliterated it with machine-gun fire. "Man, was I relieved," Kerry went on. "I could actually feel my heart tremble."

Even as bullets peppered the water near them and a poisonous snake was shot to pieces in front of their very eyes, the two VC didn't move. Medeiros recalled that they "stayed perfectly still. Frozen as if they were in shock. They may have had a weapon or bomb. We tied a rope to them and searched them, but they continued to play dead. It was bizarre. It was scary." Eventually, Kerry and Medeiros were able to get the two unarmed VC onboard PCF-94. One of the captives had a gaping gash on his leg. "It looked like our boat had run over him while he was trying to escape," Medeiros explained. "He needed medical help, although he wasn't bleeding that much." Kerry inspected the wound, wanting to apply first aid, but the femur was completely exposed. "Everything got cut off by our propeller," Kerry recalled. "All the muscle, tendons, everything, right to the bone."

Meanwhile, other VC started firing on the Swifts. Quickly, Kerry ordered the searchlight shut off.

Forced to abort the mission to establish a listening post, Kerry ordered PCF-94 to head directly to a Coast Guard cutter so the two prisoners could receive proper medical attention. "The two men refused to talk," Medeiros said. "Other lieutenants may have had them shot on the shore. But Kerry wanted them alive. It was the humanitarian side of him." The other PCF went on to capture three other VC in the same vicinity. It was all an evening's work in Vietnam.

Chapter Thirteen

The Medals

"The greatest honor history can bestow is the title of peacemaker," proclaimed President Richard M. Nixon upon his inauguration on January 20, 1969, a month before he would authorize the secret bombing of Cambodia and less than three months before U.S. troop strength in Vietnam hit its peak of 543,400 servicemen. Nixon had won the White House largely on his campaign pledge to forge "peace with honor" in Southeast Asia, but the first few months of his presidency sparked a flare-up in combat intensity, throughout the region and in every branch of the U.S. armed forces.

The upswing in battlefield ferocity was particularly noticeable in the Swift boat community, and certainly to Lieutenant (j.g.) John Kerry and the men of PCF-94. Between January 30 and March 13, 1969, the Swift crew would execute eighteen missions in and around the Mekong Delta river system; in just one hellish eight-day stretch, their boat would participate in some dozen vicious firefights. Before the winter was out, Kerry would have a Silver Star, a Bronze Star, and two more Purple Hearts to add to the front of his uniform.

The action really began to heat up for PCF-94 on February 18, when Kerry's Swift set out from An Thoi bearing mail for the LST that supported them, as well as for a squad of Navy SEALs who were headed up the Bay Hap River to Cai Nuoc to check out the proposed area for their future base camp. The elite commandos Kerry respected so much had been playing an ever larger role in Operation Sealords since the previous autumn. Indeed, Kerry and his crew would quickly be reminded of the SEALs' preeminence. On that mail run, after a pleasantly restful passage through the Gulf of

Thailand, the men aboard PCF-94 felt relaxed and confident by the time they pulled alongside the LST in the afternoon. The Swift crew were looking forward to a good night's sleep followed by some not-too-taxing operations over the next few days. But when they arrived at the support ship, the SEALs informed Kerry and his men that they wanted to get to Cai Nuoc that very night, so somebody would have to give them a ride up the river.

The other two Swifts working off the LST were already at the village, so Lieutenants Bob Hildreth and John Kerry and their crews were tapped to run their elite guests up the Bay Hap without delay. At least PCF-94 was still ready for duty; Hildreth, a New Jersey native, and his men had been out on patrol most of the day, and had already put their boat, PCF-72, to bed. Needless to say, they were not very pleased to have to head out again. "When you got back from a mission, you were exhausted," Del Sandusky recalled. "You'd have been running on adrenaline. You felt sleep-deprived. But when the order came to go back out, there wasn't much you could do. You had to repump the adrenaline."

Kerry took the lead, and as the two Swifts entered the river he noticed that the fish stakes that traversed the entire mouth of the Bay Hap were open in only two places, which was unusual. "Passing through fish stakes was always done with the breath held tightly and with the gonads drawn firmly up into the stomach," Kerry noted in his diary. "It is a favorite ambush spot because the VC know that the boats can only go through the opening that exists, or they must come to a dead stop and cut new ones—something they are obviously loath to do."

Hildreth told Kerry to take the left opening. He did and waited for an explosion as PCF-94 roared between the sticks protruding from the water a few feet off both sides of the gunwales. His Swift sailed through in silence, however, and Kerry was about to write the trip off as a milk run. But as he turned to watch Hildreth take his boat through the same hole, an earsplitting boom suddenly shook the river and a geyser of water shot up next to PCF-72. At the same instant the banks burst with flames in their direction, and Kerry watched in horror as a rocket streaked past his bow and exploded on the opposite bank. "We got bracketed with five B-40 rockets," Hildreth recalled. "We could actually see them go by. Then they started hitting us with small-arms fire." The Swifts' guns barked back as PCF-94 moved quickly to clear the ambush area. Another mine explosion then rocked the river just off Kerry's bow, inundating PCF-94 with water. "Luckily, we didn't feel anything, though—just the spattering of spark on the windshield," Kerry recorded. "Hildreth called me on the radio and said that he couldn't get maximum [rpms] out of his boat but that he had no major damage. For most of the remaining three miles to the village we fired recon fire on the banks and thought about the stupidity of what we were doing."

Once they made it to Cai Nuoc, they looked over their boats and saw just how close they had come to disaster. PCF-94's lifelines had been shot in two and there were several bullet holes through its hull, not to mention one in its battle flag and another in the pilothouse right above Kerry's head. Hildreth's PCF-72 had been equally lucky. Its flagstaff had been shot in half, and some of the bullet holes in his boat looked like high-caliber stuff. "We were even more pissed off because the ambush had obviously been meant for the boats that were already at Cai Nuoc and that would have been exiting the river had we not made our journey in with the SEALs," Kerry reported. "No one enjoyed taking potshots for someone else. This was rather a personal thing. We had sprung an underwater mine devised for somebody else."

That evening at Cai Nuoc, Hildreth and Kerry met with Michael Miggins, the U.S. Army advisor of the ARVN in the region, and asked him to call in some artillery to cover the Swifts operating in the area. Unfortunately Miggins and his Army cohorts instead brought up all manner of questions regarding who had jurisdiction over artillery in that zone. They never did figure out whose permission one had to ask to fire there, so in the end it was determined that, despite the deadly enemy threat, they did not have the authority to exchange tit for tat in that area. In any case, the Swift skippers had no desire to head back at night, considering the reception they had been given on the way in. In addition, they thought it might be wise to run a sweep through the zone the next day to see if they could find out what kind of emplacements the VC had been shooting at them from. Thus they spent the night at Cai Nuoc, letting the South Vietnamese patrol the river in the vicinity of their boats.

As they watched some Vietnamese children milling around the Swift boats, Kerry supposed that any one of them could have "lobbed some plastic in one of the windows." But he acknowledged that it would have been so out of character for this village that he stopped worrying about the unlikely prospect of anti-American sabotage. In fact, the place seemed so benign that some of the PCF crewmen wandered around the village's main street for a while, smelling the locals' food cooking over open fires and checking out how the Vietnamese lived after dark. "Going up the river we would always smell wood burning," Hildreth recalled. "And I recall all the outhouses along the canals. And at dusk the dogs in these villages were always barking, sounding an alarm. It was eerie."

Meanwhile, in the command shack from which the Army advisors operated, the rest of the visiting Navy men followed the incoming reports of Viet Cong movements detected by an American plane then flying reconnaissance over the Delta. It appeared that the guerrillas *were* circling Cai Nuoc, but the Swift boat crews were so exhausted they managed to get some sleep anyway. "On dangerous evening Special Ops missions, none of us really slept," Sandusky recalled. "We would try to establish one hour nap cycles. But it was near impossible. It was not easy to fall asleep when you knew your life was in danger."

The next day was wasted trying to get the South Vietnamese troops to conduct an organized sweep of the area. They went ashore in the wrong places several times and when they found the right ones, they refused to hike any farther than a thousand yards inland in fear of what might lay beyond. Instead, "they looted hooches as they went and came back to our boats laden down with chickens and clothing, and they refused to bring in people they found in the areas [where] they shouldn't have been," Kerry scoffed. "All told, it was a terribly effective way to risk one's life in the middle of a river."

Late that night the Swifts left the unhelpful allies back in their village and returned to the LST from what was supposed to have been a quick shuttle run up the Bay Hap River. Once they got there, Kerry and his crew received their orders for the next day's mission. PCF-94 and five other boats were to travel some fifteen miles up a river into VC territory where the American flag had not been shown in more than two years. "The canals in this area were really small," Hildreth recalled. "Branches hung over the canals. Wires were often across the water. We would cut them. Most of the people we met had never met an American before."

Lieutenant Hildreth was the senior Swift officer on this high-risk raid, but his PCF-72's radio wasn't working too well, so Kerry took on the duties of the mission's officer in tactical command. Only later would it occur to them just how dangerous it had been to send a boat with a faulty radio on that unpredictable a mission, but there had been no time to fix it and no other PCFs available. About four o'clock in the morning of February 20, the six Swifts converged toward their destination all the way around the tip of South Vietnam and then north up its South China Sea coast to the mouth of the Bo De River. They arrived a little after daylight to find the rocket-launching *St. Francis River*, a landing ship medium, rocket (LSMR), already there and awaiting orders. PCF-31's Lieutenant Skip Barker radioed over his opinion that it would be pretty stupid to shoot into the area before the Swifts entered the river, but the LSMR went ahead and unleashed a barrage of her mighty firepower into the Bo De and "softened the area up for us," as Kerry put it. The one thing the ship's rockets were sure to do was let the Viet Cong know the Navy was coming. "I couldn't get over how disorganized the raid instructions were," Kerry recalled. "It was mind-boggling."

After the LSMR bombardment, the Swift boats formed up in two lines of three and continued up the river. Memories of the Bo De massacre still made the crews nervous, so they had entered the waterway expecting the worst. They had all felt a little more secure once their support helicopters arrived on the scene just after the PCFs got to the entrance of the river. "Helos upset the VC more than anything else that we had to offer and any chance we had to have them with us [was] more than welcome," Kerry declared.

The trip up the Bo De proved one of the most beautiful transits Kerry had ever seen. The sun had just come up and the water reflected its dancing golden light everywhere, even on the smoke rising from the sterns. It seemed hard to believe that lurking somewhere in the mangrove or down at the mud line along the embankment there might be someone intent on shooting them. But no one did, and they droned on for miles through alternating flat plain and heavy jungle, passing houses along the way that struck Kerry as some of the most prosperous he had seen in South Vietnam. "Occasionally one of the boats would call and say that it had spotted people running off into the field, but they didn't fire," Kerry noted. "It was strange—we were traversing a Free-Fire Zone and our orders were to destroy all the hooches and sampans that we could find."

Take the fight to the enemy's backyard, Operation Sealords ordained, and thus the American Swift boats roamed the waterways of the Mekong Delta shooting up everything in sight. But on this raid no one was convinced, no matter what the orders said, that they were in a stronghold of the Viet Cong. "How, if no Americans had been up there in ages, did one know whether or not they could be approached by friendlier means?" Kerry wondered. "There was no doubt that after a trip through there by Swifts with guns blazing, randomly knocking down houses and sampans and shooting with an impersonality that betrayed the very reason that we were supposed to be there, more VC would have been created than would have been deterred."

And so the crews watched in shock at the contradictory nature of what they were doing as a woman ran for shelter with a child in her arms or an old man in a field ducked behind a stump. As the boats swept farther upriver, the waterway tapered down to only some twenty or thirty yards across, turning the Swifts into even easier targets. "As so many times before, the boats would disappear behind a bend until another boat rounded the turn and caught up with it, only to be lost again farther on," Kerry recalled. "Anywhere, a mine and a rocket could have wreaked havoc and probably would have resulted in guns from one PCF shooting at another."

When they finally reached the tiny, dilapidated outpost they had been sent to call upon, some of its inhabitants came out to greet them with the traditional outstretched hand; however, this was not a welcoming handshake but indicated "What have you got to offer me?" The Americans presented gifts of official PsyOps packets containing a small yellow-and-red South Vietnamese flag, some soap, and a fishhook or something likewise deemed appealing to Vietnamese peasants by the Pentagon. Then, feeling awkward at having nothing else to offer the villagers, many of whom looked as though they could use a doctor, and frustrated that they couldn't even communicate with the people because none of the crew spoke Vietnamese, the representatives of the U.S. Navy got back on their boats and left. "They were all so malnourished," Del Sandusky recalled. "Obesity was not the problem. They made us feel guilty leaving because they were clearly starving. We weren't. When we were hungry, we could always grab a bologna and cheese sandwich. The average annual wage for a Vietnamese male was $100 a year. But in those little villages the men made nothing close to that. They were surviving on fish heads and rice. And both were in short supply."

One of the mission's support helicopters had been hit by small-arms fire during the trip up the Bo De and the rest had returned with it to their base to refuel and get the damage inspected. While there the pilots found that they wouldn't be able to return to the Swifts for several more hours. "We therefore had a choice: to wait for what was not a confirmed return by the helos [and] give any snipers more time to set up an ambush for our exit or we could take a chance and exit immediately without any cover," Kerry recorded in his notebook. "We chose the latter."

One of the small spotter planes that flew strafing runs was in the air at the time and lent the Swifts some support on their way out of the treacherous river. Kerry had hoped to complete the mission without firing a single shot, ardently wanting to demonstrate that perhaps more could be achieved by approaching remote areas quietly and actually doing something to win the hearts and minds of the people there than by shooting their way into the Delta hamlets whether or not there was anything to shoot at. In the absence of the helicopters, however, it would be just too chancy to move out without reconnaissance fire, and so the Swifts broke the silence that might have proved so valuable.

Just as they moved out onto the Cua Lon, at a junction known for unfriendliness in the past, kaboom! PCF-94 had taken a rocket-propelled grenade round off the port side, fired at them from the far left bank. Kerry felt a piece of hot shrapnel bore into his left leg. With blood running down the deck, the Swift managed to make an otherwise uneventful exit into the Gulf of Thailand, where they rendezvoused with a Coast Guard cutter. The injury Kerry suffered in that action on February 20 earned him his second Purple Heart—and thus the second third of his ticket out of Vietnam (although not an ironclad rule, it was standard operating procedure for any U.S. serviceman who received three Purple Hearts to be moved out of the combat zone into a softer desk job).

The cutter's radiograph machine was out of order so the next day Kerry returned to An Thoi to have his leg X-rayed. While there he also had it out with Commander Charles Horne, who ran the base's Swift boat operations. "I tried to explain to him that we were risking the lives of Americans in an effort that was going to give us minimal return unless we got some support down here, and that even then it was a dubious enterprise," Kerry recalled. "I tried to point out that men were losing any idea of accomplishment when they spent hours in a river being shot at, and then came out to find that they could go back in the next day and that it was just as dangerous and that no territory had been secured. In fact, the only territory that we ever secured was that which stretched two thousand yards on either side of the river—namely, the range of the guns. And the idea that we were ever going to make these people our friends, when we went through shooting up their homes and cutting off the rivers [that were] their . . . livelihood—Christ, if they weren't VC before we went in, they were bound to be by the time we got finished." Commander Horne just agreed with Kerry and said he was just following orders that came from Saigon by way of Captain Hoffman, and that Kerry should therefore try to talk with him about all this.

Kerry's wound was not serious enough to require time off from duty to mend. Meanwhile, rookie Gunner's Mate Third Class Frederic Short was now taking David Alston's place manning the twin .50-caliber mounted guns as the South Carolinian recuperated from his January bullet wounds. Raised in North Little Rock, Arkansas, as a boy Short became a fan of the 1950s television programs *Navy Log*, an anthology series about Navy life, and *Victory at Sea*, a twenty-six-part documentary about World War II naval operations. Short enlisted in September 1967. After boot camp he was assigned to Amphibious Construction Battalion 1 at Coronado and then sent to Vietnam in February 1969. "Wow, was it wild from the start," Short recalled thirty-four years later. "I was a blank slate. Everything in the Mekong Delta was new to me. Kerry, Sandusky, Medeiros, and those guys had already learned how to survive. I was like a new child, having to learn *fast*."

Short felt lucky to join the talented crew of PCF-94, who would need all their combined skills and experience to make it through the harrowing next month alive. According to the invaluable handwritten personal log Michael Medeiros kept of every mission he went on in 1969, between February 8 and March 13 PCF-94's Officer in Charge John Kerry conducted daily river raids on Swift boat missions 81 through 98. All eighteen of these outings were risky, if variously intended. Sometimes the orders were to entice Viet Cong guerrillas out of hiding; at other times the crew gave C-rations and first aid to hungry and sick South Vietnamese civilians. On many occasions PCF-94 transported Navy SEALs on covert missions to search and destroy VC strongholds in the Ca Mau Peninsula or along the Cambodian border. Of all the patrols charted in Medeiros's log, however, Mission 91 on February 28 stands out. In the margin next to his entry for that date, Medeiros recorded that PCF-94 had killed nine Viet Cong and captured two others (although the official U.S. Navy report on the action reported ten VC fatalities). And it was on that day that John Kerry won the Silver Star, the Navy's third-highest combat award.*

The last day of February started out pretty much like any other morning for the crew of PCF-94. Along with two other Swifts, they were to transport thirty-odd South Vietnamese Popular Forces (PF) troops to a dangerous river on the southernmost tip of the Ca Mau Peninsula. Still called the Ruff Puffs by American servicemen despite having been split off from the French-created Saigon government's Regional/Popular Forces by President Ngo Dinh Diem in 1954—and then from his Civil Guard into the People's Self-Defense Force in 1956, before being renamed the plain Popular Forces in 1964—the PF was a village-based militia of poorly paid South Vietnamese volunteers who preferred staying closer to home than they could in the ARVN. Assigned to aid in the destruction of the Viet Cong infrastructure in the interests of "pacification," as the U.S. military now called it, the PF comprised nearly 152,000 troops. "The Ruff Puffs were usually around a small village," Commander George Elliott recalled. "They usually had an ARVN first lieutenant as their advisor, sometimes a sergeant or captain. We used them mainly for intelligence information. They often knew where the VC were camped. Some of their information was accurate but a lot of it wasn't. Sometimes they also patrolled the rivers."

Lieutenant Kerry was put in tactical command of the high-risk operation to deliver the Ruff Puffs plus a handful of U.S. explosive experts to a designated spot along the Dong Cung River in the peninsula's An Xuyen Province. The area was said to be thick with mangrove that afforded complete protection to VC snipers. Kerry, tired of being shot at, was thinking hard about how to win an ambush. "We were basically inserting South Vietnamese troops at a certain point in this river," Medeiros explained the mission. "And then they were going to do a sweep. After they checked for VC

**In the U.S. Navy's Order of Precedence, the Silver Star comes after only the Medal of Honor and the Navy Cross.*

we were supposed to pick them up and then take them back to their village. While they were hunting around onshore, we on the 94 and the 23 boat[s] were going to go farther up the river to act as a blocking force in case something came downriver. We would be like a blocking force for the troops that were doing the land sweep."

Deciding precisely where to insert the Popular Forces that day proved easy, as on their way upriver the three Swifts were ambushed and taken under fire by small-arms weapons. Realizing just how plump a sitting duck PCF-94 had become, Kerry made the snap decision to beach his boat on the bank at the exact spot where the ambush was coming from. Kerry gave the order: "Turn 090. All boats turn 090. Charge the beach." The officers of the other two Swifts, Don Droz and William Rood, followed suit. "His philosophy was it's stupid to just sit in the river and fire back," Medeiros recalled of his officer in charge. "For one thing, we never knew whether we killed any VC or not. When fired upon he wanted to beach the boat and go get the enemy." The Swift immediately spotted a VC guerrilla with a weapon on the shoreline. "We caught him totally by surprise," Medeiros remembered. "He just kind of looked up and one of our boats opened fire and took him out." They beached the Swifts right there and began searching the dead man's body for useful information; in that remote part of the Ca Mau Peninsula, logic dictated that where there was one enemy insurgent there must be dozens more. The PF troops fanned out into the mangrove swamps to look for the VC while PCF-94 headed farther upriver to block for them and to investigate shots they had heard. The Swift got only some two hundred yards from where they had killed the VC lookout when a rocket was fired at them, which blew up their portside windows.

Once again Kerry ordered helmsman Del Sandusky to drive PCF-94 right into the point of ambush. The crew immediately spotted a "spider hole" emplacement, the narrow, VC version of a foxhole where the guerrillas often stashed their ceramic crocks full of rice and other staples. "We called them spider holes because what they basically would do is duck down and pop up and shoot at you," Medeiros detailed. "Then they would duck back down and you couldn't really see them from the middle of the river." Standing in this particular hole, to the horror of the Swift crew, was a VC guerrilla holding a B-40 rocket launcher aimed right at them. With the grace of god, he was more startled than they were. The enemy, staring at them, fumbled around with his rocket-propelled grenade in hopes of blasting the American boat into oblivion, but he was too close to PCF-94 to arm his weapon in time. Suddenly panicking, he started to run away from the river. Tommy Belodeau was manning the Swift's M-60 machine gun and managed to hit the fleeing foe in the leg. "The guy fell down, but he got back up with the B-40 rocket launcher in hand," Medeiros related. "And then I saw him make a left-hand turn, and he ran down this little trail. In fact, I remember grabbing for the M-16 that I had next to the bulkhead. I picked it up and tried to take a shot at this guy, but I had the safety on. And by the time I got the safety off he had already made the turn. He was now out of view."

While Medeiros tried to draw a bead on the guerrilla, Kerry had already jumped off the boat with an M-16 and lit off after him down the trail. His crew's perceptions of their skipper's courage under fire at that moment remained vivid—and admiring—even decades later. "As a rookie, I was shocked when Kerry beached the boat," Fred Short marveled. "He saved the day and our lives. The VC had set up a double ambush for us. If we had gone upstream, there was another ambush waiting to happen." Kerry's decision to go charging after the VC guerrilla caught Sandusky by surprise. "What Kerry did was against the rules," a grateful Sandusky reflected decades later. "We had been taught at Coronado that we weren't supposed to become jungle fighters. But thank god he did."

"Well, we had an agreement," Medeiros explained of his own heroic reaction in the instant when it counted. "If he went on the beach, I went on the beach with him. I was his radioman. I grabbed a PRC-25, which we called a Trick 25, an M-79 grenade launcher, and an M-16. And, of course, my .38. I didn't, however, have time to grab the radio. With my adrenaline racing, I started following him off the boat. So I was right behind him." Belodeau soon followed, while the other three crewmen manned the Swift—Sandusky was at the helm, Thorson on the .50-caliber machine gun, astern, and Short in the gun turret with the mounted twin .50-calibers. "As the VC guerrilla got twenty or thirty meters down the path, just about in front of a lean-to, the [future] senator shot the guy," Medeiros related. "He had been standing on both feet with a loaded rocket launcher, about to fire. He fell over dead."

Kerry and Medeiros approached the sniper with caution, made sure he was really dead, and then searched his corpse. After they confiscated his B-40, they quickly headed back to their boat. Medeiros radioed the other two Swifts and Coastal Division 11 to let them know PCF-94 was going to make a sweep of the area in hope of ferreting out more Viet Cong. Kerry, Medeiros, and Belodeau then began combing the terrain inch by inch, their awareness at high alert for booby traps, *punji* stakes, snipers, or anything else they could imagine the VC coming up with. Kerry lamented that there was no helicopter cover. In the process they stumbled upon a motherload: the foundation for a new insurgent village. Camouflaged contraband matériel lay everywhere and they confiscated everything from VC flags to American-made 20-mm shell casings and sewing machines of potentially nefarious intent. The suspicious goods in hand, the trio from PCF-94 burned the tiny hooches they had found the stuff in. In the distance they could see VC running toward a tree line out of range, looking for cover in the U Minh Forest. Meanwhile, the PF soldiers began hiking their way through the tall elephant grass. Along the way they engaged in firefights that claimed the lives of two more Viet Cong operatives. "Spider holes were all over the shore," Fred Short recalled. "The VC were all over the place." But after February 28, there were nine fewer of them, thanks to the uncommon bravery of John Kerry and the rest of the crew of PCF-94.

One lighthearted result of the February 28 incident was that Mike Medeiros had lived up to a nickname he'd been given earlier. Medeiros earned the moniker on a previous river mission when he was ordered to search a Vietnamese encampment along a canal. Because the water was so shallow, PCF-94 couldn't beach. So Medeiros was ordered to wade in the muck and have a look around. Wanting to be prepared in case he encountered VC, Medeiros put on every armament he could think of, including forty-five grenades hanging from his belt, bandoliers, and ammo rounds, and an M-16. He had on a helmet, flak vest, and heavy Air Force jump boots with a knife strapped along the side. Unfortunately, he was so weighed down with armaments that he started to sink in the mud, right up to his knees. Just as he called for a rope, VC guerrillas started firing at PCF-94. Medeiros was in the unenviable position of being in the cross fire. The guns of PCF-94 pacified the VC fire and Medeiros was pulled back on the Swift. "We just howled with laughter," Sandusky recalled. "We started calling him

Duke." But in the February 28 incident he really did behave like John Wayne. "We still called him Duke," Sandusky explained. "Only we now meant it out of respect."

Shortly after the February 28 incident, Kerry was granted two days' leave in Saigon, which he scheduled so that he could meet up with his old friends David Thorne and Giles Whitcomb. "I stayed with Giles in his wonderful apartment in Saigon," Kerry recalled, "and then we all went out for dinner at the Rex Hotel. Then boom—Admiral Zumwalt got hold of me, sent me a message saying, 'You're expected for breakfast tomorrow at the admiral's residence. We're flying you back to An Thoi with the admiral, and he is going to award you the Silver Star, and you have to be there.' My reaction was, 'Oh, shit, my day in Saigon is over and I have to go back.'" Leading Petty Officer Del Sandusky remembered the day Kerry received the Silver Star with about as much enthusiasm. "Because Zumwalt was a three-striper, we had to dress up in our clean, starched uniforms. We didn't really want to," he noted. "John's Silver Star was the highlight of the day, but we also got Commendation Medals. But at the time the whole ceremony was an annoyance. We were used to wearing scruffy, faded jungle greens. The ceremony was an interruption in our lifestyle. Quite a few of us bitched that it was all bullshit. But it was John's day and he deserved that medal. So in the end, we took the whole thing pretty seriously."

The decision for awarding the Silver Star was made by Zumwalt. He had dozens of Swift boats getting shot at and he wanted to make sure his junior officers were given proper honor. Zumwalt had at first put Kerry in for the Navy Cross, but that award would take time to clear with the Navy bureaucracy back home. The Silver Star, by contrast, could be given quickly. It was an "impact" award, given shortly after an action to lift morale. "The Silver Star was rare in Vietnam," Commander Horne, who was in Cam Ranh Bay the day of the ceremony, recalled. "Most naval officers were on great big ships. They never saw the kind of action that would earn such an awesome honor. The Silver Star meant you had an unbelievable moment of combat heroism. Although what Kerry did was dangerous, it was also extremely brave." In addition to Kerry's Silver Star, PCF-94's performance on February 28 also earned Bronze Stars for Tommy Belodeau and Mike Medeiros and Navy Commendation Medals with Combat V Devices for Del Sandusky, Fred Short, and Gene Thorson. "The ceremony was meant to be a morale booster," Commander George Elliott recalled. "We were trying to pay tribute to Kerry and the others for going above and beyond the call of duty. The Silver Star is always a big deal."

Kerry felt strange at breakfast that early March morning and even more so as he sat in a military plane with Admiral Zumwalt and Captain Roy Hoffman on their way to An Thoi for the awards ceremony. As much as he wanted to seize the opportunity to complain about Operation Sealords to the top brass, his good manners and common sense kept him mum. "With that crowd, you talked when you were talked to," Kerry explained. "I didn't know what the hell I was doing with them, eating breakfast. The admiral asked me questions about how we were doing. I gave my opinions. At that point we were little lieutenants, and there was a pretty big line between you and the admiral."

That afternoon Vice Admiral Zumwalt himself pinned the Silver Star on Lieutenant John Kerry's chest. Photographs of the occasion show Kerry staring ahead like a good soldier, with a serious mien befitting the solemnity of the event. The names of the Navy men killed that year in Vietnam had been read off the rolls and Kerry professed to remember little else about the ceremony. With his bright new medal gleaming from his chest, his mind wandered away from himself to memories of Dick Pershing. For a moment, standing tall there in his crisp Navy uniform, he choked up.

The citation Kerry received along with his Silver Star for valor was remarkable both for its glowing account of his bravery that day and for its readable, decidedly unmilitary prose. It read:

> *For conspicuous gallantry and intrepidity in action while serving with Coastal Division ELEVEN engaged in armed conflict with Viet Cong insurgents in An Xuyen Province, Republic of Vietnam, on 28 February 1969. Lieutenant (junior grade) KERRY was serving as Officer in Charge of Patrol Craft Fast 94 and Officer in Tactical Command of a three-boat mission. As the force approached the target area on the narrow Dong Cung River, all units came under intense automatic weapons and small arms fire from an entrenched enemy force less than fifty-feet away. Unhesitatingly Lieutenant (junior grade) KERRY ordered his boat to attack as all units opened fire and beached directly in front of the enemy ambushers. This dar-*

> *ing and courageous tactic surprised the enemy and succeeded in routing a score of enemy soldiers. The PCF gunners captured many enemy weapons in the battle that followed. On a request from U.S. Army advisors ashore, Lieutenant (junior grade) KERRY ordered PCFs 94 and 23 further up river to suppress enemy sniper fire. After proceeding approximately eight hundred yards, the boats were again taken under fire from a heavily foliated area and B-40 rocket exploded close aboard PCF 94; with utter disregard for his own safety and the enemy rockets, he again ordered a charge on the enemy, beached his boat only ten feet from the VC rocket position, and personally led a landing party ashore in pursuit of the enemy. Upon sweeping the area an immediate search uncovered an enemy rest and supply area which was destroyed. The extraordinary daring and personal courage of Lieutenant (junior grade) KERRY in attacking a numerically superior force in the face of intense fire were responsible for the highly successful mission. His actions were in keeping with the highest traditions of the United States Naval Service.*

In 1996, when John Kerry was running for reelection to the U.S. Senate against William F. Weld, reporter David Warsh of the *Boston Globe* wrote an article raising the question whether Kerry had won his Silver Star for shooting a wounded, unarmed man in the back. A brief brouhaha ensued: Did Kerry commit a war crime? But Warsh's piece grew from a single offhand remark from a single source—Tom Belodeau, who died in 1997. Medeiros, Sandusky, and Thorson, who had all been there on PCF-94 with Kerry on the February 28, 1969, mission, vigorously disputed Warsh's wispily based charges. "The VC guy was a lethal threat," Medeiros asserted. "He still had the B-40 rocket launcher in his hands." Fred Short certainly agreed. "Those reporters who wrote that garbage weren't there in Vietnam," he noted in a 2003 interview. "They don't have a clue. Kerry saved our lives. The guy was dangerous, and there were others waiting." Bristling at the *Globe* article as an attack on his integrity, Kerry held a press conference on a Boston pier in front of the U.S.S. *Constitution* and declared that questions about his war record "dishonor the men I served with, they dishonor the U.S. Navy, and the men here who made this award." There backing him up was none other than Admiral Zumwalt himself, who had flown in from South Carolina to show his support. Commander George Elliott was also there to stand by his former officer, as was Tom Belodeau.

The day after the action that won him the Silver Star, Kerry's PCF-94 was ordered back to the Bay Hap River, but not a man on the boat felt good about it. "Morale was just beginning to reach the lowest depths," Kerry wrote in his journal for March 1, 1969. "People were risking their lives and just not seeing any results for the effort. No one, from engineman to officer in charge, could find the rationale for letting the enemy have the first devastating shots at us in the rivers, and without exception they always did." As Kerry saw it, he had simply been lucky that the VC rocket had missed his Swift, and that he had thought quickly enough to beach it. He could not stop wondering: Instead of one VC with a B-40 in the spider hole, what if there had been three, or five, or ten? He knew the answer, of course: in all likelihood Western Union would have been delivering death notices to the families of the men under his command, not to mention to his own parents.

Adding to its crew's unhappiness was PCF-94's grueling schedule. Medeiros's log through the first two weeks of March 1969 confirms that the Swift had been sent on river raids nearly every day, with little time off in between. And every one of those raids had proved harrowing. Of course, Kerry was not alone in his concerns about the foolhardiness of what Operation Sealords had turned into. Among his allies on the point was his friend and fellow lieutenant Don Droz. After one typically ill-conceived mission in early March, Kerry and Droz decided to double-team a higher-up and made an appointment to see Commander Adrian Lonsdale in An Thoi. Tall, thin, and distinguished looking, Lonsdale was Captain Hoffman's operational commander for the Cau Mau Peninsula all the way to the Cambodian border. A public relations specialist, he ran the Coastal Surveillance Center at An Thoi. Once in Londsale's office, the junior officers requested permission to speak frankly, and got it. Droz went first: "Sir, what really is the point in going into these rivers the way we're doing?"

"We're proving to the VC that the rivers don't belong to them," Lonsdale replied.

"Are we really? Christ, sir, we're just going in there and getting shot at and that doesn't prove that the rivers don't belong to Charlie," Droz continued. "Hell, all we do is go in and make a lot of noise and come out and what have we gained? Have we secured any territory? Have we made any friends? Can we dare to go in any safer the next time? On the contrary. We've probably made it a hell of a lot more dangerous, and we're not the ones who choose where and when the shooting will take place. All we do is get the shit kicked out of us, get some men wounded or killed, and come out to sit around and go back in again. That seems crazy to me." Clearly, a lot of frustration had been building up inside, and once the dams were breached there would be no holding back the young man's emotions and ideas.

"Well, what do you want me to do—what can I do?" Lonsdale asked.

"Why can't somebody talk to Latch [Captain Roy Hoffman] and make him see what's happening?" Droz wondered. "I can't believe that nobody else sees what's happening."

"I have spoken to him," Lonsdale claimed, "but he wants it the way it is. There's nothing I can do."

Then Kerry jumped in. "Sir, I can't believe that these people can't be made to listen to reason. I mean, let's look at it seriously," he said. "It's not as though the men are afraid or chicken to go into the rivers. It's not that they're not willing to risk their lives, or that they don't agree with the principle of what's being done over here. It's just that they want to have a fair chance to do something that brings results and what they're doing now isn't bringing them anything. If we were to have some support, something that would guarantee that we were gaining something but for a country with all the power that we have, we're making men fight in a fashion that defies reason."

Lonsdale just nodded at Kerry, so he continued: "I mean, sir, what we need are some troops to sweep through the areas and secure them after we leave; otherwise we're just going to be shot to hell after we go through, and there'll be nothing gained. It doesn't make sense to have the boats shoot up an area, then disappear not to return for a couple of weeks [to] that place. What the hell does this mean to the people who live there?"

The commander continued to just look at Kerry. Then he said: "I've asked for these things, but there just isn't that high a priority down here. We can't get those kind of things."

Kerry and Droz were dumbfounded. "Sir, I don't see how you can ask American troops to risk their lives when the priority in that area isn't high enough to warrant their getting certain support," Kerry protested. "I just don't think that's right. If this area belongs to the Vietnamese, then why can't we let them fight for it? But why do we have to fight ourselves for something that admittedly doesn't mean enough to the American command for them to lend it a little support?"

Lonsdale didn't really answer the lieutenants' questions, instead telling them that he was only doing what he was told and that he couldn't fight it. Like Commander Charles Horne, Lonsdale was career Navy, and Kerry and Droz realized that as such, it would be asking too much of him to take issue with Captain Hoffman's pet project, Operation Sealords. But the Swift skippers pressed Lonsdale to fight harder for the support their missions needed. Kerry opted to implore him for a policy change one last time. "Sir, all we need is to give the men a little more of a chance," he began. "They're willing to fight because they're here and that's what they have to do, but they want to fight with some prospects of getting somewhere. Why can't we have some helos that are waiting, always on the ready for us, so that when we get fired on first, if we must, we can at least expect some support and hope to get some of the enemy and make them actually worry about what we are doing? That way we might be able to catch some of them, or get some weapons or something. We could at least have a reason for going in there."

After the meeting Kerry recorded in his journal: "Commander Lonsdale said something—I don't remember what—but he said that they would try to get some helos and other support, if possible. That really got us nowhere but I felt better for having spoken about it. It wasn't really his fault. He had a job and he wanted to keep it, and he had already been threatened to be fired if he didn't keep doing what 'Latch' wanted. He was in a difficult position. But then, so was Lieutenant Commander George Elliott, who was the commander of our division. He listened to all of us and he was willing to take 'Latch' on. I remember several conversations he had with the old man. Some of them got to a shouting level. But he stood up for us, which is one of the things a commander is supposed to do. As a result I think that everyone worked for him and worked hard."

Nothing happened as a result of the meeting with Lonsdale, however, and Kerry, Droz, and their PCF cohorts resumed getting shot at with pronounced regularity. At least they had found some Vietnamese soldiers who weren't quite so loath to do any soldiering: the Nung, an ethnic minority group concentrated in North Vietnam and in the South's Central Highlands. Many Nung in the latter region fought with U.S. and ARVN troops against the Viet Cong as mercenaries in the CIA's Civilian Irregular Defense Groups. On March 12, PCF-94 and three other Swifts were assigned to transport a squad of Nung mercenaries up a small canal and drop them off at a designated spot. The plan was for the Nung to sweep through the area, which was said to harbor a Viet Cong tax station, then rejoin the Swifts on the other side of the canal a mile or so away. As the mercenaries made their way into the jungle, Kerry took PCF-94 upstream to investigate an area he and his crew had raided several weeks earlier to see how things had turned out there after the destruction wrought upon the vicinity. To Kerry's amazement when they drew near they could smell food cooking on open fires; the people had already moved back in, and rebuilt their fish stakes to boot. "It seemed to me to be testimony to the futility of what we were doing," Kerry opined.

After some more reconnaissance the Swifts moved back out of the canal and over to the rendezvous point on the other side of the narrow little waterway. Called the north/south canal for lack of any other distinction, it joined the Cua Lon and Bay Hap Rivers to become a very long, straight, and thin canal along which VC ambushes had already taken their toll. Fish stakes placed at varying intervals required very careful navigating through a space narrower than a single-lane road. Lieutenant Larry Thurlow's PCF-53 went through first, whereupon a mine exploded off its right bow, giving the boat a good shake but causing no real damage. PCF-94, meanwhile, had stayed on the beach just down the canal to wait with the other Swift for the Nung troops to emerge from the jungle. When the crews heard the explosion, however, they rushed to their boats and then upstream to help. Kerry's was the first Swift to reach Thurlow's, which came at them from the other direction until they met bow to bow on either side of the fish stakes. As they approached, the Swifts came under automatic weapons fire from the left bank. Indeed, PCF-53's injured bow gunner was so distracted by pain that he didn't notice Kerry's boat coming and came within a hairbreadth of pumping PCF-94 full of machine-gun fire. "For some reason," Kerry exhaled, "he looked up at the last moment and relaxed his finger on the trigger."

PCF-53 was in good hands with Larry Thurlow, among the toughest and ablest Swift boat officers in Vietnam. Born October 23, 1942, in Garden

City, Kansas, Thurlow spent his childhood moving around the Sunflower State, following his arthritic father's job posting as a biology teacher and basketball coach. Western Kansas was for landlubbers, though, and young Larry Thurlow had dreams of the open sea. "It might sound ridiculous," he admitted, "but it was my love of fishing that got me interested in the Navy." Upon graduating from high school, Thurlow enlisted in the Navy and was sent to OCS in Newport in November 1965. After a stint in Long Beach aboard the amphibious assault ship *Princeton*, he went to Vietnam, serving in-country from July 1966 to December 1967. Like so many other young officers, Thurlow wanted command of his own Swift boat, so he did the usual drill of training at Coronado, followed by SERE. On April 5, 1968, he returned to Vietnam, assigned to Coastal Division 11 based in An Thoi. "It was an education from day one," Thurlow recalled. "All the officers were volunteers, and most of the enlisted men were too. So the *esprit de corps* at An Thoi was high, considering the dangers we faced."

After serving as officer in charge of PCF-10 from April to September 1968, Thurlow was reassigned to train would-be Vietnamese sailors. "George Elliott said to me, 'If you want to be a teacher back in Kansas, this would be a good job for you.'" That made sense to Thurlow, who had a wife and two children back home and did indeed plan to follow in his father's footsteps. Thus Lieutenant Thurlow took on the difficult task of training South Vietnamese sailors how to operate and maintain Swift boats. "I had to write reports that said Vietnamization was working but in my gut I knew it wasn't," Thurlow related. "The Vietnamese crews I tried to train were born into the war. They decided not to risk their lives. We wanted to get the bad guys and get it over with. They were willing to let the war drag on. They didn't want to get their heads blown off."

By March 12, 1969, Thurlow was off Vietnamese teaching duty and back in command of PCF-53, which was to participating in Mission 97 along with Lieutenants Droz, Kerry, and Bill Rood. Thurlow's opinion of Kerry seemed mixed, but he admitted "John was sharp as a tack." They shared respect for each other regardless of their differences. "But he came from a background most of us couldn't understand," Thurlow recalled. "And he was both distant and foolhardy."

The disparate Swift lieutenants wouldn't have time to worry about personality conflicts on the north/south canal, however. Just as PCF-94 turned parallel to Thurlow's boat, another deafening explosion boomed through the narrow waterway, and for an instant Kerry thought it was all over for him and his men. Thurlow's boat "rocked over on its side towards us and smashed against our gunwale with a thud," Kerry reported in his diary. "I could just see the men on the stern of his boat get thrown up in the air in a somersault as the engine doors sprang open from the concussion. Then one of the crew limped around the side and was engulfed in the smoke. Everything went black—I couldn't even see across my own pilothouse through the mud and the smoke."

When the air cleared, the Swifts were still under fire and Kerry could hear the distinct clatter of AK-47s heading their way. Suddenly, piercing through all the noise like a siren, Kerry heard the voice of PCF-23's Lieutenant Rood screaming over the radio. "This is 23: I've lost my eye—I can't see." His pilothouse windshield had shattered, spraying bits of glass into Rood's eyes. (Rood's loss of vision proved mercifully temporary and thus no obstacle to his later career as a Chicago editor.) By the time the other Swifts got to PCF-23, "Rood's crew had already had him bandaged up," Thurlow detailed. "We thought he'd lost an eye, but it turned out he only temporarily lost his vision." Kerry immediately radioed PCF-43's Droz to get his boat up into the area to help, and fast (Droz had remained back at the rendezvous point to wait for the Nung troops, but they still had not shown up). Kerry vaguely recalled saying something inane back to Rood like "This is 94: have you actually lost your eye or can you just not see out of it?" By the time he should have answered, PCF-94 was slowly passing the open pilothouse door of the 23 boat, and through it Kerry could see Bill Rood standing in a shocked-pale daze as his driver put a battle dressing over the lieutenant's damaged eye. "I have never been so scared in my life—and furious, because it seemed so stupid for a young guy to lose an eye," Kerry wrote. "Droz had arrived by now and he started to pump mortar into the area from which the AK was coming."

Even in the midst of this absurdity, there was a moment of comic relief. During the firefight PCF-94 nearly lost its mascot. A few weeks earlier, Kerry and his crew had been in a village where they came across a puppy some of the locals were about to kill and eat. The Americans immediately traded some meal rations for the little dog and adopted him, naming the mutt V.C. Displaying great fortitude, V.C. soon took to standing in the bow of the Swift, with his muzzle sniffing in the air, barking at villagers along the riverbanks. His only drawback was his inability to be housebroken. That afternoon when the mine exploded just off PCF-94, to the crew's horror the concussion blew open the two big engine cover doors, catapulting V.C. into the air and overboard. After the ensuing firefight the crew fell silent. "Nobody was wounded," Kerry explained, "but V.C. was gone." A few minutes later, however, the Mekong Delta river featured a scene straight out of a Walt Disney movie, as the men of PCF-94 heard a yap, yap, yap, and turned to see their puppy on another Swift's deck, barking to come home.

Thurlow towed Bill Rood and his boat to an LST, whence the temporarily half-blind lieutenant medevaced out. "The crazy thing was that it would take about three hours to complete the medevac process—get the helos called up, get Rood out where they could pick him up, and then from there to the hospital," Kerry continued in his journal. "Rood tried to tell us that he didn't need a medevac but that he wanted to stay and fight. In fact the whole way out his crew had a hard time keeping him quiet, and it was hell to get him off the radio." Rood liked to call Kerry "Ichabod," after the similarly tall and lanky protagonist of Washington Irving's 1820 story "The Legend of Sleepy Hollow," Kerry had noted, "and he'd keep hopping on the phone and say[ing] 'Hey, Ichabod, I'm all right—let's go in there and get 'em.'" Only when the morphine kicked in did Rood finally lie down and take it easy.

Kerry and Droz stayed put on their PCFs, not out of choice but because they couldn't leave without the Nung troops who were still ashore, not only out of sight but now out of radio contact. "This was absurd," Kerry wrote afterward, "because not knowing where they were going to come out could trigger anybody's finger, seeing as we were all jumpy anyway." When the Nungs finally did appear on the bank, it was fortunately far enough from where they had been expected to emerge that they were clearly visible and there was no danger of anyone's accidentally getting shot. The mercenaries then came sprinting down to the waiting PCFs, shouting that someone had been chasing them the whole way. "The sergeant who was in charge of them had an old man with him as a hostage, and we hustled him aboard, too," Kerry reported. "Then we screamed balls-to-the-wall out of there."

Once they were back out on the main river, the Swifts contacted their LST, whose officer suggested that Kerry and Droz return to the canal and redeposit the troops on the shore near the site of the ambush on the PCFs. The crew were exhausted and not much in the mood to go back, but the specter of Bill Rood's bandaged eye made them feel obligated to. Angry and full of adrenaline, Kerry concurred. So the boats turned around and headed back, and as they reentered the canal they powered their engines down to minimum revolutions to keep the noise as low as possible as they glided up the narrow waterway.

They dropped off the Nung troops and moved off only a little, where they waited in case the ground squad called them for assistance, although Kerry didn't know what help the Swifts could provide, beyond mortar fire. At that point, the Nung's Green Beret advisor, who had remained aboard PCF-94, turned to Kerry and said, "I wouldn't be surprised if our prisoner tries to escape or if he falls on some pungi [*sic*] stakes in there." Kerry noted in his journal, "For the first time I realized that they were planning to take the old man in with them to act as a human mine detector, and also because he supposedly knew the area." Apparently, the old man had come out with some useful information while he had been on the boat. "The certainty of his trying to escape or of his falling on some pungi stakes didn't mean anything to me until later, when I had thought about it and it dawned on me what they meant to do."

The Swifts waited on the canal for about an hour, Kerry recorded, "and then the very tired group of mercenaries made their way back to the boat. They related to us how they had seen armed VC walking around with impunity, but that because the VC had been well equipped and it was late in the day they [had] refrained from making contact.

"I noticed that the prisoner wasn't with them and the Green Beret winked at me when I asked, and said that he had tried to escape. With a smile on his face they boarded the boats and we left the area. For a few nights I wondered about the man—not what, but how, they had done it, and then one night we got drunk back at the base and I found out that he had been knifed by one of the Nung and then sliced up and left with a note of warning to the VC."

The very next day after the canal ambush, March 13, would prove among the worst and best John Kerry spent in Vietnam. By the time it was over Kerry would have earned a Bronze Star for bravery plus his third Purple Heart, and unbeknownst to him at the time it would begin his journey out of Southeast Asia. At six o'clock that fateful morning, Kerry's PCF-94 and Lieutenant (j.g.) Rich McCann's PCF-24 left their LST to join three other Swifts for Operation Sealords Mission 98, a raid on the Bay Hap River. The day dawned gray and sunless, making forming up particularly difficult. The humid air hung stagnant and their radars worked spottily, if at all, through the unusually dark morning. PCF-94 was particularly far from tip-top shape. "All the windows on my boat and on [Larry Thurlow's 53] boat had been blown out in the ambush, and water slopped into the main cabins as the Gulf of Thailand sloshed us back and forth," Kerry recalled. "Finally we started to head in toward the beach, [using the] compass and general direction only again, because all the radars were in such bad shape. As it started to lighten a bit I noticed that we were actually moving in two echelons with Rich McCann leading the left one. Each boat had its quota of Chinese-Cambodian [Nung] mercenaries on board and speed was reduced because of the extra weight."

McCann was born June 6, 1944, in Teaneck, New Jersey. His father worked at an advertising agency in New York City, a specialty of his, in fact, was writing slogans and jingles. His most glorious moment was penning the popular margarine commercial "Everything Is Better with Blue Bonnet on It." Young McCann lived in New Jersey until he was sixteen years old. The family then moved to Boston. He attended Providence College, spending his summers in Newport at a Navy ROTC program. His first assignment was aboard the U.S.S. *Hyades*, a refrigeration ship. "We were a floating grocery store," McCann recalled. "We carried everything: steaks, ice cream, milk, eggs, you name it." It was a wonderful billet. He was a gunnery officer dividing his time between Norfolk and the Mediterranean.

Vietnam, however, was his ultimate destination. After training at Coronado, he went in-country in May 1968. Based in An Thoi he served as officer in charge of first PCF-11, then PCF-21. His craft was now PCF-24. "What I learned on Swift boats," McCann recalled in 2003 from his home in Virginia, "was that when your vessel is being fired at, all distinctions—wealth, education, looks—disappear." Smart and soft-spoken, McCann thought of John Kerry as one of the more aggressive sailors he encountered in South Vietnam. "I can still remember John carrying into the officers quarters at An Thoi a claymore mine he had captured," McCann remembered. "I told him to get out of here with that thing. He was aggressive. Which was a good thing, for the most part, when you were in a river with him."

Near the entrance to the Bay Hap, McCann's boat developed engine trouble and had to stop. "We transferred his troops to our boats and continued into the river," Kerry reported, his PCF-94 now so weighted down with men and gear it looked like a World War II Higgins boat heading for the Normandy beaches on June, 6, 1944. Aboard were Special Forces teams of both Green Berets and SEALs they were to transport into enemy territory. "Just past the river entrance we found a deserted sampan anchored near the muddy shore," Kerry continued. "One boat covered us while we investigated it. Everything was suspicious, but we couldn't find anyone near it. We blew it up with a well-placed concussion grenade." Years later, McCann marveled at how the Special Forces were unafraid to enter no-man's-land in an effort to "secure an area."

Kerry, who had hunted deer and quail as a boy, enjoyed shooting at stationary targets—if decidedly not at people—and he knew something about snipers, including that he had some on board. By March 1969 the U.S. military had sent numerous marksmen to Vietnam who could hit a bull's-eye from a hundred yards away with their scoped rifles. For instance, U.S. Marine Chuck Mawhinney of Lakeview, Oregon, was credited with 103 confirmed kills and another 216 probable ones. Getting to know some of these crack American snipers and watching them shoot a snake out of a tree a football field away gave Kerry only more reason for concern about their VC counterparts. "Going up the rivers we were big, fat, defenseless targets," he explained. "Anytime we were in a river or canal I felt the scope of some VC guerrilla aimed at my neck."

Gunner's Mate Fred Short, who also knew a good bit about firearms and how to use them, harbored similar constant worries about the camouflaged Viet Cong snipers. Using the jungle as cover, they would spy on the PCFs, closely studying the vessels' routines in order to devise the most effective ambushes upon them. "The VC would never stick it out and fight unless they were backed by the North Vietnamese Army," Short recalled. "For example, if we always came in the river on the left side, the next time they would have a mined ambush. They knew that Americans were creatures of habit. They always watched what we were doing. In the Ca Mau

Peninsula there were only two villages friendly to U.S. [forces]; the rest was a free fire zone. The enemy was all around us."

For the first time during Kerry's many trips up the Bay Hap, the river was choked with early-morning traffic. He could not understand where all these people came from, or why, seeing how often his and other Swifts had moved up and down the usually empty waterway with their guns blazing into its banks. In this free fire zone, after all, anything moving along it was fair game. Kerry and his men had never really shot at anything specific in the area but used their booming machine guns to frighten off would-be ambushers. "It made me wonder again," Kerry reflected on the sudden river traffic, "how the hell they classified an area a free fire zone." Kerry's friend Wade Sanders, in an unpublished memoir, cut to the crux of the problem the PCFs were facing: "The difficulty was that we were enforcing rules on a people who had moved freely for thousands of years without those restrictions and in any event, all of them weren't aware of them."

Swarms of fishing boats, of all shapes and sizes—including round bathtub-shaped vessels made of woven reeds with tar bottoms—were bobbing around. The sampans and motorized junks on the water that morning were crowded with old men and women and the usual smattering of children. The sight made Kerry feel once again like an invader in a foreign culture he knew little about. Two years later, when a *New York Times* reporter asked him how he had *felt* about being in Vietnam, he would reply: "My mother was actually born in France and when we lived there I used to play in the old German bunkers outside my grandmother's house. From listening to her stories, I got a vivid impression of what it was like to live in an occupied country, and when I went ashore in those [Vietnamese] villages, I realized that's exactly what I was in—an occupied country." In those early morning hours on the Bay Hap, as a member of the U.S. occupying force, had Kerry felt like enforcing the no-movement rule to the letter, or had it been only half an hour earlier, he could have ordered his men to shoot every Vietnamese on the sampans. The dead would have been matter-of-factly written off as VC trafficking goods during the U.S.-ordained curfew time. As Sanders summarized for all Swifties in his memoir: "I literally had a license to kill."

Instead, Kerry slowed his boat down so that its wakes wouldn't swamp the locals' shallow-drafted river craft and proceeded with caution and guns cocked in the event a grenade or anything else unfriendly suddenly came their way. "They looked at us and we at them—each staring with mistrust and fear," he remembered of the expressions on the Vietnamese faces. Finally PCF-94 arrived at the opening to the small Dong Cung canal, where they were to rendezvous with the Swifts skippered by Lieutenants Skip Barker, Don Droz, and Larry Thurlow, who would join them bearing South Vietnamese PF troops they had picked up in Cai Nuoc.

Wearing a bandanna around his head, Barker arrived leading the other boats, and the Swifts got in line headed for the tiny entrance to the Dong Cung canal. At its mouth the PCFs stopped and waited anxiously while Barker's boat cut the wire connecting the fish stakes to make a hole wide enough for the Swifts to pass through. "Only a week earlier I had personally cut that same line," Kerry related. "But then we had learned not to place any particular suspicion in any quarter. Everything could turn around and kill you, and any omen was suspect. And then an innocent fisherman could easily have replaced it."

The canal was so narrow that the first two boats roiled the entire waterway around the fish stakes, sending the others following in their wakes rolling wildly from one side to the other. "It was like being in a large bathtub with waves of water pouring over the sides," Kerry described the scene. Overburdened by the additional troops they had taken aboard from the engine-troubled PCF-24, Kerry's Swift could get up to only about ten knots, making it impossible to plane across the surface of the water. The boat strained to get up on the plane but could just not surmount the extra weight on its stern. Fortunately, every so often the rebounding wakes from the boats in front of them would combine into just enough of a wave to pick PCF-94 up and send it hurtling forward down the surges. "One moment the faces of scared South Vietnamese soldiers, huddling together on the aluminum fantail waiting for the boom of a rocket or the mesmerizing clack-clack of the Chinese Communist AK-47s, would loom up close to you and each of them became a person, frozen in your situation," Kerry recorded, "and then, as the quartermaster pulled wildly back on the throttles and slipped behind to a reasonable distance, they again became a group of soldiers, impersonal and foreign."

As PCF-94 twisted and turned up the river, its crew occasionally losing sight of the other Swifts around the waterway's sharp turns, the U.S. Special Forces captain with Kerry in the pilothouse glanced over at him knowingly.

"Fucking crazy," the captain laughed, then went back to scrutinizing the banks intently for any sign of movement. But none appeared, in part because the mangrove rose so thick above them on both sides that they could barely see through it. "Christ, they can hear us coming for miles," he pointed out, "and I can't remember any fuckin' thing in the history of war that runs like this—taking friendly boats smack into VC territory so that they can be shot at." Then, "with a sigh that said 'shit,'" as Kerry put it, the captain returned to staring out the pilothouse door. "We spoke vehemently for a few minutes of the seeming stupidity" of sending friendly troops into the "enemy heartland without attempting to conceal themselves, parading indelicately as targets in a shooting gallery," Kerry noted in his journal.

Slowly, PCF-94 approached the same area where they had been ambushed a few weeks before and began to look for the place where they were supposed to deposit the Nung troops. Lieutenant Skip Barker led the way upriver, seeking to drop off the South Vietnamese troops at a point some two thousand yards past the position of the Nung mercenaries. "The Nung were different from the Vietnamese," Barker remembered. "They had broad shoulders, were heavyset, and had square features. The Vietnamese were angular; the Nung were much more muscular."

Kerry found the dropoff point, and the boats nosed ashore into a small clearing to begin moving the troops onto the banks. A few minutes after all the Nung were ashore, the Swift crews heard a loud but muffled explosion inland. A superior officer's voice came over the radio asking, "Can you come back in here and pick up a body? I've got one of my boys killed by a booby trap."

Kerry nosed PCF-94 back into the bank, where, along with Mike Medeiros and a few others, he went onshore. They took with them a couple of U.S. military-issue rain ponchos with them to carry back the mercenary's remains. Kerry secured his boat and followed the others ashore, where "the Nung seemed to be wandering around almost aimlessly, unaware that one of them had bought the ticket," he noted in his journal. Their leader came up to him and told him where the dead body was; about fifty yards down the path. "I remembered easily who he was," Kerry wrote of the dead man, Bac She De, "the loud boisterous, fate, impish man who was something of a ringleader among the Nung and who had endeared himself to everyone by his funny face." As the Americans started down to where the

Nung mercenary had died just moments earlier, Kerry asked: "Is it bad?" The Nung leader replied "that you could put him in 'a bucket,'" Kerry wrote. "I walked more carefully, looking where each step went so that I wouldn't trigger another trap. I was also almost fearing that Bac She De might suddenly appear and scare the living shit out of me. There is nothing worse than approaching a dead body when you don't know exactly what you're going to find."

Right where the Nung leader had indicated, Kerry and the others came upon the crumpled remains of the mercenary. "I never want to see anything like it again," Kerry confided to his diary. "What was left was human, and yet it wasn't—a person had been there only a few moments earlier and . . . now was a horrible mass of torn flesh and broken bones; bent and bloody, limbs contorted and distorted as if they could never be alive. Most of his stomach was hollowed out and there was a huge hole that went through his mouth and nose to the other side. I didn't really want to look and so I concentrated on looking right through him, avoiding contact with any personality."

As two enlisted men bundled Bac She De up, Kerry wrote, "I looked at the small green sack that had been booby-trapped, and was awed by visions of the blast that must have ensued when it had been grabbed at. Half a hooch had been blown out by the concussion and there was a frightening hole in the ground." The situation was too dangerous for further on-site reflection, however, and the Navy men started back to their boats. "We had to extricate this body by carrying him in a poncho," Medeiros remembered of the hard duty. "I was carrying the front end and somebody else was carrying the back. We had to go through a mangrove swamp, tramping through all this mud. I had an M-79 [grenade launcher] strung over my back, my radio was inoperable, and we had to deposit this poncho on the stern of the boat."

As the landing party turned back up the path, "light AK-47 fire suddenly started to rake us from a field over the trees to the left," Kerry recorded.

> *Everyone dove instantly into the ditch by the side of our dike and I landed in water and mud up to my waist with the muzzle of my M-16 firmly planted in the crap. A whole line of mercenaries had already formed in the ditch, all shooting madly at what seemed like nothing. However, the whiz of bullets over our heads that was visually nothing was clearly lethal. . . . And Bac She De lay in front of*

> *us crumpled in the poncho while this holocaust went on. His feet were sticking out of one end and I couldn't take my eyes off the boots—one going one way and the other the opposite direction—and the whole thing just silhouetted where he had been dropped suddenly when the shooting began. The alive shooting over the dead to remain alive.*
>
> *I was amazed at how detached I was from the whole scene. I just lay in the ditch, not firing because I wanted to save ammo and because I couldn't see what I was firing at, and I thought about what was happening in New York at that very moment, and if people really felt that I was doing something worthwhile while they went down to Schrafft's and had another ice cream sundae or while some fat little old man who made another million in the past months off defense contracts was charging another $100 call girl to his expense account. And then, when the shooting stopped, I came back to where I was.*

Suddenly it grew quiet, and everyone stayed low in the ditch while a call was placed to the LST to report what had happened and to request some helicopters to support their sweep through the area. The crews back on the Swifts had no idea what was going on ashore and stayed beached where they were, waiting for instructions. Lieutenant Commander George Elliott aboard the LST requested the helicopters, but the word came back that none were available. "I swore," Kerry declared. "We had been promised that when the shooting got heavy [that] we would have helos to help us out. But the headquarters just said that they were otherwise disposed and we would have to do the best we could."

Unrattled, Elliott told the command that his Swift boats were all right and generally played the whole thing down. He proceeded to send a couple of small flanking groups out on either side to feel out the enemy. They immediately met and expended a huge volume of fire and some more AK started to come their way. By this time Kerry had called the South Vietnamese out of their insertion area and asked them to come down to their position to help. Their insertion had been a travesty anyway.

The Navy SEAL team that had been designated to deploy in the area had gotten lost and stumbled onto a friendly outpost; a firefight mistakenly ensued, leaving one of the SEALs hurt and several of the South Vietnamese badly wounded. In the course of their sweep the South Vietnamese troops actually had flushed out some VC, but as no one had been there to intercept them, nothing had come of their efforts. "I understood from one of the Green Berets that in the cross fire civilians were killed," Medeiros recalled, "mostly by their own people. I'm getting this all from secondhand knowledge, because the Green Berets said that when they opened fire the South Vietnamese troops jumped over this wall—women and children huddled there, mainly because they were interrogating them. And unfortunately a lot of them were taken out by their own people. That's the story I was told; John and I were moving the body to the boat, so I really don't know."

Using a walkie-talkie after the radio went dead, Kerry contacted Del Sandusky, who was running things on PCF-94 while its skipper was ashore. Kerry told his second in command to take the Swift downstream and beach it at the spot he could determine was parallel to theirs by the gunfire. Then Kerry and Medeiros sneaked up onto the dike and pulled the poncho down with them into the ditch on the other side. A crewman from a different Swift dragged Bac She De's body through the knee-deep mud as the group slogged through the mangrove toward the river. "Larry and I both felt that we should be back with our boats in the event the VC tried a charge or something; since we knew what [the] territory in front of us looked like, it just made sense," Kerry noted.

The PCFs had beached almost directly behind them, but their ooze through the mud proved so arduous it took another ten minutes of hard work before they could haul the mutilated corpse of Bac She De on board, much to the disgust of the crewmen. After a quick rinse in the canal and a few moments' rest, the other Swifts joined PCF-94 bearing the South Vietnamese PF troops they had collected. Together the boats proceeded fifty yards downstream, where a bigger clearing in the mangrove allowed the Swifts to train their twin .50-caliber machine guns more accurately at the area whence the earlier firing had come. Meanwhile, Elliott had moved his troops downstream on a parallel course through the mangrove cover. "With a few small flanking teams covering them on either side, we powwowed there to decide the next move," Kerry recorded.

The mission had ordained that the Nung troops make a sweep for Viet Cong over several thousand yards before being picked up in another canal on the other side of the jungle. "When the Nung were on board Swifts, the VC stayed away," Fred Short recalled. "They were terrified of them." Yet, after several hours, the ground troops had still gotten nowhere. The AK-47 fire had proved the enemy's presence, and the paid-by-the-kill mercenaries wanted to sweep the area and find them. The South Vietnamese PF troops, however, were far less motivated to risk their lives. "We wanted the Ruff Puffs to join us in order to create a force that definitely outclassed what we assessed the firepower of the enemy to be," Kerry wrote in his notebook. "They unloaded from the Swifts but then froze on the bank and refused to go any farther. What followed was disgusting and disheartening. At first they refused to go in because they said they would be fighting under an American commander and that this wasn't their bag. But I thought of the other times that we had gone in here, and that had to be bullshit."

Mike Miggins, the U.S. Army advisor to the local ARVN, told Kerry the problem was most likely that the PF didn't want to fight alongside the mercenaries. Then the Ruff Puff's leader got on the radio with the ARVN district chief and spent the next half hour apparently arguing about something in frenzied Vietnamese. An agitated Kerry and crew just stood by helplessly, thinking about poor Bac She De and how they had almost joined him, as the ARVN bureaucrats on hand engaged in a political debate over the angles of the present situation. Having had enough of this, the mercenaries' leader "called his perimeter men and told them to fake a firefight, hoping this might excite our Viet colleagues into battle," Kerry reported. "The next few minutes were filled with tremendous thunder, as though the whole war was being fought in front of them, but it was to no avail. Some of the South Vietnamese got up and looked around and some others cocked their single-fire rifles, but by and large they just milled around. Then they just walked back and got onto the boats and sat down."

While the "Cream Puffs," as Kerry liked to call the PF after that, scampered back to the boats, a few of the bravest made little forays around the area in front of them, where they turned up a substantial cache of Chinese-made grenades, ammunition, and one or two land mines. One group found some wire leading down into the water and pulled it up slowly and carefully. No mine was attached, but there was a string of batteries at the other end. "They were American-issue, and we realized that the VC had been planning a reception for us in this very spot should we have tried sometime in the future to make a landing," Kerry recorded. "I could just see us beaching and unloading troops into a volley of claymores and underwater mines."

When AK-47 fire started to rake them again, Miggins and several of Kerry's men finally got so angry they leaped off the Swift and joined the mercenaries to start a sweep of their own. The impromptu squad moved through a string of surprisingly prosperous huts and fields and then back on small paths toward a large open field. Kerry suddenly found himself walking through tall grass with a few of the Nung, unable to communicate. "I don't think I ever felt so alone," Kerry remembered of that moment. "I could just see VC jumping out on us and these guys running away or all of us getting rubbed [out]. I breathed a lot easier when we joined up with [the unit leader's] number-two man, who had been on another path over to the right of me. We found a small, little hut on the outskirts of the field, and there we regrouped and rested until moving back toward the boats. The hut was full of chickens and rustic implements and a small pot was cooking on a fire that was still hot. As usual, the occupants were nowhere to be seen, having probably run away as soon as they heard the boats congregating in their area earlier in the morning."

The Nung blew up some huge bins of rice they had found, as it was assumed, as always, that these were the local stockpiles earmarked to feed the hungry VC moving through the Delta smuggling weapons. "I got a piece of small grenade in my ass from one of the rice-bin explosions and then we started to move back to the boats, firing to our rear as we went," Kerry related. "Once on the boats we moved full-speed out of the zone, as we had been there for a dangerously long time now and the chance of something greeting us on exit was pretty good. My eyes still smarted and my throat was dry because Miggins had let off a gas grenade into a hooch and the wind had shifted and blown it down on us. I don't know how one of us wasn't shot by his own men when I consider how randomly we had been walking around in there—shooting and blowing things up."

Every man on the mission fully expected to get hit on the thin stretch of the Dong Cung canal on the way out, but somehow they all made it to the Bay Hap unscathed. The Swifts rounded the point toward the village of Cai Nuoc, anxious to get rid of their "Cream Puff" passengers, out of the river, and back to the LST. For the first time that Kerry could remember, however, there were no village children there to greet them; in fact, the pier was practically deserted. "Had we been paying attention," Kerry noted, "the obvious clue that something was up would have hit full force in the face, but because we were all pissed off and anxious to clear out we just unloaded and moved away."

Almost casually, as though their day's work was over, the Swifts formed up and headed out from the village. The Swift boats had gone about half a mile when the blast came. Right where they had been hit on the earlier mission, a mine went off directly beneath PCF-3 just off Kerry's port side. The Swift lifted about two feet up out of the water, engulfed in mud and spray, then settled, rocking so hard from side to side the boat started zigzagging from the banks to the middle of the river. Everybody on board PCF-3 was wounded. "At the same moment, we came under a hail of small-arms fire from both banks," Kerry recorded in his journal. "I turned the boat into the fire on the left with the intention of trying to get the troops ashore on the outskirts of the ambush, but Sandusky, who was driving the boat and who had his eyes glued on the crippled 3 boat, pointed out to me how badly hit they had been. We veered back toward her then and tried to provide cover from the engaged side. Suddenly another explosion went off right beside us, and the concussion threw me violently against the bulkhead on the door and I smashed my arm." At that instant, Army Lieutenant Jim Rassman, who was on PCF-35, was blown overboard, although nobody knew it. Kerry wrote, "We continued sidling up to the 3, and as we came closer I could see that her twin-.50 mount over the pilothouse had been completely blown out of its stand and had landed on the gunner. No one was moving on the stern. [PCF-3 crewman] Ken Tryner, on his first real river expedition, was kneeling dazed in the doorway with a small trickle of blood down his face, aimlessly firing his M-79."

Larry Thurlow had maneuvered his PCF-51 over by this time, and he hopped aboard PCF-3 to offer assistance. The boat was a shambles but they were still shooting too hard to assess any damage. Meanwhile, Lieutenant Rassman, several hundred yards in back of them, was receiving sniper fire from both banks. "Someone on the fantail must have noticed Jim swimming in back of us ducking against the fire that was trying to pick him off because I suddenly heard the yell of 'man overboard' and looked back to see the bullets splashing in the water beside him," Kerry reported. "We turned around with the engines screaming against each other—one full astern, the other full forward—and then charged the several hundred yards back into the ambush where Jim was trying to find some cover. Everyone on board must have been firing without pause to keep the sniper heads down."

Kerry ran out of his pilothouse and, thanking god the scramble nets were over the bow, struggled to get Rassman on board. "It must have looked like a comedy," he recalled. "Jim was exhausted from swimming and my right arm hurt and I couldn't pull very hard with it. Everyone else was firing a machine gun or something, except for Sandusky, who was maneuvering the boat trying not to run over Jim but also trying to get near him as quickly as possible. Christ knows how, but somehow we got him on board and I didn't get the bullet in the head that I expected, and we managed to clear the ambush zone and move down near the 3 boat that was still crawling [on] a snail-like zigzag through the river." For pulling him out of the water, a hail of gunfire causing him to risk his life, Rassman put Kerry in for the Bronze Star.

Thurlow was now struggling to get PCF-3's wounded gunner out of his hole and onto the deck. Suddenly the damaged Swift ran aground hard on a shoal on the right side of the river, sending Thurlow somersaulting into the water. At the same moment, the Swifts came under fire from the right side again, and Kerry remembered thinking that was it—they were going to get completely cut off and annihilated in a cross fire. Spontaneously, however, every boat there stood its ground and filled the entire right bank of the river with .50-caliber, M-79, M-16, and any other firepower any of them had, while one of the Swifts moved in and retrieved Thurlow, who had picked himself up out of the mud. PCF-94 then moved in and attached a line to the wounded boat's stern to try to tow PCF-3 out, but the tether snapped. Kerry went in again to put another line on, and this one held. "We managed to get her clear of the kill zone," he exulted. "Finally, the tumult subsided." The wounded were transferred to another of the Swifts, which set off at full speed with a cover boat to take them out to the LST to be medevaced.

Thurlow remained on the crippled PCF-3 and the other boats slowly began a procession out of the Bay Hap River, working desperately to keep the damaged Swift afloat. Only one of its engines was functioning, and the overboard discharge pipe had been blown apart inside the engine room, filling the hull with water fast. Even shutting down the one good engine didn't help much, and it became a race against time to get PCF-3 out of the river before it sank. One of the Swifts that had gone out with the wounded was on its way back with a damage-control team from the LST, but if they didn't get back soon there would be no boat to pump. Just as they reached the mouth of the river, PCF-94 and the other healthy Swift positioned themselves on either side of their endangered cohort to hold it up between them on two lines while a bucket brigade of tired and bloody sailors kept bailing for all they were worth, managing to keep PCF-3 afloat despite the water pouring into it. And all this while Bac She De's body lay in a poncho on PCF-94's fantail, next to a pool of blood where the corpse's head had slipped out of its shroud.

Finally the LST damage team arrived with the pump and the "bucket brigade" plopped down, exhausted. Once they caught their breath, the men all sat around and smoked and talked about what they had just gone through. As usual, in such circumstances, their chatter was animated as each man, exuberant to be alive, described how close the shells had fallen and everything else he had seen over the past few hours. One of the Nung troops still aboard, however, displayed no such elations. "Bac She De's closest buddy, who looked like a mouse and whom we called that, was crying now because it had all caught up with him," Kerry related sadly. "I have never seen eyes so wide and so full of tears, and since I couldn't say anything to him I just patted him on the arm and nodded stupidly and then went away to sit down and enjoy the peace and quiet."

After what felt like an eternity, the Swifts neared the LST and one of the other boats came alongside. The helicopter bearing the wounded was on the landing deck waiting for them to arrive with the body. PCF-94's crew transferred the corpse onto the free Swift and Kerry left Sandusky in charge of his own as he went in to have his gashed arm looked at. "Mouse came in with us because he didn't want to have to leave Bac She De, and it didn't occur to any of us at the moment that this would be worse," Kerry recorded. "We pulled alongside the LST and the helo started its engine and everyone stared over the side. I was determined that they wouldn't have the satisfaction of glaring at the body of this fighter and so when the body bag was lowered down we put Bac She De in it, poncho and all, and zipped it up. My God, the look on Mouse's face as he saw Bac She De, friend and tough mercenary of only a few hours before, zipped into a nondescript Marine-green rubber bag and hoisted away forever. The tears ran down his face. How I hated the war and everything about it at that moment—more than any other."

Kerry and the other wounded men received medical attention aboard a Coast Guard cutter, which was the closest ship capable of treating them. In addition to getting his arm patched up, Kerry, who had suffered a slight concussion, also had the bits of shrapnel and rice extracted from his backside.

Along with a third Purple Heart for the injury to his right arm, Kerry was also awarded a Bronze Star Medal for his bravery in the line of duty that March 13 on the Bay Hap River. He had certainly earned it, as had Larry Thurlow.

Admiral Zumwalt himself signed Kerry's Bronze Star citation, which read:

> *For heroic achievement while serving with Coastal Division ELEVEN engage in armed conflict with Viet Cong communist aggressors in An Xuyen Province, Republic of Vietnam, on 13 March 1969. Lieutenant (junior grade) KERRY was serving as an Officer-in-Charge of Inshore Patrol Craft 94, one of five boats conducting a SEA LORDS operation in the Bay Hap River. While exiting the river, a mine detonated under another Inshore Patrol Craft and almost simultaneously, another mine detonated wounding Lieutenant (junior grade) KERRY in the right arm. In addition, all units began receiving small arms and automatic weapons fire from the river banks. When Lieutenant (junior grade) KERRY discovered he had a man overboard, he returned upriver to assist. The man in the water was receiving sniper fire from both banks. Lieutenant (junior grade) KERRY directed his gunners to provide suppressing fire, while from an exposed position on the bow, his arm bleeding and in pain and with disregard for his personal safety, he pulled the man aboard. Lieutenant (junior grade) KERRY then directed his boat to return and assist the other damaged boat to safety. Lieutenant (junior grade) KERRY's calmness, professionalism and great personal courage under fire were in keeping with the highest traditions of the United States Naval Service.*
>
> *Lieutenant (junior grade) KERRY is authorized to wear the Combat "V".*

By any standard, John Kerry had become a bona fide war hero. When the Commander of Coastal Division 11, Charles F. Horne, recommended him for this Bronze Star on March 23, he pointed out that the twenty-five-year-old lieutenant (j.g.) had previously earned two Purple Hearts (on December 2, 1968, and February 20, 1969) and the Silver Star Medal (on March 6, 1969). In fact, Kerry had joined Mike Bernique and Larry Thurlow as one of the most decorated officers in the brown water navy. Yet he had also become a more "uncommitted soldier" than ever in the wake of the combat experiences for which he had earned a chestful of shiny medals and the horrific memories that came with them.

Chapter Fourteen

The Homecoming

Every St. Patrick's Day is a good time to drink. But on March 17, 1969, it felt downright wonderful to be a certain U.S. Navy lieutenant (j.g.) serving on a Swift boat in South Vietnam. The officer in charge of PCF-94—John Kerry—was very much in a mood to party with his colleagues in An Thoi. "All the Swift officers present went to their footlockers and pulled out clean jungle greens and put them on for the trip into the beach. Then, with a huge Irish flag, which the commander had produced from somewhere—nobody knew where—they massed on one of the boats and proceeded at full throttle the few hundred yards into the island where the Coast Guard maintained its headquarters and where they were briefed before their missions." Somebody even broke out a bottle of Mekong, an inexpensive Thai liquor that tasted like Kentucky mash bourbon.

When the boat got there, the Irish flag was transferred from its staff to an oar pressed into service as a standard; under its colors the twenty Irishmen-for-a-night paraded past a group of gaping South Vietnamese to the bungalow that housed the An Thoi Officers' Club, which included a dormitory. Some of the officers stationed at the austere facility were outside playing volleyball and interrupted their game to caution this erstwhile Celtic cortege against wrecking their club. Kerry and company then disappeared into the long barroom and gleefully proceeded to ignore the warning.

Among the first to get out of control was the local U.S. Army representative on An Thoi, who had the stressful job of keeping tabs on the site's prison, which held as many as thirty-five thousand captured Viet Cong.

The prisoners kept escaping into the interior of the island and joining the local cadre; according to one veteran infantryman cited in Michael P. Kelley's exhaustive 2002 guide *Where We Were in Vietnam*, late in 1968 more than thirty internees of the An Thoi POW camp dug an escape tunnel six feet deep and 150 yards long using soup spoons. They hid the entrance under a bed pallet and scattered their excavated dirt during their daily exercise time in the prison yard, just like in the 1963 World War II movie *The Great Escape*, which had given the idea to a captive North Vietnamese officer—he had seen the film while attending college in the United States.

Because of all the An Thoi escapes, no one really knew how many VC were on Phu Quoc. Often, however, warnings of an impending attack came in, sometimes forcing the Swifts out in the middle of the night to man positions surrounding the base and wait to provide flares or mortar the perimeter. The PCF crews took the procedures seriously after a practice drill saw one of the Swifts' flares land inside the POW camp and burn down a number of tents, creating substantial havoc. The Army's monitor at the An Thoi prison thus had a lot to unwind from, and got drunk so early on St. Patrick's Day that it wasn't long before he thought he saw Viet Cong outside the O Club. He had just picked up a wine bottle and gone to the window to throw it at the enemy when another officer tossed the Army man himself through the screen instead. Upon that, the quietest of all the Swift skippers in An Thoi let loose with a long, one-note howl that could have carried across the Grand Canyon. The party was on. "Dinky," as everyone called the well-loved Don Droz, brought out a razor and some shaving cream, "and we had a ceremony to formally shave off his Fu Manchu mustache as per the orders that Captain Roy Hoffman ('Latch') had sent down in a message," Kerry recorded in his notes. "Dinky put shaving cream everywhere and he finally found his lips with the razor and all of a sudden his mustache was gone and we all drank a toast to 'Latch,' who had such wisdom to order such an important thing in the middle of this war in the middle of the godforsaken Mekong Delta."

After that everything became a blur. "We fought imagined VC and charged a few people and sang and shouted and probably we all wanted to cry but it was one hell of a good time," Kerry reported, rivaled only "by the party a few weeks earlier when I had gotten so drunk that I passed out and woke up at three in the morning spread-eagled on the pier and spent the rest of the night freezing in the galley of the Coast Guard cutter tied alongside."

Kerry's full-bore merriment that St. Patrick's Day owed to more, however, than the rare respite from the dangerous daily grind of Operation Sealords. Four days after taking his third hit of shrapnel, the commander put in for a transfer out of Vietnam under the instruction that allowed thrice-wounded active-duty personnel to request reassignment—in Kerry's case to "duty as personal aide in Boston, New York, or Washington D.C.," according to the form forwarded on his behalf to the Navy Bureau of Personnel at 7:42 A.M. on March 17, 1969.

The getting was good that March. As the month progressed, it seemed that every few days another Swift boat sustained serious damage from enemy fire or underwater mine. In essence, the PCFs had become Admiral Zumwalt's riverine battleships. Guns always blazing, they had become the heart and soul of the Navy's pacification efforts. As the Swift boats were sent deeper and deeper into the treacherous VC territory, their crews began to rack up more and more medals—and death certificates. This had caused Swift boats to go looking for VC deep in the hidden reaches of the Mekong Delta. Suddenly, being a junior officer assigned to Swift command was the easiest way to earn these medals—and death certificates—in Vietnam. On March 21, for example, two rounds of 75-mm recoilless fire smashed into PCF-21 and nearly sank it. Before that, PCF-32 had hit a three-hundred-pound underwater mine, leaving the entire crew wounded.

Given the "new reality" of Operation Sealords, all the junior officers could do was perform well on river raids, stay alert, complain to Zumwalt or other top brass, and work together as a cohesive unit. As they were jockeyed about between An Thoi and Cat Lo and Qui Nhon and Cam Ranh Bay and Danang, the junior officers corresponded with one another. These letters constitute the most important primary source of what it was like to be a Swift boat officer in Vietnam. Stationed at Cat Lo, Wade Sanders often wrote letters of encouragement, humor, and despair to John Kerry. "John—you really have my sympathy," Sanders wrote him that March. "Everyone here at Coastal Division 13 was sorry to see you go, but it is all part of the usual confused manner of doing things. Just as you left with little reason, the 6 boat, which was traded for the zapped 51, has been snatched from our grip and is being presented to the Vietnamese. It wouldn't have been so ludicrous if it had been taken when it first got here, but they waited for us to perform nearly a major overhaul, spending much money and sweat before it was snatched. Time and effort wasted—familiar words. . . ." The letter went on:

John, for God's sake be careful and preserve yourself. I know you well enough to know you'll do your best but never forget there is so much more to be experienced, and so much more to living.

A little madness has settled upon us all here. Several rather embarrassing incidents involving drunk OINCS [sic] *has shaken each of us. . . . It's getting shaky here, believe me.*

I had a "Sea Lords" raid scheduled last week, and I couldn't sleep until nearly 4 AM, two hours before I had to get up. I laid on my back, with eyes wide, and was just plain scared. We're all just plain scared. . . .

Once again, old friend, keep your head down and enjoy what bit of living you can, it's all precious and each minute is unique.

Wade

Sanders, like his friend John Kerry and tens of thousands more young American servicemen, was growing up fast in Vietnam. The brutality of their experiences brought a sudden maturity and appreciation for life itself and their firsthand knowledge of the South Vietnamese people brought compassion and respect for other cultures. Lieutenant Skip Barker, for example, had grown tired of reading Vietnam War policy books. He turned to anthropology to try to make sense of the culture gap between Americans and the Vietnamese. He shared with Kerry what he discovered in Robert Ardrey's *African Genesis* and Desmond Morris's *The Naked Ape*. "In a nutshell," Barker later noted, "I concluded we weren't of the same tribe."

Over and over, the U.S troops in Southeast Asia saw, heard, and felt the myriad reminders that this was not the Good War or the Great War fought by their fathers and grandfathers. In Vietnam, they learned that their government's motives were far murkier, and its goals insubstantial, if not suspect. Paying any price and bearing any burden "to assure the survival and the success of liberty" looked rather less lofty from the view of those actually on the ground meeting the hardships to support any friend and oppose any foe. And increasingly the questions nagged: Whose liberty, and by whose definition? The world seemed to have grown so much more complicated in the mere eight years separating the inaugural addresses of Presidents John F. Kennedy and Richard M. Nixon.

To begin with was the massive toll taken on the entire region by the indiscriminate daily U.S. bombings. "When Lyndon Johnson decided to cut back the bombing of North Vietnam in November 1968, the Joint Chiefs reluctantly agreed after Secretary of Defense Clark Clifford assured them that the strikes could be redirected against Laos," wrote William Shawcross in *Sideshow*. "The statistics help tell the story. In 1968, 172,000 sorties were flown against North Vietnam and 136,000 against Laos. In 1969 the bombing halt reduced sorties against the North to 37,000—the attacks in Laos rose to 242,000. In January 1969, when Nixon and Kissinger arrived in the White House there was capacity to spare for Cambodia."

Barely a month after Nixon's inauguration, the North Vietnamese launched their "high point" campaign of rocket and mortar attacks on some hundred South Vietnamese cities and towns. The Communist offensive claimed the lives of 1,140 Americans and many more South Vietnamese. "Nixon then took the first step in the old scenario for extended war," as Michael Maclear put it in his 1981 *The Ten Thousand Day War: Vietnam: 1945–1975*. The President approved Operation Menu: a series of secret B-52 bombing strikes against the supposed North Vietnamese "sanctuaries" inside Cambodia. The raids represented one of the first actions under the new U.S. administration's policy of shielding its withdrawal of American ground troops with escalated military attacks from the air.

Operation Menu began March 18, 1969, and during the next fourteen months 3,630 B-52 sorties dropped more than 100,000 tons of bombs within five miles of Cambodia's border with South Vietnam. Kerry and the other Swift crewmen running the rivers near Cambodia were stunned at their government's concentrated bombing campaign directed into a neutral country. Every time they had so much as fired an M-16 round into the Cambodian underbrush it seemed the U.S. State Department had received a complaint from Prince Norodom Sihanouk. Yet the United States was dropping bombs with stark regularity into South Vietnam's neutral neighbor, and the international media was ignoring the sorties.

Even more disturbing in retrospect was how little of a threat the supposed Viet Cong in these supposed Mekong Delta "sanctuaries" of theirs actually posed. As Neil Sheehan revealed in his biography of John Paul Vann, a longtime U.S. advisor in Vietnam, *A Bright Shining Lie*, Vann "discovered that many of the 2,100 hamlets the [Hamlet Evaluation System] listed as under Viet Cong ownership in IV Corps in February 1969 (another 2,000 hamlets were listed in varying degrees of Saigon control) were actually held by half a dozen guerrillas." Nevertheless, as Sheehan continued, "Richard Nixon's 'de-Americanization . . . with all deliberate speed' was not proving cheap in American lives. During 1969, 11,527 U.S. servicemen perished in Vietnam [9,414 of them in combat]. . . . In all, nearly 21,000 Americans were killed in Vietnam during Nixon's presidency and about 53,000 seriously wounded, more than a third of the total U.S. casualties." At the same time, Michael Maclear observed of Nixon's "Vietnamization" period, "South Vietnamese military losses would rise fifty percent to more than 250,000; civilian casualties—including deaths—would also rise fifty percent to 1,435,000." And that wasn't counting the Cambodian peasants killed by the Menu bombings. And all the while, Kerry's PCF-94 and its fellow Swifts in An Thoi went on dropping U.S. Navy SEALs off along the Cambodian border to destroy Viet Cong base camps along "Bernique's Creek." Kerry described one such mission in detail in his "War Notes":

> *The banks of the "Creek" whistled by as we churned out mile after mile at full speed. On my left there were occasional open fields that allowed us a clear view into Cambodia. At some points, the border was only fifty yards away and it then would meander out to several hundred or even as much as a thousand yards away, always making one wonder what lay on the other side, always aware of the ease with which the VC were using the sanctuary of this imaginary line to protect their smuggling and rearming efforts.*
>
> *On occasion we had shot towards the border when provoked by sniper or ambush but without fail this led to a formal reprimand by the Cambodian government and accusations of civilian slaughters and random killing by American "aggressors." I have no doubt that on occasion some innocents were hit by bullets that were aimed in self-defense at the enemy, but of all the cases in Vietnam that could be labeled massacre, this was certainly the most spurious.*

When we arrived at the prearranged drop point, the boats slowed down and one of them nudged into a bank. The SEALs jumped ashore and within seconds disappeared into the darkness. They called quickly on the radio to check communications and then there was complete silence except for the drone of the diesels.

To try to convince the VC that we were on a normal patrol, we continued upriver occasionally slowing down as we had at the drop point, and then we sat at the Vietnamese outpost at the very top of the creek—at a fork whose one arm went directly into Cambodia and whose other arm became the Vinh Te Canal and a route to Saigon for the small craft. We always stopped there to rest and grab a drink or relieve ourselves or something. It was a chance to relax and open [our] flak jackets, have a smoke and brace up for the trip back down the river. I used to try and catch twenty minutes "shut-eye" if I could, but it wasn't easy to sleep up the river. Some of my men would never sleep and one of them didn't even like to move around the boat.

From the outpost we would move down the river to Tra Phu. If no one shot as us—and they usually didn't at night—the trip would still provide its own excitement. Running the rivers at night was something like flying a plane on instruments. The only difference was that we didn't have any instruments. When it was pitch-black and there was no moon, the banks would blend in with the water and looking through the glass windows in the pilothouse, it was exceedingly difficult to find a horizon and tell where one was in the river. One could feel a graceful motion from the calm cut of the boat through the water, and this only lent further to the feeling of vertigo that sometimes caught the drivers. At one point we careened off a bank thinking that we were in the middle of the river.

Finally we decided that the only safe way to negotiate the river at the speed that provided an extra margin of safety was for me to hang out the door on whichever side of the river we were covering and warn Ski [helmsman Del Sandusky] when the bank appeared to be looming too close for comfort. And so we would move downstream at about twenty knots, this figure hanging out the door with an M-16, pretending to survey the countryside for enemy movement, while actually turning to the helmsman and saying, "Better come right a little"—"Okay, now, we're too close"—and all the while praying that we weren't going to hit something under water.

We arrived at Tra Phu and parked the boats opposite the village. Then, with the engines shut down, we would listen to the radio signals that told us the inserted SEALs were safe. One click, which would answer with a voice response, meant that the group downstream was all right. They would check every half hour, and in the meantime we would keep continual watch in both directions for a red flare shot into the sky, which meant they needed immediate extrication.

A group of the Vietnamese junks pulled in alongside us and we sat there listening to their small engines burp into the quiet of the night. Across the river, perhaps fifty yards away, we could see the villagers moving around in their huts, silhouetted by the fires that burned throughout the small town. The peculiar smell of their wood burning blew across us in wisps. All over Vietnam this smell had been the same. It was a smell that brought to mind poverty and dirty food and the ground to sleep on at night and it made me feel very clean and out of place.

There were a great many mosquitoes and despite the fact that it cooled down at night they kept biting away. There was nothing worse than lying in a bunk waiting for sleep to carry one away when this grating buzz swept by an ear and perked one into an annoyed alertness. For some reason, the idea of lying back and letting the goddamned insect bite away was like walking willfully in front of VC fire. It was just one more pestilence to fight and it had to be fought. Besides, I hadn't taken my malaria pills for weeks and I was convinced that the one mosquito that I let get away with it was going to be the wrong kind.

One of the men occasionally popped a flare to see if anything or anyone was sneaking up in front of us, and the junk would be suddenly as bright as daylight. Slowly the flare would drift down on its parachute and then extinguish on the ground in a ballet of flickers with huge palm leaves forming the backdrop and stage. Every half hour the radio would click once or twice and the clicks would remind us that out there in the dark our men were creeping around, trying to find the enemy.

At about three in the morning a call came in requesting us to pick up the group upstream. Somehow they had compromised their position and staying there was only wasting time. We aroused the sleepers and moved quickly up to [the SEALs'] position. As we neared them, a red light flashed intermittently to bring us in on their position and we beached a few yards from where they were grouped.

The sergeant in command of the group wanted us to take them to another area where they could perhaps perform a "body snatch." At first I didn't know what he was talking about, but he quickly elucidated. They wanted to be left off near a small village which they would enter quietly, sneak into a hooch, and kidnap someone that they thought was good for information. Often the individuals in the villages would volunteer information in the hopes of making what they were sure would be an ordeal easier. Essentially, this is what the SEALs were looking for.

We moved back up the river to an area from which lights had been coming a few hours earlier. There the SEALs disembarked again. We moved the boats to the first bend in the river and there we sat with everything that made noise turned off. It was eerie. The slightest movement was heard across the water. The bruising of an M-16 against the bulkhead seemed to carry for miles. And in front of us in the brush, we kept hearing movement of some kind.

About an hour passed and then we heard the noise of the SEALs special boat moving slowly towards us—its red signal light flashing intermittently to identify it. We were all amazed at how little the sound of the engine would carry, and frightened when we thought that his engine, compared with ours, had been especially designed for clandestine missions and was supposed to be particularly silent. This merely confirmed the fact that anyone in the river knew when we were as much as three miles away. No wonder we never caught any contraband.

The SEALs came alongside and loaded their gear aboard the Swifts. They had a young man with them who apparently had a good deal of information about VC movements and who had volun-

teered to come with them to the headquarters. The SEALs' boat was tied alongside, and we left the area to pick up the group downstream. When they came aboard it was amidst a barrage of swearing because one of the Vietnamese had lost his way and led them into a mud swamp in which they had been forced to spend most of the night, because of a number of noises near them, which they interpreted as an enemy force larger than their own. All in all, it had been a long and profitless night but no shots had been fired and it was better to be tired than dead.

But after that mission, PCF-94 was turned over to a new lieutenant.

Because Kerry had earned his third Purple Heart, he was headed home. After a final uneventful patrol, he was transferred to Cam Ranh Bay to await further orders. "I spent probably five or six days sitting around there," Kerry recalled. "Then I got assigned to Brooklyn."

Decades later, on June 17, 1995, at a ceremony where a Swift boat was being put on permanent display at the Naval Historical Center in Washington, D.C., Kerry reflected not on the tragedy of Vietnam, but on the fellowship. After years of painful memories, bright spots burst forth. "We sunbathed and skinny-dipped," he recalled of those days. "We traded sea rations for fresh shrimp, and left our Vietnamese recipients of Uncle Sam's technology grinning from ear to ear as they believed they got the better deal. We happily basked in wide betel-nut smiles. We glorified in shouts of 'Hey, American, you number one,' and we casually brushed off taunts of 'Hey, you number ten.'" He recounted with fondness the youthful, boyish moments when they lobbed raw eggs at one another, engaged in epic flare fights, with a life raft even once catching fire; and the difficulties of house training V.C., his onboard puppy. He called the men "brothers in combat" and he missed them.

Quickly Kerry said his good-byes to Vietnam, making telephone calls to Wade Sanders and Skip Barker. On a flight back to the United States, Kerry, dressed in his Navy khakis, slept. Eventually he woke up from a nap to find himself at McChord Air Force Base in Tacoma, Washington. He studied the Lockheed C-141A Starlifters on the runway, ready to head back to where he had come from. As he boarded a commercial flight bound for San

Francisco, he received strange stares from the civilian passengers. "It was as if I was an alien being," Kerry recalled. "There was no thanks for being in uniform. Only glares." He was flying down to California to console his sister Diana, who was torn apart by the suicide of the mother of a close family friend. "So I spent one night in San Francisco paying my respects," Kerry recalled. "Then flew back to see Julia in New York."

When John Kerry finally landed at Kennedy Airport in New York he was greeted by an ecstatic Julia Thorne. "I went to pick him up," she recalled. "It was a packed airport, [but] in those days you could go to the gate. People were pouring off the plane in Hawaiian shirts and leis, some seemed drunk on mai tais. Then, emerging from the crowd in his dress blues and his white hat, came John Kerry. He was bandaged, some of it was sticking out, and nobody was paying attention to him while I was sitting there going, 'Everybody stop and look at this man.' Part the seas and say, 'This is your veteran coming home from serving his country.' But nobody cared. Nobody gave a goddamn. Nobody gave a damn at all. Nobody."

Hugging each other, they rocked back and forth and cried. It had been decided that he would live with Julia, his fiancée, at her fourth-floor apartment on East Seventy-seventh Street (his parents were now living in Groton, Massachusetts). With little money to draw upon, they pulled together like a family. "We started to live together," Julia explained. "We cooked at home, we went to parties together, when he wasn't on duty, we could be human beings." For children of privilege, they now found themselves on a shoestring budget. Sometimes Kerry headed over to Elaine's on Second Avenue, forging a friendship with *Paris Review* editor George Plimpton; another time he hooked up with journalist Pete Hamill at the Lion's Head on Christopher Street in Greenwich Village. To the outside world Kerry seemed happy and content. "This is the period when I most remember safekeeping John's ghost," Julia recalled. "The Vietnam ghosts. The nightmares. John would wake up in sweats. He would wake up and he'd get out of bed. He was always in combat, in an emergency situation. He was saving men. It was never anything about him—it was saving the boat and saving the men."

For almost a year, in the middle of his sleep, Kerry would have a Vietnam flashback, and Julia would comfort him, stroking his head and reassuring him that everything would be all right. While John and her two brothers had been in Vietnam she had felt sorry for herself, fretting over the disease of loneliness. Every evening from November 1968 to March 1969 was one of constant worry. She used to think it would have been better to have lived at a military base with Navy wives—people to share the waiting. Now, due to John's nightmares, she understood in a more pronounced way the anguish of the Vietnam veteran. "John was experiencing a level of pain I had never known," Julia recalled. "And he really had nobody to talk it out with. He could maybe have talked with Dick Pershing—but Dick was dead. There was nobody in his immediate vicinity that he could share this with. When he wasn't working his Navy job in Brooklyn, he just immersed himself with my friends, who really knew nothing of his pain."

Nightmares aside, good fortune had found Kerry when he was assigned to serve as a special assistant to Admiral Walter Schlech Jr., commander of the Military Sea Transportation Service, U.S. Atlantic Fleet, headquartered in Brooklyn. Kind, caring, and considerate, Admiral Schlech had amassed a sterling military record, winning the Silver Star Medal with Gold Star, the Legion of Merit and the Commendation Pendant with Combat V, among dozens of other honors. Born on July 8, 1915, in Brooklyn, Schlech was roughly the same age as Kerry's father. Mentoring came naturally to him. After graduating from the Naval Academy he had been assigned to serve on the battleship *Arizona* as turret officer and Assistant Navigator just three years before it was bombed at Pearl Harbor. Starting in January 1940, Schlech reported to Submarine School in Groton, Connecticut. During World War II he became a war hero, participating in five submarine patrols in the Pacific Theater (Aleutians, Truk, Palau-Puam areas, and northern Japan).

Sharing Silver Star status helped Kerry bond with the paternalistic Schlech who was the father of four children. But Kerry also quickly learned to appreciate the international relations aspects of Schlech's intrepid career. Ever since he had become a rear admiral in July 1964, Schlech traveled the world. He regaled Kerry with stories about his time spent with the Navy's first squadron of Polaris missile firing submarines at Holy Loch, Scotland, and his dogged efforts implementing the U.S. Military Assistance Program in Ankara, Turkey. His billet back in Brooklyn was essentially a last post to serve before retirement.

After only a few days of working for Admiral Schlech, Kerry asked for permission to tell his commanding officer about Operation Sealords. "I had a lot to get off my chest," Kerry recalled. "I wanted him to understand what I had experienced." He told Schlech how he had warned his superior officers in Saigon of the inevitable catastrophic consequences of continuing the kind of raids the Sealords program was engaged in. He even tried to alert Admiral Zumwalt of the dangers—to no avail. It was Kerry's prophecy that the U.S. Swift boats in the Mekong Delta were due for a VC ambush of a ferocious intensity not experienced since the Tet Offensive. He also felt that, without a more professional approach to the problem of Delta pacification, the Navy's riverine vessels would only suffer losses more serious and unnecessary than those they had already sustained. With great emotion Kerry recounted to Schlech how PCF-51 lost six men in a short four-month span. "He listened," Kerry recalled. "But he didn't fully agree with my bleak assessment of the war. He thought things were a little screwed up, but he didn't want to talk about it. I talked with him mainly about *his* war—World War II."

Kerry had not personally known the men killed on PCF-51 that March and April while on patrol. That, unfortunately, was not the case for the OIC of PCF-43. On April 12, two weeks after Kerry had left Vietnam, Don Droz, one of his closest friends from Coastal Division 11, was killed in action on the Duong Keo River while on PCF-43. The tragedy was the result of a surprise heavy weapons ambush by the VC. The vessel was completely destroyed; it had been carrying 800 pounds of high explosives. "It was broke in half" is the way one eyewitness explained the sight. "It was beyond repair." A recounting by Lieutenant (j.g.) Peter Upton titled "Death of the 43" uses phrases like "ripped asunder," "gaping wound," and "raked the deck" to describe the damage done by the VC attack. As the Western Union telegram reported, Droz's Swift boat had come "under heavy rocket and automatic fire on operations."* He left behind his wife, Judy Huddart-Droz, whom he had married in Pennsylvania only a short time before coming to Vietnam, and a newborn daughter.

Only two or three weeks before his death, Droz had received orders which were to send him to Dartmouth College in New Hampshire for a two-year tour as an ROTC instructor. Like Kerry, the eloquent Droz had often written and spoken of his violent objection to Sealords and to the

**Others wounded in action included Seaman Third Class Art Ruiz, Seaman Michael Sandlin, Seaman Third Class Robert Lowry, Seaman William Piper, Gunner's Mate Rickey Hinson, and Lieutenant (j.g.) Peter Upton.*

entire war. "Can you possibly imagine how much I've been thinking of you these past few days since returning from R & R?" Droz wrote his wife just a day before he died. "Everything about you is now so vivid in my mind. I could just die when I think that I must live without you for even two more months. It's incredible how much two little months can mean. I love you and Tracy so much that all else in life seems truly secondary. I almost constantly think of life in Hanover. Do you have any idea how wonderful you are to come home to?"

Words cannot express the outrage Kerry felt when he learned that Droz had been killed on the Duong Keo by a claymore mine and a fusillade of rockets courtesy of approximately seventy-five VC guerrillas. He wanted to grab a megaphone and denounce the Vietnam War publicly as a sham. He had risen in life—through education and the military—by playing, for the most part, by the rules. But what was happening to the Swifties in the Mekong Delta was, to his mind, a grave government mistake. Grateful to be alive—and to have gotten out of Vietnam with three Purple Hearts, a Silver Star, and a Bronze Star—he knew he had an obligation to speak out. His conscience was haunting him. He felt muzzled and frustrated. Part of him loved the U.S. Navy, particularly the Swift boat community and great officers he had encountered like Captain Allen Slifer, Commodore Charles Horne, Commander George Elliott, and now Admiral Walter Schlech. His heart was still in the Navy. But due to his firsthand experiences in Vietnam his head told him that the Nixon administration's foreign policy was dangerously wrongheaded. Everyday good sailors like Don Droz, a family man of the first order, were being sacrificed for Nixon's ego trip that he would not be the first U.S. president to lose a war. "Don's death cut like a knife," he recalled. "It was all so senseless."

Kerry had found out about Droz's death via an April 28 letter from Skip Barker, who was still at An Thoi. Reflecting on why he chose to write Kerry about Droz, Barker spoke of their intellectual kinship and a shared outrage about sending sailors up canals to get shot up. "Besides the complete loss of PCF-43, 5 boats had been hit by B-40s, two of these two rounds," Barker recounted to Kerry.

Every boat had numerous bullet holes—AK-47 and .30-caliber and every boat had blood on her decks. That night we had a taped debriefing in which each officer gave his account of the action. At the end, Captain Hoffman made his remarks. He said that we had taken a few licks—but we were the victors in the battle of the Duong Keo. We had "stayed in there and would return. Charlie, having hit us, had run." He said he "regretted terrible the loss of our men and that Don Droz was a personal friend of mine. That Don was a real fighter—and there was no question but that when he went down—he did so fighting." The next morning we picked up the Marines and continued operations—that is those boats that could still get underway. We finally ceased this part of our insanity on 22 April . . . The only good news I have is that Hoffman leaves country in 2 weeks—end of tour. His replacement couldn't be as bad though we expect little change.

Barker's letter hit Kerry like a ton of bricks. For a few days he wandered around in anger, not quite sure at whom or what. He heard from a friend in the Mekong Delta that Commander George Elliott of Coastal Division 11 was furious at Captain Hoffman about Droz's death. Apparently they got into a dispute over Operation Sealords. "I was trying to weigh values of the risk versus rewards," Elliott recalled. "Hoffman wanted to send men into the U Minh Forest. It never did any good. There was a series of things we didn't see eye-to-eye on. He thought I was being too cautious and slow." A few days after their confrontation, Elliott, who lost the policy debate, was transferred from An Thoi to Coastal Division 15 at Qui Nhon. He had been essentially demoted. "There were no rivers to go into at Qui Nhon," Elliott recalled. "Occasionally there was a VC incursion but it was a more placid post. We called it Peace Corps. The Swifts under my command did coastal patrols."

The death of Droz created a wave of angry sentiments from the Swift boat community. Lieutenant Larry Thurlow, for example, summed up the feeling many Swift boat officers had to Captain Hoffman: "That man wanted in the worst way to be admiral—that's what Sealords was about." Lieutenant Jim Galvin of PCF-22, an Annapolis classmate of Droz's, was ambushed at almost the exact spot where PCF-43 sank, just a few weeks later. "I only got a flesh wound," he recalled. "But I was lucky. The VC's trajectory was just a little off." Lieutenant Stephen Hayes, for example, wrote extensively about the accident in his diary. A few days after the ambush he had gone on a mission past the "charred remains of the 43 boat." The only word he could find to describe it was "grisly." Frustrated at the carelessness of

Operation Sealords, he took aim at Captain Hoffman with his pen. "Latch had the gall to call it a victory!!!," he wrote. "Such a tragedy that these operations are run by bunglers and hard core military men who apparently get some strange sense of satisfaction out of playing around with other men's lives. What is it? The sense of importance. The excitement of danger? It's just not worth it."

Proud of his Swift boat service, Kerry now felt duty-bound to denounce Operation Sealords raids like the ones in which Droz was killed and Galvin was wounded in as ludicrous missions aimed at sacrificing the best Americans to satisfy a president's geopolitical ambition. Every week that Nixon was in the White House the ire and frustration in Kerry grew. Why, he wondered, didn't the American citizens rise up and demand an end to the war. He was shocked that a Harris poll taken shortly after Droz's death found that only 9 percent of U.S. citizens were prepared to accept a peace settlement in Vietnam if it meant communist control of the Southeast Asian country. "By the time I came back from Vietnam I was livid at the Nixon administration because they claimed they had a secret plan for peace," Kerry recalled from his Boston home in 2003. "One of the things I remember saying when I got back was 'They kept their promise. They kept it a secret.' I was angry about it. Vietnamization had become this bogus process of continuing the war in surrogate form. I didn't think throwing people at the Hamburger Hills or Hill 81s was a realistic solution."

The frustration Kerry felt in April 1969 came rushing back to him in 2003. A number of times when talking about Droz's death he would raise points about U.S. policy made by Arthur M. Schlesinger Jr. in *A Bitter Heritage*, Townsend Hoopes in *Limits of Intervention*, and Neil Sheehan in *A Bright Shining Lie*. "Firsthand experiences taught me that the Vietnamese people weren't hard-core Communists," Kerry went on. "They weren't hard-core anything. They were opportunists. Unless the government in South Vietnam was willing to make their lives better, which they didn't seem to be doing, it was going to be very difficult to advance the American cause. Which is exactly what happened. . . . I think I learned very quickly how screwed up it was. And part of what prompted me to start publicly protesting the war was a frustration with a conventional strategy that was almost dumb, a sort of calculated draw fire but not really accomplishing anything but losing American lives. I wanted to secure territory. I wanted to change people's lives. We didn't seem to be doing that except by violence. And losing Droz . . . well, it ticked me off."

In a later interview, he mused that he also lost his St. Paul's School friend Stephen Kelsey in Vietnam along with Dick Pershing, John White, and Bob Crosby. "Strangely," Kerry stated, repressing tears, "some of the people who were most involved in my life in both high school and college got killed." About a month after Droz's death, on May 15, Kerry received news that a column of PCFs had been attacked in the Bay Hap River, with two more fatalities. To Kerry the war had now become personal. Compartmentalizing his emotions from his sense of reason became more and more difficult. Perhaps Navy Lieutenant L. J. King from Houma, Louisiana, who spent thirty-three months in combat zones, summed up the feelings of Kerry and thousands of other servicemen best: "Vietnam disgusted me because I lost several of my friends."

For the most part Kerry's desk job in Brooklyn involved answering correspondence, setting up interviews, and reviewing documents. Always at Admiral Schlech's side, Kerry accompanied him on trips to such far-flung locales as Greenland and Panama. Photographs from 1969 always show him holding the admiral's black briefcase, attentive to his superior officer's every request. He was grateful not to be one of the 543,000 troops still in Vietnam—or one of the almost 12,000 sent home that year in body bags. From his post in Brooklyn, however, Kerry met fellow Vietnam veterans, many of them psychologically devastated by the same kind of recurring nightmares he was experiencing. GI guilt was widespread. Stories were circulating in the press of GIs raping and murdering Vietnamese women or torching entire villages with a click of a cigarette lighter. After World War I, doughboys came back "shell-shocked." In World War II the GIs were inflicted with "combat fatigue." But the grunts who fought in Vietnam were said to have suffered from post-traumatic stress disorder (PTSD). One veteran explained, "PTSD meant you no longer lived in the world of stars and stripes, you lived in the world of scars and stripes. Suddenly, you returned from Vietnam and were known as a baby killer."

Historian Marilyn B. Young, in her landmark *The Vietnam Wars: 1945–1990*, comes as close as anybody has in explaining the rage veterans felt toward American society. She points out that a crucial deficiency of the 1960s and 1970s was that the U.S. government had no program to prepare veterans for "the shock of return." While in the Mekong Delta or along the DMZ GIs always dream of ending their tour, of coming home to a national embrace. But when they got home nobody was there to greet them. They were shunned as if they harbored a contagious disease. "Later, many veterans would tell stories of having been spat upon by antiwar protesters, or having heard of veterans who were spat on," Young wrote. "It doesn't matter how often this happened at all. Veterans *felt* spat upon, stigmatized, contaminated. In television dramas, veterans were not heroes welcomed back into the bosom of loving families, admiring neighbors, and the arms of girls who loved uniforms; they were psychotic killers, crazies with automatic weapons."

One Navy veteran, John McCain, who had been shot down over North Vietnam and was held as a prisoner of war in Hanoi for five and a half years, much of it in solitary confinement, couldn't believe how badly American society treated his "brothers." When he became a U.S. senator in 1987, he recounted a story that illustrated the indifference—even contempt—granted Vietnam veterans by average citizens. "One of my most unfavorite stories is about the guy that lost an arm," McCain said. "He was in a San Francisco bar and a young woman walked up to him and said 'How'd you lose your arm?' And he said, 'Vietnam.' And she said, 'Good!' "

It's important to stress that not all Vietnam veterans suffered from PTSD; in fact, only about one fourth who served believe they had developed psychological effects from combat. And about 70 percent of all Vietnam veterans loathed the antiwar movement. There were prowar veteran groups like Vietnam Veterans for a Just Peace, B-52s for Peace, and Vietnam Veterans for Nixon. Oddly, a group of right-wing prowar hawks—including Senators Barry Goldwater, Richard Russell, Strom Thurmond, and Russell Long—echoed some antiwar veterans when they claimed if the U.S. government didn't have the courage to "win the war," disengagement made perfect sense. In June 1969, when Nixon announced that he was withdrawing 25,000 soldiers from Vietnam, doves claimed *not enough*, while hawks protested the reduction as akin to appeasement. Meanwhile, the deaths of GIs continued. From Nixon's inauguration to the departure from Vietnam that August of the Ninth Infantry Division, more than 7,000 U.S. troops had died in Southeast Asia.

* * *

Throughout the summer of 1969, Kerry grappled with his conscience. As a highly decorated sailor, with a cushy assignment with Admiral Schlech, part of him decided to not make waves. Just put in your last year, he thought, then enroll in law school. He learned about two antiwar groups—the Vietnam Moratorium Committee and a new reconfigured New Mobilization Committee to End the War in Vietnam—but refused to join. But the nightmare, guilt, and anger wouldn't subside. Constantly he found himself seeking out other Vietnam veterans to share experiences. As Robert Muller, a disabled founder of Vietnam Veterans Against the War (VVAW), once explained: "You don't go to war, come home, and not talk about it." And the poet W. D. Ehrhart, speaking for thousands of vets, including Kerry, summed up the moral dilemma they all felt when he stated: "I want it to have been worth *something*, and I can't make myself believe it was." *New York Times* reporter James Reston, sensing the growing veterans' movement, wrote quite perceptively that President Nixon "has been worried about the revolt of the voters against the war . . . but now he has to consider the possibility of a revolt of the men if he risks their lives in a war he has decided to bring to a close."

An important turning point in Kerry's life occurred in October 1969, when his sister Peggy—then actively protesting the Vietnam War—volunteered her brother to fly former Robert F. Kennedy speechwriter Adam Walinsky around the state to deliver antiwar addresses. Walinsky was somewhat of a legend in Kennedy circles. A gifted wordsmith and policy advocate, he had served in the Justice Department helping draft the Immigration Act of 1965, the Civil Rights Act of 1964, and the Economic Opportunity Act of 1964. When Robert Kennedy became U.S. senator from New York, Walinsky became his key advisor on everything from Vietnam War dissent to the creation of the Bedford-Stuyvesant Restoration Project.

Although Kerry flew all over New York with Walinsky, he refused to take part in the so-called Moratorium Day Demonstrations on October 15, declining to show up at massive antiwar protests in either Central Park or Boston Commons. But he carefully monitored the event, stunned that New York Mayor John Lindsay ordered flags at half-staff as tribute to the men killed in Vietnam. Millions across America—including the United Automobile Workers and the Chemical Workers of America—endorsed the Moratorium. And his main political heroes—Allard Lowenstein, Ted

Kennedy, and Eugene McCarthy—vigorously took part in the mass protest. In Washington, D.C., Coretta Scott King held a candlelight vigil while word spread that thousands of U.S. soldiers in South Vietnam were donning black armbands in solidarity with the Moratorium. Even San Clemente, California, where Nixon's Summer White House was located, attracted dozens of Marines from Camp Pendleton to protest the war. "The Moratorium convinced me that I had to speak out," Kerry recalled. "If you loved America you knew we had to extricate ourselves from Vietnam at once."

Kerry was inspired by Walinsky's antiwar beliefs and the power of the Moratorium, which *Life* magazine deemed "a display without historical parallel, the largest expression of public dissent ever seen in this country." More than anything else he wanted to publicly speak out against the ongoing war in Vietnam. When President Nixon declared on November 3 that "the great silent majority" of Americans were opposed to immediate withdrawl, Kerry felt as if a gauntlet had been thrown down. Six days after Nixon's silent majority speech, 1,365 GIs signed a petition that ran in the *New York Times* proclaiming: "We are opposed to American involvement in Vietnam. We resent the needless wasting of lives to save face for the politicians in Washington. We speak believing our views are shared by many of our fellow servicemen. Join us!"

But in Kerry's case "joining" was not that simple. He was still enlisted, wearing a naval uniform, and working in Brooklyn at the behest of Admiral Schlech. He had, however, a plan. Quietly, with the help of Julia, he had evaluated the possibility of running for Congress from Massachusetts. Over the years he had lived in so many different places in the state that he felt he could just pick his hometown. He decided on the Third District, which stretched from Fitchburg to Newton. His rationale was simple: the current congressman, seventy-one-year-old Democrat Philip J. Philbin, was a hawk. Kerry would challenge him for his party's nomination as the peace candidate. But before any of this was possible he would have to be granted an honorable discharge from the Navy six months early. He decided to petition Admiral Schlech, to tell his boss that his conscience dictated that he protest the war, that he wanted out of the Navy *immediately* so he could run for Congress. Admiral Schlech consented to his request and on January 3, 1970, the U.S. officially issued an honorable discharge. "It was surprisingly easy," Kerry recalled. "I went and saw the admiral and told him my conscience was bothering me, that I had to speak out against the war. He was a great man. He simply said 'I understand.' "

As promised, Kerry went to work, but another challenger had entered the race: the famed antiwar Jesuit Reverend Robert F. Drinan. Father Drinan's peace platform was deeply rooted in the Roman Catholic Church. "My feeling was that even if the objective was good, namely to stop the Communists, it couldn't be justified by a massive violation of international law," Drinan recalled in 2003. "I publicly said that all of the American bombing clearly violated all the rules of war in Geneva and also Catholic teaching over the years of unjust war." In 1969, in fact, while Kerry was in Brooklyn, Father Drinan, dean of Boston College Law School, had traveled to Southeast Asia on an "observation tour" and ended up writing a small book, *Vietnam and Armageddon*. Deemed a troublemaker by the White House, Father Drinan was placed on Nixon's famous "enemies list" and the FBI constantly monitored his antiwar activities. Upon meeting Kerry he was immediately taken with his poise, charisma, and commitment to end the Vietnam War. "I thought that he was actually a better person than I to run for Congress," Drinan recalled. "But commitments had been made and money had been raised for me. He was just too late."

With very little money John Kerry and Julia Thorne left Manhattan and moved to Waltham, Massachusetts, home to both Brandeis University and Bentley College. They rented a small downtown apartment next to a gas station; a far cry from the exquisite Robert Treat Paine estate that graced the town. "It was a dumpy little apartment," David Thorne recalled. "It was our campaign headquarters."

On February 21, 1970, Kerry took his campaign to become the Democratic congressional nominee from the Third District to a caucus at Harvard University. In preparation of his visit, the *Harvard Crimson* published a story headlined "John Kerry: A Navy Dove Runs." Kerry had made a personal visit to the *Crimson* building and talked for a couple of hours with reporter Samuel Z. Goldhaber. The resulting story essentially mocked Kerry for carpetbagging. But the candidate's voice came out forcefully against the Vietnam War. He lambasted, in particular, House Armed Services Committee Chair Mendel Rivers while praising the fiercely antiwar House member Allard Lowenstein. He supported a "volunteer army" and envisioned the United Nations expanding its influence on the U.S. military.

He was extremely hard on the CIA. "Everybody who's against the war is suddenly considered anti-American," Kerry complained. "But I don't think they can turn to me and say I don't know what's going on or I'm a draft dodger. . . . I should be at law school but the problems are too great to sit back and watch them go by."

By all accounts Kerry delivered a brilliant speech at the Saturday caucus. Instead of discussing Newton's sewage problems or Fitchburg's lack of snowplows, Kerry hammered away on the need to disengage from Vietnam. "I'm not saying that there's any special virtue in what I did because it takes just as much courage not to do—not to fight in this war—but I do want to tell them what it's like to walk amidst death in a country that wants only life and to have to shoot at people with whom we have no quarrel," he intoned, the caucus clinging to his every word. Worcester's *The Evening Gazette* reported that the impassioned Kerry spoke in "faintly Brahmin tones with Yale overtones."

Given the power of Kerry's antiwar speech, in which everybody from conservative philosopher Edmund Burke to novelist Joseph Heller was evoked, it took the more than 1,000 caucus members a surprising four ballots to finally decide on Father Drinan. The Jesuit activist would now run against Philbin in the primary election. Kerry, the underdog candidate, politely withdrew from the campaign. "When John got up to concede at the caucus I was there reporting on the event," George Butler recalled. "There was a glass on the podium, and I saw John looking at it. Suddenly he picked it up, held it in front of the audience and got their attention. And then he said: 'We may not agree on everything, but at least we're drinking from the same glass.' Father Drinan had the same glass and it just caught everyone's imagination right away. People started laughing. He lost, but he won because he really grabbed people's attention." An impressed Drinan told the press that Kerry was "a very good kid."

Stepping aside with dignity, Kerry, with David Thorne helping, went on to chair the Drinan campaign. But Kerry's mind—and heart—that spring were, for the most part, focused on personal matters. On May 23, 1970, John Forbes Kerry and Julia Stimson Thorne got married in a large outdoor ceremony in Bay Shore, Long Island. The society wedding, widely covered in the East Coast press including the *New York Times*, was held at the estate of her grandparents. Over 350 people attended the reception, including John's old Yale crowd who served as groomsmen: Danny Barbiero, Harvey

Bundy, Fred Smith, David Thorne, and Jack Pershing, who attended on behalf of his brother. It was an ecumenical service with both an Episcopalian minister and a Catholic priest accepting their vows, which on Clare Booth Luce's advice they had written themselves. The wedding dress that Julia wore—ivory colored, with clusters of red roses and green leaves—was two centuries old, worn at the 1786 marriage of her ancestor Suzanne Boudinot (Alexander Hamilton had been in attendance). A casual flip through their wedding album, which Julia keeps in Montana, tells the story: "Papa" Kerry posing with Admiral Schlech, Lanny Thorne in his Marine whites, David Thorne pushed into a pond, and a search committee rushing around trying to locate John so he could cut the cake. When the reception was over, the newlyweds were helicoptered from the estate.

From Long Island, John and Julia flew to Jamaica for their honeymoon. The Pershing family had given them private run of their Round Hill estate for a week. George Butler had also just gotten married, to Victoria Leiter, and they likewise honeymooned in Jamaica. Like most Caribbean newlyweds they snorkeled and windsailed and drank rum. But even there, Vietnam didn't leave John's consciousness. While at Kennedy Airport en route to Jamaica, Kerry purchased a copy of *Life* with a feature story titled "Our Forgotten Wounded." Laden with a dozen photographs of forgotten Vietnam veterans dying in understaffed and ill-equipped VA hospitals, the exposé made Kerry sick to his stomach. In Jamaica he could talk of little else. He had visited a few VA hospitals in the New York area—including a poorly run one in the Bronx—and he had come to the conclusion that it was a national disgrace. The *Life* article confirmed his hunch. Over and over again, he would read out loud the testimony by a former Marine—a man paralyzed from the neck down in a mortar explosion during the Battle of Khe Sanh—who spoke of the crude unsanitary conditions of his VA hospital. "It's like you've been put in jail," the Marine told *Life*, "or been punished for something."

Upon their return, the Kerrys gave up their Waltham apartment for more spacious quarters: a modest two-story Cape Cod saltbox house with green shutters at 33 Tavern Road, just off Highway 128. John immediately had bookshelves constructed in the basement, removed the mildew and cobwebs, and set up an office that would serve as his "nerve center" until 1972, when they moved to nearby Lowell.

Upon returning to New York for a week, the Kerrys set upon themselves the goal of helping other Vietnam veterans cope with PTSD and help end the war. Kerry, in fact, was invited to appear on *The Dick Cavett Show* to calmly promote veterans' rights and denounce the war. His performance was superb. Overnight Kerry became a celebrity. "We lost our privacy," Julia recalled. "We lost our freedom. Privacy is the greatest freedom in this world. You no longer have that the minute you're in the public eye. People are looking to you as a role model. You have a responsibility to them to act appropriately. Not to be fake, to be responsible. But I was proud of John. He was doing a tremendous job of championing veterans' rights."

The afternoon following the Cavett show, for example, they left Bloomingdale's department store in New York at rush hour. The Kerrys were standing on a corner waiting to cross the street. "All of a sudden these two cars stopped and people got out," Julia recalled. "And they started shouting 'Hey, John Kerry.' Traffic stopped and everybody was spontaneously applauding him in the street. It was this moment of, you know, 'Oh my god, I am married to a superstar.' But it's not easy. You become in demand. Our little world was over. The newlyweds were over."

It was while serving as chairman of Drinan's congressional campaign that Kerry first came to the attention of VVAW. He joined up immediately. Founded in 1967 by six New Yorkers, VVAW had a singular mission: to end U.S. military involvement in Vietnam (later Cambodia and Laos) through educating the public. During early antiwar demonstrations, the founders had marched under the banner of Veterans for Peace, an organization composed primarily of World War II veterans. Veterans for Peace took out a full-paged in the *New York Times*, marched in "peace parades," and lobbied U.S. senators to end the war. But they had no clear impact on Johnson administration foreign policy. Recognizing the need for a protest organization whose charter members had been shot at in Vietnam, and therefore had a wrenching eyewitness perspective, Jan Crumb, VVAW's principal founder, gained momentum by handing out mimeographed flyers on Greenwich Village street corners.* "Jan Crumb had originally conceived of

---

**Wanting to be a writer, Crumb adopted the pen name Jan Barry. Both names appear in news accounts and other documents of the era, causing historians confusion. This book will use the name Jan Barry hereafter, except in quoting other scholars, such as Gerald Nicosia.*

the organization as more of an educational tool than a political group," historian Gerald Nicosia explained in his *Home to War: A History of the Vietnam Veterans' Movement*. "For one thing, Crumb was against the wearing of military uniforms to impress the public with the veterans' military service. He saw VVAW as a group of *civilians*, albeit civilians who had gained a vital experience in war, and he felt they could do the most good on talk shows and in debates with Congressmen and State Department people."

Nicosia's *Home to War* makes it clear how many antiwar and counterculture organizations were vying for media attention with acronyms like MOVE, LINK, and CCI. But it's VVAW that had the greatest impact. Whether it was a senator putting one of its statements in the *Congressional Record* or its ability to frequently get veterans on TV's *David Susskind Show*, VVAW became a force to reckon with. It quickly gained a reputation for fearless street theater and grassroots organizing. While Kerry stayed aware of VVAW activities, joining the organization after his Jamaican honeymoon, it was not until Labor Day 1970 that he was catapulted into a prominent role. It came about when he spoke at Valley Forge, Pennsylvania, on September 7, as part of Operation RAW, an acronym for Rapid American Withdrawal, and "war" spelled backward.

Angered by the My Lai massacre, and the shabby medical treatment vets were receiving, Operation RAW was conceived by Al Hubbard, an African-American firebrand organizer who felt a genuine affinity for the Vietnamese people. With his lean looks, emphatic speaking style, and bushy mustache, Hubbard, among other things, was a tireless avatar of Black Power. His heroes were Malcolm X, Bobby Seale, Huey Newton, and Eldridge Cleaver. Tired of VVAW's just handing out leaflets and spinning the media, he preached the gospel of *direct action*. The idea behind Operation RAW was for Vietnam vets to march eighty-six miles between two Revolutionary War sites—Morristown, New Jersey, and Valley Forge, Pennsylvania—engaging in guerrilla theater along the way. The vets, including 110 who had earned Purple Hearts, would march about twenty-five miles a day. Most veterans who participated in the event came from the New York area, but others were bussed in from such Midwest cities as Chicago, Kansas City, and Milwaukee. The spectacle of this ragtag band of ex-soldiers was bound to get the media's attention. And it did.

The scores of newspaper articles about the march from Morristown to

Valley Forge are colorful, if somewhat childish. The Philadelphia Guerrilla Theater, for example, was brought into the march to play the role of captured soldiers. They were gagged, hands tied behind their backs, and forced to march. Girlfriends of the antiwar protesters were organized as Nurses for Peace. The music of the Doors and Creedence Clearwater Revival blasted out from portable tape players. Peace signs were omnipresent. Marijuana was in the air. Skinny-dippers frolicked in the Delaware River. Their long hair, ripped jeans, army-surplus-store canteens, and toy guns gave VVAW the look of a ragtag band of Haight-Ashbury refugees. Bitter that African Americans were carrying the burden of combat warfare in Vietnam, Hubbard enlisted black veterans to participate in the Labor Day weekend procession so it wasn't "a Caucasians only affair." Along the marching route, veterans would shout out phrases like "Kill him!" and "Cut his belly open!" for dramatic effect. "I think we raised some questions," Hubbard told the press after the march. "I don't think we've converted anyone. I think we've caused them to think a bit, and I think that's all we set out to do, is to make them think."

When the two hundred vets finally arrived at Valley Forge chanting "Peace now!," they were greeted by a cheering crowd of over one thousand. Just down the road, however, was a counterprotest held by the Douglas MacArthur Post of the Veterans of Foreign Wars. Conflict was in the air. From a makeshift stage on the very ground where General Washington and his Continental Army had nearly frozen to death in the winter of 1777–78, a revolving door of antiwar speakers took to the microphone. Al Hubbard denounced "international racism" while author Mark Lane spoke of GI revenge. Actor Donald Sutherland read passages from Dalton Trumbo's antiwar novel *Johnny Got His Gun* while actress Jane Fonda, standing on the bed of a pickup truck, denounced the Nixon administration as being a beehive for cold-blooded killers (one journalist, taken with the power of Fonda's rhetoric, dubbed her "the next Susan B. Anthony"). Then there was the Reverend James Bevel of the Southern Christian Leadership Conference—Martin Luther King Jr.'s mentor on the virtue of Ghandian nonviolence—who charged the United States with committing genocide in Southeast Asia, and Allard Lowenstein, who used sardonic humor to win the crowd over.

But it was John Kerry who stole the day. Unknown by most of the veterans, except for those who caught him on *The Dick Cavett Show*, Kerry delivered a spellbinding address reminiscent of his class oration at Yale University, peppered with Kennedyesque cadences. "We are here because we above all others have earned the right to criticize the war on Southeast Asia," Kerry shouted into the microphone. "We are here to say that it is not patriotism to ask Americans to die for a mistake or that it is not patriotic to allow a president to talk about not being the first president to lose a war, and using us as pawns in that game." The crowd went wild. Some veterans in the audience, unable to participate in the march from Morristown because they were in wheelchairs, thrust their fists in the air as spontaneous signs of empowerment. With his sister Peggy at his side, Kerry, wearing an army jacket, spoke like a man possessed. One VVAW leader elbowed Peggy and said, "Whoa, man, whoa!! Your brother looks like Abe Lincoln and sounds like Jack Kennedy." By the time he hopped down off the pickup to thunderous applause, he was the new leader of VVAW by popular default. Barry, Hubbard, and others would continue to do the organizational work but Kerry was now the celebrity vet, the highly visible voice of the antiwar cause.

Of all the things that happened in 1970, it was the fact that Spiro Agnew, the vice president of the United States, had deemed antiwar protesters as "criminal misfits" that gnawed at Kerry. How could Agnew say such a thing about men who were paralyzed from the neck down or missing three limbs because they served their country? How could veterans be rotting away in VA hospitals without proper care? Why wasn't Congress holding Nixon accountable for widening the war into Cambodia and Laos? Kerry, the "uncommitted soldier," was energized on the bed of a pickup truck at Valley Forge. While epiphany is too strong of word, it certainly proved to be an emotional turning point. He was not just a spokesperson, but a new man. From Valley Forge onward he was a committed antiwar activist, one who would spend much of his free time in VA hospitals interviewing men about what had happened in Southeast Asia.

Chapter Fifteen

The Winter Soldier

They called themselves the "winter soldiers" and John Kerry came to hear their stories. Sponsored by VVAW for three days, January 31 to February 2, 1971, more than one hundred veterans met at a motor lodge in downtown Detroit to hold a public "winter soldier investigation" about atrocities they committed—or witnessed—in Vietnam. These veterans adapted their name from the Revolutionary War pamphleteer Thomas Paine, who wrote in "The American Crisis" on December 23, 1776: "These are the times that try men's souls. The summer soldier and the sunshine patriot will, in the crisis, shrink from the services of their country; but he that stands it *now* deserves the love and thanks of man and woman." Paine's inspiring words were written to rally the downcast men in General George Washington's ragged army, to implore them to stand tough and fast even though the harsh snows of winter were upon them. Meanwhile, the catalyst for the winter soldier investigation was the series of November 1969 stories that *New York Times* investigative journalist Seymour Hersh had written about the My Lai massacre.* "My reports took on a life of their own," Hersh recalled. "The veterans in Detroit, independent of me, decided to speak up independently and tell their own stories. I had nothing to do with it."

Just how important the Hersh articles and the trial of Lieutenant William Calley were to VVAW is evident in a statement made in Detroit by Lieutenant Bill Crandell. Six foot tall, with shoulder-length brown hair and

---

**On March 16, 1968, the soldiers of Charlie Company, 11th Brigade, Americal Division, murdered 347 Vietnamese civilians in the hamlet of My Lai, in the village of Son My, under the command of Lieutenant William Calley. In March 1970, fourteen officers were charged with covering up the incident and Calley was convicted of premeditated murder in 1971.*

a bushy mustache, Crandell hailed from Sylvania, Ohio. His introduction to the military began as an ROTC recruit at Ohio State University, then the largest Army college program in the country. Eventually, he was sent to Vietnam where he served as an infantry platoon leader in 1966–67. While in-country he witnessed—or participated in—the killing of Vietnamese unlucky enough to be in free fire zones. His conscience haunting him, he became part of the Buckeye Army of Liberation, along with Art Flesch and Bill Bayer; their antiwar antics got them tossed out of an Upper Arlington, Ohio, Independence Day parade. "We intend to demonstrate that My Lai was no unusual occurrence," Crandell announced on the first day of the public Detroit investigation. "We intend to show that the policies of the Americal Division which inevitably resulted in My Lai were the policies of other Army and Marine divisions as well. We intend to show that the war crimes in Vietnam did not start in March 1968 [when My Lai occurred] or in the village of Son My or with one Lieutenant William Calley. We intend to indict those really responsible for My Lai, for Vietnam, for attempted genocide."

Detroit was selected to hold the VVAW hearing because it was an industrial mecca in the heartland. So many of the men killed in Vietnam had come from blue-collar families and this was a way to honor them. But there were other reasons. The United Auto Workers, for the most part, were firmly in the antiwar corner. There was a large Vietnamese community living across the Detroit River in Windsor, Ontario, which VVAW wanted to engage in a healing dialogue. Canada was also the home to approximately sixty thousand draft resisters. Another consideration was African-American participation. Racial strife had reached dangerous proportions in Vietnam, particularly at the air base in Danang. Detroit was Motown—home of Malcolm X's black nationalism and Berry Gordy's Hitsville USA—and a vortex of the civil rights movement. It was a good place for VVAW to show concern for racism in the U.S armed forces. Congressman John Conyers of the Fourteenth District in Detroit strongly supported VVAW. His personal assistant at his Detroit office was none other than Rosa Parks, of Montgomery bus boycott fame. "We didn't want it to be a coastal event," Bill Crandell offered. "We didn't want the media to write it off as just another Berkeley or Columbia antiwar event. We wanted to hold the investigation in the heart of a working class area."

John Kerry was pleased that VVAW alluded to Thomas Paine in their manifesto. And he enjoyed getting to know committed antiwar activists like Bill Crandell and New York's Scott Moore. But he was terribly uncomfortable with antiwar activists who quoted Vladimir Lenin or Mao Zedong for inspiration. Dissent, Kerry believed, was part of the American grain—communism or socialism were not. And Paine, Jefferson, Lincoln, and King were to his mind the guiding intellectuals of the antiwar movement. Not Karl Marx or Che Guevara. He preferred the law-and-order approach of the Bernard Russell Peace Foundation's *Against the Crime of Silence* to the anarchist street theater of Abbie Hoffman or the Berrigan brothers. Far from being part of the "new left," in fact, Kerry was uncomfortable with groups like the Yippies, SDS, or the Black Panthers. He was a Kennedy Democrat, not an anarchist radical. And while Kerry thought the U.S.-declared free fire zones, B-52 bombing raids, defoliation campaigns, and search-and-destroy policies in Vietnam all morally reprehensible, he refused to mount a soapbox and detail atrocities he witnessed in the Mekong Delta at a forlorn motel on the corner of West Grand Boulevard and Third Avenue in Detroit. "John came to watch, learn, and listen," his friend George Butler recalled. "He was not ready—or willing—to break down in Detroit or anywhere else. Yet, he felt empathy for the veterans who confessed to war crimes."

There is not a known written record of what John Kerry did or said in Detroit, just eyewitness recollections and a handful of black-and-white photographs. While in Detroit he read Jonathan Schell's *The Village of Ben Suc* (1967), a brutal account of how a Vietnamese community was destroyed by U.S. armed forces. Butler, however, is the key link in the event, for Detroit had become his home. Instead of enlisting in the military, like his other college friends, Butler had joined VISTA in the fall of 1968, after the Detroit race riots. Frightened by the urban violence all around him in the Motor City, Butler bought a German shepherd and shotgun for protection. He rented an apartment on Oakland Avenue and started a grassroots community newspaper. "It was like Dodge City," Butler recalled. "VISTA paid me $3 a day to live in a combat zone. Detroit was seething with anger. People were changing. 'Throw out the wealthy. Throw out the rich. Throw out the establishment. Blood on the streets. Revolution in America.'"

Kerry was vehemently opposed to such overwrought rhetoric. Always well groomed, he was adverse to the cultivated sloppiness of professional peaceniks. He was short-fused when it came to blowhard apparatchiks like Rennie Davis discussing showdowns at the barricades. He was against extremism of any kind, and he had cultivated views on Vietnam that he thought were pragmatic. Bolstering his viewpoint was the fact that the U.S. senators he respected most—Ted Kennedy and George McGovern—believed as he did. Yet he was not dogmatic about the war—or anything else, for that matter—so he arrived in Detroit for the winter soldier hearings seeking new answers to the Vietnamese quagmire. "My main objective was to educate myself," Kerry recalled. "I arrived in Detroit a student. I knew what was wrong with Sealords. But I didn't know what had happened at Khe Sahn or Hamburger Hill. You might say I was on a fact-finding mission."

Kerry and David Thorne checked in at the motor lodge not far from the Fisher Building, headquarters to General Motors. In an act of sincere generosity, Senator Eugene McCarthy, they learned, had rented rooms for some of the veterans so they could take daily showers. A local Catholic organization, through the intervention of antiwar activist Father Daniel Berrigan, arranged for other out-of-town veterans to take over a six-bedroom house. The meeting room at the motel was filled with long conference tables with microphones scattered about. Wandering around were young men dressed in Army fatigues with long hair. Many looked disheveled. One special consideration had been made: plywood wheelchair ramps were erected. "The key to the winter soldiers investigation was gaining the trust of the veterans," Crandell recalled. "We spent a lot of time first checking credentials. Then once the veterans got to Detroit, we let them all know it was an obligation to tell the truth." The big news in Detroit at the outset of the investigation was that the radical icon Jane Fonda had arrived. Outspoken on civic issues dear to her heart, Fonda had become a favorite pinup of GIs due to her role in the 1968 film *Barbarella*. That was soon to change: she would earn the moniker "Hanoi Jane," for her unconscionable visit to North Vietnam in 1972.

Butler, who had great contacts with the *Detroit Free Press* and *Detroit News*, served as a sort of VVAW press liaison for the event. Veterans flashed their DD-Z14 cards as bona fides. Some men took to the microphones and started confessing to crimes they committed in Vietnam at the behest of the U.S. government. With their tormented stories and emotional breakdowns, the winter soldiers investigation was incredibly powerful. "It was a mind-blower," Butler recalled. "It was extraordinary. With my background I couldn't imagine a soldier saying, 'I've done wrong.' I couldn't imagine people expressing hatred of the war so articulately—in my opinion what ended the war in Vietnam was the veterans. Once the troops spoke out against it you couldn't fight the war much longer."

It's quite unsettling to read transcripts of the winter soldiers investigation.* Veterans from all branches of the military spoke up. With great solemnity, in an atmosphere that resembled an Alcoholics Anonymous meeting, veterans stepped up to a microphone in the front of the conference room. Fairly well organized in terms of chronology—from boot camp and degradation to combat atrocities—the horror stories the veterans spun were chilling. The speakers all shared one attribute: a search for forgiveness through repentance. Joe Bangert, for instance, recounted how an officer at Camp Pendleton used to keep a pet rabbit, which was essentially his mascot. "He has this rabbit, and then a couple of seconds after just about everyone falls in love with it . . . he cracks the neck, skins it, disembowels it. . . . That's the last lesson you catch in the United States before you leave for Vietnam."

The stories quickly got much worse. Vietnam veterans spoke of killing gooks for sport, sadistically torturing captured VC by cutting off ears and heads, raping women, and burning villages. The amount of sadism revealed was numbing. Special Forces SP/4 Steve Noetzel recounted how North Vietnamese prisoners were shoved out of U.S. helicopters to their deaths. Air Force Cavalry Regiment Sergeant Michael Hunter told of GIs gang-raping women outside of Hue. Pilots complained of having to napalm villages. Most confessions, however, revolved around random machine-gunning of peasants in villages. It was all part of the U.S. government's win-the-war mandate. And even though the Pentagon had recently claimed the United States had not invaded Laos, the testimony in Detroit told a vastly different story. The collective effect was devastating to the senses. "Its power lay in its very boredom," Peter Michelson of *The New Republic* recalled. "[F]or one kept being jolted by the recognition that multiplied accounts of murder and bestiality were boring. But as the accounts slogged on, the very commonness, the quotidian character of atrocity, identified itself as the core of dehumanization that accounts for war crimes."

**The most easily accessible account is Vietnam Veterans Against the War,* The Winter Soldier Investigation: An Inquiry into America's War Crimes *(Boston: Beacon Press, 1972).*

Just as he had in the Mekong Delta, Kerry circulated among the crowd, asking questions and making vague suggestions about a follow-up event that spring in Washington, D.C. All around him was anguish. Burned-out veterans, tears rolling down their faces, were sharing nightmares. Nerves were shattered. Everybody was on edge. Laughter was not heard. One insensitive comment could be like a trip wire—kaboom. This was not like the O Club at Cam Ranh Bay, where officers bragged of winning medals and getting girls. It was a macabre feast of disassembling. Something deeply disturbing had happened to these men. This frayed generation was going to pieces. Just two years earlier, Kerry had been interviewing soldiers about the horrors of the Bo De River. He had heard tales of atrocities committed in the Ca Mau Peninsula. Now, in front of his very eyes, he was witnessing the psychological damage immoral U.S. policies had done to a generation of GIs. "There was a lot of stuff that I hadn't heard," Kerry told historian Gerald Nicosia. "There was a lot of rough stuff out there, and it blew some of my images. I mean it shattered some of my perceptions."

Writing in his journal, Kerry philosophized about the timeless nature of the war in Vietnam. World history, it seemed, was like one endless panoramic of Picasso's *Guernica*, where the sword always ruled. "War has always contrasted the real with the absurd—unless someone was in love with what they were doing and then there was a reason for everything—even dying," Kerry wrote. "But not many people went to Vietnam because they wanted to fight or because they felt in the slightest that they were fighting to save a country or a people—making a country 'safe' for democracy. It was a tour of duty—a one year long absence from all that made sense and all that one wanted to do. And so, it was that absurd it struck even harder. One moment there was beauty and the silence and the next moment there was the macabre and chaos. The days went by as though in a dream—and only a dream could soften what one felt."

What differentiated Kerry from many of the other winter soldiers was that he had not lost his faith in the armed services. Even though he had left active duty in January 1970, he was still in the Naval Reserves. He considered Commodore Horne, Commander Elliott, and Admiral Schlech mentors. What Kerry thought was missing from the Nixon administration's approach to the war was compassion for the field soldier and sailor. They were coming back blown to pieces while President Nixon kept talking about

Vietnamization. It seemed incredible to Kerry that the gallant soldiers convening in Detroit were painted as anti-American while men like National Security Advisor Henry Kissinger were touted as patriots. At the very least, Kerry thought, these veterans needed to be heard—they had earned that right the hard way, on the battlefield. Democracy, after all, was supposed to be a dialogue. The injustice of it all gnawed at him. "I wouldn't put Vietnam away," Kerry recalled. "It was a question of responsibility, of keeping faith with why I survived, why I was lucky enough to come back—and others didn't. I took on a deep responsibility on a very personal level of making sure guys like Dick Pershing and Don Droz didn't die in vain. We had to make something positive out of all this absurdity. I couldn't just leave it there like some kind of negative, horrible, dark moment."

What infuriated Kerry the most was the fact that Vietnam veterans had to fight for recognition and legitimacy in America. After all, the combat, pain, suffering, loneliness, cold nights, hot days, and heartache were the same as they'd been in Lexington, Concord, Iwo Jima, or Normandy. And, in many ways, it was far worse. Battles like Gettysburg and D-Day only took a few days. Because of the helicopter, the combatants in Vietnam often would finish one battle and be flown immediately to the next. In Vietnam it wasn't if you were going to be wounded, it was when and how badly. Some Vietnam veterans, like Lieutenant David Christian, thought the World War II veterans could have helped the Vietnam veterans reintegrate into American society. "The older generation, our fathers, let us down," Christian, who would become a military analyst for Fox News, recalled. "They treated us, their sons, as if we failed. If only they could have seen the gallantry of their sons on the battlefield."

Christian was the most decorated officer of his age in Vietnam. His platoon, known as "Christian's Butchers," had taken the fight directly to the enemy. They also waged a clandestine war inside Cambodia. Badly wounded on his first tour of duty in 1968, he was sent home to recover. Once healed, he was sent back in 1969, only to get terribly burned by napalm and shot up. "We had run out of ammunition," he recalled of the firefight, "so we were reduced to arming ourselves with flares." In the coming years, he underwent thirty-three operations for injuries suffered in Vietnam. "While Kerry and I disagreed on some political issues we saw eye to eye on one key point," Christian recalled. "Vietnam veterans were getting shafted."

Instead, many of the so-called greatest generation questioned the patriotism of VVAW. While Kerry met wheelchair-bound men who had to carry their legs over their shoulders because their stumps were still raw, some whose spines had been shattered like glass, and still others who had been reduced to defecating in plastic bags attached to their lower torsos, these veterans couldn't get respect. The generational schism was wide. Heroes like David Christian never got embraced by either the Pentagon or the liberal press. Many World War II veterans treated Vietnam vets as social misfits because they did not *win* their war. Yet, in truth, the GIs of Vietnam never lost a major battle. Essentially, American society treated Vietnam veterans as "losers." In a culture that only embraces winners, they were marginalized. The media didn't want to run stories on their true battlefield heroism. As antiwar activist Ron Kovic, who was paralyzed from the waist down after being shot in the shoulder in Vietnam, used to say, "We never got our parade."

For all the explosive drama at the Detroit motel, the national media aired only a few stories on the event. CBS reporters filmed the public confessions but they never aired on the evening news. A *New York Times* stringer covered the spectacle, but a story of merit never appeared in print. The best press VVAW got was a one-hour NBC special titled "The Vietnam Veterans," which aired March 16. VVAW, however, received a significant endorsement from Hugh Hefner of *Playboy*, who devoted a full page out of his magazine to an endorsement of the organization—its haunting visual was a stark photograph of a coffin with an American flag solemnly draped over its sides. Under it, in bold print, it warned: "And more are being killed and wounded everyday. We don't think it is worth it." Thousands of veterans joined VVAW because of the *Playboy* piece, which even included a cut-out membership coupon. "Hefner was enormously helpful," Crandell recalled. "He put our cause in front of an incredibly large male audience."

SDS leader Tom Hayden was disappointed—but not surprised—that the mainstream media backed away from covering the Detroit event. "The winter soldier investigation was one of those places you weren't supposed to go if you wanted to be taken seriously in American life," Hayden recalled in a 2003 interview in Cambridge, Massachusetts. "It was off limits. The veterans were supposed to have developed amnesia. But they were horrified by what they had done. They were traumatized by Vietnam. Now, as they tried to talk about what happened they were traumatized a second time, called liars, Communists, malingerers. They were told that they had hallucinated their stories." What Hayden encountered in Detroit were not radicals, but frightened men in need of therapy and counseling. "They were desperate for somebody to listen," he recalled. "They were reaching out to each other for reassurance, comfort, and help."

Hayden was not the only antiwar activist concerned about the lack of media coverage. Given that so many newsbreaking stories were told, including front-page-worthy accounts of the Nixon administration's secret war in Laos, it seemed to many veterans that a conspiracy was underfoot. Paranoid about FBI infiltration, they naturally feared the worst: that the media was too co-opted by the U.S. government to speak the truth. "Speculation about the poor exposure was rife," historian Andrew E. Hunt wrote in *The Turning: A History of the Vietnam Veterans Against the War*. "Some veterans believed the American public simply did not want to hear that atrocities were commonplace in Vietnam." The question Kerry faced, therefore, was: How to get more press? Following the lead of Martin Luther King Jr., Kerry understood that VVAW needed to take its fight to the streets of Washington, D.C., to parade up and down Constitution Avenue in the shadow of the Lincoln Memorial and the U.S. Capitol, hurling peace anthems like thunderbolts. He recalled how in 1932 the so-called bonus army, World War I veterans demanding government assistance to help alleviate their suffering, had caused a major stir. But now was not the time to press the issue. The leaders of VVAW needed to call an executive meeting in New York and regroup. On February 2, as the hearings wound down, Kerry said good-bye to his friends—he knew he would see them soon. "You couldn't help but feel numb and sickened by these men's testimony," Kerry recalled. "It was brutal; I sat pained hearing their agonies. It made me physically ill. It was all so graphic and ugly."

Besides Kerry, who didn't play a major role in the Detroit event, the VVAW member most concerned with the poor press coverage was Larry Rottmann. Born on December 20, 1942, in Jefferson City, Missouri, he was perhaps the most artistically inclined member of VVAW. Before he went to Vietnam in March 1967 as a second lieutenant in the Army's 25th Infantry Division, he had earned a literature/journalism degree from the University of Missouri. Based out of Cu Chi, he had seen combat during Operation

Junction City and the Tet Offensive, earning a Purple Heart and the Bronze Star. "I thought the war was stupid," Rottmann recalled. "There was no way of rationalizing any of it. But I loved Vietnam, the people were incredible."

When Rottmann returned home in March 1968, he thought about putting the war behind him, but he couldn't. Six foot tall with shoulder-length blond hair and a beard, his first public event was speaking in front of the Springfield, Missouri, Chamber of Commerce. He had jotted down poems on three-by-five note cards and read them out loud. "I felt that the war was destroying America," he recalled. "The U.S. was the best country in the world. The war was hurting us. I couldn't not act. Protesting the war needed to be done by people—veterans—who were willing to stand up and be counted for. I wasn't obsessed by Vietnam. I was simply concerned that we might not survive as a nation if we didn't get out. It was destroying my entire generation."

Disdainful of groups, Rottmann decided to join VVAW as a last resort. He became its press officer, arranging for media interviews, cultivating reporters, and always trying to advance the antiwar message. He quickly noticed that Kerry had star potential. "A lot of veterans were inarticulate, angry, scruffy young men," Rottmann recalled. "Kerry was the opposite. He had medals, education, the initials, and Lincolnesque stature. I tried to steer the media away from John, at times, to get profiles of Native Americans from Oklahoma or antiwar mothers who had lost their sons. I did pretty well. But Kerry usually was the one people wanted to meet."

With Kerry and Rottmann getting more and more involved with VVAW, the group was gaining new friends. Senator Mark Hatfield (R–Ore.), who had gone to Vietnam back in 1945 to help Ho Chi Minh, then an ally against Japan, entered portions of the winter soldier testimony into the *Congressional Record* in April 1971 and asked that official hearings be held about the conduct of U.S. forces in Vietnam. "These veterans showed the diversity in the antiwar movement," Hatfield recalled. "They weren't just college students complaining. They weren't armchair politicians. They were articulate and specific about what had happened to them in Vietnam. They were asking: What is our national interest? All of these veterans gave witness that the domino theory was ridiculous. We had lost Vietnam. But we were still playing the game that we hadn't lost." For supporting the winter soldiers, Hatfield caught flack from fellow Republicans, particularly Nelson Rockefeller. "He told me I was terribly naïve about communism," Hartfield recalled. "I told him I had been following Vietnam closely, since 1945, and he didn't know what he was talking about."

But by far the most helpful new assistance to the VVAW movement came from Senator George McGovern (D–S.Dak.), who was deeply moved by the testimony of veterans. Nicknamed "Senator Dove" by many of the veterans, he monitored what had transpired in Detroit very carefully. Pacifica Radio had taped the entire investigation and McGovern ordered a transcript. He was worried that, ever since the President inaugurated the bombing of Cambodia, Nixon was carrying on a secret war against dissent groups like VVAW. "Once I read the testimony of these young veterans in Detroit, I knew Nixon would do anything to discredit them," McGovern recalled. "I realized they were going to need politicians and reporters to stand up for them, to vouch for the authenticity of their combat experiences."

On the popular culture front, the winter soldier hearings also strengthened VVAW support. Jane Fonda—who met her future husband Tom Hayden in Detroit—now personally adopted the group as her leading cause. Just before the investigation, she had organized "Acting in Concert for Peace," a show that included performances by actor Donald Sutherland and comedian Dick Gregory. She also financed the 1972 documentary *Winter Soldier*, produced by the Winterfilm Cooperative. Folksinger Phil Ochs, who had played a free concert on the evening of the hearings, started performing pro-vet songs on stages around America. The talented songwriter Graham Nash, who had penned the classic peace ballad "Teach Your Children," had come to Detroit at the request of Fonda. Stunned by the cathartic confession made by ex-Marine Scott Camil—who described how GIs raped a Vietnamese woman and knifed an old Vietnamese man—he wrote a song. Titled "Oh! Camil (the Winter Soldier)," the protest song appeared on Nash's *Wild Tales* album and included the following haunting verse: "Oh! Camil, tell me what your mother say / when you left those people out in the fields / rotting along with the hay? Did you show her your medals? Did you show her your guns?"

Although the American public and media were not yet paying much attention to the winter soldiers, the Nixon White House was. Presidential aide Charles Colson, in particular, was worried. Intelligence information had leaked out of Detroit and he smelled a future public relations debacle in the making. If these so-called winter soldiers were anointed with credibility, his boss's entire war strategy could go tumulting into the abyss. Anxious to be proactive, Colson disseminated an internal memo, "Plan to Combat Viet Nam Veterans Against the War," outlining a way, through FBI wiretaps and surveillance, of discrediting VVAW. Do it by attacking each veteran's biography. "The men that participated in the pseudo-atrocity hearings in Detroit," Colson wrote, "will be checked out to ascertain they are genuine Viet Nam combat veterans."

The winter soldier hearings clearly showed that the Vietnam War had damaged a generation. As Philip Caputo so elegantly put it in *A Rumor of War*: "We left Vietnam peculiar creatures, with young soldiers that bore rather old heads." In his classic work of new journalism, *Dispatches*, Michael Herr wrote that Vietnam was what his generation had "instead of happy childhoods." Clearly the atrocities revealed in Detroit—plus the revelations about the government's invasion of Laos—increased popular disillusionment with the war. Only two months after the winter soldier hearings, even leading Republicans like Michigan Congressman Gerald Ford and Tennessee Senator Howard Baker had grown sickened by the Nixon administration's widening of the war into Laos. In fact, no less than five resolutions were introduced to hamper the President's enlarging the war. And the protests grew in frequency. On February 8, demonstrators invaded the South Vietnamese embassy in Washington, D.C., while on March 1 the Weathermen blew up a restroom in the U.S. Capitol in outrage over U.S. involvement in Laos. "Winter Soldiers was a small stone thrown hard and far out into the large pond—America's perennially overconfident, jingoistic consciousness," Gerald Nicosia wrote in *Home to War*, "and it would take a long time for all the ripples to reach shore."

If the winter soldier hearing was a stone dropped into a pond, then their next planned action was a boulder crashing down off a cliff. The idea that 1,000 Vietnam veterans, with John Kerry as their leading spokesman, would descend on Washington, D.C., for five days, was decided at a February 18 VVAW executive meeting in New York. "We were trying to determine what our next step was," Jan Barry (formerly Jan Crumb) recalled. "Detroit gave us momentum but we now had to keep it going." Tensions flared at the VVAW meeting—the gap between moderates and extremists was large. Insults were tossed about in vicious fashion. The anger in the air was tangible. As the day wore on, Kerry had heard enough. "Kerry was quite something," Barry said. "In a roomful of angry veterans Kerry stood up and suggested we all go to

Washington. It was essentially decided then and there. He had done something positive with their anger." They would call the demonstration Dewey Canyon III. (Dewey Canyon II was a scheduled second invasion of Laos, which got scrapped largely because of the winter soldier hearings. It was renamed Lam Son 719, with South Vietnamese troops—not Americans—doing the bulk of the fighting.)

Word of Dewey Canyon III sent a shock wave that landed right on President Nixon's desk. The White House disinformation campaign intensified at once. Rumors aimed at discrediting many of the winter soldiers were unleashed. As Dewey Canyon III approached, Nixon grew worried—and defiant. "He thought it was an outrage," Johnson administration Attorney General Ramsey Clark recalled. "They didn't think of the consequences of trying to stop a veterans' antiwar march. Nixon was just too full of resentment that soldiers, *his* soldiers, had turned against him." The president's first reaction was to have the Interior Department order it illegal for the veterans to camp out on the Mall. On April 7 Nixon delivered an address aimed at lulling the antiwar distress that was fanning out across America like a virus. "Tonight I can report that Vietnamization has succeeded," Nixon told the nation, accompanied by a promise to end the war. "I expect to be held accountable by the American people." The address fell flat. With newspaper headlines announcing new revelations about Lieutenant Calley at My Lai, plus ongoing speculation about U.S. military operations in Laos, Nixon was unable to turn the media to his side. *Time* magazine, for example, wrote off his rhetoric as a "foxhole speech, digging in tenaciously in defense of his existing position."

The Lieutenant Calley situation had Nixon in a vise. Many in the military were furious that Calley was being court-martialed. At Fort Benning, Georgia, where Calley was being held in a stockade, about one hundred supporters, most of them GIs, shouted loudly for his immediate release. Meanwhile, the White House received a quarter million letters demanding that Calley be honored, not imprisoned. It was as if the Vietnam War itself were being placed on trial. The Defense Department, however, knew that Calley was guilty and *had* to be punished. One internal Pentagon memorandum summed up the political reality quite succinctly: "Based on the evidence produced in the public trial . . . Calley fully deserves to be punished, morally and legally." VVAW, by contrast, claimed that the high-ranking officials of the Johnson and Nixon administrations were the ones who should be punished. Men like Lieutenant Calley, they brazenly charged, were merely following orders. "With the conviction of Lieutenant William L. Calley the real dilemma of my generation has finally been brought unmistakenly home," Jan Barry wrote in the *New York Times*. "To kill on military orders and be a criminal, or refuse to kill and be a criminal, is the moral agony of America's Vietnam War generation."

Meanwhile public support for the war was slipping. Approximately 43 percent of the American people polled were opposed to Nixon's Vietnam policy. His silent majority was becoming a slim majority. Exacerbating the situation from Nixon's perspective was the fact that many Republicans were abandoning him. Republican doves in Congress like Mark Hatfield and Jacob Javits were growing in stature. "Laos is one more straw—and a substantial one—on the camel's back," Congressman Peter Frelinghuyen (R–N.J.) declared. "Most Americans—myself included—have come to feel that this war has gone on too long." Even members of his own administration thought Nixon was bungling the war. Cocktail party gossip around the capital was about "Nixon's quagmire" and "Nixon's war."

Intensifying Nixon's public relations problem was that the media refused to give Vietnamization the benefit of the doubt—which was time for it to work. President Lyndon Johnson always argued that he had fought hard for common people, pointing to his Great Society programs as evidence. But Nixon was seen as void of compassion, a law-and-order man insensitive to the plight of U.S. veterans and minorities. "We saw him as a monster," gonzo journalist Hunter S. Thompson recalled of the liberal press mentality circa 1971. "He had no redeeming value." Nixon felt the press was breathing down his neck—and it riled him to no end. In machine-gun-like fashion Nixon could rattle off the names of all his enemies in the press, including John Osborne of *The New Republic*, Richard Rovere of *The New Yorker*, and George Frazier of the *Boston Globe*. When Henry Kissinger telephoned the President about how much he liked the April 7 speech, Nixon blurted out, "Screw the Cabinet and the rest. . . . I'll turn right so goddamn hard it'll make your head spin. We'll bomb the bastards off the Earth."

Nixon's speech only egged VVAW onward. VVAW press releases announcing the five-day "limited incursion" of Washington, D.C., by over 1,000 veterans attracted widespread media speculation. Liberal columnist

Mary McGrory, for example, interviewed Kerry about the upcoming protest, which was rumored to include mock search-and-destroy operations throughout the metropolitan area. Kerry complained to McGrory about how he had been ordered to fire on villages supposedly VC friendly, how Admiral Zumwalt and General Abrams tried to persuade naval officers to burn hooches, and how the U.S. government wasn't taking proper medical care of veterans. "The My Lai court-martial, Kerry thinks, forced Americans to put aside their myths of GIs giving out chocolates and salvation," McGrory wrote. "The dissenting veterans are pledges to non-violence during their 'peaceful incursion.' Nobody can call them 'effete snobs,' but somebody in the Administration will have to figure out a policy of handling the thorniest problem in protest that has arisen since the War began."

As Dewey Canyon III approached, the twenty-seven-year-old Kerry crisscrossed America, raising money for the event. *Boston Globe* columnist Thomas Oliphant wrote a sympathetic article on April 15 about how Kerry was raising bus fare so veterans from Chicago, New Orleans, Denver, and Oregon could make the long journey to Washington, D.C. A group of veterans in Boston dressed in combat fatigues held a preliminary "search-and-destroy" protest on City Hall Plaza that attracted national attention. It was just a taste of what was to come. Representatives Bella Abzug (D–N.Y.) and John Dingle (D–Mich.) praised VVAW before Dewey Canyon III. Meanwhile, Nixon claimed that only about 30 percent of VVAW were *real* veterans, the rest impostors.

The headquarters of the Washington VVAW chapter was a frenetic beehive of activity that April. Everybody was working the telephone banks, getting ready for the winter soldiers to descend upon the nation's capital in droves. Meanwhile, the Nixon administration made a desperate, last-minute attempt to prevent VVAW members from camping on the Mall, placing wreaths at Arlington National Cemetery, or returning their medals on the steps of the Capitol. Using Attorney General John Mitchell as his front man, Nixon wanted to shatter VVAW before the first veteran was lifted off a Greyhound bus and placed in a wheelchair. Representing the Nixon administration in the courts was C. Patrick Gray, while the plaintiffs (veterans) hired Ramsey Clark. The Justice Department, who feared 5,000 veterans were coming to town, was ordered to turn hostile—that is, use intimidation tactics. Exacerbating Nixon's fears was that the day after Dewey Canyon III was to end, an even larger antiwar group, calling themselves the People's Coalition for Peace and Justice (PCPJ), was scheduled to arrive in Washington with an estimated 200,000 protesters. While VVAW was asking permission to camp on the Mall, PCPJ put in for Rock Creek Park. Nixon worried he was going to be surrounded by hostile forces like Lincoln was during the Civil War. "The administration is frightened about what we have to say," Kerry told the *Washington Post*. "Frightened that we will name war crimes; that we will tell the people that Vietnamization is not working."

There was something unjust about the Nixon administration's obtaining an injunction to forbid the veterans from camping on the Mall. After all, all of these men had served their country in Vietnam, slept in jungles and elephant grass, on barracks and boats. In 1969 Reverend Ralph Abernathy had led a "poor people's march" to Washington, D.C., and had been granted permission to camp on U.S. government property. Why weren't the veterans allowed? As the veterans started arriving in West Potomac Park, near the Lincoln Memorial, where a registration booth for early arrivals had been set up, they denounced the injunction to the press. "It's really ridiculous," VVAW leader Tom Butz declared. "We could have a wild rock festival, or hours of speeches, but they tell us we can't quietly and peacefully go to sleep." Reporters started making unflattering comparisons of Nixon's treatment of Vietnam veterans to President Herbert Hoover's decision to call out the cavalry to chase the bonus army from the Mall in 1932. "It apparently never occurred to many layers of public relations wizards at the White House that had they let the veterans alone the affair might have been a washout," Mary McGrory wrote. "But through harassment, the Administration has guaranteed maxim impact."

Just as the VVAW members were arriving in the capital, New York Senator Jacob Javits held a private dinner at his home where Secretary of Defense Melvin Laird was the guest of honor. It was a carefully staged event. Javits had eight other leading Republican members of the House and the Senate also stop by. After dinner, while cigars were passed around, Pennsylvania Senator Hugh Scott dropped the gauntlet. "You don't see any hawks around here," he told Laird in a no-nonsense fashion. "The hawks are all ex-hawks. There's a feeling that the Senate ought to tell the President that we should get the hell out of the War. We just can't hold the line any longer on numbers. The President must think in terms of finality. He must make some formula that clearly indicates the end of American participation in the war." Also at the dinner was Alaska Senator Ted Stevens, a World War II veteran known for being fiercely prowar and anti-Communist. "I come from the most hawkish state in the union," Stevens said. "I ran [last time] as a hawk. I can't do it again in '72."

The Sunday before the Dewey Canyon III protest began, Kerry appeared on NBC's *Meet the Press*, along with Al Hubbard. Poised and clearly comfortable under the camera's glare, Kerry complained about how the Department of Defense was trying to undermine VVAW. When asked about the winter soldier phenomenom—i.e., confessing to atrocities—Kerry boldly proclaimed that he *had* violated the understood law of warfare. "Yes," Kerry told moderator Lawrence E. Spivak. "I committed the same kinds of atrocities as thousands of others in that I shot in free fire zones, fired .50-caliber machine bullets, used harass-and-interdiction fire, joined in search-and-destroy missions, and burned villages. All of these acts are contrary to the laws of the Geneva Convention, and all were ordered as written, established policies from the top down, and the men who ordered this are war criminals."

Kerry had his work cut out for him that week in Washington, D.C. Emotions were high. Many veterans arrived on Harley-Davidsons shouting epithets at the White House. Chants like "I love my country, but fear my government" were unleashed from veterans who congregated in Lafayette Square Park. Up and down Pennsylvania Avenue could be seen Volkswagen vans with bumper stickers that read, "Question Authority" or "Give Peace a Chance."

A goal of Kerry's over the past year had been convincing VA hospitals to provide drug rehabilitation programs. Too many veterans, from Kerry's perspective, were trying to numb their emotional pain with narcotics. Whenever Kerry raised the issue of narcotics use, however, with a drug-dependent veteran, he was shot down. "Being a leader of VVAW was like organizing anarchists," Jan Barry recalled. "Kerry was constantly calming down veterans who wanted participatory democracy—nothing more. They weren't into niceties and negotiations." Peggy Kerry, who spent that entire week in Washington, D.C., helping her brother, recalls that he was "in nonstop motion," always trying to "make peace with all parties."

Just a day after the *Meet the Press* taping, Kerry and Barry were brought over to the State Department to discuss the VVAW. "What was amazing is the way Kerry got through doors," Barry recalled. "He had connections. He was articulate in these situations and got the top department people to listen." And, as always, Kerry was keeping notes. Later that year, Macmillan would publish a book called *The New Soldier*, written by John Kerry. It was a hodgepodge of veterans' oral histories and black-and-white photographs of the April antiwar happening. Today it reads—and looks—like a cultural relic from a distant era. It was compiled by Kerry's two most stalwart friends—David Thorne and George Butler—who practically never left his side during the weeks leading up to Dewey Canyon III.

On Monday, April 19, Kerry was in the front row of approximately eleven hundred veterans who marched from the Lincoln Memorial across the Potomac River toward Arlington National Cemetery. It was a heartbreaking sight. Many of the veterans were in wheelchairs or on crutches. They were searching for dignity. They had chosen this day because it was the anniversary of the Battles of Lexington and Concord—the opening salvo of the American Revolution when "the shot heard 'round the world was fired." Mothers of veterans who were killed in Vietnam—known as "Gold Star moms"—led the procession. The main chant of the day was "Bring our brothers home!"

When the column of antiwar marchers reached the gate at Arlington National Cemetery, they were turned away, barred from entering. (Later that day, without the veterans present at the Tomb of the Unknown Soldier, Reverend Jackson Day prayed for *all* those who died in Vietnam—in both North and South. He had been a military chaplain just the day before, but in an emotional moment triggered by the Dewey Canyon III he quit. "I knew in my heart the killing had to stop," Day told the press. "Redemption was needed.") Although the White House had been doing whatever they could to sabotage the efforts of VVAW, nobody had predicted a blockade at a military cemetery. Kerry, who was planning on placing a wreath on Dick Pershing's grave, was understandably livid. "It's too bad," he shouted, "they lock the gates where are brothers are buried." Al Hubbard, sensing the marchers' anger, grabbed the megaphone denouncing the "insensitivity" of the White House. A U.S. government helicopter flew over the cemetery, causing the enraged veterans to flip it their middle finger. Little did they realize it was President Nixon flying above Washington, D.C., checking out the street theater so he could decide how best to counter the situation.

The Nixon administration's strategy remained based on discrediting the veterans as being frauds. As White House aide Dick Howard told reporters, VVAW was a "group that are apparently not veterans, who are apparently trying to compensate for some guilty feelings . . . by participating in this organization." An internal memorandum written by White House Chief of Staff H. R. Haldeman to Charles Colson came right to the point: "The President should know that we are continuing the effort to discredit VVAW." It was one thing to try to stop them from sleeping on the Mall. But to disallow veterans, many disabled, from praying at the grave of a son, brother, or friend was by any standard of decency just plain wrong. If it were true, as Nixon saw it, that Washington, D.C., was being invaded, then a tactical mistake had been made. By turning the veterans away from the cemetery he had made martyrs of the winter soldiers. "It was a terribly stupid and crass thing to do," Mary McGrory recalled. "Only Nixon would have turned wounded soldiers away from visiting a military cemetery."

Thwarted at Arlington, the protesters headed for the U.S. Capitol. The group was joined by Representatives Pete McCloskey (R–Calif.) and Bella Abzug. Jan Barry was there to present Congress with sixteen VVAW demands. Meanwhile, the Mall was starting to look like a mini-Woodstock. Veterans were blasting Phil Ochs's "Draft Dodger Rag" and Crosby, Stills, Nash, and Young's "Ohio" from eight-track players inside customized vans. Banners reading "End the War" were displayed. Toy machine guns were handed out. Disabled veterans pointed wooden crutches at pedestrians pretending they were gunning them down. The grown men, acting like little boys, went ratatat-ratatattata. Anne Price, a Gold Star mom, turned herself into a moving billboard with a sign that said, "A Birthday Not to Be." Neil Olsen of Russell, Pennsylvania, publicly mourned his son's death in Vietnam by playing "Taps" all day long. In one widely publicized exchange, a woman who was a member of the Daughters of the American Revolution confronted a Vietnam veteran. "Son, I don't think what you're doing is good for the troops," she said. To which the veteran replied: "Lady, we *are* the troops."

More than any other VVAW member, Kerry was running all over town, meeting with Interior Department officials regarding camping permits in Potomac Park and spinning top-tier reporters on the righteousness of the antiwar cause. Although he slept on the Mall, he used the Georgetown home of Oatsie and Robert Charles as a place to conduct business. (The Charleses were prowar Republicans, but they let Kerry use their home because John and Julia were friends of their daughter.) Some in VVAW were annoyed at Kerry for avoiding the grunt work of envelope stuffing and Porta Potti cleaning. There was jealousy that he was garnering all the media attention, and he was lampooned behind his back as an "establishment pig" and "pretty boy." Kerry paid any slights—both real and imagined—little mind. Desperate to keep Dewey Canyon III nonviolent, he ran from flash point to flash point, trying to quell veterans who were spouting revolution. For the operation to be successful, Kerry believed, VVAW had to be embraced by at least some of Nixon's silent majority. A mantra for Kerry throughout the week was: "I am determined to work with the system. We are totally nonviolent and nonaggressive."

But as hard as he tried, Kerry couldn't possibly speak on behalf of all disconnected veterans. For example, a blowup occurred after he had VVAW leader Jack Mallory fill in for him at a Common Cause rally. Mallory denounced Democratic liberals as lackeys who supported Nixon's imperialistic policies. When word of Mallory's grandstanding reached Kerry, he went ballistic. "When I'm out sticking my neck out for this organization," Kerry shouted, "you're blasting away at everyone's politics and espousing revolution." Throughout the week, in broken-record fashion, Kerry was urging everybody to believe in the "system," to lobby politicians and vote in the next election. Such polite platitudes obviously did not sit well with certain outraged veterans, who were enjoying standing in front of the Justice Department giving the Nazi salute and shouting, "*Sieg, heil!*"

The difference of approach between Kerry and the more radical VVAW wing became glaringly clear. Kerry and Hubbard clashed over everything from pamphlets to transportation. But, in truth, it was Kerry's more diplomatic, nonconfrontational approach that was starting to pay off. Things, in general, were starting to look up for VVAW. On Tuesday, April 20, approximately two hundred VVAW members attended Senate Foreign Relations Committee hearings in which Senator George McGovern took the lead in demanding the United States pull out of Vietnam at once. McGovern's advocacy signaled a shift in momentum.

It seemed as if a brutal showdown between the Nixon White House and VVAW would take place. But two White House advisors, Counsel John Dean and speechwriter Patrick Buchanan, wisely prevailed on the President to ease up on the hard-nosed tactics. As Dean wrote in an April 21 memorandum to

Haldeman and John Ehrlichman, Nixon's domestic policy adviser: "The policy—which the VVAW are totally unaware of—is that there will be *no arrests* made of VVAW who violate the order and it has been clearly and unequivocally given to the appropriate authorities." Decades later, Dean recalled that Nixon was "worried to death" about the VVAW. "He tried to pretend that the protest didn't bother him," Dean said. "But every half an hour—literally—he wanted an update."

Later in the week the second wave of antiwar protesters, from PCPJ, was going to flock to Washington. Buchanan advised via a memorandum getting tough on that group and leaving the veterans alone, particularly because they had "an articulate spokesman"—i.e., Kerry. Handcuffing a man in a wheelchair, Buchanan reasoned, was akin to beating up an old woman. It was bad politics. "Kerry's group were veterans, so they deserved some respect," Buchanan recalled. "We had real nut balls coming in a few days. I was raised in Washington, D.C., and my dad had told me about the bonus army. Arresting veterans—like bonus marchers—was horrible public relations. So I urged the President to leave them alone."

Suddenly the White House resistance lifted. "Don't bust the Vietnam veterans on the Mall," Nixon instructed Ehrlichman. "Avoid confrontations." Unbeknownst to another contingent of veterans who marched back to Arlington National Cemetery, the likelihood of a showdown with the authorities had been greatly diminished. This time they were allowed in the cemetery, the superintendent now welcoming them with open arms. A VVAW guerrilla theater search-and-destroy mission held on the Capitol steps went off without a hitch.

All of these small victories served to empower Scott Camil of VVAW, a group leader since the Detroit winter soldier investigation, who believed in direct confrontation. That evening a VVAW fund-raiser was held in Georgetown, hosted by the genial Senator Philip A. Hart (D–Mich.), known for his conscientious support of civil rights and environmental protection. It was an odd event. Blue-collar veterans with mud on their boots—men who were sleeping on the Mall—walked around on the plush oriental rugs, with old *Harper's Weekly* covers framed hanging on the wall. They were mixing with lobbyists and politicians in Brooks Brothers suits. A buzz occurred when Senator J. William Fulbright made an appearance. Eyeing him from afar, Camil took the opportunity to pigeonhole the Senator and accuse him of being a hypocrite. "Here he was, bragging about how much he was against the war," Camil recalled. "So I just went up to him and said, 'Look, the Gulf of Tonkin [Resolution] gave the President the power to do what he did. You voted for it. Why did you vote that way?' "

Ambushing Fulbright at a cocktail party was hardly the way to get the senator's genuine attention. Kerry, by contrast, circulated around the party with grace and dignity. His nonconfrontational, pleasant demeanor impressed the Arkansas senator, who had already heard good things about Kerry from his colleague Ted Kennedy. The next morning Kerry received a telephone call from a Fulbright aide. "The senator wanted me to testify in front of his committee the following day," Kerry recalled. "It was extremely short notice, but of course I said yes." Immediately he started working on what would clearly be the seminal speech of his young life. Luckily he had his speech file with him. He started reworking the remarks he had delivered at Valley Forge a year earlier.

Kerry spent most of the day pulling his most impassioned thoughts together. Dividing his time between VVAW headquarters, several media events, and the Mall, he made a series of afternoon telephone calls to friends, seeking advice and guidance. One such friend was Adam Walinsky, the former Robert F. Kennedy speechwriter who had inspired him to speak out when they met during the 1969 Moratorium. Meanwhile fifty veterans had marched on the Pentagon, hoping to goad the security police to arrest them—instead they were just ignored. At 4:30 P.M. the police, because of the latest decision—this time from the Supreme Court—in the legal struggle between VVAW and the White House, were scheduled to evict veterans from the Mall. The hour came but no police. "Camping?" one police officer is reputed to have said. "I don't see any camping."

Earlier in the week the Nixon administration had tried to frighten the veterans, to intimidate them with the possibility of arrest. VVAW had clearly called their bluff—they stayed firmly planted at West Potomac Park next to the Lincoln Memorial. As the courts kept flip-flopping over whether the veterans could camp on the Mall, Mike Oliver of VVAW issued a heartfelt letter to the local police. Addressed to "our brothers in blue," the letter pleaded with the police to understand they were combat veterans, not a cabal of wandering hippies or potheads. All over District of Columbia police stations, the VVAW friendship letter was posted. That, coupled with the fact that most of the veterans were courteous, had brought the police emotionally over to the side of the protesters. When asked to arrest two double amputees for trespassing during Dewey Canyon III, Washington Police Chief Jerry Wilson said, "I won't do it. I just won't arrest them." After all, if Reverend Abernathy had been allowed to have a poor people's march in the capital and his followers had been granted permission to sleep on the Mall, why was not the same privilege granted to men who had lost arms and legs in Vietnam? The Nixon administration's early tactics of intimidation had clearly backfired. "Scared?" one ex-Marine from Connecticut fired back at a journalist's question. "Are you kidding? I was scared when the Russians' 22s were coming over, but scared of park police?"

What an odd sight these winter soldiers must have made to the tourists who had come to the Mall to see America's great monuments honoring Washington, Jefferson, and Lincoln. Here were long-haired men, with beards, chanting Beatles songs like "Come Together" and waving the American flag upside down (the battle sign for distress). Bedrolls, canteens, and banners were scattered about. The wind swept mimeographed antiwar flyers in the air. In a sense, the veterans were collectively a living monument to such high-water marks of American dissent as the Underground Railroad and the Selma–Montgomery March. As Gerald Nicosia inventories in *Home to War,* these war veterans were deeply embedded in U.S. tradition, wearing the insignias of some of America's finest fighting units: the Screaming Eagles (101st Airborne), the Big Red One (First Infantry Division), and the Black Horse Regiment (11th Armored Cavalry). With their Woodstock smiles, love beads, peace-sign flashes, blue jeans, frumpy hats, leather sandals, and toy plastic guns, they seemed more like followers of the Grateful Dead than men who had killed for their country. "What was truly amazing," Kerry recalled in a 2003 interview, "was that we were setting up tents, camping, at almost the exact same place the Vietnam War Memorial was dedicated in 1982."

Taking a break from speechwriting, Kerry wandered around the Mall talking to his "brothers." The one-hundred-plus men of the Massachusetts contingent on the Mall was the largest of any state, and Kerry was one of them. If they got handcuffed, he was going to get handcuffed with them. Years later, when critics accused Kerry of using Dewey Canyon III to further his career, it made no sense. There were easier ways to become a congressman or senator in 1971 than getting a rap sheet with the police. Realizing that morale was waning, Kerry telephoned his friend Peter Yarrow—who was in New York—for help. "These guys needed their spirits lifted," the folksinger recalled. Yarrow contacted the road company of the Broadway musical *Hair*, who were on tour in the capital. Within a few hours, they showed up as promised to entertain the vets. "They blasted into singing 'Let the Sun Shine' and 'Aquarius,'" Yarrow said, "and all the men's anxieties dissipated." At 5:30 P.M. Ramsey Clark took to the makeshift stage to make an announcement about the legal struggle. "Stay on the Mall, don't sleep, and the government won't arrest you; or sleep on the Washington Mall and the government will arrest you." It was a perplexing court decision. Mary McGrory, who was covering the possible showdown, interviewed Kerry about why President Nixon had resorted to playing brinksmanship with decorated Vietnam veterans. Speaking on behalf of the VVAW, Kerry's response was simple: "He didn't understand us, and he doesn't understand the country."

As darkness fell, the police refused to make any camping arrests. Senator Edward Kennedy showed up unexpectedly, signing autographs, patting backs, and showing his complete solidarity with the Vietnam veterans. Ever since 1966, when he wrote an antiwar article for *Look* about his journey to Vietnam, Kennedy had been a virulent critic of U.S. policy in Southeast Asia. He thought Vietnamization would never work because the Vietnamese were "indifferent" to both their own plight and the virtues of democracy. When trying to educate himself about Vietnam, Kerry paid close attention to what Kennedy said or did. "It was the first time I really spoke to John," Kennedy recalled about the Dewey Canyon III campout. "He peeled off to the side and we talked about the veterans' movement. I was impressed by his deep commitment and unshakable intelligence."

In the evening, Kerry stood on a platform clutching a megaphone and delivered a spellbinding antiwar speech. Some active GIs from nearby Quantico, Virginia, were in the audience cheering him on. "He was filled with passion," Ramsey Clark recalled. "He was setting the tone for the speech he delivered the next day to the Senate Foreign Relations Committee."

Kerry would stay up most of the night, dozing for a few hours on the Mall. At the crack of dawn he headed over to the VVAW office to shave. Like in Vietnam before a mission, insomnia haunted Kerry. His mind was racing. He was practicing his speech in his half-sleep. In the morning he met with some VVAW leaders to discuss holding a candlelight vigil in front of the White House that evening, and then went to the Supreme Court, where veterans were singing "God Bless America" and were begging to be arrested. (The police were refusing.)

More than any other journalist, Tom Oliphant of the *Boston Globe* had gotten to know these veterans as individuals. He knew what high schools they had gone to and how many Purple Hearts they had in their fatigue jacket pockets. While the head of the Veterans of Foreign Wars was deeming VVAW "Communist controlled, motivated and funded," Oliphant was writing stories about soft-spoken, quiet veterans like VVAW organizer Arthur "Bestor" Cramm of Massachusetts who personified decency. And out of all the veterans swirling about the Mall, Oliphant had figured out John Kerry was the one to follow. "He was in the zone," Oliphant recalled. "It was a truly amazing performance that week. He was going from veteran to veteran, trying to keep them happy. He essentially succeeded—for a week."

Kerry's star had risen even higher once it was reported that Al Hubbard, his sidekick on *Meet the Press,* had *not* been an Air Force captain in Vietnam as he claimed. He had been a staff sergeant, one who had never seen combat. For three years Hubbard had been lying to whoever would listen. He had left the Air Force in 1966 as an instructor flight engineer. His time abroad—between 1963 and 1965—had been at Tachikawa Air Force Base outside Tokyo. "I allowed this lie to continue because I realize that in this country it has been very important to have an image," Hubbard offered by way of excuse on NBC's *Today*. "That is compounded if one is black and attempting to do something. I don't justify the lie. I'm trying to explain it. Those people on the Mall know nothing about this discrepancy and they'll be as shocked as anyone else. But they'll accept it because they're telling the truth."

Hubbard's exposure as an impostor gave credence to the administration's claim that VVAW was rife with GI pretenders. The pressure was thus on for Kerry to deliver a knockout speech in front of the Senate Foreign Relations Committee on Thursday, April 22. The entire reason for marching on Washington on the heels of the winter soldier investigation in Detroit was to garner widespread media attention. The week had been pretty good, with the exception of Hubbard. But as Kerry approached the Dirksen Senate Office Building, with Oliphant at his side, he knew this was his one opportunity to make the VVAW case public, to reach into the living rooms of the average American TV watcher and say, "Wake up." The stakes were high. A strong performance was mandatory. He had confidence in himself but was slightly queasy.

Over the years he had learned that he did best under extreme pressure. In baseball terms, he was a clutch hitter. Since 1965, VVAW had been trying to bust open the door of the so-called establishment. It had taken out a full-page ad in the *New York Times*, participated in moratoriums, and held public investigations of atrocities. But the group was always on the outside looking in. Even if Kerry and Barry had just spoken at the State Department, it was behind closed doors; nobody else heard their concerns. By contrast, Kerry's testimony before the Foreign Relations Committee would be delivered for all Americans to hear. Kerry was testifying not as a lone wolf with a belly full of gripes, but as the embodiment of a kinetic grassroots veterans' movement that was directly challenging the government's foreign policy. In true democratic fashion, he was taking his dissent directly to the American people, challenging Congress to end its complicit behavior and end the long national nightmare of Vietnam. His credibility, like most of the legitimate winter soldiers, was unimpeachable: he had killed VC, so he had the right to explain why killing more had to stop. As he entered the Senate hearing room, he realized that history would judge Dewey Canyon III as "the moment the soldiers tried to stop the war."

In pure political terms Kerry's statement put forth three primary aims: U.S. disengagement from Southeast Asia by December 31, an end to Vietnamization, and proper funding of veterans' hospitals. His two hours of heart-wrenching testimony were stunning. Even the more hawkish senators present, Stuart Symington and Clifford Case, were awed by his oratorical command. Reporters used words like "eloquent," "moving," and "mesmerizing" to describe his presentation. Only *Newsweek* took a swipe at his performance. "Some of his rhetoric was exaggerated and irrational," the newsweekly claimed, "but there was no arguing with the conviction with which he spoke for the marchers." In coming weeks, he would get parodied by Garry Trudeau in his *Doonesbury* comic strip and lambasted by William F. Buckley in the *Washington Star* and the *Boston Globe*. But the glowing reviews far outweighed the bad. "His credibility was unassailable," Robert Healy enthused in the *Boston Globe*. "Articulate and direct. He pulled it all together."

That afternoon Ted Kennedy was not in the chamber but he heard the totality of Kerry's testimony on National Public Radio and was taken aback.

It was the most elegant testimony he had ever heard a young man give—on any issue. "His presentation made a mark on my psyche," Kennedy recalled in 2003. "Sometimes words are like poems or songs or prayers. They touch a part of your heart that cuts through the bureaucratic haze or tangle of legal briefs. His words—coupled by a flawless delivery—carried people away on a cloud of hope. It was a master stroke. From that moment onward, I knew John Kerry was a potent force for powerful change. He had taken a front seat in the hard-fought effort to bring our boys home, to cut our losses in Vietnam and get out."

Although by anybody's standard Kerry did a superb job testifying, not all of his friends—or former crewmates—were happy. As could be expected, when Tedd Peck, fully recovered from his combat injuries, living in San Francisco, and working as a salesman, turned on the TV and saw Kerry's face, his "stomach turned." Bill Zaladonis of PCF-44, who was living in Pittsburgh, couldn't believe his eyes when he saw Kerry on national television denouncing President Nixon. "I just about fell out of my chair," he recalled. "Shortly thereafter he tried to get me to join VVAW. I told him no way. I didn't care for what he was doing. The only thing he was right about was that after Tet we should have gone in and closed the deal. We had already kicked the North Vietnamese in the butt. But Washington let us down." The loyal James Wasser of PCF-44 caught part of Kerry's testimony on the nightly news. "I love John, but I was pissed," Wasser recalled. "I was living in a small one-room apartment in Illinois. I saw him on the tube and was outraged. 'What the hell are you doing?' I shouted. At that point I thought what he was doing was dead wrong."

Decades after Kerry's testimony, Michael Medeiros remembers the shock he experienced when he found out his old PCF-94 skipper had joined forces with "hippie veterans." He had just registered for classes at San Jose State University. He had missed out on the whole winter soldier hearings and the Dewey Canyon III protest. Wandering around his university bookstore, he stumbled across an offering for a sociology course, *The New Soldier*, by John Kerry. The cover showed disenchanted veterans with the American flag upside down—the distress code. "I felt disappointed and betrayed," he recalled. "I was still staunchly prowar. I couldn't believe the flag upside down. Now, with 20/20 hindsight, his position makes sense."

Mike Bernique, who was busy taking classes at the University of Chicago, was particularly disgusted by Kerry's antiwar stance. "I was appalled, angry but not surprised," Bernique recalled in a 2003 interview. "I thought he was behaving like an opportunist." The Kerry he remembered, and admired, was the fighting man who when a fellow officer said, "Let's hit the river," would immediately push the throttle. "John was a very courageous man," Bernique went on. "If you wanted somebody to watch your back, John was the man. He was one hell of a leader. But when you risked your life on a day-to-day basis, and then had these little anti-American turds criticizing what you did, it made me angry. I refused to have anything to do with John or VVAW."

But Kerry did have allies from the Swift community. Del Sandusky of PCF-94 remembers being in Norfolk, still on active duty, when Kerry's face appeared on CBS News. He was proud that Kerry had the courage to take on the Nixon administration. "I felt he was right," Sandusky recalled. "It took guts to stand up to those guys in Washington. Too many things had gone sour in the war. If I could have I would have been standing right next to John, I would have joined his protest." Likewise, Drew Whitlow of PCF-44 rejoiced at Kerry's brazen actions. "He did what needed to be done," Whitlow recalled. "Us regular guys had just become pawns on a chessboard. We were treated like lepers, spat on. John brought us attention. He shed light on the lies of Vietnam. Before his speech most Americans had taken us for granted." Among the Swift officers Kerry had served with, Skip Barker and Wade Sanders cheered him on. Each man, in separate interviews, said he was "proud" of Kerry on April 22 for having the "boldness" to look the U.S. government in the face and say, "You're wrong."

On Friday, April 23, the final public act of Dewey Canyon III was its most successful guerrilla theater action to date. A makeshift wooden and wire fence had been erected around the Capitol. "We felt, 'Those motherfuckers are putting this fence between us and that Capitol, the Congress,'" Jan Barry recalled. "We're going to go there tomorrow with our medals—let's throw them over the fence!" Federal authorities were concerned that the scheduled May Day Tribe—the antiwar convocation of the PCPJ, which had nothing to do with VVAW—was going to attract up to 200,000 protesters, in the largest antiwar rally in United States history. That Friday, Congressman Jonathan Bingham, whom Kerry first met at Yale University back in 1965, held hearings with former intelligence officers regarding the distortion of top-secret memos concerning the Vietnam War. Senators McGovern and Hart also held hearings on atrocities by U.S. soldiers in Vietnam while about 800 veterans marched up to the barricades around the Capitol and threw their medals back. "It was a day filled with drama," McGovern recalled. "When a proud soldier is forced to throw back his medals then you knew something was terribly amiss in America."

The White House was worried sick about the medal-returning ceremony. Their main fear was that VVAW was going to abandon the Capitol and hurl them over the White House gate instead. President Nixon and his advisors considered having a U.S. military representative accept the medals in front of the White House—such a gesture would ensure there was no violence or wild TV images. But historian Tom Wells, in *The War Within*, explains that General Don Hughes, Nixon's chief military aide, found the idea repulsive. Meanwhile, word of the medals "throwaway" jarred veterans worldwide. Many were insulted by the prospect. "I did not admire the throwing of medals on the steps because I did not believe it was appropriate when so many brave men and women had sacrificed in order to get those same medals," then POW John McCain recalled in 2003. "John and I later became great friends. But I never addressed this one issue with him directly."

While McCain's sentiment was held by many sailors, particularly careerists, other active duty GIs cheered VVAW onward. In *Who Spoke Up? American Protest Against the War in Vietnam, 1963–1975* (1984), Nancy Zaroulis and Gerald Sullivan detail how much internal sabotage was going on within the Navy. In 1971 alone, there were 488 cases reported (191 sabotage, 135 arson, 162 wrongful destruction). Stories of "fragging" also became widespread. Angry GIs sought revenge on officers and men in their platoon or unit. Widespread mutiny was feared by the Nixon administration.

So, as their closing salvo VVAW, in a carefully planned action, had 800 veterans congregated near the Capitol's front steps. Jack Smith of West Hartford, Connecticut, a Marine Corps veteran, read a statement explaining why men who earned Purple Hearts and Bronze Stars were now giving them back to the government. For over two hours, men hurled their medals and ribbons over the fence toward a statue of John Marshall, the first Chief Justice of the United States. Dramatically, veterans from all branches of the armed services broadcast their names, units, and citations, and then rid themselves of their mementos in disgust.

Words can not properly describe the chilling effect the event had on the speakers and participants. Each soldier had his own horror story, which brought him to this precipice. As an antiwar action—or a piece of street theater—it was a powerful demonstration. But it was more than that. The bitterness and rage exhibited by these soldiers ripped at the nation's conscience. Anybody who heard, for instance, Paul F. Winters pray for forgiveness as he hurled a Silver Star, Distinguished Cross, and Bronze Star over a fence and then watched him limp away was forever scarred by the memory. Some men, however, were not quite so dramatic. They gave only their first name and a calm statement: "Robert, New York, and I symbolically return all Vietnam medals and other service medals given me by the power structure that has genocidal policies." Others vented their spleens, which were bursting with defiance: "Here's a bunch of bullshit," one veteran shouted as he hurled a handful of medals. Folksinger Peter Yarrow said, "There had never been a more symbolic gesture ever committed in American history: this was raw despair."*

Over the years some have raised the question as to why Kerry chose to dispose of his ribbons, not his medals. Critics saw it as trying to have it both ways. It gave credence, they believed, to what A. J. Liebling of *The New Yorker* once claimed of Rough Rider Theodore Roosevelt: he was "dilettante soldier but a first-class politician." Further confusing the issue was the fact that Kerry *did* throw the medals of two no-show veterans toward the Marshall statue at their request. "The point of the exercise was to symbolically give something up," Kerry recalled in his defense. "I chose my ribbons, which is what many of the veterans did." The medals he tossed had been given to him by two angry veterans who wouldn't make it to Washington, D.C.; he was merely serving as their surrogate. Before Kerry discarded his ribbons, he declared: "I'm not doing this for any violent reasons, but for peace and justice, and to try to make this country wake up once and for all." Decades later, his medals are kept in a desk drawer in his Boston study—he's proud of them.

Much of Kerry's time that afternoon was spent hand-holding with two Gold Star moms, Anne Pine of Trenton, New Jersey, and Evelyn Carrasquillo of Miami, Florida. He stood by them as they hurled medals back at the gov-

---

**Estimates vary as to how many veterans participated in the medal-throwing event; for example, VVAW claimed 3000,* Newsday *1,500 to 2,000, and the* New York Times *700.*

ernment. As a World War II veteran played "Taps," and about five hundred people gathered around, names of men who died in Vietnam were called out. Watching TV that evening was Rich McCann, who had traveled the Mekong Delta rivers with Kerry and was now a graduate student at George Washington University. "When he threw those medals over the fence, I was pretty upset," McCann recalled. "I was grappling with a lot of issues myself. It was hard to accept that I had given a year of my life for a lost cause. In retrospect, however, what he did was right."

Not all the men gave up medals or ribbons. Many chose to turn back hats, jackets, and discharge papers. A photograph taken by George Butler shows that the offerings included recruitment letters, induction papers, and discharge forms. Historian Andrew E. Hunt, in his superb *The Turning: A History of the Vietnam Veterans Against the War*, gives the rationale of several veterans for returning personal possessions. Ron Ferrizzi, for example, a Philadelphia native, disowned his Silver Star and Purple Heart against the pleas of his family. "My parents told me that if I really did come down here and turn in my medals, that they never wanted anything more to do with me. That's not an easy thing to take. I still love my parents. My wife doesn't understand what happened to me when I came home from Nam. She said she would divorce me if I came down here because she wanted my medals for our son to see when he grew up."

For a World War II veteran, the tossing away of medals must have been a painful sight. Either these long-haired hippies were on drugs or something had happened in Vietnam that they couldn't fully understand. As the memorabilia piled up, and the media took it all in, it was clear that the antiwar movement had just turned a sharp corner. First Kerry's speech, now this. George Butler captured the emotions of the afternoon with his camera lens. Collectively his photographs speak of personal liberation. For many of the veterans the discarding of military paraphernalia set them psychologically free. It was as if the U.S. government had corrupted them, seized their moral compass with a shiny pin-on honor. "You have no idea how healing the whole experience was," Bill Crandell recalled. "It was our hour of claiming ourselves back."

Julia Kerry, who was with her husband the entire week, came to the conclusion that the veterans had been in deep depression and denial. "There was so much buried pain," she recalled. "It was numbing to witness." On that last day veterans also planted a tree, as a symbolic gesture for the preservation of life over death. "The truly impressive thing was that no acts of violence had been committed that entire week," Kerry recalled. "We had promised to be nonviolent and we were."

Just as President Nixon feared, the days following Dewey Canyon III, with tents and platforms coming down, another mass of demonstrators flocked to Washington, D.C. Unlike VVAW, they were interested in civil disturbance. The May Day Tribe—led by Dave Dellinger, Rennie Davis, and others—hoped to shut down all traffic flowing into the city, creating gridlock. They rolled logs down Canal Road, overturned trash cans, and purposefully sat in the middle of downtown streets. Tear gas hovered over Georgetown. Not since Martin Luther King Jr.'s 1963 March on Washington had so many people congregated *anywhere* for a single cause. Senator Fulbright, talking directly to May Day Tribe leaders, summed up the situation perfectly: "I can detect that some of you have lost all confidence that this system can work." The bad behavior exhibited brought the organizers negative press. "Even Mary McGrory was angry that they trashed Georgetown," Pat Buchanan later laughed. "Their behavior was awful." When a group of protesters tried to block the path of Buchanan's Cadillac, he accelerated right at them—they jumped out of the way, thinking he was crazy. "I wouldn't have run over them," he insisted. "But they thought I would."

One of the May Day protesters in Washington, D.C., was Rosemary Kerry, John's mother, who had journeyed down from Groton to be part of the historic demonstration on the Mall. (That same day over 150,000 protesters gathered in San Francisco.) She listened intently as Coretta Scott King, Ralph Abernathy, and Ernest Gruening spoke. When it came time for her son to speak, she couldn't see over the throngs of people. So she climbed a tree to get a bird's-eye view. "There she was sitting in a tree with her sneakers on, and her white hair and all," Kerry recalled. "Underneath the tree were all these kids who were smoking pot and she was getting high—inadvertently—from secondhand smoke." Kerry got to introduce the folk trio Peter, Paul, and Mary who sang "Puff the Magic Dragon" and "Blowin' in the Wind." While they were performing, Kerry remembers staring out at this immense crowd and knowing that the war would soon be over. "We now had power in numbers," Kerry recalled. "We were on the ascendancy."

Chapter Sixteen

Enemy Number One

In the course of researching his June 2003 profile of John Kerry for the *Boston Globe*, reporter Michael Kranish dug through some of the Nixon materials housed at the National Archives in College Park, Maryland. Looking for any mention of Kerry amid the boxes of White House memoranda and reels of secret Oval Office tape recordings, he struck gold. A tape from April 28, 1971, revealed that at 4:33 that afternoon, Counsel Charles Colson telephoned President Nixon to discuss the troublesome twenty-seven-year-old Vietnam veteran who had turned so passionately against the war. With his Eastern establishment and Ivy League pedigree, his gentlemanly manners, his preppy look, and a chestful of medals—including no less than the Silver Star—the former Navy lieutenant was perceived as a menace to the Nixon White House. "This fellow Kerry that they had on last week," Colson remarked to the President about the clean-cut youngster's appearance before the Senate Foreign Relations Committee. "He turns out to be really quite a phony."

"Well, he is sort of a phony, isn't he?" Nixon replied.

Colson went on to inform the president—falsely—that while the rank and file of the VVAW had been camping out on the Mall, Kerry spent the night at the posh digs of a Georgetown grande dame. "He's politically ambitious and just looking for an issue," the White House counsel continued. "He came back a hawk and became a dove when he saw the political opportunities." Nixon can be heard muttering his assent. Around the same time, Kranish wrote in the *Boston Globe*, "Colson, in a secret memo, revealed he had a mission to target Kerry: 'Destroy the young demagogue before he becomes another Ralph Nader.'"

As a result of his impressive performance in front of the Foreign Relations Committee on April 22, John Kerry had emerged from the Dewey Canyon III demonstration a media darling, and thus a bane of the Nixon administration. In *The Haldeman Diaries: Inside the Nixon White House*, posthumously published in 1994, Chief of Staff H. R. Haldeman's diary entry that April 23 mentioned the "media problem" generated by the VVAW actions and their impact on Nixon's public approval ratings. "The veterans' deal, and the coverage of it, is the cause" of the President's drop in the polls that week, Haldeman wrote, "so we've got to see if somehow we can't make the media the issue. We'll probably have to crank up the VP again and get him going on it." Indeed, Spiro Agnew would soon go full out after the press coverage of VVAW, and by extension after the group's public face—which belonged largely to John Kerry.

Stories about the antiwar ex-Navy officer appeared in *Newsweek*, the *New York Times*, and the *Washington Post*. An Associated Press profile headlined "John Kerry: A Hot Item" ran in more than a hundred newspapers across the nation. When *Boston Globe* reporter Barbara Rabinovitz visited Kerry at his home in Waltham, she found a young man in dizzying demand. "An interview with Kerry these days is punctuated with interruptions for television taping sessions, numerous telephone calls, quick snacks of ginger ale and chocolate-chip cookies," she wrote. "Gone are the fatigues he wore during the week-long demonstration in Washington. Instead he sports glen-plaid pants, wide orange and blue ties, and a blue shirt with the monogram 'JFK.' (His middle name is Forbes.)"

With Kerry as the group's spokesperson and the appeal in *Playboy*, VVAW membership climbed above thirteen thousand. Every day twenty or thirty more Vietnam vets signed up instead of opting to associate with the more hawkish Veterans of Foreign Wars or American Legion. Unlike those hoary organizations, VVAW was seen as "hip"—the "good vets," as rocker Neil Young called them. And the group's core organizers were in the protest business for the long haul. Texas coordinator Jon Floyd returned from Dewey Canyon III to Dallas with his broken-down school bus; his colleague Larry Rottmann headed back to New Mexico, distrusting mainstream politics; but Al Hubbard, although remaining VVAW's executive secretary, temporarily faded from view. Meanwhile, Rottmann recalled, "John had become perhaps our best fund-raiser. He now had name recognition."

So Kerry kept speaking to crowds to great effect. Just weeks after his Foreign Relations Committee testimony, he headed back to VVAW's New York headquarters for an executive meeting. *New York Post* reporter Helen Dudar caught up with him there and engaged the former Navy officer in a candid interview about his future. Kerry shrugged off the notion that he was politically motivated, but acknowledged how much his life had changed since April 22. "People stop me on the street and talk to me about the statement," he told Dudar. "The most amazing thing has been the reaction of men who I suppose [could] be considered hard hats. A lot of them said they were moved and changed." His growing celebrity indeed served to bring attention to the VVAW cause, but it also set waves of bitter envy rolling through the ranks of the veterans' antiwar movement. In addition, it did not go over well that when other leaders in the organization were cleaning up after the march in Washington, the Kerrys took off back to Massachusetts. Michael Kranish reported an even more telling anecdote in his June 2003 *Boston Globe* profile. According to VVAW leader Scott Camil, Kranish wrote, on one occasion a group member "had tried to reach Kerry by telephone and was told by someone, presumably a maid, that 'Master Kerry is not at home.' At the next meeting, someone hung a sign on Kerry's chair that said: 'Free the Kerry Maid.'" The Kerrys, of course, had no maid. (The culprit was Larry Rottmann, who later regretted the joke.)

Other VVAW members speculated whether Kerry's political ambitions included an imminent run for Congress in Massachusetts's Fifth District and wanted him to resign from the group if that was indeed his intention. On the eve of the march, in fact, a *Detroit News* report claimed that some of his fellow VVAW leaders were contemplating booting Kerry out for "indulging in a cult of personality" and using the veterans' movement as a "political stepping-stone." *Boston Globe* reporter William J. Cardoso, however, came to the rescue with an article refuting the Detroit paper's account. "It was a bogus story," Kerry agreed. "But some people were after me—namely, some people in the Nixon administration and FBI." He was not about to retreat, but he did acknowledge his troubles within the organization as well. "There is no question that I was having problems with the more radical elements of VVAW," Kerry admitted in retrospect. "But it was important to keep the cause front and center." To the surprise of the organization's hard-core militants, Kerry refused to quit that spring or summer.

A few weeks after Kerry's Senate Foreign Relations Committee appearance, CBS News correspondent Morley Safer ventured to New Hampshire to tape a *60 Minutes* segment on the VVAW star, who was staying at his wife's cousin's cabin in the White Mountains. According to historian Todd Gitlin in *The Whole World Is Watching* (1980), Safer had become persona non grata in the White House since 1965, when he covered U.S. Marines burning down the Vietnamese village of Cam Ne for CBS News. The segment shocked many Americans. Titled "First Hurrah," Safer's *60 Minutes* profile of Waltham's new hero aired nationally except in Massachusetts, where a Boston Red Sox baseball game preempted it. It was essentially a flattering profile of VVAW by an antiwar correspondent. When asked during the televised interview how he felt about the results of Dewey Canyon III, Kerry replied: "Hopeful, because a lot of people listened to us; depressed, because so many people who couldn't be bothered with us before suddenly became our friends and joined the bandwagon." Later, when Safer asked him flat out, "Do you want to be president of the United States?," Kerry answered, "No. That's such a crazy question when there are so many things to be done and I don't know whether I could do them."

Morley Safer came away from the interview impressed by the smooth composure of his subject. They had wandered together through the rolling New England countryside, talking about the traditional or nonviolent resistance, campus unrest, and the poetry of Wallace Stevens. They joked that the real slogan of the GI in South Vietnam was CYA—"cover your ass." They pondered what was then being called the "automated battlefield," where indiscriminating laser-guided bombs were destroying primitive villages; the incongruity was mind-boggling. "I was knocked out," Safer recalled. "I never heard somebody so articulate during all my days covering Vietnam. And the thing that struck me about Kerry was that he was a man with another life, an intellectual life. He was extremely well read. He had already taken some heat for his initials and Kennedy cadences. But he had a truly engaging quality that was just hard to describe."

The *60 Minutes* profile triggered a second wave of interest in the VVAW spokesman. The *Boston Globe* even speculated that Kerry might challenge African-American liberal Republican Edward Brooke for his U.S. Senate seat in 1972. "The idea never crossed my mind," Kerry averred years later. "The press was just starting making up scenarios." Calling Kerry a young man of "impressive eloquence," the *Globe* article pointed out that Massachusetts's impending extension of the vote to eighteen-year-olds could help catapult the antiwar veteran right into national politics. Other possibilities the *Globe* broached for Kerry included another try for Father Robert Drinan's Third District congressional seat, or even moving out of Waltham to take on either Democratic Representative James A. Burke in South Boston's Ninth District or the Republican incumbent farther south in Plymouth County's Fourth District. "Without political victory, he has in a year become a nationally recognized figure," *Globe* reporter Carol Liston wrote of Kerry. "Overnight he was able to impress the public."

Kerry also had two increasingly unimpeachable political missions: to end the dishonorable Vietnam War and to get adequate funding for the nation's Veterans Administration hospitals. He knew that he was better at speaking, lobbying, and fund-raising to accomplish these broad goals than helping start the new VVAW newspaper *The First Casualty* or acting in mock search-and-destroy missions. Yet he continued to participate in VVAW's political theatrics. Over the Memorial Day weekend he joined nearly two hundred fellow veterans on a protest march to Boston from Concord's Minute Man National Historical Park—down Battle Road, the route traveled so famously by Paul Revere to warn that the British troops were coming in the wee hours of April 18, 1775. As with the organization's 1970 march from Morristown to Valley Forge and by playing on Thomas Paine's words to name their winter soldier investigation in Detroit a few months earlier, VVAW was once again hitching its message to America's Revolutionary War. And, as it had in the nation's capital the month before, the irony level again rose along with a controversy over whether the freely assembling citizen soldiers would be permitted to camp on the nine-hundred-plus acres of parkland at Concord's North Bridge and at Bunker Hill.

Various legal obstacles were thrown in the VVAW marchers' path by local authorities. The Lexington Board of Selectmen, for example, barred the veterans from camping on the town's Village Green or in its tour park, on the grounds that "no good purpose could be served" by it. Father Drinan raised his clear voice stridently in favor of the protesters, demanding at a press conference/rally that their campout be allowed to proceed and disdaining efforts to stop them as "wrongheaded attempts" by local politicians controlled by the Nixon administration. Reminding the citizenry of the significance of the Battle of Lexington—at which some seventy colonial minutemen had taken on ten times that many British redcoats in the Revolution's opening salvo on April 19, 1775—Drinan quoted Captain John Parker's determined instructions to his men: "Stand your ground. Don't fire unless fired upon, but if they mean to have a war let it begin here."

The first night of Operation POW—so named to suggest that all Americans had become prisoners of the Vietnam War, as well as to express solidarity with the 339 known U.S. POWs held captive by the North Vietnamese—went smoothly. Kerry bivouacked with his peers at the North Bridge in Concord that evening. The next morning, May 29, the veterans would start their march to the Boston Common, reading the stanzas of Henry Wadsworth Longfellow's poem "Paul Revere's Ride" in reverse order, as they retraced the Revolutionary silversmith's route from Concord through Lexington and Charlestown to the Common. The intended point was that, just as during the Revolutionary War the British had committed war crimes in the North American colonies, the United States was now guilty of doing the same in Southeast Asia. "We made the connection because we were all patriots," Kerry explained. "Paul Revere, Samuel Adams, George Washington—they were all trying to tell the American people the truth. They were trying to stop British oppression. As veterans, we were defining our duty as going beyond the actual war. We also had to struggle at home to stop U.S. soldiers from dying senselessly in Southeast Asia."

As they marched slowly toward Boston, the veterans would act out search-and-destroy missions, hand out leaflets to spectators, and sing such antiwar anthems as John Lennon's "Give Peace a Chance" and Country Joe and the Fish's "Feel-Like-I'm-Fixin'-to-Die Rag." The lyrics of the latter song, written by Country Joe McDonald, lacked the philosophical intensity of Bob Dylan's antiwar songs, but no VVAW marcher had trouble memorizing stanzas like "Well come on mothers through the land / Pack your boys off to Vietnam / Be the first one on your block / To have your boy come home in a box." The very straightforward simplicity of the song made it easy to chant. A few of the veterans would read gut-wrenching poems they had written while serving in Vietnam (many of these would be collected in VVAW anthologies like *Winning Hearts and Minds*). Kerry's favorites were Jan Barry's "In the Footsteps of Genghis Khan" and Larry Rottmann's "Rifle, 5.56 mm, XM16E1." Julia's brother Landon wrote a popular antiwar poem, "Brothers," which he read at peace rallies. As for music, Kerry preferred Peter, Paul, and Mary's "The Great Mandala," which played out a deeply psychological exchange between a father and son over Vietnam.

These modern-day singing Paul Reveres were sounding a tocsin to the nation. The theme would continue once the marchers made it to Boston Common at the end of the weekend, where they would host a "celebration of life" picnic that local children—and Vietnamese immigrants—were encouraged to attend. The event would feature Revolutionary War fare from a fife-and-drum parade to a public reading of the Declaration of Independence. There were no discussions of mutilated bodies or napalm. It was a much more gentle spectacle than Valley Forge, Detroit, and Washington, D.C., had been. The only truly jarring moment, in fact, occurred when Jimi Hendrix's version of "The Star Spangled Banner"—his electric guitar sounding like artillery fire—was blared out of an eight-track tape player.

Toy gun manufacturers did well that Memorial Day weekend off of VVAW. Stacks of plastic guns, some with flowers sticking out of their muzzles, dotted the Operation POW encampments. At one point the vets piled all the toy guns together and then stomped them to pieces under their combat boots. All in all, with fine weather and the Red Sox's series against the Oakland Athletics blaring from transistor radios all weekend long, Operation POW had quite a festive air. The demonstrators came from area towns like Bedford and Lincoln and Cambridge and Lawrence. Many were first-time marchers; some of the veterans' supporters were as young as sixteen. "It was the first real antiwar incursion into the suburbs," noted VVAW organizer Bestor Cramm. "It was a good opportunity to get our message to the people who weren't classic liberals. We were received very well. Veterans like ourselves were reviving the antiwar spirit of the 1960s. It had faded for a while. Protests like Operation RAW and Dewey Canyon III had reawakened the spirit. This was the next round."

Cramm had risen quickly to a leadership position in the veterans' movement. Born in New York City, raised in Westchester County, and educated as an economic major at Ohio's Denison University, he had enlisted in the Marines and completed a full tour of duty in Vietnam from 1968 to 1969. He had trained as an engineer, and woke up most mornings in-country to another long day of sweeping for mines and detecting booby traps. Combat "found" Cramm, as he put it, and seeing dead comrades became a grotesque reality. Just as disturbing, while in Vietnam he learned firsthand that the U.S. government was pursuing numerous policies that flew in the face of the Geneva Conventions. Before his tour was out, he had concluded that the war was not "justifiable." As a result, Cramm became the first active-duty Marine to put in for conscientious-objector status. His case never made it to court, however, because his contract with the U.S. government expired before his filing could wind its way through the Pentagon bureaucracy. Around the same time Cramm, who was then living in North Salem, New York, saw the VVAW's lone-coffin advertisement in *Playboy* and immediately signed up. "My first VVAW action was Operation RAW at Valley Forge," he recalled in a 2003 interview. "I wanted the war to end. I didn't shave or cut my hair for two years."

By the end of 1969 Cramm had gotten married, moved to Cambridge, and become a full-time VVAW organizer. It was at the New England regional office, at 65 Winthrop Street in Boston, that he connected with Arthur Johnson, who was then working for the Legal Service Project, an effort aimed at finding lawful ways for GIs to renege on their U.S. military obligation—fast. At that point there were some 2.7 million servicemen in the armed forces, and the lingering Vietnam War—with its napalm and Agent Orange and other toxic effects—was turning out to be more than many of them had bargained for. As a Navy officer who had served in Vietnam, Johnson wanted to "help guys who had gone AWOL and had no legal recourse." He wanted to save other men from becoming "beasts" and "monsters" and "killers"—all words veterans had used to describe themselves in Detroit. Although Johnson would eventually earn a law degree from Northeastern University, at the time of Operation POW he provided simple "legal counseling."

In early 1971, disaster struck the Winthrop Street office. For the most part the branch had focused on typical grassroots political organizing: fundraising, pamphleteering, debating issues on radio talk shows, and staging search-and-destroy missions. The New England operation differed from other VVAW regional offices, however, in its concentration on helping GIs get out of the service. In fact, the small Boston office pioneered the use of the law to aid conscientious objectors. That, plus its close proximity to downtown Cambridge and the Harvard University campus, made the Winthrop Street office, in the eyes of the White House, a beehive of subversive antiwar activity. Then, in the middle of the night after a slow, quiet day, a bomb exploded in the office. Fortunately, nobody was hurt. "Our suspicion was that it was done by the government," Cramm stated later. "It destroyed files and cabinets, but in truth there wasn't anything of importance in the office." Apart from unnerving the building's tenants, the attack accomplished nothing. "We never found out who did it," Cramm continued. "But believe me, it was a warning." And it wouldn't be the last: on the eve of the May 1971 veterans' march in Boston, Cramm received several threatening telephone calls, including one that promised him "bullets" if the demonstration proceeded.

Another Operation POW ringleader was Chris Gregory. Born in Maine and raised in New Jersey, he had emerged as an important player in the Boston VVAW office. "It was," he remembered, "a boiler-room operation." From 1964 to 1968 Gregory had served as a medic in the Air Force, but in 1971 he was a philosophy student at Brandeis, and willing to put his life on hold for the antiwar movement. John Kerry and he often visited VA hospitals together, growing despondent at the negligence they found, filing complaints about the poor conditions, and lobbying the facilities to establish drug rehabilitation clinics. "These hospitals weren't taking care of paraplegics and quadriplegics," Gregory explained. "John was very patient and good with these guys. They had deep psychological problems caused by Vietnam. He listened carefully to them. He cared deeply. And he got a great deal out of these exchanges. It energized him in his fight."

With the possible exception of Massachusetts Institute of Technology linguistics professor and noted antiwar activist Noam Chomsky, the star of the Memorial Day veterans' march was undeniably the photogenic twenty-seven-year-old Kerry. Over the course of that weekend, he posed for group pictures in Concord, signed autographs for a high school civics class in Lexington, and tossed a football around with strangers at the Boston Common. Young women flirted while discussing fund-raising with Kerry, as an eager recording-studio producer trailed behind him trying to catch his interest in cutting a "hit record." Henry David Thoreau's influential 1849 essay "Civil Disobedience" was read aloud as the demonstrating veterans found a kindred spirit in its author. Thoreau, after all, had spent a night in jail for refusing to pay his taxes on the grounds that he did not want to support the Mexican-American War. Good cheer filled the air as Kerry linked arms with his fellow vets—including the indomitable Al Hubbard—and chanted, "Bring them home; bring our brothers home." Cramm remembered finding himself impressed that Kerry eschewed grandstanding that weekend, wearing a camouflage jacket and marching in lockstep with his brothers like a regular foot soldier instead. "All these vets were warriors," Cramm recalled. "But Kerry was a listening warrior. He heard what everybody was saying. He was extremely thoughtful. I remember thinking about him that he was the guy you wanted with his finger on the trigger. He knew when to shoot and what targets to hit."

His burgeoning fame did not, however, shield Kerry from being arrested. At two o'clock in the morning of May 30, while the veterans were asleep on the Village Green of Lexington, a phalanx of local and state police in riot gear suddenly appeared in the soft crescent moonlight glowing on the old battlefield shrines. The marchers were all being charged with trespassing. "We had defied their ban on camping on the battlefield," Kerry explained. "They were waiting to get us." The police arrested 441 demonstrators camping on the green. Kerry and the other war veterans were read their rights, put on school buses, and hauled off to the local Department of Public Works garage. A few were handcuffed, although most were not. The Massachusetts police had done what the District of Columbia authorities had proved afraid to do: bust VVAW. Selective in their arrests—refusing to detain paraplegics or local residents known to them—the police wrote the protesters up for violating park rules and fined each of them five dollars. Charges for disturbing the peace were dropped, but the process of arraigning all the putative marchers took hours nevertheless. Some rancor flared up at Lexington Town Hall, just twelve miles from the renowned Walden Pond, where Thoreau wrote his transcendentalist prose. "They had wanted us to camp at the dump," recalled Chris Gregory of the transparent attempt to discredit VVAW. "Of course we refused. By arresting us they introduced a local controversy into the national controversy. But the police treated us gently. They took us to a garage where snowplows were usually kept. They gave us blankets and we went to sleep. It wasn't too bad."

Thanks to the negative media coverage of the arrests, the VVAW bust actually backfired on the Lexington Board of Selectmen, particularly Chairman Robert Cataldo. "We were elected to represent the town," Cataldo recalled in a 2001 *Boston Globe* interview. "This was a highly charged issue.

I went to Concord the night before and saw what was going on: drugs, liquor, the whole bit." Many residents of Lexington and Concord, however, were so incensed by the police action that they got together to raise the money to post bond for the self-proclaimed "army of peace." This was not all that surprising, of course, given Massachusetts's long history of embracing dissent. Even pro-Nixon citizens felt sympathy for veterans being arrested because they wanted to camp in a historic park. "The citizens of those communities thought the mass arrests were callous and unfair," Kerry recalled. "They were on our side because they thought we had the constitutional right to march. They were tremendous. The vast majority of the people ended up very mad at the Board of Selectmen for what they did."

In the end, the organizers of Operation POW regarded the heavy-handed official reaction to their protest as a public relations coup for VVAW. Nobody, it turned out, seemed to like seeing young war veterans in wheelchairs being treated as common criminals. The *Boston Globe*'s article on the arrests quoted several defiant demonstrators, such as Ed Lloyd of Chicago, who proclaimed, "We've already begun the Battle of Lexington—the whole country knows it." Kerry, who had denounced the excesses of the May Day Tribe, tried to instill in his cohorts the public relations advantages of the nonviolent approach. "I had learned from Dr. King," Kerry explained. "Peaceful demonstrations make the opponent look bad." The District Court judge, in fact, took notice of the vets' gentle demeanor. To the surprise of the Selectmen, he issued a statement that he had been "impressed with the behavior of the defendants."

After they were released, the protesters continued marching to Boston, single file and even more fired up, courtesy of the ham-handed arrests. Like the untested militiamen of 1775 and the civil rights Freedom Riders of 1961, the antiwar veterans of 1971 had, at least for that weekend, achieved the status of martyrs in a righteous cause. When they arrived at the Common some three thousand supporters were gathered to greet them. At the main rally, even Kerry's star was eclipsed by the éminence grise of the antiwar movement: former U.S. Senator Eugene McCarthy. Whether it was by paying for protesting veterans' rooms at the motel in Detroit or by championing improvements to VA hospitals in the U.S. Senate, the grand old Minnesota Democrat proved himself, with the possible exception of Massachusetts Senator Ted Kennedy, perhaps the most powerful ally VVAW had. Sporting a tacky red, white, and blue necktie that fine day, McCarthy proclaimed that he, too, had come to Boston to "spread the alarm" like Paul Revere. "The men who are fighting this war have come to call this country to a judgment of itself," McCarthy declared. "One of the marks of a first-class nation is what it does to satisfy its own conscience. . . . We are in that process."

Because Operation POW had been garnering a lot of media attention even before the event, Cramm had arranged for New York PR man Hart Perry to film the VVAW's Memorial Day march. Throughout the weekend the same crew that had already notched the 1970 Best Documentary Oscar, for director Michael Wadleigh's three-hour *Woodstock*, captured in cinema verité the mass arrests, the spirited march, the collegial sing-alongs, and the fiery speeches of Operation POW. The resulting color footage was separated from its audio tracks into ten-minute reels that remained stored away in their cans for the next twenty years. During that time, after a stint in Vermont running children's camps, Bestor Cramm had moved to Great Britain to study filmmaking at the West Surrey College of Art and Design. By 1985 he had returned to the United States and formed a small film company, Northern Lights Production, but he never forgot about the Operation POW reels he had commissioned in May 1971. Then, in 1995, a grant from the Lexington History Project enabled Cramm to acquire the rights to all the Hart Perry footage of the VVAW Concord-to-Boston march.

From the uncut vintage film, Cramm put together *The Unfinished Symphony*, largely about Operation POW. It was released in 2001 and shown at that year's Sundance Film Festival. Set to composer Henry Gorecki's Symphony No. 3, the *Symphony of Sorrowful Songs*, the hourlong movie opens with John F. Kennedy's 1961 inaugural speech, which had stirred an entire generation to engage toward public service. The film shows how Vietnam veterans shared a common sense of societal dislocation and the search for spiritual renewal. With their fashion-defying appearances—peace medallions and Army surplus clothing—they all believed that redemption came from speaking out, in letting one's conscience dictate one's life. Co-director Cramm's cinematic summary of the Vietnam veterans' antiwar movement closed with ghastly footage of American B-52s carpet-bombing Southeast Asian villages. *The New Yorker*'s review of *Unfinished Symphony* called it "a lucid analysis of an incisive political stroke and a sweeping threnody for the lives blasted and lost in America and Vietnam."

* * *

After Dewey Canyon III, Julia Thorne Kerry became so worried about her husband that she took to sleeping with a hundred dollars in cash under her pillow for his bail bond. "I knew it was just a matter of time," she recalled. "The FBI and others were after him. My job was to stay behind the scenes to bail him out. . . . His boldness had really annoyed the White House." When the inevitable finally occurred during Operation POW in Lexington, the Associated Press ran a photograph of Kerry, with his hands behind his head and a dark shadow over his face, being hustled off to the makeshift police holding station as if he were Lenny Bruce. With his camouflage jacket and long sideburns, the wire-service perp-walk photo almost made the descendent of John Winthrop look like a real outlaw. But Kerry took pride in the image and what it represented. Just as his opponents within VVAW were grumbling that he was "too elite" to speak for the antiwar veterans' group, here was the AP wire with the evidence that Kerry was genuinely putting himself on the line for the veterans' movement. "Nobody could ever claim John wasn't very cool in high-pressure situations," Chris Gregory remembered. "He was so much more mature than the rest of us. We were veterans, but we were still in our twenties. John operated with wisdom. And he did so knowing VVAW had been infiltrated. People would come to our campsites selling drugs, but we knew they weren't real dealers. They were informers." David Thorne recalled that after Dewey Canyon III whenever he talked with Kerry on the telephone he could hear a "weird clicking noise" on the line. "He was tapped," Thorne observed, "no doubt about it."

Kerry was not, of course, the only protester the Nixon administration was after. In an August 16, 1971, memorandum, presidential counsel John Dean would broach the matter of "how we can use the available federal machinery to screw our political enemies," formalizing an already tacit policy with the creation of a team of cloak-and-dagger specialists—including former CIA operative E. Howard Hunt and former FBI agent G. Gordon Liddy—ordered to conduct a smear campaign against a whole range of antiwar activists. A White House "enemies list" was compiled by an aide to Charles Colson, naming two hundred U.S. citizens deemed unfriendly to the administration—among them actors Gregory Peck, Paul Newman, and Bill Cosby; football star Joe Namath; journalists Daniel Schorr, James

Reston, Tom Wicker, Marvin Kalb, Garry Wills, and Morton Kondracke; and Kennedy and Johnson administration alumni Robert McNamara, Sargent Shriver, Arthur Schlesinger Jr., and Theodore Sorensen. John Kerry never made this "enemies list"—officially, at least. The idea behind the list was to make life difficult for opponents of the President's foreign policy via intimidation tactics such as personal surveillance, telephone taps, and tax audits.

But the Nixon administration was also getting the heat turned up on them. In mid-May, for example, the Senate began debating the Hatfield-McGovern amendment, a peace proposal that, if passed would have ended U.S. military expenditures in Vietnam by December 31. While the amendment failed 55–39, the writing was on the wall. Nixon's foreign policy was under siege. To Nixon, the most troubling aspect of the snowballing dissent was that previously hawkish senate Democrats—like Stuart Symington and John Stennis—were becoming doves. Everywhere he looked, Nixon discovered former allies now determined to rip him down. "Southern Democrats like Sam Ervin and John Stennis had even seen the light," Kerry enthused. "They were now ready to curb the power of the chief executive and end the war in Vietnam."

The next month brought even more bad news for the White House. On June 13, 1971, the *New York Times* published the first in a series of excerpts from the Pentagon Papers, a compilation and analysis of government documents regarding the Vietnam War ordered by Johnson-era Secretary of Defense Robert McNamara. The top-secret Pentagon study filled forty-seven volumes chronicling U.S. involvement in Vietnam from 1945 to 1968. Disgruntled National Security Council analyst Daniel Ellsberg, a former Marine and liaison officer at the American embassy in Saigon, had copied the massive document, which he had helped put together, in February 1971, and then turned it over to *Times* reporter Neil Sheehan. It was a stunning moment in journalism. "Sheehan's very readiness to entertain the notion that Americans might have committed war crimes, and that the war itself might be a crime, already stamped him as having one foot in the movement," Ellsberg wrote in his memoir *Secrets*. "Before the night was over, I had described to him the McNamara study and told him that I had it, all of it. I told him of giving it to [Senator J. William] Fulbright, and where that stood, and that [Senator George] McGovern had agreed to use it, then changed his mind." The *New York*

Times had no such qualms, and put the Pentagon Papers in print. It amounted to a leak of epic significance. "Nixon was furious," Kerry recalled. "The heat got turned up on guys like me considerably."

In *The Haldeman Diaries* entry for April 23, 1971, President Nixon's chief of staff wrote of that week's VVAW demonstration in Washington: "We got into quite a discussion of the media problem; they're really killing us because they run the veterans' demonstration every night in great detail, and we have no way to fight back. It's a tough one, and we've been trying to figure out some ways of getting back at it. One thought we're going to try is to have Mr. Rainwater of the [Veterans of Foreign Wars] call all the network presidents and ask if they'll assure him of equal time for veterans in favor of the Administration's position on the war next week. . . . In the meantime we're getting pretty well chopped up."

Not long after John Kerry led the Dewey Canyon III protest in Washington, D.C., another group of young ex-GIs was being invited by the White House to serve as alternative youth-culture representatives. Among them was Armistead Maupin, who would go on to become America's most popular gay novelist on the strength of his *Tales of the City* series. The son of a conservative descended from Huguenot general Gabriel Maupin, who operated a saloon in Colonial Williamsburg, Armistead Maupin grew up in Raleigh and then attended the University of North Carolina at Chapel Hill, where he befriended George Butler and thereby indirectly entered John Kerry's vast circle of friends. "George used to tease me shamelessly for being such a young Republican," Maupin recalled. "We formed our own eating club—the Society for the Preservation of Buck Taylor's Mutton and Shoats—modeled after an Ivy League club."

Under intense pressure from his domineering father, Maupin entered law school after graduation, but flunked out after only a year. He then spent a brief stint at Raleigh's WRAL-TV, working for an executive named Jesse Helms before joining the Navy. After attending the Newport Officer Candidate School (like John Kerry), Maupin found himself stationed in Charleston, South Carolina, attached to the destroyer tender *Everglades*, which sailed to Naples and Malta. "Can you believe my ship was named after a swamp?" Maupin would later joke. "That says something, doesn't it?" While deployed in the Mediterranean, Maupin befriended his ship's captain, Emmett Tidd, who a year later would rise to the post of chief of staff under Admiral Elmo Zumwalt. Tidd asked the aspiring writer to join him in Saigon as a protocol officer. At the time Maupin was still in the closet, and he speculated later that he gladly went to Vietnam as a way of demonstrating his manliness. "My mother said I had a Lawrence of Arabia complex," he recalled. "She didn't know how true that was."

Among Maupin's duties as Zumwalt's protocol officer was squiring officers' wives to such tourist sites as Hue, the ancient royal capital, and Saigon's Notre Dame Cathedral, constructed between 1877 and 1883. Bored with his job as a tour guide, he sought a new assignment as a communications officer with the River Patrol Force. He was stationed at Chau Doc, a small village near Cambodia. "It was very exotic," Maupin said, "a tiny place with heavy French influences and little straw houses on stilts." His primary function in Chau Doc was keeping the U.S. Army from accidentally shooting at the U.S. Navy vessels that suddenly appeared in the Mekong Delta river system. Maupin's most lasting memory of this assignment was the sound made by what American troops had dubbed the "fuck-you lizard," also known as the "re-up lizard." The reptile's mating call sounded like a drawn-out "fuck you." Maupin relished how the lizards always seemed to start singing just as senior officers were delivering the duty, honor, country spiel to their men.

After his year in Vietnam, Maupin returned to South Carolina, where he got a job as a reporter. His stories garnered fan letters from all over the Piedmont region. "I've been seeing your fine hand at work in the pages of the *News and Courier*," his old boss Jesse Helms, who was already preparing for a 1972 U.S. Senate run, wrote to him on February 9, 1971, "so obviously my wish for your success has come true." Shortly thereafter, Maupin got a phone call from Melville L. Stevens, a recently retired lieutenant working in the White House. "They were distressed by the Vietnam Veterans Against the War, and wanted a counterpropaganda effort," Maupin explained. "John Kerry had gotten under their skin, so Nixon's dirty-tricks squad decided to orchestrate a concerted strategic assault on him. They wanted my advice on how to do it. I thought about it for a while."

At the time, Maupin was disturbed by the tone and tenor of Operation Dewey Canyon III. "The whole thing was a carnival that seemed to be aimed at two and a half million Vietnam veterans, negating everything we'd worked for and fought for," he told the Associated Press then. "In their desire to end the war at any cost, a lot of people appeared to be trying to portray all veterans as scag-heads and radicals, apparently hoping to use this as a lever against the Administration. It seemed all wrong, and I wanted to do something about it."

Shortly after Kerry's appearance before the Senate Foreign Relations Committee, Maupin came up with an idea: why not organize a group of U.S. veterans who were *pro*-Nixon and *pro*-Vietnamization, and have them return to Southeast Asia on a goodwill mission? "I'll find the veterans to do some good in Vietnam," he wrote to Admiral Zumwalt on June 3, "if you can give us a project and get us over there." On June 28 Maupin received a positive reply. "I appreciate your concern for the continued progress of the Vietnamization effort and your offer to return to Vietnam," Zumwalt wrote back. "I have asked Rear Admiral Marshall of [Observation Post-]44 to work with the Office of the Secretary of Defense on what support we may be able to supply, and I have asked Rear Admiral Rauch . . . to identify a few of the numerous worthwhile projects that so desperately need the participation of more personnel to allow them to be a significant contribution to the people of Vietnam."

The White House thus hatched a plan to recruit ten all-American Vietnam veterans, including Maupin, to go to the village of Cat Lai, fifteen miles east of Saigon and build a twenty-unit housing project.* "The idea was for us to construct a block of houses so disabled South Vietnamese veterans could live cheaply and comfortably," Maupin explained. "It was a PR stunt to both promote Vietnamization and discredit Kerry." "We wanted to demonstrate our sincere concerns for Vietnamese veterans," one of the selectees, Charles P. Collins III of Dallas, Texas, informed the *Washington Post*, "and we wanted to counteract the negative image of American Veterans as presented by a portion of the media. John Carey [*sic*] doesn't speak for all three million veterans. The media takes a small minority of veterans and implies that the majority are drug addicts or war criminals. I don't feel bad about anything I did over there."

**In addition to Armistead Maupin, the veterans chosen for the Cat Lai project were Lou Abad, John Butler, Charles P. Collins III, Zeph Lane, Jesse Leadbeter, Jack Myerovitz, Thomas M. Neilsen, Kaki Reese, and Rick Will.*

On July 3, 1971, Maupin and the other nine young administration envoys found themselves aboard a C-140 military transport flying out of Alaska for Vietnam, along with a load of helicopter parts and a ton of blood. The village of Cat Lai—home to a South Vietnamese Navy base and several small U.S. advisory units—was just a kilometer away from a Viet Cong stronghold. "The whole point for the White House was to show these fine young anti-Kerrys doing good deeds as part of Vietnamization," Maupin recalled. "We were a high-profile, right-wing Habitat for Humanity club." Taking a swipe at the counterculture back home, Maupin began referring to the housing project as the Cat Lai Commune. In the end, the ten veterans erected a rickety concrete-and-cinder-block apartment complex that looked like a particularly ugly and poorly constructed American motel. "We knew nothing about construction," Maupin explained. "Nothing fit its frame; everything was off-kilter."

Upon the anti-Kerry squad's return to the United States, the then proud Goldwater conservative wrote a ringing endorsement of Nixon's Vietnamization program for William F. Buckley's *National Review*, headlined "The Ten Vets Who Went Back." In rather breathless but elegantly balanced prose, Maupin limned the do-good mission. "We lived in shelters for two months—without rank, pay or guns," he wrote, "to prove a point to ourselves and, perhaps, to our countrymen." He also imagined the future prosperity of Vietnam: "Someday I'll go back to Cat Lai to see the houses. I'll arrive at dusk when the sky over Saigon is amber and people are coming home from the rice fields. There will be children then, laughing and running and playing in the mud by the road. Their clothes will be white, not olive drab, and the watchtower on the edge of the village will be empty."

Years later, Maupin remembered the thrill he felt when Priscilla Buckley, the *National Review* publisher's sister, told him, "Bill read your piece and really enjoyed it." At around the same time, Maupin secured a job as a reporter for the Associated Press in New York, so he loaded up his Opel GT with glee and set off for the Golden State. He stopped in Clinton, Iowa, a town of twenty-five thousand on the Mississippi River, to look up his friend Tom Nielsen, a coparticipant in the Cat Lai Commune stint. "I was having dinner with Tom and his folks when the telephone rang," Maupin related. "It was Bob Haldeman, inviting Tom to the White House. When they heard I was there they invited me, too. Nixon wanted the whole Cat Lai

Commune group of ex-vets with him as a counterpoint to Kerry and the VVAW."

Maupin immediately telephoned the San Francisco AP bureau to let his new office know he would be late reporting in for his job because President Nixon wanted to see him. "They weren't impressed," he laughed, looking back. "You could hear their minds asking: 'Who is this little fascist?' "

On October 26, 1971, limousines bearing the Cat Lai goodwill ambassadors pushed their way through a crowd of protesters to the White House grounds; the anti-Kerry crew were ushered into the Oval Office at 12:20 P.M. for a scheduled twenty minute meeting with the President. Counsel Charles Colson and Deputy Assistant to the President Alexander Butterfield sat in on the conversation, which proved pure vintage Nixon, veering between astute geopolitical analysis and achingly awkward attempts at one-of-the-guys banter. "My chief memory is having to put Nixon at ease," Maupin recalled. "I was a twenty-seven-year-old closeted gay veteran. Of course, Nixon didn't know. So he tried to talk guy talk and was just miserable at it. I remember almost cringing when he said, 'Isn't it amazing how sexy those little Vietnamese girls look on bicycles with their dresses flapping?' "

As fine a humor writer as Armistead Maupin turned out to be, the transcript of the White House tape recording of this meeting reads even funnier than does his remembrance of Nixon's small talk. First, the President explained why he had not yet ended America's military involvement in Vietnam. "Now, let us see what would happen if we were to get out just a little bit too soon—a month too soon, three months too soon, four," Nixon supposed. "One of the things that would happen certain is that everything that *you* have served for and some people have died for would be lost. But let's forget that; what's it do to *them*? You could be sure that in a matter of a few months what had happened to occur in North Vietnam would be unstoppable in South Vietnam."

Upon that, Nixon remarked what "good people" he found the South Vietnamese, and that's when he edged into the bizarre. After one of the vets remarked how the South Vietnamese women who cooked and cleaned at the U.S. bases would warn the Americans when VC activity "was going to get hot and we might get in trouble," the President of the United States made his comments about the women's dresses.

After the young veterans nervously laughed, Nixon continued: "It's quite a sight. They told me when I was there in '56 that a Vietnamese mother tells her daughter that she is to carry herself like a swan. And I don't mind saying, just among our . . . and I'm not an expert on this thing, but the Vietnamese women are actually not all that attractive. But I have never seen clothing that does more in, shall we say, a spectacular way than it does for the Vietnamese. But you all know that!"

Taking the lead out of the painful small talk, Maupin said, "We've come under some criticism with the press. We're made to look like bozos in magazine articles. It's a burden, and we've got to work on that." Nixon lauded those present for doing just that, and agreed about the negative coverage of their humanitarian effort. Maupin pressed on: "We called ourselves the 'Cat Lai Commune' when we were there, partly because we thought if we gave ourselves a leftist-sounding name, it might attract a little more attention. And we did, toward the end; the *New York Times* came out to see us. We came back from Saigon to show Gloria Emerson, the correspondent, the site of the commune . . . and she talked to everybody. She spent the whole day trying to get us to indicate that the reason we had come back was that we were guilty because the war crimes we had committed—that we felt some kind of mea culpa in the situation."

At that, the President sneered, "Yeah, of course. . . . She is a total bitch. She's been writing that stuff for years in the *Times*—totally inaccurate, totally distorted." (When Maupin ran into Emerson at a book festival in the 1990s, he told her what Nixon had said. "She howled with laughter," he remembered. "It was a badge of honor.")

After their White House session, a press conference was arranged for the Cat Lai Commune volunteers. Maupin served as lead spokesperson, praising President Nixon for his Vietnamization program. "The most immoral thing you could do in a war," the twenty-seven-year-old Navy veteran proclaimed, "is to pull props out from under people." While Kerry and his VVAW cohorts were singing John Lennon's "Give Peace A Chance," Maupin and the other Cat Lai veterans, were spouting slogans like "All Vietnam veterans aren't potheads" and "Building instead of bombing." Meanwhile, in addition to getting his article published in *National Review*, awards began pilling up in recognition of Maupin's civic contribution. Among the honors were the Freedom Leadership Award from the Freedoms Foundation at Valley Forge, presented for his volunteer actions demonstrating exceptional individual achievement and for serving as a role model for all Americans, and a commendation on behalf of the Veterans Administration.

Over the next few years Maupin remained proud of his unofficial role as an anti-Kerry mouthpiece for the Republicans. In fact, his status remained such at the White House that Maupin was invited to sit in the Presidential Box right behind Julie Nixon and David Eisenhower at Nixon's 1973 inaugural-celebration concert in Washington. "The Republicans played bad rock to prove they were hip," Maupin remembered of the event. "And I'll never forget Mamie Eisenhower sticking her fingers in her ears to drown out the noise."

Even though he considered President Nixon a "dorky, insecure man," Maupin prominently displayed the photo taken of him with the commander in chief that day in his San Francisco apartment. Only later, in 1974 when he came out of the closet, and started bringing men home, did Maupin feel ashamed of having once served as a prowar flunky for Nixon. "The guys I brought home from the bars freaked out at my picture with Nixon," Maupin observed. "Their entire attitude toward me changed. They figured I was Jeffrey Dahmer or something." After a few such incidents the picture came down; Maupin eventually became a best-selling novelist. And over the years he learned to respect John Kerry for the courage he had shown in speaking out as an antiwar veteran. "People are shocked when they learn I worked for Nixon," Maupin averred of his political progression from right to left. "Kerry, as it turns out, was dead right about Vietnam. I'm now a non-doctrinaire practicing humanist. These days, of course, I'm a lot more antiwar than Senator Kerry, so I hope he takes a tougher stand against Bush's invasion of Afghanistan and Iraq. Just as important, he also needs to lead on the issue of allowing gay people in the military."

Creating the Cat Lai Commune turned out to be the most benign of the Nixon administration's attempts to discredit VVAW. Over the next year the Kerrys would be followed and their telephones tapped, presumably by federal agents. Perhaps even more menacing was the FBI's hiring of William Lenner to infiltrate the antiwar veterans' organization. A strange, shady character, Lenner wormed his way into heading VVAW's Arkansas-Oklahoma regional office although, according to the *New York Times*, he was a paid informant tasked with discrediting the group. By 1972, the *Miami Herald*, reported, VVAW had turned into an "organization thoroughly infiltrated by police and federal informers before and during the [Republican] Convention." As Kerry saw it, the only crime most VVAW leaders were guilty of was refusing to have amnesia when it came to what had transpired in Vietnam.

John Kerry remained in demand on the lecture circuit throughout 1971. A Boston TV station invited him to appear on a program called "The People Speak Out on the War"; the other guests were Common Cause President John Gardner, former U.S. Ambassador to Japan Edwin O. Reischauer, former U.S. Army Lieutenant General James Gavin, and President of the National Council of Negro Women Dorothy I. Height. When the discussion turned to Vietnam veterans, Kerry offered that U.S. servicemen were coming home from Southeast Asia "brutalized." A few days later, in a speech before the Massachusetts Political Action Committee in Cambridge, he announced that he would soon return to Vietnam to study U.S military operations—provided that the U.S. government did not deny him a visa. That trip never came off, but two decades later, as a U.S. senator, Kerry would travel to Vietnam on seven different occasions to search for information on remaining American POWs and MIAs and to work to establish better trade relations with the burgeoning free market economy there.

Perhaps Kerry's favorite memory of his antiwar activities in 1971 was the one commemorated in a photograph he later displayed on a wall in his U.S. Senate office. Taken by George Butler, the image shows Kerry at a New York antiwar rally, wearing his favorite leather bomber jacket and standing next to a granny-bespectacled John Lennon wearing his trademark Greek fisherman's hat.

It was all thanks to Kerry's old friend Danny Barbiero. After having helped Kerry with his 1970 congressional campaign, Barbiero had "dropped out" for a year in the Virgin Islands. "It was good down there for post-Vietnam syndrome," he explained. "I lost touch with John for a while." Barbiero returned from the Virgin Islands to work as a recording engineer with his musician brother Michael. The Yale graduate and former U.S. Marine platoon commander found himself at home as an assistant engineer at Media Sound, a studio used by some of the biggest rock 'n' roll performers. It was through Barbiero's contacts that Lennon had been brought to the 1971 antiwar rally at New York's Bryant Park.

Just two years later, in the summer of 1973 while the Watergate hearings were captivating the country, Lennon would ask Barbiero to engineer his *Mind Games* album. "I was just in awe of him, and he knew it," Barbiero remembered of that job with the ex-Beatle. "So we spent the whole first session just talking. We didn't turn on the equipment. He came in the next session with a piece of chalk and wrote on the wall, 'wall,' and wrote on the tape machine, 'tape machine,' and he wrote on the glass, 'glass,' and I looked at him like he was crazy. He said, 'Dan, I just want you to know what everything is so that we can get to work.' We then had just a great recording session."

At the Operation POW march to Boston Kerry had led the ragtag crowd of veterans in singing Lennon's "Give Peace a Chance." Just a few weeks later he got to introduce the icon at this much bigger rally. "Lennon had seen my Senate Foreign Relations Committee speech on TV, and he had heard some of my antiwar statements," Kerry recalled of that day in Bryant Park. "He liked what I was saying. Our government was giving him some flak because of his antiwar statements. So he asked me to be the guy to introduce him at the New York event. I met him ahead of time. We just hung around and talked. I love the picture because I love John Lennon."

By early June, the Nixon administration unleashed a series of public initiatives beyond the farcical Cat Lai Commune in its ongoing effort to discredit the antiwar veterans' movement in general and John Kerry in particular. On June 2, a White House front called Vietnam Veterans for a Just Peace (VVJP) challenged Kerry to a debate. Formed shortly after Kerry's Senate testimony, the group claimed it already had some 5,000 members in a statement by one of its leaders, Bruce Kesler. "We demand that the mass media of America, which has helped paint us as bloodthirsty murderers by giving undue prominence to 1,000 out of 2,500,000 of us, give ample time and space to clear our name and the conscience of the American people," read a VVJP press release picked up by the Reuters wire service. "We are forced by the abuse poured upon us to rebut the many absurd or baseless charges made against us so that we may resume our lives knowing that our sacrifices were not in vain." Kerry cleverly fired back that he would relish debating any member of VVJP "anytime, anywhere," including Richard Nixon himself. Not wanting to look outgunned, Kesler retorted that Kerry's "face has monopolized the airwaves for too long." To which Kerry scoffed: "Anyone can hold a press conference. Until they become active, they don't shake me up that much." A few days later in

Cambridge, Kerry derided VVJP for "mouthing" Nixon's bankrupt Southeast Asia policy.

The White House effort against Kerry didn't end there. The same day as the VVJP press conference, Vice President Spiro Agnew went after him, just as Haldeman had suggested. Speaking from Paradise Island in the Bahamas, Agnew first slammed the news media for making Kerry their darling. Then he sneered that Kerry had hired former Robert F. Kennedy speechwriter Adam Walinsky to ghostwrite his Senate Foreign Relations Committee testimony for him. The *Boston Globe* immediately sprang to Kerry's defense, avowing that he had written his remarks himself. The *Globe* claimed Agnew—who had hired wordsmiths Patrick Buchanan, William Safire, J. C. Helms, and Herbert Thompson to write his speeches—had unfairly maligned Kerry. The next day Walinsky also chimed in on Kerry's behalf. "I can understand how the vice president could fail to comprehend how the Vietnam experience could produce such bitter eloquence from a man of twenty-six [sic]," Walinsky told reporters. "But the fact is, that's what produced it—not some fancy speechwriter."

The source behind Agnew's allegation that Kerry had used a ghostwriter turned out to be Nixon-boosting *Detroit News* columnist Jerald F. ter Horst, who had made the charge in the course of comparing Kerry unfavorably with another young Vietnam veteran—Melville Stevens of Hanford, California. What ter Horst failed to mention was that Stevens was a paid operative of the Nixon administration, working for Chuck Colson. Details like that could get in the way of the White House's quest to convince the public that John Kerry was a phony.

As if the Cat Lai Commune and VVJP weren't clumsy enough, on June 7, with the backing of the White House, the Veterans of Foreign Wars trotted out its own Swift boat officer from the Navy's Coastal Division 11 in Vietnam: John O'Neill of San Antonio, a Naval Academy graduate who had just been released from active duty. With his aggressively fit physique and close-cropped hair, O'Neill had struck Chuck Colson as the ideal pro-Nixon to go head-to-head with John Kerry. Although the two twenty-something vets had never met each other in Vietnam, they were both river-running in March 1969. Publicly O'Neill lambasted Kerry for speaking "for no one except himself and his embittered little group of 1,000." By contrast, O'Neill claimed to have the over 500,000 Vietnam veterans who had joined

VFW on his side. "The President does our talking for us, as with most Americans," O'Neill said. "Mr. Kerry certainly does not."

In many ways Colson had picked the right fellow. Patriotic to the core, O'Neill spoke about his experiences in the Mekong Delta with pride. He truly believed in the U.S incursions into Cambodia and Laos. He felt President Nixon was an outstanding commander in chief. And he thought the whole winter soldier investigation was bunk. "I served in Coastal [Division] 11 for a year," he said. "I never saw one war crime committed by allied forces. I served for much of the previous two years in water adjacent to Vietnam. I never saw one war crime committed by allied forces. This is not to say that there are no war crimes committed in Vietnam. I saw kidnapping of minors or assassinations utilized almost daily by the Viet Cong forces in the area. Even among the allied forces, there are certainly war crimes. In the city of Boston last year there were 129 murders. Any group or city has psychotics. To say murder is part of the public policy in Boston is a lie. To say war crimes are commonly committed in Vietnam as a matter of policy is a lie."

VVAW held its national meeting in St. Louis on June 4 through 6, 1971. It was a "where do we go from here" session as well as a power struggle. For many in VVAW, Kerry had become a hero; others wanted him ousted. The media's attention to Dewey Canyon III had given some VVAW members big heads. In the end, however, Kerry received the full endorsement of the national organization to continue. The real question at St. Louis wasn't whether VVAW wanted Kerry but whether Kerry wanted to be associated with VVAW. As Rusty Sachs, a Kerry supporter, later noted, VVAW had become populated by "extremists" and "radicals." Kerry, like Sachs, wanted no part of them. "There were divisions in St. Louis," Kerry recalled. "It was polarizing. But I decided to work with the organization for a while longer. It was important not to lose sight of bringing home our soldiers."*

With a vote of confidence behind him, Kerry started publicly debating O'Neill about the war. On June 16, they each gave speeches at the closing session of the U.S. Conference of Mayors. Kerry's main point: the Nixon administration should withdraw *all* U.S. troops by December 31. From that moment on, offers for Kerry–O'Neill debates came pouring in. At first,

---

**At the meeting the executive committee also decided to hold a large VVAW "action" that Fourth of July in St. Louis.*

Kerry was ready to do each one. But in discussions with his wife, Julia, and his brother-in-law David Thorne, he realized that he was falling into a trap. He had become, in Mary McGrory's words "a household name." VVAW had proved its strength by holding a peaceful march in Washington, D.C. that past April. VVJP was nothing, a clownish paper organization concocted by the White House to counterbalance him. Nobody had ever even heard of O'Neill. So Kerry played cat-and-mouse games, picking and choosing the venues where he appeared with O'Neill. He turned down, for example, offers to debate O'Neill on *60 Minutes* and on an NBC special hosted by Edwin Newman.

Instead, Kerry said yes to Dick Cavett, who was known to be antiwar. The format would be a sit-down conversation, as on Kerry's previous appearance on the late-night talk show. With his message well-honed, and a gift not to freeze when the camera light went red, Kerry triumphed over O'Neill. "Kerry's arguments were more sound," Cavett recalled. "He understood the importance in TV of speaking calmly and staying on message." Years later Cavett, giving his alert staff credit for booking the debate, remained "knocked out" by how good Kerry was on television. Cavett allowed Kerry to "educate" his audience about the tragedy of U.S. involvement in Vietnam. While O'Neill spoke in a clear, strident fashion, Kerry was scholarly without being a know-it-all. "He was so clearly intellectual and sensitive," Cavett went on. "We knew that when he left the set the second time it wasn't the last we were going to hear from him again."

On the heels of his successful thwarting of O'Neill on TV, another group went after Kerry: the wives of POWs. Led by Patricia Hardy of Los Angeles, they charged Kerry with using antiwar issues to further his political career. When a *Boston Globe* reporter caught up with Kerry, he was saddened that these angry wives had taken issue with him. It hurt. But he kept to his point. "What bothers me most is that action must be taken now in the Paris peace talks," Kerry stated. "We need a date set, for total withdrawal from Vietnam."

The rest of 1971 was one of nonstop locomotion for Kerry. In Boston he debated John O'Neill yet again on WBZ television with John Cole as moderator. During a trip to Washington, D.C., four wives of POWs heckled his press conference. But others, like Shirley Culberton of McLean, Virginia, whose brother was a POW, defended his honor. "I think it's a shame that

Mr. Kerry is being maligned by a group who are being used by Mr. Nixon and won't acknowledge it," she said. At another time, Kerry arrived at a rally at the Plymouth, Massachusetts, Superior Court to support the need for Vietnam veteran drug rehabilitation centers, and was met by a prowar group. He refused to shun controversy—for example, he took up the cause of Joseph "Bert" Westbrook, an Army major who earned three Silver Stars and one Bronze Star in Vietnam but was now demanding that he be released from further military service because he had become a conscientious objector. And he appeared at a winter soldier event held at Faneuil Hall in Boston, where Edward Kennedy, George McGovern, and Daniel Ellsberg participated. More than 50,000 people attended. "We both laughed at the fact that the President of the United States was after John," David Thorne recalled. "Government agents used to shadow John around at rallies. It was just unbelievable. He was as patriotic a guy as Yale University and the U.S. Navy ever produced. In some ways their hounding him was flattering. But it was mainly disturbing. I had come from a Republican family. Patriotism was in my veins. So we just thought because the war should end didn't mean we deserved to have our basic rights infringed upon."

Sometimes Kerry, often accompanied by Julia, would travel around the lecture circuit with another VVAW member in tow. Because he wrote war poetry, Larry Rottmann, for example, was booked to share the stage with him on a few occasions. He has fond memories of their being together in Missouri. "In our travels in the early seventies we talked about the future," Rottmann recalled. "We talked about what we could do with our lives. Kerry had decided that politics was his best way to express his concerns, mine was education. We weren't just concerned about Vietnam. What we cared about was America's values." In May 1971 the two got together in Albuquerque. Rottmann, who lived in nearby Corrales, had been arrested for handing out leaflets and had been locked up in the Bernalillo County Jail for three days. "Once I was out of jail, he stayed at my house," Rottmann recalled. "We had some laid-back times." They spoke about the poetry collection Rottmann had co-edited, *The First Casualty*, and they gardened, trout fished, and attended the dedication of the first Vietnam War memorial, erected at Angel Fire.

The stunning memorial at Angel Fire is located on the outskirts of Taos. It is second only to the Vietnam Veterans Memorial in Washington,

D.C., as a sacred spot for survivors of the war. How the memorial came into existence is a moving story unto itself. On May 22, 1968, a prominent physician in the area, Victor Westphall, learned he had lost his son in a Vietnam firefight. In deep mourning Westphall—who died in 2003—decided to be proactive. He commissioned Santa Fe architect Ted Luna to erect an angelic monument to the Vietnam veterans, perched high atop a mountain with a spectacular view of the serene Moreno Valley. A vast gull-like structure with curving walls and sweeping arches, the memorial was completed around the time Kerry was to testify before the Senate Foreign Relations Committee. Moved by Kerry's impassioned remarks, Dr. Westphall invited Kerry to be the principal speaker at the dedication on May 22. Front-page stories in both the *Santa Fe New Mexican* and *Albuquerque Journal* soon trumpeted Kerry's heroics in both the Mekong Delta and the Mall.* "It was quite an honor," Kerry recalled. "This was the first effort to pay proper respect to the Vietnam veteran. It was—and is—a beautiful, majestic memorial. My speech talked about the need to pay homage all across America for the Vietnam solider."

Julia enjoyed being on the lecture circuit—particularly because it exposed her to the Rocky Mountain states. She fell in love with lonely cities like Billings, Montana, and Denver. "I remember landing on a bluff in the middle of nowhere," Julia recalled of the intrepid schedule Kerry kept through 1971. "It was the first time I was in the *real* West." In one typical week for her husband, he would speak in six different states. During the first week of October, for example, he lectured in Washington, D.C., Kansas, Colorado, Nevada, California, Illinois, and Massachusetts. Sometimes folksingers Peter Yarrow or Tom Paxton would perform before he spoke to help draw larger crowds. "John and I saw the world from a similar perspective," Yarrow recalled. "I was trying to get artists involved in stopping the war and so was John." Occasionally, however, Kerry was met with angry protesters at his campus gigs. Lecturing at Georgetown University, for example, he was met by shouts of "Fight liberalism" and "Trash John Kerry." His lecture at Fort Hays State College in Kansas, however, covered by *The National*

**A visitors center, telling the story of Vietnam veterans, and a chapel were also eventually built at Angel Fire. Over the years, CBS reporter Charles Osgood would regularly tell the TV audience the story of the moving monument in the New Mexico hills.*

Observer, was more typical. Kerry spoke for fifty minutes, then answered questions. He chided the students for not doing more to protest the war. Apathy, he said, was the agent of selfishness. "Now is the time to keep the pressure on," he implored. "There is a sad history in this movement of what comes of not keeping the pressure on."

Imbued with an old-fashioned belief that words swayed people, throughout 1971, Kerry polished his speaking style. He continued to worry that VVAW was becoming too radical. He heard through the grapevine that a dozen or more members were planning on locking themselves inside the torch-carrying arm of the Statue of Liberty. The notion appalled him. In the same way he thought it unwise for VVAW members to sue President Nixon to end the war—it was a waste of time. He had also come to believe mock search-and-destroy theater was *passé*. His views were maturing. His solutions to Vietnam were the same as those of Senators Edmund Muskie of Maine and George McGovern of South Dakota—not SDS's Rennie Davis. The power of VVAW, he believed, was no longer centralized out of New York; it was now in branch offices in cities like Cincinnati, St. Louis, Denver, and Dallas. The issue dearest to his heart, however, remained the "atrocious conditions" of VA hospitals. At the same time, his concerns grew to encompass such domestic issues as increased unemployment, failed schools, dilapidated public housing, and environmental degradation. He cheered for the Ellsbergs and Berrigans but his *admiration* was for men like Ted Kennedy and Mark Hatfield. He wanted to join their ranks.

Given this maturation it only made sense for him to resign from VVAW even as membership topped the 20,000 mark. In a November 10 letter housed at the VVAW Papers in Madison, Wisconsin, Kerry quit, politely noting he had been proud to serve in the national organization. His reason was straightforward: "personality conflicts and differences in political philosophy." In two days VVAW was meeting in Kansas City and he would be a no-show. A clear indication of his clean break with the organization was that he immediately changed his stump speech from "Vietnam Revisited" to "America Revisited." And he started settling old scores.

William F. Buckley, a fellow Skull and Bonesman, derided Kerry in a syndicated column. In these columns, published as "Kerry I" and "Kerry II," Buckley tried to make mincemeat out of the young antiwar activist. The situation deteriorated to the point where a debate was in order. On

November 14 Kerry and Buckley squared off for fifty-eight minutes on the latter's TV show, *Firing Line*. Buckley began by introducing Kerry as the man who "stepped aside in his congressional race in favor of Father Robert Drinan, who was elected, some say, on a day when God was out to lunch." The barbs continued to fly, but tempers never flared. In a role reversal Kerry championed Nixon's recent visit to China deeming it "a brave, bold gesture," while Buckley bemoaned the visit as a treaty violation with Taiwan. Perhaps their sharpest point of departure occurred over GIs and drug use. Bemoaning the "deplorable" VA hospitals, Kerry asserted that New York alone had between 15,000 and 20,000 veteran heroin addicts; he blamed the dependency on Vietnam. "Oh, come on," Buckley retorted. "The reason they take drugs is because drugs are fun."

Kerry understood Buckley's point about drugs. Occasionally during his travels around America in 1971, usually with either Julia or a fellow veteran, Kerry had smoked marijuana. Because he always played athletics, the idea of putting smoke in his lungs was essentially anathema to him. He never smoked pot while at Yale or in the Navy. But during his stint as a leader of VVAW, he occasionally indulged. "Yeah, I smoked pot when I came home from Vietnam," Kerry noted in a 2003 interview. "I didn't mind getting high. I certainly enjoyed it. But I didn't like the out-of-control component. I like being alert. So I tried it a few times, but I didn't touch it after 1972."

But what Buckley didn't understand, Kerry believed, was that many veterans were using heroin to escape from the horrors of Vietnam. These veterans felt, as one journalist phrased it, that they were the "butt end of a bad war." Starting at the winter soldier investigation in Detroit, Kerry had befriended Yale University psychiatrist Robert Jay Lifton, who was doing pioneering work in PTSD. During the early 1950s, Lifton had been an Air Force psychiatrist serving in Japan and South Korea. He understood Asian culture. In a distinguished career that has spanned over fifty years, Lifton wrote detailed reports, for example, on Hiroshima survivors. Although Lifton hadn't yet written his landmark study on Vietnam veterans, *Home From the War*, he expressed his findings on the lecture circuit and in magazine articles. In head-on fashion, Lifton explored such explosive issues as the "heroin plague," CIA complicity in selling drugs, and why soldiers "in the rear" were most likely to become addicts. What Kerry learned from Lifton was something that he also understood intuitively: that many of the

GIs on drugs were victims. While Kerry himself encountered virtually no drug use while skippering PCFs-44 and -94, he did meet addicts in VVAW. They were searching for help. Drug rehabilitation programs for veterans, to Kerry, were a humane solution to the problem. Calling veterans "potheads," "dope fiends," or "losers," he believed, showed a coldhearted disregard for the valor those men had displayed in Southeast Asia. He felt society had an obligation to help these troubled patriots get over their dependency.

In December rumors were again flying in Massachusetts about Kerry's political plans; the *Boston Globe* reported that he was going to challenge Congressman Harold Donohue in the Fourth District seat in Massachusetts. On December 12, after a television interview, reporter Gordon Hall, who was a regular contributor to the Sunday *Boston Herald American*, caught up with Kerry. It was a setup. The well-bred Kerry, who turned twenty-eight the day before, was not yet used to this sort of tabloid assault. After some phony friendly banter, Hall put Kerry on the defense about his stint in VVAW, an organization supposedly loaded with Communists. "I don't like Communists," Kerry exclaimed. "In fact, I hate them. I hate all totalitarians. I'm totally dedicated to representative, pluralistic, free democracy. If a country is taking care of the needs of its people, there is absolutely no reason to fear an alternative philosophy." Hall kept pushing, and Kerry kept rising to the bait. He was forced to defend why Communists were allowed to march in VVAW. "When a decision is democratically made to march to protest the war these groups decide to come along, to participate in the demonstration," he answered, "you must understand that they are not coming along with us. We are not in any way going along with them." Eventually, as Hall got rude, Kerry, angry, ended the interview.

Politics has been called a blood sport in America—it is. In many ways Kerry had walked into its buzz saw. He had become the symbol of the veterans movement and also a target of the White House. He had absorbed attacks on his character from Richard Nixon, H. R. Haldeman, Charles Colson, VVAW organizers, and William Buckley. *Doonesbury* had twice mocked him. Yet he felt good about his accomplishment: Vietnam veterans now had an antiwar voice in America. "In my mind the veterans were the least publicized, most important group in the antiwar movement," Boston University professor of political science Howard Zinn recalled. "The students and priests got most of the headlines. The veterans took away from

Nixon the idea that the peace movement was against GIs. Because veterans—and GIs out in the field—were now demanding immediate withdrawal from Vietnam, it was difficult for the White House to characterize them all as Communists or anti-American zealots."

That Christmas season Kerry had a decision to make: Was a career in public service worth the cost? Searching for the answer, he took long solitary walks along the beach. Fame, he understood, had its toll. He may have become a household name but it came at a price. Throughout 1971 his pace had become a sprint. He had little time for reflection. If in Vietnam the sound of mortar fire at midnight used to jangle his nerves, now it was the telephone that never stopped ringing. What he was most grateful about was retaining his longtime circle of friends: Harvey Bundy, Danny Barbiero, David Thorne, George Butler, Fred Smith, Landon Thorne, among others, were always by his side. So, of course, was Julia, and his sisters and brother. "I had real friends who stuck with me through thick and thin," Kerry recalled. "Collectively they gave me great strength."

As Christmas approached, Kerry spent an afternoon at his parents' home in Groton. It was a time of inventorying for John. He read some of the letters he had written his mother and father from his tours of duty. How innocent he had been, how naïve. His new book, *The New Soldier,* was selling modestly well. On December 7 in Manhattan, George Plimpton, editor of *The Paris Review,* had thrown a book party for Kerry, packed with high-profile editors, antiwar reporters, and progressive socialites. "When I had heard John testify, I was terribly moved," Plimpton recalled. "He was so damn good. I knew him—and Julia—from Elaine's. I liked him enormously. And I thought *The New Soldier* was fantastic. It caught the pathos of the veterans' movement brilliantly. The photographs in particular. I urged John, that evening, to write a memoir of his Mekong Delta experiences. Something that would make the hair stand up on our necks. He told me he was already considering it."

Now, on the encouragement of Plimpton and others, he was reading his old correspondence about the brown water navy in Vietnam. Even though he had told his parents he was an "uncommitted soldier" from aboard the U.S.S. *Gridley* and was essentially against the war, his prose—even after Dick Pershing was killed—smacked of hopeful defiance. He read how excited he was about seeing his first Swift boat glistening in the waters outside of Danang and cringed. That letter had been written three and a half years before, but it seemed like an eternity. So much had happened to him that he hardly recognized the young man he encountered in his *Gridley*-era letters, still basking in the glow of Skull and Bones and fanciful rides through Europe in Harvey Bundy's Austin-Healey. Like many young sailors, he had expected to see the world in the Navy and he had: a rum-soaked night in Honolulu, port call in Subic Bay, a dance club in Australia. A quick glance at his maroon-and-black covered *Gridley* yearbook, which his mother kept, had photographs of him slurping up spaghetti noodles in a timed race and laughing with his men. Those were better, more carefree days. Even San Diego now glowed with nostalgic delight. For some reason he always remembered himself bicycling around town feeling free, his future ahead of him like a cloud. But his cloud had turned dark. Thunder had struck him in November 1968 when he went in-country. He had never expected war to be pretty. He actually enjoyed firing guns. The coastal waters of Vietnam were utterly gorgeous, and even the muddy rivers of the Mekong Delta harbored a kind of murky enchantment. But then he started seeing dead bodies. Then he had to kill. Then he learned that his government was telling lies. He suddenly learned a new vocabulary in Vietnam: atrocities, napalm, Agent Orange, Nung—somehow these monstrous words had not been in the lesson books at Coronado. When he thought about his old crewmates on PCFs-44 and -94, which was often, a smile always came over his face. These men were all his "brothers." And sometimes the lighthearted moments of Vietnam—water balloon fights and shrimp boils—almost numbed out the bad. Almost. But no matter how hard he tried, the nightmares lingered.

Somehow in Kerry's upbringing, he had developed an intense disdain for injustice. His heart was always with the underdog. For the past year he had worked as tirelessly for the cause of veterans' rights as any man in America. The entire year had become a blur of protest speeches and antiwar demands. He was tired yet, in an odd way, invigorated. Every time he talked to a man with no legs because he had stepped on a DMZ land mine or had to hold the hand of a Gold Star mother who would never get to see her son become a full-fledged man, his insides swelled with certified rage. He was not unique in this regard. By anybody's standard, the anger index in almost any Vietnam veteran far exceeded that of the average citizen. But unlike many of his colleagues in VVAW, he didn't know how to drop out from society. He had a type-A personality. By instinct and inclination, he was a fighter who favored relentless pragmatism over anarchistic fury. The key skill that he had developed in his twenties was seeing the hole in the opposition's defense. He had learned to find the opening as surely as NFL running backs Gale Sayers or Franco Harris. His critics saw this as opportunism. He saw it as pushing his objective over the line and winning. Because he had a fundamental belief in American democracy, he knew a single person could make a difference—especially if they had a forum, like a seat in Congress. While his main political goals were ending the Vietnam War, bringing back his "brothers," and improving VA hospitals, he knew that was only the start. His entire life, he decided that Christmas, would be like his father's career: devoted to public service. There was, however, a major difference. While his father was best thinking about policy, he wanted to initiate it. Deep down it probably made more sense to enroll in law school and perfect his understanding of the U.S. Constitution and habeas corpus. But he was not living in sensible times.

Two Christmases earlier he had decided to retire early from the Navy and run for Congress. He had lost to Father Drinan, who had since become a friend. In the past two years, he had learned much, much more about leadership. In Vietnam, on a Swift boat, he had given the orders. As a VVAW leader, he had to compromise—but only in nuances. He had, in his mind, a clear sense of direction for America. He knew that in Congress he would have to hone both skills. The Nixon administration—with the rare exception of opening the door to China and vigorously supporting NATO—was opposed to his principles. At twenty-eight, even with a White House dirty-tricks squad smearing his name, he was prepared to enter the great arena of electoral politics once again—this time in earnest. No bowing out early. As he headed to Groton House on Christmas Eve, he told his wife he was ready to once again fight for a seat in Congress. He was going to put the memoir on hold. She was not surprised.

Chapter Seventeen

Duty Continued

The presidential election year of 1972 was among the most politically charged in American history. Across the nation, the Vietnam War was the lightning-rod topic; around Boston, so was John Kerry's political future. Sharing the stage with former Democratic Representative Al Lowenstein at an event in Cambridge on January 10, Kerry remarked that he "was not that concerned with the presidential race" because "politics at the local level is where the action is." When pressed by a local reporter whether that meant he would be seeking elective office, Kerry admitted that he was 98 percent certain he would run for Congress—from either the Fourth District against the venerable Harold Donohue, a Democrat, or the Fifth District against the well-connected F. Bradford Morse, a Republican.

What Kerry had realized by the break of the new year was that if he wanted to launch his career in Massachusetts politics with a seat in Congress, he and his wife would have to move. Their Waltham home was in the Third District, which the silver-tongued Father Robert Drinan represented in the House. Kerry and the Roman Catholic politician-priest had become friends in early 1970, when Kerry withdrew at a caucus gathering in favor of Drinan. When it came to the need for an immediate U.S. withdrawal from Vietnam—and most social issues—the two men spoke with essentially the same voice.

So the Kerrys went looking for a new place to live, eventually buying a home they liked in Worcester, the Fourth District hometown of the ailing seventy-year-old Harold Donohue, then the third-ranking majority member of the House Judiciary Committee. Over his twenty-six years in Congress the unpretentious, hardworking Donohue had come to be known as a classic Massachusetts constituent service politician. While hardly outspoken on the war or any other issue, he had been hawkish on Vietnam. Kerry determined to go after him on style rather than substance, as the two were basically in agreement on the latter. Thanks to the widespread media coverage of his appearances as a VVAW leader, Kerry had become nationally known as a vociferous public critic of the Vietnam War and the Nixon White House. Donohue, by contrast, was seen as a quiet, Truman-era keeper of the status quo in Congress—a well-meaning man who didn't like to make waves. Kerry planned to capitalize on this generational difference.

Along these lines, Kerry made two harsh anti-Nixon speeches that February. He delivered the first right in Donohue's backyard, at Worcester's prestigious Holy Cross College, founded by the Jesuits in 1843. Before an audience of some two hundred students Kerry called for the "dumping of Nixon." Two days later, speaking to a Young Democrats group in Providence, Rhode Island, he explained what he meant. He wasn't merely advocating voting Nixon out of the White House at the polls in November; he wanted the President driven from office *now* for having widened the war into Cambodia and Laos. "There are grounds to impeach the President," Kerry intoned, and not just because U.S. troops still weren't fully withdrawn from Southeast Asia, as Nixon had promised in the last campaign. Now, Kerry complained, "the President said he would not obey the law." It was an over-the-top statement, and it got a lot of applause.

What came across clearest, as Boston's *Record American* headlined, was that John Kerry was "Running for Something." But practically as soon as the Kerrys prepared to move into their new Worcester house, the news reached them that Congressman Bradford Morse was giving up his Fifth District seat to take a high-ranking job at the United Nations. An unexpected opening had been created for Kerry. The Fifth District was his home of record while at Yale and in Vietnam. His justification for the move at the time was at least plausible: for the past ten years his parents had lived in nearby Groton, also in the Fifth District. Eager to capitalize on Morse's departure, the Kerrys didn't take long to decide to move again, to Lowell in Morse's district. "Voila!" the *Boston Herald Traveler* gibed that April 5. "Kerry miraculously discovered that he would rather live in Lowell and represent the people of the Fifth District in the halls of Congress. If he doesn't stop

House hunting soon, he'll not only need a campaign manager, but a full-time real estate agent."

Decades later, Kerry admitted that he deserved the criticism he got for his brazen district-shopping. "I was perceived as a carpetbagger," he noted. "I was young, a novice to politics, and you make those kinds of mistakes." Working-class Lowell, to the east on the Merrimack River, stood at the core of the regional economy. Named after Francis Cabot Lowell, whose Boston Manufacturing Company ran the first mill in the country capable of turning raw cotton into finished cloth, the city had retained its reputation as a textile manufacturing center. Over the years communities of Greek, Polish, and French Canadian immigrants had sprung up in proudly blue-collar Lowell, which aggressively rejected the Brahmin culture prevalent in neighboring prep school towns such as Andover and Groton. In his 1950 autobiographical first novel *The Town and the City*, Lowell native Jack Kerouac captured his hometown with "the river coursing slowly in an arc, the mills with long rows of windows all a-glow, the factory stacks rising higher than the church steeples."

After formally launching his congressional campaign in the Fifth District, Kerry spoke at an antiwar rally at Boston Common on April 13. He used the opportunity to flog Nixon and endorse George McGovern's bid for the White House. "President Nixon has committed himself to do everything the people of this country elected him not to do," Kerry shouted from the podium. "His secret plan for peace . . . is still secret." A few cries of "Right on!" and "Power to the people!" echoed back, but the *Boston Globe* wrote the rally off as "listless." The newly declared Fifth District candidate quickly realized that he was going to have to do more than denounce the Vietnam War to get elected to Congress.

Many of Lowell's residents were what would later become known as Reagan Democrats and, like their homegrown Beat novelist Jack Kerouac had before his death in 1969, many of them supported U.S. involvement in Southeast Asia. The factory and construction workers who populated the Irish and Greek bars on Lowell's Moody Street, for example, made no secret of what they thought of the pot-smoking hippies they saw protesting the war on the news damn near every night; in the tough old mill town those snotty college kids looked like a bunch of degenerates out to undermine the American way. Few of the local veterans of World War II or the Korean War cottoned much to the notion of throwing away one's combat medals. These Lowell citizens liked Kerry the Silver Star winner and welcomed him into their community; they felt otherwise about Kerry the protester and treated him as a Eugene McCarthy wannabe. Sensing the resentment from this crucial constituency, the new candidate began making speeches all over the Fifth District about the need for economic renewal. At the Lowell Technological Institute's commencement ceremonies on June 12, for example, Kerry offered the crowd of four thousand a cliché-ridden address full of economic proposals and employment statistics. "Surely a nation that can send a man to the moon can now find a way to send a man back to work," Kerry proclaimed before going on to criticize how deeply the welfare system had suffered under the Johnson and Nixon administrations.

The shift in focus proved a smart political move. Unemployment was sky-high in Lowell and similarly blue-collar Lawrence. What's more, while the Fifth District included 90,000 registered Democrats and only 44,000 Republicans, another 77,000 of its voters were registered as independent of party affiliation. The candidate who would become Kerry's Republican opponent, former State Representative Paul Cronin of Andover, was assiduously courting these independent voters. A bespectacled, bookish moderate Republican blessed with an Eisenhower-like smile, Cronin was formidable in an understated way. It was clear that whichever contender came across as more likely to bring jobs to the district would win. Cronin's close ties to the Nixon White House—via his friendship with Charles Colson, who hailed from Massachusetts—made the Republican's potential slugfest with John Kerry a highly anticipated prospect. It was one of the few 1972 congressional races bound to attract national attention.

First, however, Kerry had to win the Democratic primary by defeating nine other candidates, most notably State Representative Paul Sheehy. Kerry's staff numbered forty people, and they directed the more than 6,000 grassroots volunteers to ring doorbells, hand out pamphlets, and hold rallies. Many of them were young antiwar activists from all over New England who had flocked to Kerry not for his economic proposals but because of his high-profile dissent with VVAW, which had made him a hero to many in America's youth culture. His five-story campaign headquarters, located inside an old bank building in downtown Lowell, had become a beehive of antiwar activity. "I flew in from Germany, where I was working for the

Olympics Committee," Kerry's sister Diana recalled. "It was just amazing how many young antiwar people had been galvanized by my brother. They were all over Lowell."

Another clear asset Kerry enjoyed was friends in high places around the country, who would do wonders for the political neophyte's fund-raising efforts. While there was some money to be gleaned in the tonier of the twenty-six communities of the Fifth District's Middlesex County, there was a lot more in New York, Los Angeles, and Washington, D.C., where Kerry had esteemed individuals virtually lining up to host fund-raisers for him. Novelist Kurt Vonnegut Jr., folksinger Judy Collins, and pop vocalist John Denver all journeyed to either the Fifth District or Boston to campaign on his behalf. George Plimpton, who just five months earlier had thrown a book party for Kerry, now held what writer Tom Wolfe might have called a "radical chic" affair to raise money for his Massachusetts campaign. Skull and Bonesmen mingled with rich young faux-hippie women in miniskirts and celebrities including journalist Pete Hamill, folksinger Buffy Saint-Marie, composer Leonard Bernstein, and movie director Otto Preminger. At the proper moment Plimpton gave a blast on his bugle, announced that a true Vietnam War hero and the next congressman from Massachusetts's Fifth District was about to speak, and turned the evening over to Kerry. "He spoke in a low voice," Plimpton recalled in an interview shortly before his death in 2003. "You had to strain to hear what he was saying, but the passion was unmissable."

Just as the Kerry campaign was gaining stride, however, a suspicious incident occurred that may have been attributable to the White House Plumbers, a clandestine unit run by E. Howard Hunt and G. Gordon Liddy under the direction of John Ehrlichman aides David Young and Bud Krogh. Recalling the September 1971 break-in at the office of Daniel Ellsberg's psychiatrist in Los Angeles, not to mention the far more momentous June 17, 1972, break-in at the Democratic National Committee headquarters at Washington, D.C.'s, Watergate office complex, a break-in of a different kind took place that September 18 in Lowell. "It was a strange affair," Kerry recalled. "It wasn't Watergate; it was more like a Watergate in reverse."

At one thirty that Monday morning John Kerry's younger brother, Cameron, and Thomas Vallely, whose father was a superior court judge, had received an ominous telephone call at Kerry's campaign headquarters on Central Street, where they had been working late. The caller claimed that the thirty-six telephone lines the campaign was going to use later that day to urge voters to the polls had been "sabotaged." The pair of impetuous twenty-two-year-olds immediately raced outside to the vacant single-story building next door, where the telephone bank was located. The entrance was locked, so they forced open a door to get inside and see if anything was amiss. Within minutes they found themselves surrounded by police officers. "We were trying to stop the crime from happening," Cameron Kerry explained years later. "But in truth we had been set up—duped."

The incident had really begun on August 30, when John Kerry had written to the New England Telephone and Telegraph Company to note his concern that his office's lines were not secure. "Without passing any personal indictments I think it is quite within reason to believe someone might tamper with the lines," Kerry had written. Clearly, his campaign was quite aware of possible "dirty tricks" involving telephone tapping or line-cutting, so when Cameron Kerry and Tom Vallely got the heads-up phone call, they took the bait. After the young men were apprehended, the Lowell police officers read them their rights, put them in a squad car, and took them to the local jail. The pair were charged with breaking and entering and eventually released without bail. "At 2:00 A.M. I received a call telling me my brother was in jail," Kerry recalled. "I immediately headed down to the police station to pick them up." The Central Street block also housed the campaign headquarters of State Representative Anthony DiFruscia, who called the pair's unauthorized entry another Watergate break-in. Furious that his little brother had been set up, a livid John Kerry denounced the curious incident as "harassment" and a "smear" against his family's name. "It was bogus," Kerry recalled. "They had no flashlights or tools. They were just responding to a tip. They had been duped."

On the positive side, as the election approached, Kerry's campaign proved stellar at fund-raising. According to the *Boston Globe*, his war chest of $279,746 was the national high for a Democratic congressional candidate in 1972. In this fiscally flush mood, instead of adopting, say, a Peter, Paul, and Mary antiwar song, his campaign appropriated FDR's rousing old standard "Happy Days Are Here Again" for its theme. Yet no matter how hard Kerry tried to focus on domestic issues, his services in the Vietnam War and subsequent protest against it remained his true calling cards in the eyes of the press and the public. When Boston's WLVI-TV hosted a debate between the Fifth District's two congressional candidates, John Kerry and Paul Cronin, the war dominated the discussion. Kerry alone snagged the next day's headline, with his demand that the U.S. withdrawal from South Vietnam hinge upon the return of all POWs.

By early October a *Boston Globe* poll showed Kerry with a two-to-one lead over Cronin, 50 percent to 24 percent. But trouble was brewing. An independent candidate in the race, Roger Durkin, who fancied himself a conservative Democrat marque, had logged a respectable 7 percent in the *Globe* poll. Meanwhile, national surveys indicated that if trends continued, Democratic presidential nominee George McGovern would be clobbered by incumbent Richard Nixon by a margin of thirty percentage points on November 7. As Election Day neared, Kerry, feeling victory was in sight, called in the state's popular second-term Senator Edward Kennedy to campaign for him. "At this point I saw Kerry as the rising star in Massachusetts politics," Kennedy recalled. "I would have done anything to help him. He was one of the most impressive, tough, yet warmhearted men I had ever met."

On the eve of Kennedy's arrival in the Fifth District, however, the ultraconservative *Lowell Sun* newspaper—pushed by its owner, Clem Costello—slammed Kerry in a blistering editorial about his recent book *The New Soldier*. More than thirty thousand copies had been sold and the collection of VVAW speeches had gotten generally decent reviews, but *The New Soldier* had turned into a political tar baby for John Kerry. "Its cover carried a picture of three or four bearded youths of the hippie type carrying the American flag in a photo resembling remarkably the immortal photo by Associated Press photographer Joe Rosenthal of U.S. Marines raising the flag on Iwo Jima after its capture from the Japanese during World War II," the *Sun* editorial carped. "The big difference between the two pictures, however, is that the photo on John Kerry's book shows the flag being carried upside down in a gesture of contempt that has become synonymous with the attitude of youth groups protesting not only Vietnam but just about everything else there is to protest in the United States. These people sit on the flag, they burn the flag, they carry the flag upside down, they all but wipe their noses with it in their efforts to show their contempt for everything it stands for."

Senator Kennedy's visit, during which he addressed large crowds in both Lowell and Lawrence on behalf of the young congressional candidate, should have been a triumphant moment for Kerry. Not only was his friend a Kennedy, but he ranked ahead of even Daniel Ellsberg on Nixon's notorious "enemies list"—a true badge of honor in liberal Massachusetts in those days. A decade earlier, Kerry had been driving his VW bug around the state, campaigning for Kennedy's first run at the U.S. Senate in 1962; now the scion of Massachusetts's first Democratic family was stumping for *him*. Try as he might to enjoy the honor, however, Kerry could only feign delight as the *Lowell Sun* editorial continued to eat at him. "Costello, along with his brother, was a pioneer in negative tabloid assaults," Kerry recalled. "He was going to screw me. It was that simple." He could hardly believe that his hometown newspaper could deem him, John Kerry, a Silver Star winner, unpatriotic. He had *bled* for his country. He had *fought for the flag*. Now, a single, if controversial, photograph George Butler had taken during Dewey Canyon III was threatening to deflate, and possibly derail, his congressional campaign. Kerry had expected hardball. What he got, however, amounted to character assassination. "The *Lowell Sun* really did a number on Johnny," former Massachusetts Governor Michael Dukakis recalled. "They had gone after both Paul Tsongas and myself. But what they did to Johnny was worse. They were rabidly right-wing and prowar. Once they decided to go after you, look out." Practically every day until the election the paper ran an anti-Kerry message of some sort. "Suddenly my opponents were questioning my integrity," Kerry still fumed decades later. "They were trying to paint a picture of me as some wild-assed, irresponsible, un-American youth. It was infuriating."

Seizing upon the flag-cover controversy for his own political benefit, Roger Durkin, the independent candidate, actually filed suit to be granted the legal right to use Butler's photograph in his campaign's newspaper advertisements. After the courts turned him down, Durkin took out anti-Kerry ads in the *Lowell Sun* with the headline "Censored." He derided Kerry as nothing more than "a dressed-up Abbie Hoffman" and to remind voters of the cover of Kerry's book took to wearing an American-flag pin on his lapel. Meanwhile, two members of Nixon's cabinet—Elliot L. Richardson of the Department of Health, Education, and Welfare and John A. Volpe of the Department of Transportation—came to the district to campaign for Paul Cronin. Nixon's son-in-law Edward Cox also came to the district, positioning himself a true clean-cut representative of the Vietnam generation. The national GOP very much wanted to see Kerry go down—hard. Even his *National Review* nemesis William F. Buckley Jr. weighed in, using a *Boston Globe* column to call Kerry a "young aristocrat" who on April 22, 1971, had delivered "the most irresponsible speech in recent congressional history." Buckley also dubbed Kerry an advocate of "enthusiastic surrender" in Vietnam.

David Thorne, Kerry's brother-in-law and campaign manager, recalled how the opposition suddenly turned up the heat on the Democratic candidate. "We tried to contain the attacks," Thorne said. "But it was a difficult, uphill battle through a minefield of dirty tricks." The Kerry camp issued press releases pointing out that the upside-down flag was the international signal of distress and trying to explain the image's symbolic use on the cover of *The New Soldier*. But Kerry was outgunned. Ever since VVAW protesters had disrupted the Republican National Convention in Miami that summer, the FBI seemed to have grown increasingly obsessed with discrediting the antiwar veterans, particularly their charismatic new spokesperson Ron Kovic, a U.S. Marine from Wisconsin who had been paralyzed from the chest down in Vietnam. The impassioned Kovic had gained widespread notoriety when he and two other veterans in wheelchairs had disrupted Nixon's acceptance speech at the convention. Just before the showing that got them escorted off the convention floor in their wheelchairs, CBS News correspondent Roger Mudd had conducted a short interview with the now twenty-six-year-old Kovic. "I'm a Vietnam veteran," the ex-Marine had yelled. "I gave America my all, and the leaders of this government threw me and others to rot in their VA hospitals. What's happening in Vietnam is a crime against humanity." That summer Kerry had attended the Democratic National Convention—also held in Miami—but he watched the GOP spectacle on TV. He cheered on his fellow vet Kovic. "He was a great, courageous fellow," Kerry praised Kovic, "a man of deep moral convictions and an uncompromising disposition."*

Roger Durkin did all he could to tie Kerry to the most radical Kovic-like antiwar element in the minds of the Fifth District's voters. Although running as an independent, Durkin ardently supported the administration's

---

**Kovic's 1976 memoir* Born on the Fourth of July, *which director Oliver Stone made into a motion picture, remains an antiwar, pro-veterans-rights classic.*

Vietnam policies. Speaking at the Polish-American Hall in Lowell, he lambasted Kerry's book as a "misuse of the flag" by "Yippie-type people," thereby equating the Democratic congressional candidate with the anarchic Youth International Party and its wild-left leaders Abbie Hoffman and Jerry Rubin. Desperate to counter such cheap shots, Kerry had his campaign manager go after Paul Cronin as a chicken who had avoided military service altogether. "He was one of these guys who loved war," Thorne quipped of the thirty-two-year-old Republican opponent, "even though he never took the time to wear a uniform."

The ad hominem attacks against Kerry—the rich and radical maverick—worked. On October 28, right on the heels of the book-cover controversy, the *Boston Globe* ran a story headlined "Kerry's Lead over Cronin Dwindling." The reason for the tumble was clear: "negative reaction to the 28-year-old front-runner." After all, the *Lowell Sun* had published two separate four-day series of articles pounding Kerry, one for fund-raising outside the district and the other for maintaining friendships with radicals. The *Sun* even published a cartoon titled "The Rape of the Fifth," showing Kerry with his foot on a prostrate woman's buttocks as he grabbed Wall Street and Hollywood cash in his hands. The shift in momentum away from Kerry was palpable in the last days before the election. At his rallies what David Thorne described as "professional hecklers" took to interrupting many of Kerry's speeches with shouts of "Traitor!" or "Loser!" Leaflets claiming Kerry was a drug addict were mysteriously distributed around the Fifth District. Cameron Kerry, who had lost much of his zest for the campaign after his arrest, was being followed. John Kerry couldn't speak into his telephone without the line clicking every few seconds, and his wife had to put up with rude remarks from strangers at the local grocery store and post office. The opposition had decided the best way to destroy Kerry was to deem him insincere. "I thought, welcome to Nixon's America," Julia Kerry recalled. "Our basic rights were being trampled."

The final straw fell on November 4, just three days before the election. Fighting to hold the line, Kerry had notched a couple of last-minute coups: he got to announce that he had won the support of Mayor John Buckley of Lawrence, and Peter, Paul, and Mary materialized, guitars in hand, to motivate the youth vote by performing "Blowin' in the Wind" and "If I Had a Hammer" at Kerry rallies. Just as the Democrat began to feel he would pull it off, Roger Durkin dropped a bombshell. The independent quit the race and threw his support to Republican Paul Cronin. "The polls showed I was the spoiler," Durkin told the press, insisting the Nixon administration had nothing to do with his last-minute decision. "My continuance in the race would assure Kerry the election. . . . In the interest of the many citizens who feel as I do—that it is important to defend [against] the dangerous radicalism that John Kerry represents—I am announcing my withdrawal from the race."

Kerry smelled a rat in this stunner, but there wasn't much he could do about it. Rumor had it that Durkin had made a deal to have his entire campaign debt wiped clean if he endorsed Cronin, but it was never proved. A ticked-off Kerry tried to pretend Durkin's sudden move didn't faze him. "I don't know the real reasons for his getting out," he offered. "There have been a lot of attempts lately to affect my candidacy, and this latest one doesn't surprise me." Ever since he had given his antiwar speech at Valley Forge Kerry had been fighting "the Power," as counterculturists used to call what the less hip still knew as "the establishment." But the multipronged smear campaign against his congressional bid in 1972, with its unrelenting attempts to paint his antiwar beliefs as anti-American, was wearing on Kerry. "But I kicked it into fifth gear," he remembered, "and did whatever last-minute things I could think of to win. I was proud to be an antiwar veteran demanding our troops be brought home. Their attempts to discredit me only goaded me onward."

Despite his best efforts to turn the tide, Kerry lost the race to Paul Cronin by a margin of nine percentage points. Like a good politician and a gentleman, Kerry gracefully conceded to his Republican opponent quickly. Early TV news analyses out of Boston attributed his defeat to his VVAW past, as misrepresented on the cover of *The New Soldier* but also including his high-profile role in Dewey Canyon III and its medal-throwing demonstration. In truth, however, the press pummeling he took for his antiwar activities is what really did his campaign in. "Saying that he received bad press is like saying that Anne Boleyn was scolded by Henry VIII," remarked former Kerry campaign worker Michael Pollak. "At times it was difficult to tell if Kerry was running against Cronin or the *Lowell Sun*."

Kerry suggested as much in his concession speech. With Julia and his mother and father by his side in the conference room at Andover's Rolling

Green Motor Inn, Kerry faced his thousand or so glumly disappointed campaign volunteers and launched into one of the sharpest antiwar perorations he had ever delivered. "I just don't believe that we lost to a tangible opponent," Kerry declared to his troops. "What we lost to is the very kind of fear and kind of things that we were running against. I know now that what we were standing for was right. We came together last March around an idea—that the war in Vietnam, which is not over, should be. . . . If I had it to do over again, I'd be in Washington with the veterans tomorrow." Then, drawing up all the conviction he could muster, he claimed that he did not feel at all bitter. He also made it clear that losing would not make him give up or go away. "John Kerry will be here and he will continue the same battle tomorrow that he fought throughout this race," he stated. "If you don't think you can walk out of here with your head [held] proud, then you're not part of the Kerry team."

With that he stepped away from the microphone, hugged his wife, thanked his parents, and high-fived Peter Yarrow before going around to thank everybody in the room individually. Nobody—not even the *Lowell Sun*—thought this was John Kerry's political swan song. Even to him, almost as disturbing as his own loss was the news that President Nixon had not only been reelected, but had trounced George McGovern in the biggest landslide in U.S. history. And in Massachusetts' Fifth District, McGovern did worse than any Democratic presidential candidate ever."I think McGovern hurt us a lot," Kerry said later. "I know he carried the district, but he set a climate. People were willing to believe that I would open up our country's defenses. Nobody wanted to believe my positions on tax reform." Instead, once again the American people preferred to buy into Nixon's claim that "peace was at hand" in Vietnam. For four years Kerry, along with millions of others, had put his heart and soul into trying to oust Nixon—and had failed. Poet and Vietnam veteran W. D. Ehrhart, in his 1995 memoir *Busted*, summed up Kerry's feelings on Election Day 1972 perfectly: Nixon and Agnew and the extended war in Southeast Asia "stuck in my throat like a stick sharpened at both ends." Meanwhile, the *Boston Globe* reported that Nixon went to bed satisfied that night only after hearing that Kerry had lost in Massachusetts—the only state McGovern won. As angry as he was over the smear campaign directed against him, he blamed himself for not having been smarter. He knew his transparent "carpetbagger" switch from Worcester to Lowell had hurt his chances. And he had counted too much on the youth vote. "John had run a terrific campaign," McGovern recalled. "But the backlash against me was like a tidal wave. He had spoken the truth about Vietnam, and some people were afraid of the truth."

Naturally, his defeat triggered some postmortem resentment in Kerry. His sister Peggy, his wife, Julia, and his brother-in-law David Thorne all mentioned, looking back, how glum he acted after he lost the election. Retreating to the confines of the rented Lowell house, Kerry read novels and built model airplanes and ships to distract himself from thinking about the immediate future. He also painted a little. But it did him no good to brood. It made more sense to ponder his next move. He had some big decisions to make by Christmas. First he had to pay off his campaign debt of about $43,000. Then he would buy a house in Lowell—the one in Worcester had already been sold—to accommodate his new primary goal to start a family with Julia. His unemployment problem was solved when Kerry was hired as a New England regional coordinator for the Recovery and Developmental Program of the Cooperative for American Relief to Everywhere (CARE). It was his season of thinking about the future. His contemplative mood often brought him to one of Lowell's Catholic churches—Immaculate Conception or St. Michael's—to sit by himself and pray. That August the ambitious young councilman, Paul Tsongas, made an appointment to see Kerry. He had one question: Was Kerry going to run again for Congress in 1974? If he was, Tsongas would not seek the Democractic nomination for himself. Forced to decide, Kerry came back to Tsongas with a definitive "No." He was going to apply for late admission to law school. "I had already taken all the law boards," Kerry recalled. "So I applied to Harvard, Boston University, and Boston College. I was extremely late. Only BC would entertain a late application." It proved to be the ideal school. Boston College, located in Chestnut Hill, was not too far from either his house in Lowell or his parents' place in Groton.

In the process, Kerry grew recommitted to the pledge that echoed through his Vietnam journals from 1968 and 1969, to "never forget." He swore to stay a watchdog over the Vietnam War until it finally and truly ended, and never to forget what he had seen in and learned from it. Not a day went by that he didn't think about Dick Pershing, Peter Johnson, John White, Steve Kelsey, Don Droz, and Bob Crosby. When Kerry visited California late in 1972, he made a point of calling on the distraught Judy Droz, who was herself studying law at the University of California at Berkeley. "We both felt that we would not rest until the war ended and the last POWs were brought back," Don Droz's widow recalled of their conversation. "We looked at each other in the eyes and knew what the world had lost in Don. The loss allowed us to be close." She and Kerry also shared a profound sense of shock and disbelief as they and the rest of the nation watched the reelection-emboldened Nixon unleash what history would know as the "Christmas bombings," of an impoverished Third World country that posed America no real threat.

Nixon had set his National Security Advisor Henry Kissinger loose to practice "jugular diplomacy" on his North Vietnamese counterpart Le Duc Tho at the Paris peace talks toward the end of 1972. What this translated into was intense U.S. pressure on the North Vietnamese to stop stalling the negotiations. To underline this point, for starters Nixon saw to it that the Pentagon immediately delivered vast amounts of military equipment to President Nguyen Van Thieu of South Vietnam. "Matériel worth some $2 billion was flown to South Vietnam from such American aid recipients as Taiwan, South Korea, and Philippines, which were to receive more modern weapons in exchange," Stanley Karnow wrote in *Vietnam: A History*. "The program, completed in six weeks, gave the Saigon regime the fourth largest air force in the world." Then, in a far more brazen move, Nixon ordered thirty-six thousand tons of bombs dropped on North Vietnam. As historian George Herring noted in *America's Longest War*, the Christmas bombings—dubbed Operation Linebacker II—exceeded the total tonnage of bombs used against the North "during the entire period from 1969 to 1971." The December 18–29 air assault left behind vast devastation throughout North Vietnam, particularly in sections of Hanoi and Haiphong. During this campaign, fifteen U.S. B-52s and eleven other planes were shot down out of the sky. More than sixty of their crew members were killed in the Linebacker II bombing raids, with another thirty-plus captured and held as POWs. An international outcry against the bombings of Hanoi erupted around the globe. Massive anti-American protests took place in London, Berlin, Rome, and Amsterdam. British Labour Party leader Roy Jenkins spoke for many when he denounced the Christmas bombings as "one of the most cold-blooded actions in recent history." Writing in the *New York Times*, columnist Anthony Lewis called it "a crime against humanity."

Kerry, back home in Lowell, entirely agreed with the criticism. In fact, he was flabbergasted by what he saw as the "monstrous brutality" of the Christmas bombings. Many other Americans felt the same way; Nixon's public approval ratings plummeted to a lowly 39 percent. In Hanoi alone only 1,318 North Vietnamese civilians had been killed in the raids. But as Herring pointed out in *America's Longest War,* the bombing did force North Vietnam back to the negotiating table in Paris. Peace accords were signed and entered into force on January 27, 1973. (For their diplomacy Henry Kissinger and Le Duc Tho would be jointly awarded the Nobel Peace Prize, which the latter declined.) All of America, John Kerry included, cheered when 591 U.S. POWs were released that spring and returned home to great fanfare. Ever since he had arrived in the Mekong Delta, Kerry had heard harrowing stories about how these servicemen were beaten, tortured, and starved. They had become *his* heroes; young Americans, like himself, who were being inhumanely treated by the vicious North Vietnamese. Their liberation was liberation to his mind and heart. Yet the war continued. "The peace treaty did nothing for Saigon," John Negroponte, Kissinger's deputy in Paris, later conceded in Michael Maclear's *The Ten Thousand Day War.* "We got our prisoners back; we were able to end our direct military involvement. But there were no ostensible benefits for Saigon to justify all of the enormous effort and bloodshed of the previous years."

As Kerry watched Richard Nixon deliver his second inaugural address on January 20, 1973, on television, he grabbed a yellow legal pad and started jotting notes. In a spontaneous form of therapy, he filled notebooks with his critiques of the President's policy mistakes. Some of what he scribbled that day made it into a long opinion piece Kerry polished for the *Boston Globe*; most of it joined his other unpublished journal entries in a box. Two paragraphs read:

> *The truth we are afraid to face is that, the signing of the Paris Peace Pact notwithstanding, we are right back where we started from in 1954—with an even less determined position. In 1954 the South and the North had withdrawn to both sides of the DMZ. There was a promise of an election within two years and the prospect of real self-determination for the Vietnamese. Today, more of Vietnam is under communist control than it was when we first started fighting.*

The troops have not withdrawn to either side of the DMZ and there is no guarantee of elections. The only guarantee is that North and South will continue the struggle and that despite our statements of support, we are actually powerless to safeguard Cambodia, Laos, and Thailand.

Over $200 billion later, over 55,000 dead Americans later, over 400,000 wounded Americans later, over millions of dead Vietnamese later, over millions of Vietnamese refugees later—after all that the too often repeated and just as often ignored litany of the war can call to mind—Vietnam and Southeast Asia [are] in exactly the same situation [as], if not worse than, when we became involved.

And yet it was clear in the wake of Nixon's inauguration and the Paris peace accords that the antiwar movement had lost its momentum by early 1973. The fact was that the brutal Christmas bombings had worked in bringing the North Vietnamese back to the bargaining table in Paris, which led to the cease-fire agreements. More important in the eyes of the American people, the same day the pacts were formally signed, January 27, Nixon ended the U.S. military draft; two months later, on March 29, the last American combat troops left Vietnam, and three days after that the U.S. POWs were released. Of course, the fact also remained that in the course of Nixon's presidency close to 21,000 Americans died in Vietnam, with another 53,000 badly wounded—amounting to more than a third of total U.S. casualties in Southeast Asia.

In truth, however, unlike in 1968, Nixon had not won reelection on his Vietnam policies. Instead he capitalized on just how sick the American public had grown of even hearing about the war, pro or con, and beat McGovern on bread-and-butter economic issues, along with charges that the Democrat favored "amnesty, acid, and abortion." In any case, Nixon's overwhelming electoral victory, combined with Kissinger's tenacious diplomacy, succeeded in temporarily demoralizing the left. In their hearts many liberals believed that the Nixon administration would soon violate the Paris agreements. But it was hard to protest something that *might* happen. Still, some antiwar activists, including movie star Jane Fonda and her future husband, SDS cofounder Tom Hayden, opted to visit Hanoi to protest the Christmas bombings. Others, such as MIT linguistics professor Noam Chomsky and 1969

Nobel Prize in Medicine Laureate Salvador Luria, wrote open letters to the *New York Times* insisting that the entire Paris peace accords process had been a ruse designed to provide a fig leaf to cover continued U.S. bombing raids in Southeast Asia. Democratic U.S. Senate Majority Leader Mike Mansfield of Montana, who denounced the Christmas bombings as a "Stone Age tactic," thereby tacitly cautioned that the Nixon administration seemed stuck on the notorious 1965 suggestion of former U.S. Air Force Chief of Staff General Curtis LeMay: "My solution to the problem [of North Vietnam] would be to tell them frankly that they've got to draw in their horns and stop their aggression, or we're going to bomb them back into the Stone Age." Meanwhile, Nixon started sporting an American-flag pin on his lapel and lashing out at organizations like VVAW. In a May 24 speech to a group of just returned POWs, the President asserted that the time had come for Americans to "quit making national heroes out of those who steal state secrets and publish them in the newspapers."

Nixon was referring, of course, to Daniel Ellsberg and the momentous Pentagon Papers, but the Kerrys continued to face regular harassment for their antiwar activism at the local level. After the *Lowell Sun* had slurred John Kerry as un-American, their house instantly became a target for egging and their car tires for flattening. But these incidents went beyond juvenile pranks and minor vandalism. Just after John started law school at Boston College, Julia Kerry had given birth to the couple's first child, Alexandra, on September 5, 1973, and only a few days after they brought the baby home from Emerson Hospital in Concord, their house was terrorized. "One night somebody threw rocks through our windows," Julia recalled. "One of them came crashing through and landed in my child's bassinet. It missed her head by three inches. We had so much vandalism on our property that we had to put Plexiglas up on the windows, and that is a horrible way to live. It taught me to be frightened. I'm not saying the Nixon administration orchestrated these attacks. They probably didn't. But they had created a climate whereby it was okay to haze us—to intimidate us. We had been battered. It's one thing if you lose an election because the populace doesn't like you and it's honest and open. It's another thing if you lose because you've been vilified and publicly abused. It's a whole different way to lose a campaign."

But out of the ashes of defeat the Kerrys had the satisfaction of watching the Nixon administration crumble. Just weeks after their daughter was born, federal officials revealed that Vice President Spiro Agnew, John Kerry's primary White House attacker and one of the few people in it untainted by Watergate, had accepted bribes while governor of Maryland from contractors in exchange for his help in procuring them state work. The federal investigators also announced they were still looking into Agnew's tenures as Baltimore County executive, Maryland governor, and vice president of the United States. To avoid winding up in prison, on October 10 Agnew resigned from office under an agreement with the Department of Justice. Pleading no contest in federal court to the charge that he had evaded income tax payments on $29,500 he had received as governor in 1967, Agnew was fined $10,000 and put on three years' probation. House Minority Leader Gerald R. Ford succeeded Agnew as vice president that December. "I didn't see it as a personal victory that my major detractor had fallen," Kerry noted in retrospect. "It was more a matter of 'thank god for America.' Many of these Nixon people, Agnew being just one, had absolutely no regard for our Constitution. They were criminals, really, and we as a nation needed to purge ourselves of them."

John Kerry spent the next decade pretty much in political exile. From the time he started law school in the fall of 1973 to his successful campaign for the U.S. Senate in 1984, he compiled a modest record of somewhat unfocused achievement. He put in a stint as a Saturday evening talk-show announcer on Boston's popular WBZ radio program, *The Jerry Williams Show*. The experience forced him to deal with the politics of the state. "I would read newspaper articles and then discuss them on air," Kerry recalled. "It was a good education. People would call in with complaints and I would try to deal with them." In July 1974 he was named executive director of Mass Action, a private organization dedicated to improving his home state's government. For a while he did local TV news. From time to time, the *Boston Globe* would speculate that Kerry might run for secretary of state or for Congress again, but he never did; taking Cronin on again just didn't interest him, and neither did any state office. He had decided to pay his dues to the party for a while. When *Globe* reporter Crocker Snow Jr. caught up with him in November 1974, he found a second-year law student content with that and being an attentive father. Under the headline "Once a Hot Political Property," Snow's story centered on his subject as the VVAW maverick who had completely faded from public view. The interview offered a few insights into Kerry's thinking on such diverse topics as the merits of Lyndon Johnson's presidential press secretary Bill Moyers running for the White House himself and Aleksandr Solzhenitsyn's teaching Americans the concepts of freedom. Most interesting, however, was Kerry's comment regarding why he was out of politics for the time being. "When I was running for office, Nixon was still President, Agnew was still Vice President, and our involvement in Vietnam was still a big issue," he told Snow. "Now the whole context of things has changed."

An unusual aspect of Kerry's education at Boston College was that he was allowed under state law to prosecute misdemeanor offenses, even though he hadn't yet passed the bar exam. "This gave me practice in front of juries," Kerry recalled. "It was invaluable." After graduating from Boston College Law School in May 1976, Kerry took a job as a prosecutor on the staff of Middlesex County District Attorney John J. Droney in East Cambridge. His first trial was a high-profile rape case, which he won. "Droney gave me the opportunity to excel," Kerry recalled. "He had a great street sense and I learned a lot from him. Unfortunately, he had a neuromuscular disorder at the time and wasn't in his full stride."

Many political observers expressed shock that the ardent champion of the underdog had not chosen to work as a defense attorney instead. They clearly did not understand what made John Kerry tick. Kerry by nature was a prosecutor. His entire public stance during Vietnam was based on holding the U.S. accountable for its Vietnam policy. He was now going after rapists, robbers, and rogues. "I gave up politics for a while," Kerry recalled. "The whole national tenor in 1976 was to put Vietnam behind us." The Democratic nominee for president, former Georgia governor Jimmy Carter, who had previously praised Nixon's Southeast Asia policy, had now dubbed the war "immoral" and "racist." When Carter was elected president, the Kerrys were preoccupied with moving into a bigger house in Newton. On New Year's Eve Julia Thorne Kerry gave birth to their second child, Vanessa, once again at Emerson Hospital in Concord. They were thrilled when, shortly after Carter's inauguration, the President issued a blanket pardon for draft dodgers. "Reconciliation was in the air," Kerry believed. "A healing was occurring."

One Massachusetts politician who understood the "new," post-Vietnam

John Kerry was Governor Michael Dukakis, who admired the young prosecutor's political zeal when it came to crimebusting. Back in 1970, his father, Richard Kerry, had been the lone delegate to the Massachusetts Democratic Convention from Groton. "It was a mid-May morning," Dukakis recalled. "I drove up a gravel drive to meet Richard, who was standing in front of his barn. To my surprise I didn't have to even utter a word. He said, 'You don't have to spend any time with me, I'm with you,' then he sort of waved me away." The next spring, in April 1971, Dukakis and his thirteen-year-old son went to Washington, D.C., for Dewey Canyon III. "Johnny was just spectacular that weekend," the future governor remembered. "My son was totally taken with him." So as Dukakis rose up the ranks in Massachusetts politics he stayed in touch with Kerry, who would be elected his lieutenant governor in 1982. In that latter post Kerry focused on federal-state relations and anticrime measures. "No matter the situation," Dukakis noted, "Johnny always brought up the plight of veterans. It was his special calling, even though the Vietnam War was over."

Kerry's workload increased after he was promoted to first district attorney in Middlesex County in January 1977, but he continued to make time for the men he had served with in Vietnam. The brown water navy had been a close-knit community, and Kerry kept in contact with his fellow Swift boat veterans. Through his friends Wade Sanders and Skip Barker, he had heard that PCF-71 had been sunk by a B-40 rocket, but recovered; that a Swift convoy on a harassment-and-interdiction mission had captured two tons of VC ammo, and that an underwater mine had blown a gaping hole through PCF-40, killing a crew member. Naturally, he felt particularly curious about what had later happened to the two boats he had skippered. PCF-44 had continued to get shot up, but had managed to avoid any more serious firefights. As for his other Swift, back in April 1970, PCF-94 had been patrolling a Giang Thanh canal when the boat was ambushed by Viet Cong firing B-40 rockets. Although PCF-94 suffered serious damage, the VC attack was eventually thwarted by the crew's gallant return of fire and some helicopter support; one man had been badly wounded. After that, the only thing Kerry could find out about his boats was that word had it both PCF-44 and PCF-94 had been turned over to the South Vietnamese government in 1973, along with most of the other still seaworthy Swifts. All anyone seemed to know about the PCF program was that its training center had been moved from Coronado to Mare Island, California. Meanwhile, SwiftShips in Berwick, Louisiana, was no longer getting government contracts to manufacture the 50-foot aluminum boats.

Early in 1978, *Boston Herald American* reporter Peter Gelzinis interviewed Kerry for a feature story. The office walls at the Middlesex County assistant district attorney's office were adorned with Vietnam memorabilia. Accompanying Gelzinis on the assignment was Richard Thompson, veteran *Herald* photographer renowned for his cut-to-the-chase, no-nonsense directness. The still chain-smoking Dickie Thompson had earned his outspokenness, having served as a cameraman in legendary director George Stevens's motion picture photography unit during World War II. He had, after all, been on hand to capture the liberation of Paris and the fall of Berlin on film. The much younger Gelzinis naturally would never have dreamed of telling the senior photographer how to behave, or for that matter how to dress, on an interview. So when the *Herald* duo—Thompson wearing a rumpled sweater and pants with a cigarette hole in them—entered Kerry's office, the photographer spoke his mind. In reference to the interviewee's April 1971 appearance before the Senate Foreign Relations Committee on behalf of VVAW, Thompson said, "Even if you live to be a hundred years old, you will never see that moment of glory again. Never. You had the whole country down on its knees." Then, making his hand into a fist for emphasis, he added: "You stood Congress on its ear. You were a hero, then. Now, you've become a politician."

It was a tough note on which to start an interview, but Kerry didn't flinch from this broadside from a veteran twice his age. "How many heroes do you know who are still alive?" he shot back. "At some point, we all have to grow up. We all have to make choices, and I didn't want to live the war for the rest of my life. It was never my intention to become a professional veteran."

Kerry, who had been spearheading state efforts to curb crime, had also clearly been trying to carve out a meaningful post-Vietnam role for himself in public service. Under his leadership the district attorney's office had grown from fifteen part-time lawyers to over one hundred. But Thompson's blunt remarks struck a chord in Kerry. He knew perfectly well that no matter how many drug dealers he put away or federal grants he procured for the

Commonwealth of Massachusetts, he would still be known as the Vietnam vet who had wondered out loud, "How do you ask a man to be the last man to die for a mistake?" As an awkward silence fell over his office that January day nearly seven years later, Kerry quickly moved to fill it. "I'd prefer to think of myself as a political activist, rather than a politician," he said. "I don't see myself as a politician, do you?"

In the resulting article Gelzinis noted, "Those 'ghosts' or 'shadows' hovering about John Kerry simply can't be ignored." After briefly recounting that his father had been an Eisenhower-era career diplomat and how enamored the young Kerry had been of John F. Kennedy, Gelzinis revealed that he had pushed his subject to defend Camelot. "All the mistakes notwithstanding," Kerry had replied, "I think the Kennedy Administration symbolized a degree of commitment and purpose that was very exciting." After that, he detailed how much he enjoyed living in Newton, raising two girls, playing hockey once a week, and flying rented biplanes at a local airstrip whenever he got the chance. Kerry also admitted that he worked twelve to fourteen hours a day and felt not the least bit of shame at harboring political ambitions.

What did perturb Kerry was the way journalists tended to paint his desire for a career in public service as self-serving and his politics as too radical. Instead, he insisted that even during his VVAW days he had been operating out of his deep love for his country. Kerry felt that his protests—like those of earlier Massachusetts rebels such as James Otis, John Adams, and Paul Revere—sprang from his ardent belief that American-style democracy remained the best possible form of government. "I was never outside the system," Kerry maintained to Gelzinis. "Working with the Vets, I spoke at about 60 colleges and in just about every state. The message was always the same—that you should try and commit yourself to the system and to making it work. I never advocated going outside the system. Even submitting to arrests in a passive demonstration was done to make a statement within the establishment."

Clearly the pair from the *Boston Herald American* had pushed Kerry's buttons. At that point in the interview he became quite animated, wanting to get the record straight. He knew where his antiwar activism had come from—it came from his anger at what he had seen firsthand of the disastrous effects of U.S. policy in Vietnam. His journals proved it, full as they were of his thoughts on the absurdity and pointless agony of the war as he was fighting it. "If I had been calculating, I would have kept my mouth shut when I got out of the service and run for office on my record," Kerry protested to Gelzinis and Thompson. "There I was, a decorated veteran, a Yale graduate—I could have gone the traditional road and probably been in Washington now. You know, not too many people question the ambition of the folks who run the mammoth conglomerates, or the ambition of an actor who wants to make a lot of money, or a writer who wants to be a success. But when ambition is applied to the political realm, it suddenly becomes evil. You become suspect."

At the end of his article, Gelzinis recalled something Kerry had said in reply to Thompson's pointing out that his April 22, 1971, speech before the Senate Foreign Relations Committee had marked the highlight of his career. "You're right," Kerry had noted. "I may never experience another moment like that again. But that doesn't mean I can't try."

EPILOGUE

September 2, 2003 (Charleston, South Carolina)

Vietnam was the most divisive time of battle in our country since the Civil War. It was the third most pivotal experience in the century—following the Depression and World War II. Its consequences are still being felt in our foreign policy, our troubled economy, in a haunted generation, in the new generation faced with possible new Vietnams, and in our hearts and minds. And yet because we lost, many refuse to face its monumental importance.

—MYRA MACPHERSON, *Long Time Passing* (1984)

The Vietnam War was different from any the United States had fought before. Not only did much of the nation come to see the conflict as a cruel and unnecessary war of aggression, but then the United States lost the peace. As a result, the veterans who fought in Southeast Asia found themselves as unwelcome back home as they had felt in-country. That shared sense brought them even closer together than the mutual rotten experience that no one who hadn't been there with them in those steamy jungles or river deltas could ever fully understand. "Why did we come back and our brothers didn't?" remained the guilty refrain that tortured the war's survivors, observed former U.S. Senator Max Cleland. "Vietnam took the real measure of all three and a half million of us. We may not have served together in the same unit or been over there at the same time, but we understand each other in profound psychological ways. I'd give my last remaining arm to any Vietnam brother." That may sound incredible coming from a man who had lived with only one limb for decades, even one who nevertheless retained a genuinely warm and sunny disposition as well as a sharp intellect and a great sense of humor. But the singular Cleland was hardly alone in his sincere empathy for all his Vietnam-veteran "brothers."

Late in the summer of 2003 the fifty-nine-year-old junior U.S. senator from Massachusetts, John Kerry, sat at a desk in the study at his house high atop Boston's Beacon Hill, riffling through his Vietnam War files. He was searching for the long statement he had written for a memorial service held for an old Swift boat crewmate who died in 1997. Kerry and Chelmsford native Thomas Belodeau had become friends serving together in Vietnam aboard PCF-94. This was in part because the officer in charge and his young forward gunner shared a Massachusetts heritage. In addition, they and their other crewmates shared a powerful bond forged in firefights on the rivers of the Mekong Delta. "You have to understand that we lived together as closely and as intensely on fifty feet of floating armament as men can live," Kerry noted. "And we learned all there is to learn about each other." At that, he pointed above his desk to a frayed American flag tattered with bullet holes. It was the one that had fluttered from PCF-94 over Kerry and his crew through all the Viet Cong attacks they had survived together on the narrow waterways of South Vietnam through the first three months of 1969. Belodeau had been the first of the Swift's mates to pass away. "I'm sorry he's not around for Charleston," Kerry said softly. "He'll be with us in spirit, though."

The "us" Kerry referred to were the rest of the men who had served under him on PCFs-44 and -94 during the Vietnam War. Collectively called the Crewmen by Senator Kerry's staff coordinator for veterans' affairs, they were about to converge on Boston from across the United States to lend their old skipper their support. For Kerry was about to officially announce his candidacy for the presidency of the United States in Charleston, South Carolina, on September 2, 2003.

Historian Stephen E. Ambrose titled his book about the men of Easy Company who served together in World War II with the 506th Regiment of the U.S. Army's 101st Airborne Division *Band of Brothers*—a phrase that applied just as aptly to the officer and crews on both of Kerry's Swift boats in Vietnam. The men he had fought alongside were a widely diverse bunch, in background, education, interests, and, of course, politics. Some were conservatives who would vote for Ronald Reagan and George W. Bush; others were liberals who favored Jimmy Carter and Al Gore, while still others remained staunchly independent of ideological or party labels. About half of these men were aghast when their onetime officer in charge spoke out as a leader of Vietnam Veterans Against the War in 1970 and 1971. The other half, however, felt proud of their former skipper for spelling out the Vietnam War's ugly truths before the Senate Foreign Relations Committee. Over the decades, however, these ex-crewmates never let their political differences get in the way of their fellowship. Virtually all of the men who served on Kerry's Swift boats backed his bid for the White House. "We all remembered what a superb, first-rate, tough and fair commanding officer he was," explained PCF-94's helmsman Del Sandusky. "We knew the real John Kerry, the man who saved our lives and made us proud, in a profound way. That's something many of the shitheads in the press simply don't get." Kerry's second Swift's engineman, Gene Thorson, agreed, calling his lieutenant (j.g.) a "hard charger" who used to reassure his men that, "We're going to come out winners."

For more than a year before the fall of 2003, their former skipper had been acting like a presidential candidate, raising funds around the country and spending an unusual amount of time in the crucial early primary and caucus states of Iowa, New Hampshire, and South Carolina. Thus the news that he would formally announce his quest for the 2004 Democratic nomination came as no surprise. Only slightly more unexpected was the word that John Kerry wanted all his Swift crewmen to come to Charleston with him. Only one of them—Stephen Hatch—turned down his invitation, on the grounds that he was "apolitical." Yet he told the press that Kerry was "his man" for president. Bill Zaladonis, PCF-44's engineman, was unable to make the trip; both of his legs and one of his wrists had been broken in a recent Harley-Davidson accident, and he had to stay put to mend. "I got stuck in an Orlando nursing home," Zaladonis complained. "It ticked me off. I consider John a good friend. But I'll be at the next big event, no matter where or when."

Kerry's staff arranged for most of the rest of his old Navy buddies to gather at Boston's Logan Airport. A chartered plane would take them to South Carolina for the announcement, and then on to Iowa and New Hampshire for what they called "the grand tour" of Kerry's first official campaign appearances. On September 1, they began arriving in Boston in clusters—first Del Sandusky of Elgin and James Wasser of Kankakee, Illinois, followed by Fred Short and Drew Whitlow of Arkansas, and Gene Thorson of Iowa. While they waited for their flight south together in a conference room at Logan's charter terminal, the Vietnam vets discussed presidential campaign strategy with Kerry's staff. They already knew and were eager to play their role: in the words of the late Warren Zevon's ballad, they would be Kerry's Vietnam War "witnesses," and he their "mutineer." "We were ready to help John any way we could," Sandusky asserted. "We felt that by just being with him on the campaign trail would speak to his outstanding character. We'd follow him anywhere, anytime."

Following their Boston briefing, at dusk Kerry's crewmates were flown to Washington, D.C., where they stopped to pick up Michael Medeiros, en route to join them from California. Then they took off for Charleston, South Carolina, where the next morning they would stand next to Kerry on a stage set up in front of the celebrated Essex-class aircraft carrier U.S.S. *Yorktown*, which had earned eleven battle stars in World War II and another four in Vietnam, and give their old skipper their stage-side thumbs-up for all of America to see as a local military band blared out "Anchors Aweigh." Kerry's campaign staff handed each of the Crewmen a matching navy-blue polo shirt with "PCF-44" or "PCF-94" embroidered in yellow on the breast pocket.

Although his 2004 presidential campaign kickoff marked the first time Kerry had gathered his wartime troops for a group, first-class-all-the-way campaign junket, it was not the first time the Massachusetts Democrat had turned to Vietnam veterans for help in the political arena. Back in 1984, in fact, when then Lieutenant Governor Kerry hit a bump in his road to the U.S. Senate, it had been a group of veterans from his home state who catapulted him over the obstacles to victory. Kerry's Democratic primary opponent, U.S. Representative James M. Shannon—a smart, popular protégé of then Speaker of the House Thomas P. "Tip" O'Neill Jr.—had tried to challenge his military record and subsequent antiwar activism. In a debate, Shannon in effect called his opponent a hypocrite for having protested a war he had fought in, and vice versa. It proved a mistake. As the *New Yorker* political writer Joe Klein explained, "The next day, Kerry headquarters was deluged with calls from infuriated veterans. Shannon hadn't fought in Vietnam. They hadn't been so lucky—and they hadn't 'chosen' to go to war, either. In their final debate, Kerry asked for an apology, and Shannon said, 'That dog won't hunt.'"

That did it. By refusing to apologize for having insulted so many of America's Vietnam veterans, Shannon inadvertently made his situation worse. A dozen Massachusetts vets, among them Kerry's old crewmate Tom Belodeau, immediately sprang to action. Dubbing themselves "The Doghunters," they rallied to Kerry's defense. What mattered was that Kerry was a "brother"; going to Vietnam had made him one of them, Yale education and all. His Brahmin family's connections could easily have kept him safe at home, but when his country had called him to duty Kerry had shown the character to respond by enlisting in the Navy for combat assignment. Others who had gone in-country knew all that meant, and acted on it. "The Doghunters were crucially important to Kerry winning," recalled his 1984 campaign manager, John Marttila. "For the last two weeks they brought incredible passion, energy, and commitment to John's cause. They had felt insulted by Shannon. They validated John. We brought them by helicopter into every media market we could."

Individually, the Doghunters boasted unimpeachable credentials, such as Jim McDevitt, an ex-Marine who had earned four Purple Hearts, and Doug Clifford, an Air Force photographer who had become active in VVAW. "I had turned sour on Vietnam back in 1969 when I was in-country," Clifford recalled. "It was obvious that we had little support among the population. But I came back home to Watertown and moved on. I had two kids. Then this guy Shannon puts us down with his 'that dog won't hunt' bunk. It was an insult to every Vietnam vet. It was an insult to what we had done after the war. It made us realize we had a direct stake in backing Kerry. Our very essence was on the line. We weren't going to let him step on us like that."

Another of the top Doghunters, John Hurley, had befriended Kerry early in 1970. A veteran of the U.S. Army's 69th Engineers operating in the Mekong Delta, Hurley had put in his yearlong tour of duty in 1967–68 and come home distraught and disaffected. What he had seen in Vietnam, including atrocities, had convinced him that U.S. forces had to get out of Southeast Asia, and fast. Hurley entered Boston College Law School and there sought out Father Robert Drinan, who got him involved in various antiwar protests. "When I met John Kerry we became immediate friends," Hurley remembered. "We shared a common interest in politics and veterans' rights." Although Hurley never joined VVAW, he did attend its Dewey

Canyon III demonstration in Washington, D.C., and agreed with most of the group's positions. So when Jim Shannon slurred his fellow antiwar Vietnam veterans in 1984, Hurley was eager to go after the other Democrat. "What he said demeaned veterans' service," the Boston lawyer recalled. "The Doghunters took the fight right into his face. Kerry had been down by ten points in the polls when Shannon claimed service had been optional. By the time us veterans were done with him, he had lost the election."

Ever since Kerry went on to win that hard-fought U.S. Senate primary, and the general election that November against Republican Raymond Shamie, more fellow Vietnam veterans had rallied around his political career. They were with him again in 1990, when Kerry held on to his Senate seat against GOP challenger Jim Rappaport. And in Kerry's toughest reelection campaign—his 1996 battle against Massachusetts Governor William Weld—no less than retired four-star U.S. Navy Admiral Elmo R. Zumwalt Jr. himself joined the Doghunters to support his former brown-water lieutenant at a rally at Boston's Charlestown Navy Yard. Once Shannon's slur on their service got the Massachusetts vets stirred up behind Kerry in 1984, the Doghunters stayed with him, playing a principal role in getting out the vote to reelect their senator in 1990, 1996, and 2002. "Don't mess with the brothers," John Hurley summed up. "We have our own disagreements, but when somebody who never served in Vietnam questions those who did—look out."

Yet most of the Doghunters understood that everyone who had to decide what to do about Vietnam from 1963 to 1973 was, in truth, a veteran of the conflict. Vietnam was different than other wars. Even those who spent the war years protesting the Cambodian incursions or the Christmas bombings were veterans. "People noted that I was a Medal of Honor winner," recalled former senator Bob Kerrey. "But there should have been medal winners for organizers of the moratoriums or peaceful protests. It's a mistake to look at Vietnam like World War II. Our generation was forced to make decisions that other generations didn't have to. There was no ambiguity in World War II. I consider conscientious objectors and draft dodgers Vietnam veterans. But that doesn't mean those who served in uniform don't have a special affinity for each other. We do."

That bond was underscored in April 2001, when Gregory L. Vistica, an investigative reporter for the *New York Times Magazine*, wrote an article alleging that Bob Kerrey had led his Navy SEAL unit to kill thirteen unarmed civilians in the village of Thanh Phong on the night of February 25, 1969. Kerrey's fellow Vietnam vets in the U.S. Senate immediately came to his aid. Joining Max Cleland in defending Kerrey's war record on ABC's *This Week*, John Kerry—as if reading from his old combat diaries—said, "In Vietnam, the civilians were often the combatants. A twelve-year-old could walk up to a café, and did, and lobbed a grenade into that café and blew up people. . . . This was a war where the combatants were completely confused with civilians." In addition, Senators Kerry, Cleland, and Chuck Hagel penned a letter published in the *Washington Post* that read in part: "Bob Kerrey's personal and difficult disclosure last week demonstrates the courage we all have known in him for years. It also reveals the very real guilt and pain that persists among combat veterans of all war, and particularly Vietnam." John Kerry simply refused to let the media excoriate his colleague, who lost his right leg below the knee in combat. "People that criticized Bob just didn't understand the realities of Vietnam," Kerry declared. "His distinguished record speaks for itself. He was braver than brave."

As deep as the mutual fealty ran among America's Vietnam veterans in general, the bond that tied the men who served in the tiny Mekong Delta brown-water navy remained even tighter. Beginning in 1995 with a reunion dubbed "The Last River Run," held in Washington, D.C., many Swift boat veterans took to gathering every couple of years at various locations around the country to catch up and reminisce. A number of the men always brought along their old scrapbooks, full of snapshots of their lives at Danang or Cat Lo or An Thoi, and of their treks up the delta rivers to wipe out Viet Cong camps. Where World War II veterans' reunions still swung to big bands and the Andrews Sisters warbling "Boogie Woogie Bugle Boy," the 2003 Swift crew gathering in Norfolk, Virginia, set the Sheraton Waterside Hotel's banquet room rocking to the more antiwar refrains of Buffalo Springfield's "For What It's Worth," the Byrds' "Draft Morning," Jimmy Cliff's "Vietnam, Vietnam, Vietnam," Creedence Clearwater Revival's "Who'll Stop the Rain," and an obscure bootleg recording of Bob Dylan's "Hero Blues," in which a boy responds to his girl's urging him to go to war so she can boast of his bravery: "Well, when I'm dead / No more good times will I crave / You can stand and shout 'hero' / All over my lonesome grave."

Since 1996, when PCF veteran Larry Wasikowski put up his Swift boat website, many of the men who had served on them began educating themselves on the history of the brown-water navy and reconnecting with one another. Before long, Wasikowski's site boasted some 2,300 Swift vets' names along with various bits of available information such as their in-country call handles, tours of duty dates, and current addresses—and the roster kept growing via e-mailed word of mouth. (Every week Wasikowski tries to track down the approximately 1,000 Swifties not yet listed on his Web site.) The good news was, the Swift boats' story had not been forgotten. Articles in *Vietnam* magazine, Francis Ford Coppola's 1979 movie *Apocalypse Now*, and John Kerry's political career all brought public attention to the riverine part of the Vietnam War. In 1996, an updated edition of historian Thomas J. Cutler's acclaimed 1988 book *Brown Water, Black Berets*, which dealt in part with Operation Sealords, was released by the Naval Institute Press. That same year, former Swift skipper Jim Kolbe of Coastal Surveillance Task Force 115 won Arizona's Fifth District seat in the U.S. Congress, giving the Republican veteran a platform to further educate the public about the brown-water navy. In an oral history recorded by the Naval Historical Foundation, for example, Kolbe discussed the downsides of Operation Sealords, particularly working for Captain Roy Hoffman, whom he described as "an absolute mirror image" of Robert Duvall's brutish character in *Apocalypse Now*, who said he liked the "sweet smell of napalm in the morning."

John Kerry thoroughly enjoyed catching up with his fellow Swift veterans at the Norfolk reunion in March 2003. The George W. Bush administration's war on Iraq had just been launched, with the Norfolk area's *Virginian Pilot* headline for Saturday, March 22, blaring, "Ferocious Assault: Massive Air Attack on Iraqi Capital Comes as U.S. Troops Advance from South." Like most Americans, the assembled Vietnam War vets were hoping Saddam Hussein's Baghdad would be captured. Even the war skeptics among the veterans were cheering on the U.S. armed forces. After all, the youngsters they saw on the television screen had once been them. To the delight of everybody, Roy Hoffman and George Elliott—both strongly prowar—attended this reunion. Nobody held any visible grudges about the way Operation Sealords had been conducted—old wounds had been healed. Admiral Zumwalt's widow, Mouza, was also there; she was, however, concerned about the global ramifications of the U.S. preemptive strike against

Iraq. A real highlight of the weekend, which Kerry relished, was riding a Swift boat—PCF-2—in the Norfolk harbor for approximately thirty minutes. This Swift boat, one of only three in the United States, was the property of the local Tidewater Community College.* The last time Kerry had been on a PCF was in March 1969. "Boy, did that bring back memories," he exclaimed upon returning to the dock. "Memories just came rushing through me when I heard the engines roar."

Kerry could not help noticing at the reunion, however, how old his Vietnam buddies were getting. Many of them had already been struck by a medical ailment of one kind or another. Kerry himself had his prostate taken out just weeks earlier due to cancer. Even his PCF-44 boatswain's mate, Drew Whitlow, who was one of the younger veterans there, was now fifty-eight and had five children and five grandchildren. Like the other members of Kerry's two Swift crews as they headed into senior citizenship and further on with their lives, to Whitlow Vietnam was simply an experience he had endured long ago. "I just carried on after Vietnam," he explained. "I never let what happened over there catch up to me." In Norfolk, in addition to passing around the Polaroids they had taken of one another on the Delta rivers and the Ca Mau Peninsula, the Swift veterans proved even prouder to pull out fresh photos of their grandchildren for all to admire.

Kerry also had plenty of personal updates to report, and much to be proud of, beginning with his two adult daughters: twenty-nine-year-old Alexandra, a Los Angeles filmmaker, and twenty-six-year-old Vanessa, a medical student at Harvard. Although neither attended the Norfolk event, both would be accompanying their father in South Carolina when he made his momentous announcement for the presidency. Julia Thorne—who after her 1988 divorce from John Kerry had remarried, moved to Montana, and in 1996 written *You Are Not Alone*, a guidebook through depression—called her ex-husband to wish him well in Norfolk. "He's always at his best when he's with those guys," she said. "They make him capable. He really loves them all in a profound way."

**PCF-1 is housed at the Washington Naval Yard Museum while PCF-104 is at Cism Field on the Naval Amphibious Base in Coronado. In addition, two other Swifts have somehow managed to make it into Malta's defense forces.*

Kerry's second wife, Teresa Heinz—whom he had married in May 1995, four years after her first husband, Republican Senator John Heinz III, had been killed in a plane crash—agreed that the Doghunters and the Crewmen invigorated her second husband's psyche. She had seen it during the last few months crisscrossing America on behalf of his presidential campaign. Born Teresa Simões-Ferreira in Mozambique to Portuguese parents, the outspoken Teresa Heinz had inherited her late first husband's food-industry family fortune, and since his death in 1991 had worked as chair of the Heinz Endowments and the Heinz Family Philanthropies. She and Kerry had met in 1992 at an Earth Summit conference in Rio de Janeiro. A few months later, they ran into each other again at a Washington, D.C., dinner party, at the end of which Kerry had invited her to take a walk with him around the National Mall. "We ended up at the Vietnam Veterans Memorial*," Heinz recalled. "We looked up names of friends he had lost."

On the last day of the Norfolk reunion, Don Droz's captivating daughter, Tracy Droz Tragos, premiered her documentary *Be Good, Smile Pretty.*† She had been only three months old when PCF-43 had been ambushed, killing her father. Tragos was determined to honor her father, to not forget what had happened to him on the Duong Keo River. In the process of honoring his memory on film, she hoped to make a collective eulogy to *all* the children of GI fathers killed in Vietnam. From her home in Topanga, California, she started interviewing men who had known her father in Vietnam; Kerry was at the top of the list. Tragos had gotten to know Kerry fairly well because she had worked in his Senate office as an intern for a year. "Don and I went there to win. . . . We were gung ho," he told Tragos. "Regrettably, we found a difficult situation."

The title of Tragos's first-person film came from the way her father would sign off his letters home from Vietnam. Because of her obvious talent and infectious earnestness, the entire Swift boat community embraced her project. After receiving a grant from the Independent Television Service, Tragos formed the Orphans of War Foundation, an outreach program to the

---

**The Vietnam Veterans Memorial, better known as the Wall, was dedicated on November 13, 1982. Designed by Maya Lin, a twenty-one-year-old Yale architecture student, the memorial has become the most visited historic site in Washington, D.C.*

†"Brothers in Arms, *a documentary by Paul Alexander focusing on Kerry, the Crewmen, and their relationships since Vietnam, was screened at film festivals and on public television stations in the fall of 2003.*

estimated 20,000 U.S. children whose fathers died in Vietnam. You could hear a pin drop in the Norfolk conference room as the hourlong documentary played. When it was over, they gave Tracy a standing ovation. Then the Swifties—including Kerry—lined up to give her a hug. "Tracy just knocked me out," Kerry recalled. "Her film has such richness and beauty and sadness. Just seeing photos of Don on the screen and then looking across the conference at Tracy choked me up. I was devastated, but in a positive sort of way."

The Norfolk Swift-crew gathering had been a great success. It set the stage for the next one, six months later in Charleston, which was not, obviously, just another reunion for the crews of PCFs-44 and -94. Their skipper didn't have the luxury of taking a Swift boat on a harbor cruise or wandering down memory lane. He was declaring himself a candidate for president of the United States, and every one of his crewmates felt honored to be a part of the event. Kerry's good friend and fellow PCF officer in charge Skip Barker also made his way to Charleston from Selma, Alabama, for the occasion. That night a few of the Swift veterans stayed up late downing beers at the Hilton Charleston Harbor Resort bar, smoking cigarettes on the terrace, and swapping Vietnam reminiscences. The next morning at breakfast they discussed how much Vietnam had changed since they had been there at the end of the 1960s conducting high-speed river raids for twenty-four to thirty-six hours at a stretch. After South Vietnam fell to the communist North in April 1975, its capital's name had been changed from Saigon to Ho Chi Minh City, but one would never have known it listening to the Swift veterans as they spoke a little sadly of how nearly all the boats they had served on had been either retired or given away long ago to the now defunct South Vietnamese navy.

Without exception, the erstwhile riverine warriors expressed their hopes that capitalism would continue to flourish in Southeast Asia. The region's economic indicators looked positive, as did their political repercussions after the United States had ended its nineteen-year trade embargo on reunified Vietnam in February 1994, followed seventeen months later by America's extension of full diplomatic recognition to the socialist republic. Kerry had shown extraordinary zeal and confidence in helping lead the effort to make this historic reconciliation happen. "It was less about legislation and more about a lot of quiet diplomacy on the part of Senators Kerry and McCain," recalled Kerry foreign policy advisor Nancy Stetson, "including a lot of time in the Oval Office with President Clinton." For their efforts, the bipartisan team of Kerry and McCain received the 2001 Christian A. Herter Memorial Award, presented annually by the World Affairs Council of Boston to "distinguished leaders who have contributed significantly to international understanding." At the award ceremony former ambassador to Vietnam Douglas "Pete" Peterson praised the veteran senators: "Through the administrations of four presidents I have seen how these colleagues of mine . . . have guided this reconciliation process," he said. "With the aid of John Kerry and John McCain, the people of Vietnam now have a chance for a better life."

Kerry had seen working toward normalized relations with Vietnam as the final act in of his nearly thirty-year-long tour of duty. And it was paying off. According to the *New York Times*, over the course of the 1990s Vietnam's poverty rate fell from 70 to 30 percent, and the country became the world's second largest exporter of rice. Coffee production rose to the point that Starbucks stores across the United States started selling a Vietnamese blend. Some of the Mekong Delta's rivers, meanwhile, were turned into catfish-breeding farms that helped bring prosperity, if slowly, to both Vietnam and the other five countries and 250 million people that shared the Mekong River basin: Cambodia, China, Laos, Myanmar, and Thailand. The opening of Vietnam's first stock exchange in July 2000 was greeted with enthusiasm across the region. Although none of Kerry's Swift crewmen had gone back to see reunified Vietnam's recent revitalization, they were delighted to hear of it and proud that their skipper's postwar visits to the country had played a large role in Vietnam's economic resurgence in the wake of the normalized U.S. relations he and John McCain had largely brought about.

In truth, John Kerry may well have worked harder on the normalization of relations with Vietnam than any other U.S. veteran who had fought there. Starting in 1991, when Senate Majority Leader George Mitchell of Maine asked him to investigate the status of unaccounted-for POWs/MIAs in Southeast Asia, Kerry had set himself to studying every aspect of the war he had so despised. Most of the other Vietnam veterans in the Senate at the time—John McCain, Bob Kerrey, Democrats Charles S. Robb of Virginia and Al Gore of Tennessee, and Republicans Hank Brown of Colorado and

Bob Smith of New Hampshire—thought the project a difficult and sensitive job spawned, in part, by a bogus *Newsweek* cover story that promulgated the widespread belief that American POWs were still alive and captive in the former North Vietnam. Feisty Texas billionaire H. Ross Perot fanned the flames by traveling around the country haranguing the George H. W. Bush administration for forgetting about these "men who were left behind." Its obvious pitfalls aside, Kerry took on the job with zeal. "Ross was on the other side," McCain recalled, "that they still had a bunch of our men in caves in Laos and all of that. There was a real estrangement that had occurred between Ross and me for a long period of time. We just had a different opinion of the POW/MIA situation." Working closely with General John Vessey, a retired former chairman of the Joint Chiefs of Staff, Kerry and McCain were able to get over a million pages of sensitive documents declassified. "Every day I was reliving Vietnam," Kerry recalled. "I learned the family histories of hundreds of our POWs. It was haunting stuff."

It was in the course of this investigation that Kerry and McCain became close. The Arizona Republican had retired from the U.S. Navy as a captain in 1981; both his father and grandfather had been four-star admirals. A great admirer of Ronald Reagan, John McCain had won his Senate seat in 1986, replacing the venerable conservative Barry Goldwater. At first John Kerry was not, to put it mildly, his favorite colleague. McCain remained put off by the fact that in 1971, while he was languishing in a Hanoi prisoner-of-war camp, Kerry had been leading his VVAW cohorts in throwing their combat medals onto the U.S. Capitol grounds. In fact, McCain had spent five and a half years as a POW, much of it in solitary confinement. Yet the two men found some common ground when they sat next to each other during a long flight to the Persian Gulf region on a 1991 fact-finding mission. "We talked about Vietnam and shared our perspectives," McCain remembered. "We found we had a great deal in common." For his part, Kerry felt in awe of his Republican colleague. "He was my hero," Kerry averred. "Just think of what this man went through. Just read his memoir, *Faith of My Fathers*. We all look small, all of us, in comparison to John McCain. He is, to my mind, patriotism personified."

With freshman Senator Bob Smith serving as Republican vice chairman of the POW/MIA investigation, the Kerry committee would once and for all put an end to the issue. While studying the matter, Kerry made fourteen trips to Vietnam. Each one was a journey of discovery. Much about the country surprised him, not the least that the government in Ho Chi Minh City had started a War Crimes Museum in which the My Lai massacre was documented and the effects of Agent Orange were examined. It was truly surreal to be back, traveling around the countryside looking for lost comrades. Small children rushed up to him, glad to see an American. "The people have largely put the war behind them," Kerry noted. "They've been invaded for a thousand years and they somehow don't hold grudges."

An entire book could be written about Kerry's journeys back to Southeast Asia between 1991 and 1993. His first trip, in May 1991, started in Cambodia. He arrived, along with a few staffers, in Phnom Penh late one evening. The capital was cloaked in ethereal darkness. There was no traffic, just the eerie dim of a few isolated streetlights. At the time U.S. relations with Cambodia were for all practical purposes nonexistent. Even the United Nations had limited interaction with the Cambodian regime. What Kerry encountered as he toured the countryside saddened him: a notorious "killing field" where human skeletons were piled high in a macabre monument; a Khmer Rogue torture chamber constructed in an old elementary school, where bare-iron beds were caked with dried blood; and a filthy hospital where surgeons operated using tarnished instruments. Overcome by disgust, Kerry nevertheless proceeded to have a six-hour dinner conversation with Cambodian Prime Minister Hun Sen. A former Khmer Rouge officer who had lost an eye during fighting in 1974, Sen had eventually turned against the Pol Pot regime when he saw the magnitude of its genocidal aims. He had fled to Vietnam for safety but was quickly imprisoned. But when Vietnam invaded Cambodia, and Pol Pot was ousted, Sen was welcomed home with open arms. At twenty-seven, Sen was named the foreign minister, the youngest in the Pacific Rim. "It was all overwhelming," Richard Kessler, a Senate Foreign Relations Committee staffer who accompanied Kerry on the trip, recalled. "At that time Cambodia was eager to make a distinction that it wasn't tied in tandem to Vietnam. Sen claimed Cambodia was an independent, pro-American nation. He wanted to try to get the U.S. economic embargo lifted against his country. And he wanted the U.S. to end its military support of the non-Communist resistance in Cambodia." To prove his "pro-Americanism," Sen ended up sending his son to West Point and his daughter to school in Long Island.

From Cambodia Kerry flew directly to Hanoi. As they prepared to land, Bud Orr, a U.S. officer escorting Kerry and a Vietnam veteran, pointed out the various bridges the United States tried to bomb during the war and missed. Much like Phnom Penh, there were no decent hotels in town, so Kerry stayed at the Foreign Ministry Guest House. With a rattling portable air-conditioner, insatiable bed bugs, and fat mosquitoes flying around, his quarters were an uncomfortable operations base. Accompanying Kerry around Hanoi was Frances Zwenig, a Peace Corps veteran whom he had first met back in 1970 when running for Congress against Father Drinan. Smart, quick-witted, and deeply committed to reestablishing U.S. diplomatic relations with Vietnam, in Zwenig's opinion the war was not officially over. "We still had a lot of unfinished business to do," she recalled. "Until we cleared up the POW/MIA issue it was impossible for the Bush administration to embrace Vietnam in any way. And the Republican right, including Bob Smith, were convinced, wrongly so, that American POWs were still being held captive."

Kerry and Zwenig decided that the first step toward reestablishing diplomatic relations with Vietnam would be to initiate a student exchange program. An excited Kerry soon sponsored a Senate bill to fund the educational exchange, deciding upon a "sailable angle" skeptical conservatives in the U.S. Senate couldn't resist. "The Vietnamese students would come to America to study law and free-market economics," Kessler recalled. "It worked. The Republicans all agreed to push that forward." Vietnamese students started coming to America.

As Zwenig tells it that first trip to Cambodia-Vietnam "whetted" Kerry's "attitude and appetite." Within hours of returning to the United States, sitting in Los Angeles Airport, Kerry started organizing small groups of veterans to head back to Vietnam on goodwill missions. "He knew the way back was through the vets," she recalled. "That would help us with Bob Smith and other conservatives." An emotional Kerry had dedicated his 1971 book *The New Soldier* to the survivors of the "Indochina War," in hope that "forgiveness will ease the pain and understanding" to produce "a lasting peace." In the aftermath of the Gulf War, which Kerry voted against, President George Bush declared the "Vietnam syndrome" over. But Kerry believed it was not. Only until the thorny POW/MIA issue, followed by normalization of relations ensued, could the history books be closed in a meaningful way.

So the shuttle diplomacy continued. On a later visit, he got to visit An Thoi, his old stomping ground. It was a beautiful afternoon, the blue water sparkling in the sun. It was hard to imagine this placid place was once the vortex of Operation Sealords. "There had been some reports of live sightings out there," Kerry recalled. "So we went out just to talk to officials. I didn't get the chance to walk around and do all the things I would have liked to have done. But I actually found some shells in the ground in the same grating where we stood back in '68–'69. And I landed on the exact same grating that the Americans had built as an airport—it was fascinating."

The most difficult part of Kerry's job was pronouncing *definitively* that no American POWs were being held captive in Vietnam. Every time Bob Smith's Senate office got a lead about a POW, no matter how questionable, Kerry was forced to check it out. Largely because he was a famous antiwar spokesperson in the early 1970s and was therefore deemed honorable, the Vietnamese government opened all of its facilities to Kerry. The most difficult rumor to bat away was that U.S. servicemen were being held hostage under Ho Chi Minh's tomb in Hanoi. Built from 1969 to 1975 the shrine—modeled after Lenin's tomb in Moscow—was sacred to Vietnamese people. But Reuters news service had run two stories in August 1992 claiming there were U.S. POWs being held captive in the underground tunnels of Hanoi near the tomb. The most convincing of these was an interview with the Russian ambassador to Vietnam, who said he believed these grim rumors to be true. So Kerry, accompanied by Bob Smith, insisted that they be permitted to conduct an on-the-spot investigation. The two senators found themselves being escorted all over the dank subterranean underground of Hanoi. "It was weird," Kerry recalled. "We were forced to keep it quiet. But there we were inspecting these musty catacombs and crazy tunnels." Together they found no evidence to substantiate the Reuters claims—and Smith's suspicions were placated. "That was the big hurdle," Kerry recalled. "By giving us that kind of special access, it proved they weren't hiding anything. There were no POWs under Ho Chi Minh's tomb. Bob Smith was starting to come around."

Another dramatic incident occurred when Kerry went back to Vietnam with McCain. C-SPAN founder Brian Lamb went with them to film the weeklong pilgrimage. Back in May 1973, Kerry had read a first-person account of McCain's POW years in *U.S. News & World Report*. He had been blown away. When he had endured SERE training in the California desert, including a short spell in a solitary cell, he almost lost his mind. McCain had somehow managed years of such inhuman treatment at Hoa Lo prison, derided by U.S. soldiers as the "Hanoi Hilton." Now Kerry was standing with McCain at the solitary confinement cell where he had tapped out signals to fellow POWs. His North Vietnamese tormentors broke both of his arms. "Just to stand there alone in this tiny cell with McCain, just to look at this guy who was now a United States senator, and my friend, in the very place where he'd been tortured, and kept for so many years, not knowing if he might live," Kerry recalled to Joe Klein of *The New Yorker*, "we found this common ground in this far off place."

After eighteen months of hard work, the Kerry Committee issued a massive, 1,223-page report. "Nobody had the patience that John had," Zwenig recalled. "He really wanted to broker a friendship between the United States and Vietnam." During this period Kerry had carefully studied photographs, analyzed reports, and tracked down rumors. As journalist Paul Alexander noted in *Man of the People: The Life of John McCain* (2002), Kerry, with McCain's constant support, tackled every allegation presented. Together they were walking a political minefield, careful not to make a misstep. All the Vietnam veterans in America were watching them with a close eye. There was, however, something extremely reassuring about the hawk (McCain) and the dove (Kerry) determined to find closure on the thorniest issue still remaining between the United States and Vietnam. "John handled that whole deal with the utmost skill, patience, understanding, and compassion," McCain summed up Kerry's work on the POW/MIA investigation. "The contribution that he made, in my view, was vital. It led the way to the normalization process. During the hearings he was far more under control than I was. I would hear some of these outrageous allegations of government cover-ups, and Henry Kissinger as a traitor and those types of things, and I'd go ballistic. But John handled it in a mature, measured, and extremely effective manner. At the end, he got every member of the committee to sign a report that said there was no compelling evidence that there were Americans alive in Southeast Asia. It was a remarkable feat."

In February 1994 the nineteen-year-old trade embargo was lifted. And at a July 11, 1995, ceremony, President Bill Clinton, with Kerry and McCain at his side, announced normalization of relations between the two old adversaries. "This moment offers us the opportunity to bind up our own wounds," Clinton declared. "They have resisted time for too long. We can move on to common ground. Whatever divided us before let us consign to the past." Out of all the days Kerry had spent as a U.S. senator, this was his finest moment.

By the time Kerry joined his war buddies for breakfast that early September 2, 2003, morning in Charleston, he had already put in several hours of major media campaigning. Unable to sleep on the eve of his big announcement, he had stopped trying at around five o'clock, when he began to prepare for his scheduled appearances on ABC's *Good Morning America*, the *CBS Morning News*, NBC's *Today*, and CNN's *American Morning*. Avoiding the studios at the network's local affiliates, Kerry gave the interviews by remote transmission from video cameras set up by his campaign staff to show the U.S.S. *Yorktown* gleaming behind the soon-to-be candidate. The morning's hectic preparations didn't allow Kerry and his crewmates much time together in the nautically decorated hotel, where junior campaign staffers scurried about pressing red, white, and blue Kerry for President buttons and bumper stickers into every warm hand they could find.

Kerry's senior advisor Robert Shrum—a veteran Democratic political consultant who had worked on earlier White House bids for Ted Kennedy and Al Gore—was seemingly everywhere, issuing instructions and fretting about everything from lighting to wheelchair ramps. He paused just long enough to explain that the theme of the day—and perhaps the campaign—was "patriotic populism," which the Crewmen from Kerry's Swift boats were to personify. Many of the reporters trickling in to cover Kerry's announcement saw the vets' presence as politicking, a jaded view Bob Shrum was quick to dismiss. "What people sometimes miss is that John really loves these guys," Shrum pointed out. "Their collective vouching for his leadership is authentic and powerful. And, with the exceptions of John McCain and Max Cleland, Kerry has been the greatest friend veterans from *all* wars have ever had in the U.S. Senate."

Just a week earlier, in an August 25 address to a VFW group in San Antonio, Kerry had sharply criticized the George W. Bush administration's handling of its war in Iraq. To underline his qualifications to argue with the

President's policies, the Democratic senator referred to his own military service in Vietnam as a prerequisite for holding the nation's highest elective office. He also used his veteran status to promise better treatment of those who had served. "I won't just bring to that profound responsibility the perspective of [one] sitting in the Situation Room," Kerry had intoned. "I'll bring the perspective of someone who's fought on the front lines." *New York Times* political correspondent David M. Halbfinger raised the question in his article the next day of just how important combat experience was to becoming commander in chief of the modern-day United States. Halbfinger's *Times* piece—written before retired Army General Wesley Clark entered the race—quoted Kerry's fellow Democratic Senator Chuck Hagel of Nebraska opining that, "It's better not to talk too much about your military record. You should probably play that experience down to some extent—not run away from it, but don't talk too much about it. The media will draw the comparisons and distinctions anyway."

In that spirit, as reporters peppered him with questions about Vietnam on the morning of September 2 in Charleston, Kerry kept trying to shift the spotlight from himself to the assembled Swift boat veterans who had served under him in Vietnam between November 1968 and March 1969. He also made glowing references to the Vietnam vets he had served with in the U.S. Senate: John McCain of Arizona; Hagel, who had served as a squad leader with the Army's Ninth Infantry Division; Hagel's fellow Nebraskan across the aisle Bob Kerrey, who had left the Senate in 2000; and Georgia Democrat Max Cleland, who lost both legs and his right arm to a grenade blast while serving in the Army in 1968, earning both a Bronze and a Silver Star, and whose reelection bid had failed in 2002. But just by virtue of mentioning these colleagues and having his old crews by his side, and by speaking in front of the bunting-bedecked *Yorktown*, Kerry appeared to be anchoring his presidential bid to his military record in Vietnam.

Retired Navy Lieutenant Bob Emory, who had served with Kerry as a Swift skipper in the Mekong Delta, wasn't the least bit surprised to learn that every one of the men who had served under him on PCFs-44 and -94 was backing Kerry's run for the White House. Emory simply believed that good enlisted men always continue to follow the good officers who had commanded them, even after their wars are over and everybody has gone home. He could easily understand how Kerry's men felt about him, even three and a half decades later; their officer in charge had brought them through combat alive, and the debt they felt they owed him for it certainly transcended politics. "I think they probably feel that if he is committed to serving the American people to the same extent that he was committed to them when they were riding in his boat, that he's going to once again do his very best in a non-self-serving way," Emory observed. "The other interesting perspective here is that in the military environment there is no question about the objective, the focus. Some people have grander expectations even at that time, but when they're in the combat zone there is only one objective to achieve. The question that I think starts to appear in people's minds is, 'To what degree does the political process sully the noble character of the guy who used to be in combat?' Because the objectives are so ill defined—to get elected so that you can implement your programs—it all seems like a kind of self-serving pursuit, whereas most people feel that leading men in combat is the exact opposite. That's what Kerry has got to deal with: a reputation in combat where everyone respected him and was proud to serve him and experienced his integrity. His commitment to his crew was obviously exceptional. But when he runs for president, the question is: Okay, now is his commitment to those he will serve, or is his commitment to getting elected? And it's a tough call."

On September 2, John Kerry's presidential campaign kickoff ceremony started with PCF-94's gunner's mate-turned-preacher making that call in the form of a prayer. Since leaving Vietnam, David M. Alston—now a chemical operator at a nuclear power plant as well as a Baptist minister in Columbia, South Carolina—had struggled to keep God in the forefront of his life. His success in doing so showed in the number of local African-American Baptists who respected his abilities as a preacher enough to come to offer Alston their support as he opened the nationally televised ceremony for his old skipper. "Down in the Mekong Delta, we lived together, we fought together, we bled together, and we survived together," Alston intoned from the star-spangled stage set up in front of the *Yorktown*. "Whether we were Democratic or Republican was not the issue. The issues at the time were trust, courage, judgment, and character." Just as they always were in a presidential campaign, he thereby suggested.

As Alston stepped back to his chair on the stage next to the rest of Kerry's blue-shirted crewmen, the candidate in the offing looked almost stricken with appreciation for the reverend's gracious benediction. In truth, Kerry had looked a tad embarrassed to be the focus of so much fuss and bunting from the moment he had mounted the red-white-and-blue-festooned stage with the Crewmen trailing behind him. Paper fans to beat away the heat had been passed out to the crowd of more than a thousand, which included such dignitaries as Steve Cheney, a three-star general and the commandant of the Marine Corps Recruit Depot at nearby Parris Island. Kerry, in a blinding white shirt and cool blue tie, had removed his jacket against the sultry air but still looked a little overcome, perhaps less from the heat than from the sight of all the shield-shaped signs proclaiming, "America's Courage: Kerry '04" everywhere he looked. At times his long, age-cragged face looked grayer than the still Kennedy-thick crop of hair late-night TV hosts had taken to making fun of for its abundance.

Then Max Cleland rolled his wheelchair to the microphone. Sporting a Kerry for President button, he began by saluting all the veterans in the audience for their service to their country, then turned to Kerry. "If any of you have doubts about his courage," Cleland declared, "his crew is here to testify." The perspiring crowd had perked up the moment the former Georgia senator started speaking, and by the time he finished the subject of his praise somehow looked younger and stronger as he unfurled his lanky frame to bear-hug the dear friend he called "a soldier's soldier, an American's American." Kerry mentioned his Vietnam service only briefly in his ensuing formal declaration of his candidacy for president. Anything more would have been superfluous, given that a symbolic photograph is worth a thousand sound bites.

After his speech, all the assembled veterans and campaign staffers retreated to an air-conditioned Hilton conference room. Kerry and his wife, Teresa Heinz, were in high spirits, as was filmmaker Alexandra Kerry, who along with CNN and C-SPAN had videotaped the entire event. She planned to follow her father around with a camera through the rest of his run for the White House to make a campaign documentary from her insider's perspective. After a quick sandwich and a Coke, Kerry's core campaign team prepared to head to Iowa. The Crewmen would be along to lend their support at a rally that evening in Des Moines. Just as the Kerrys and Crewmen were about to leave, Max Cleland wheeled himself over. He sounded vigorous and upbeat as he started talking to John Kerry about how

Vietnam created an eternal bond among those who had served there, and how in consequence they were all truly brothers. He offered to campaign for Kerry at VA hospitals in Iowa and New Hampshire; the candidate accepted. He then told his friend how he had finally found out exactly what had happened to him that awful day in Vietnam when a grenade ripped off his legs and right arm. In 1999 Cleland had taped an interview for a History Channel documentary on combat medics in Vietnam. In it he had lavished praise on the medics at Khe Sanh who had saved his life. He also mentioned on camera that the circumstances of his accident had haunted him since, as he feared that the grenade that tore into his body had accidentally dropped from his own gear belt when he had jumped out of a helicopter hovering a few yards above the ground. He added that not knowing exactly what had occurred remained a source of guilt for him.

Shortly after the History Channel aired the combat-medics program, Cleland related, he received a call from a man named David Lloyd, the medic in Vietnam who had not only witnessed Cleland's accident but had cut his own trousers into the tourniquets he applied where Cleland's legs had once been attached to stop him from bleeding to death. Lloyd told the ex-senator that his severed legs had been "smoking from the explosion." Most important, the former medic added that he had tracked Cleland down to let him know after all these years that it had not been his but another, inexperienced GI's grenade that had accidentally dropped and blown Cleland's limbs off. Thus had David Lloyd given Cleland the greatest gift he had ever known: "peace of mind." After that, Cleland told Kerry, he would never again be plagued by the possibility that what had happened to him had been his own fault.

By coincidence, he continued, at Kerry's campaign-announcement rally that very day Cleland had run into another man who had been at Khe Sanh with him when the grenade went off: his old buddy Steve Price, who lived in South Carolina. Price had also witnessed the grenade explosion that sent Cleland's limbs flying into the air. As Lloyd and the other medics had loaded the grievously wounded twenty-six-year-old onto the medevac, the terrified Price had muttered: "There goes a dead man." Although Price had later learned he was wrong about that, and had followed Cleland's political career closely, the two men had not seen each other in more than three decades. Now, on September 2, 2003, in Charleston, South Carolina, because of John Kerry, and because *his* Vietnam experience had brought him into politics, Cleland and Price had been reunited.

With tears trickling from his eyes, Cleland grabbed Kerry's right arm and squeezed it hard. He told the newly official presidential candidate that to his mind, both Lloyd and Price were gifts to him from God. Now, Cleland wanted Kerry to share in that higher power as he went forth in his quest to make a better nation. "Brother," he said as he placed a well-worn volume in Kerry's hand, "this is my family Bible. I want you to keep it with you at all times." Near speechless with emotion, Kerry refused. But Cleland pressed on, explaining how his mother and father had read to him from this same little King James Bible when he was a boy. Then he reminded Kerry how Abraham Lincoln had found solace in the Bible during the Civil War, and how Jimmy Carter had often dropped to his knees to pray during his successful 1976 campaign for the highest office in the land.

"Max, I can't," Kerry insisted. "It's your family Bible. I'm afraid to lose it." Cleland pushed the small tome into Kerry's hands and replied, "You've got to keep it, brother. You've got to win. It's your duty." Kerry just looked at his friend for a moment in awe and admiration. Then he leaned down and embraced Cleland's ravaged frame. "I know," he whispered back. "I will. I won't let us down." He didn't have to explain who "us" was. They both knew.

Timeline

December 11, 1943	John Forbes Kerry born in Denver at Fitzsimmons General Hospital.
Spring 1962	Graduates from St. Paul's School. Begins studying Southeast Asia.
November 22, 1963	President John F. Kennedy assassinated in Dallas, Texas.
August 2–4, 1964	U.S.S. *Maddox* and *Turner Joy* allegedly attacked in the Gulf of Tonkin.
August 7, 1964	Gulf of Tonkin Resolution passed by Congress.
December 31, 1965	There are roughly 184,300 United States servicemen in Vietnam.
February 18, 1966	Formally enlists in the U.S. Navy.
June 1966	Graduates from Yale University with a bachelor of arts degree in political science and a specialization in American government.
August 22, 1966	Reports to the Naval Officer Candidate School at the U.S. Naval Training Center in Newport, Rhode Island.
December 16, 1966	Receives his commission as an ensign in the U.S. Naval Reserve.
December 31, 1966	There are roughly 385,200 United States servicemen in Vietnam.
January 3, 1967	Kerry reports to Naval Schools Command at Treasure Island in San Francisco. He takes a ten-week Officer Damage Control Course.

March 22, 1967	Reports to U.S. Fleet Anti-Air Warfare Training Center in San Francisco. Receives training as a Combat Information Center Watch Officer. He completes these courses on April 14, 1967.
June 8, 1967	Reports to the guided-missile frigate U.S.S. *Gridley*, aboard which he serves in several capacities. The *Gridley* serves mostly in and around Southern California from its base in Long Beach until February 1968.
December 31, 1967	There are roughly 485,600 United States servicemen in Vietnam.
January 30, 1968	The Tet offensive begins.
February 9, 1968	The *Gridley* departs for operations in the Pacific. The ship spends time in the Gulf of Tonkin off North Vietnam, at Subic Bay in the Philippines, and in Wellington, New Zealand.
February 10, 1968	Kerry requests duty in Vietnam. He lists his first preference for a position as an officer in charge of a Swift boat, his second as an officer in a patrol boat, river (PBR) squadron.
March 16, 1968	What would become known as the My Lai massacre occurs.
May 12, 1968	Paris peace talks begin.
May 27, 1968	The *Gridley* sets sail for the United States. Kerry arrives in Long Beach on June 6, the day after Democratic presidential contender Robert F. Kennedy is shot in Los Angeles.
June 16, 1968	Kerry is promoted from ensign to lieutenant (junior grade).
July 20, 1968	Leaves the *Gridley* for specialized training at the Naval Amphibious Base in Coronado, California, in preparation for service in Vietnam.
August 25, 1968	Begins Swift boat officer training at Coronado.
November 17, 1968	Reports for duty to Coastal Squadron 1, Coastal Division 14, Cam Ranh Bay, South Vietnam. First serves as officer in charge of PCF-44, then is moved

	to PCF-94 near the end of January 1969. As part of Task Force 115, participated in Operation Swift Raider, which involved PCF incursions into enemy strongholds and sanctuaries in and about the river-mouths, inlets, coves, and canals of coastal South Vietnam to sever enemy supply lines and communications. Kerry's Swift boats also transport U.S. and South Vietnamese troops on reconnaissance missions.
December 2, 1968	Wounded in action, for which he receives his first Purple Heart.
December 6, 1968	Moved to Coastal Division 11 at An Thoi on Phu Quoc Island.
December 13, 1968	Moved to Coastal Division 13 at Cam Ranh Bay.
December 31, 1968	There are roughly 536,000 United States servicemen in Vietnam.
January 20, 1969	Richard Nixon inaugurated as President.
February 20, 1969	Kerry is again wounded in action, and receives his second Purple Heart.
February 28, 1969	Engagement for which Kerry received the Silver Star.
March 13, 1969	Engagement for which Kerry received the Bronze Star. Kerry is wounded in the action and is awarded his third Purple Heart.
March 17, 1969	Requests reassignment out of Vietnam after sustaining three injuries.
April 11, 1969	Reports for duty at Military Sea Transportation Service, U.S. Atlantic Fleet, based in Brooklyn, New York. He is assigned as personal aide and flag lieutenant to Rear Admiral Walter F. Schlech Jr.
October 15, 1969	Kerry flies former Robert F. Kennedy speechwriter Adam Walinsky around New York State to make antiwar speeches.
December 31, 1969	There are roughly 475,200 United States servicemen in Vietnam.
January 1, 1970	Kerry is promoted from lieutenant (j.g.) to lieutenant.

January 3, 1970	Requests an early discharge from active duty, which is granted.
February 1970	Kerry drops out of the Democratic primary race for Massachusetts's Third District seat in the U.S. Congress. Reverend Robert F. Drinan goes on to win the nomination.
February 20, 1970	Secretary of State Henry Kissinger begins secret peace negotiations with the North Vietnamese in Paris.
May 23, 1970	Kerry marries Julia Thorne in Bay Shore, Long Island. Shortly after their honeymoon he joins Vietnam Veterans Against the War (VVAW).
April 31, 1970	The United States invasion of Cambodia begins.
September 4–7, 1970	VVAW march from Morristown, New Jersey, to Valley Forge, Pennsylvania, dubbed Operation RAW (Rapid American Withdrawal).
April 18, 1971	Kerry arrives in Washington, D.C., for VVAW's weeklong Operation Dewey Canyon III antiwar demonstrations.
April 22, 1971	Speaks for VVAW before the Senate Foreign Relations Committee, chaired by J. William Fulbright.
May 28, 1971	VVAW march from Concord, Massachusetts, to Boston Common begins, dubbed Operation POW.
June 13, 1971	The *New York Times* publishes the Pentagon Papers.
November 1971	The Macmillan Company publishes Kerry's antiwar book, *The New Soldier*. It coincides with his resignation from VVAW.
July 1, 1972	Kerry is transferred to the U.S. Naval Standby Reserve.
November 1972	Due to his antiwar record, Kerry loses to Republican Paul Cronin in the race for the Massachusetts Fifth District congressional seat.
December 18–29, 1972	"Christmas bombings" against North Vietnam.

December 31, 1972	There are roughly 24,000 United States servicemen in Vietnam.
February 1973	North Vietnam begins releasing United States POWs.
August 14, 1973	All United States military operations in Indochina come to an end.
December 13, 1974	Battles break out between the armies of North and South Vietnam.
April 29–30, 1975	North Vietnamese take Saigon and the war ends.
February 16, 1978	Kerry is honorably discharged from the U.S. Naval Reserve.
November 6, 1984	Having defeated James M. Shannon in the Democratic primary with the help of Massachusetts veterans known as the Doghunters, Kerry wins race for U.S. Senate seat against Republican Raymond Shamie.
July 21, 1991	The United States Senate votes in favor of opening an investigation into the existence of American soldiers still captive in Vietnam. Kerry begins making his fourteen trips to Vietnam, some accompanied by Senator John McCain.
January 13, 1993	A U.S. Senate select committee chaired by John Kerry issues a 1,223-page report based on his findings from numerous trips to Vietnam, stating that there is "no compelling evidence that proves that any American remains alive in captivity in Southeast Asia."
September 2, 2003	Accompanied by the crews of PCFs-44 and -94, John Kerry announces his candidacy for the office of president of the United States in Charleston, South Carolina.

Glossary

AK-47	Chinese-built North Vietnamese Army semiautomatic rifle
APC	armored personnel carrier
APL	auxiliary personnel lighter, a non-self-propelled barracks craft
ARVN	Army of the Republic of Vietnam
ATC	armored troop carrier
CO	commanding officer
DD	destroyer
DMZ	Demilitarized Zone
FSB	Fire Support Base
GQ	general quarters
H&I	harassment-and-interdiction
KIA	killed in action
LSMR	landing ship medium, rocket
LST	landing ship tank
MACV	Military Assistance Command, Vietnam
MIA	missing in action
NLF	National Liberation Front
NTC	naval training center
OCS	Officer Candidate School
OINC or OIC	officer in charge
PBR	Patrol Boat, River
PCF	Patrol Craft Fast
POW	prisoner of war

PRU	Provincial Reconnaissance Unit
PX	Post Exchange
RAG	River Assault Group
RAW	Rapid American Withdrawal
SDS	Students for a Democratic Society
SEAL	from "sea, air, land," a U.S. Navy commando
SERE	from "survival, evasion, resistance, escape," an advanced U.S. Navy training course
SVNAF	South Vietnamese Armed Forces
SVNAV	South Vietnamese Navy
VC	Viet Cong
VVAW	Vietnam Veterans Against the War
WHEC	high weather endurance cutter, a U.S. Coast Guard craft
XO	Executive Officer
YIPPIE	a member of the Youth International Party

Interviews

John Kerry Interviews

August 9, 2002, in Washington, D.C.
October 17, 2002, in Washington, D.C.
December 20, 2002, in Boston
January 30, 2003, in New Orleans
March 2, 2003, in Washington, D.C.
March 18, 2003, in Boston
June 30, 2003, in Boston
September 8, 2003, in Boston
September 22, 2003, in Boston

Other Interviews and Oral Histories

I conducted all these interviews between August 2002 and October 2003. An asterisk indicates that more than one interview was conducted.

David M. Alston*
Jim Bales
Daniel Barbiero*
Elliott "Skip" Barker*
Michael Bernique
Barry Bogart
Robert Brant
Patrick J. Buchanan
William F. Buckley Jr.
Harvey Bundy*
George Butler*
Dick Cavett
David Christian
Ramsey Clark*
Max Cleland*
Doug Clifford
Arthur Cramm
William Crandell
John W. Dean III
Ron Drez
Robert F. Drinan
Michael S. Dukakis
George Elliott*
Bob Emory
Gerald R. Ford
Leslie Gelb
Chris Gregory
Stephen Hatch*
Mark Hatfield
Kitty Hawks

Tom Hayden
Stephen Hayes
Seymour Hersh
Bob Hildreth
Roy Hoffman
Townsend Hoopes
Charles Horne*
John Hurley
Thomas Huston
Robert Jack
Virgil Jackson
Arthur Johnson
Dang Quang Kai
Bob Kerrey
Alexandra Kerry
Cameron Kerry
Diana Kerry
Peggy Kerry*
Vanessa Kerry
Judy Droz Keyes
L. J. King
Carter Kirk
Brian Lamb
Dorsey Lee
Calvin Leleux
Robert Jay Lifton
Armistead Maupin*
John McCain
Eugene McCarthy
Pete McClosker Jr.
George McGovern*
Mary McGrory
David McKean*
Michael Medeiros*
George J. Mitchell
Chuck Mohn
Paul Nace
Gerald Nicosia
Paul Nitze
Tom Oliphant*
Tedd Peck
Jeffery Perin
George Plimpton
Garland Robinette
Larry Rottmann
Nils Rueckert
Morley Safer
Mark Salter
Wade Sanders*
Del Sandusky*
John Shattuck
Fred Short*
Bill Shumadine
Robert Shrum
Fred Smith*
Michael Solhaug
Fritz Steiner
Nancy Stetson
Hunter S. Thompson
David Thorne*
Julia Thorne*
Eugene Thorson*
Larry Thurlow
William vanden Heuvel
Adam Walinsky
Larry Wasikowski*
James Wasser*
Drew Whitlow*
Curtis Wilke*
Perry Williamson
Peter Yarrow*
William Zaladonis*
Howard Zinn
Frances Zwenig

Notes

The notes that follow list only the main primary and secondary sources used in writing *Tour of Duty*. It's worth noting that books by Stanley Karnow, Michael Kazin, George Herring, Ronald Spector, Larry Berman, David Halberstam, John Prados, Jerry Lembcke, Todd Gitlin, and Michael Lind helped inform my general knowledge about the Vietnam War.

Prologue: April 22, 1971 (Washington, D.C.)

A full transcript of John Kerry's address before the U.S. Senate Foreign Relations Committee appears in his book *The New Soldier* (1971). Correspondence pertaining to the writing of *The New Soldier,* including Kerry's financial obligations to the VVAW, can be found in the VVAW Papers Archives Division of the State Historical Society of Wisconsin (hereafter VVAW-Wis). I also drew upon a file from Kerry's Boston archive that contains various drafts of the speech as he worked on it in 1970–71. Among all the major U.S. newspapers' coverage of Kerry's Senate testimony, I found Thomas Oliphant's *Boston Globe* articles the most detailed and precise. Oliphant's many pertinent articles about Dewey Canyon III include his "Senators Told 'We Created a Monster' " on April 23, 1971. My conversations with Oliphant afforded a vivid firsthand perspective on the scene around Kerry that day. Another veteran *Boston Globe* correspondent, Curtis Wilkie, also was helpful, explaining to me Kerry's career trajectory through his VVAW years. An important article was Peter Lisagor's "Two Men Demonstrate Power of a Single Voice," *Chicago Daily News*, April 25, 1971.

Associated Press reporter Brooks Jackson conducted incisive interviews with a number of veterans at Dewey Canyon III, including Phillip Lavoie, for valuable articles that ran nationwide. Robert Muller, who went on to become president of Vietnam Veterans of America, was quoted in *The New Soldier*; my conversations with him gave me a better understanding of the veterans' movement. Bill Crandell likewise helped educate me about the political infighting within VVAW. Several good books have been published on J. William Fulbright, including Randall Woods's excellent *Fulbright: A Biography* (1995) and William Berman's concise *William Fulbright and the Vietnam War* (1988). Also important was J. William Fulbright's September 29, 1971, letter about Kerry to Samuel Stewart, which can be found in the VVAW-Wis archives. My interviews with Mark Hatfield, Ted Kennedy, and George McGovern contributed much to my understanding of the Senate Foreign Relations Committee in the early 1970s.

Video footage of Kerry's Senate testimony is available from the major American TV networks of the time. A more easily accessible condensed version appears in the documentary *The Winter Soldier*, which can be purchased over the internet. In writing about Kerry's VVAW role during Dewey Canyon III, I found it helpful to watch tapes of his TV appearances on ABC's *Dick Cavett Show* and NBC's *Meet the Press*. The Philip Hart papers are available at the University of Michigan at Ann Arbor, including his files on the Vietnam War. Both *Time* and *Newsweek* covered Kerry's testimony (May 3, 1971, issues). I gleaned further details on the key U.S. battles Kerry discussed in his testimony from Robert Pisor's *The End of the Line: The Siege of Khe Sanh* (1982), Eric Hammel's *Khe Sanh: Siege in the Clouds, An Oral History* (1989), and Samuel Zaffiri's *Hamburger Hill: May 11–20, 1969* (1987). My friend David Christian, co-author of *Victor Six* (1990), helped me understand the differences between veterans of World War II and Vietnam.

It is impossible to express how inspiring I find Max Cleland, a genuine American hero. Throughout the writing of this book we spoke frequently; his quotes here come from our conversation of September 19, 2003. In addition to Cleland's own memoir, *Strong at the Broken Places* (2000), I found two articles about him particularly useful: Tom Baxter and Jim Galloway, "Max Returns, with Fire in His Eyes," *Atlanta Journal-Constitution*, June 16, 2003, and Peter Carlson "Political Veteran," *Washington Post*, July 3, 2003. And every American should read Lewis B. Fuller Jr.'s *Fortunate Son* (1991). It's one of the most honest memoirs ever written about duty, honor, country, and the psychological devastation of war.

For a discussion of the leading antiwar generals, see Joseph R. Conlin, *American Anti-War Movements* (1968); Kerry read this book before his testimony. Also of interest is the transcript of the General David M. Shoup–John Kerry press conference held in Room 345 Canon House (Washington, D.C.) on March 16, 1971. A fascinating recent study touching on generals opposed to U.S. intervention in Vietnam is Robert Buzzanco, *Masters of War: Military Dissent and Politics in the Vietnam Era* (1996).

Chapter One: Up from Denver

For information on the history of the Forbes family, I owe special thanks to the Captain Forbes House Museum in Milton, Massachusetts. A fine account of the life of John Winthrop and the roots of the Winthrop family appears in Francis J. Bremer's *John Winthrop: America's Forgotten Founding Father* (2003). Also useful was James Hershberg, *James B. Conant: Harvard to Hiroshima and the Making of the Nuclear Age* (1993).

A few good profiles of Kerry's early life have appeared in recent years, notably Paul Alexander, "John Kerry: Ready for His Close-Up," *Rolling Stone*, April 11, 2002; Laura Blumenfeld, "John Kerry: Hunter, Dreamer, Realist," *Washington Post*, June 1, 2003; and Michael Kranish, "John Kerry, Candidate in the Making Part 1: A Privileged Youth, a Taste for Risk," *Boston Globe*, June 15, 2003.

Many queries about Kerry's precollegiate education were answered by Bob Rettew, director of the Ohrstrom Library at St. Paul's School. He provided the St. Paul's obituary for John Walker as well as an article he had written on the occasion of Walker's departure for the National Cathedral in the Summer 1966 issue of the school's *Alumni Horae*. Also useful was August Heckscher, *St. Paul's: The Life of a New England School* (1980). Valuable information on the origins of the Vietnam War appears in David L. Anderson, *Trapped by Success: The Eisenhower Administration and the Vietnam War, 1953–1961* (1991) and in Lloyd Gardner, *Approaching Vietnam: From World War II Through Dienbienphu* (1988).

Kerry keeps his father's flight logbook in his Boston office desk drawer; the quotes came directly from it. Additional information pertaining to Richard Kerry also comes directly from his son's archive, including a 1962 résumé. Insights into Richard Kerry's political thinking can be gleaned from his *The Star-Spangled Mirror* (1990). On December 20, 2002, I spent two hours with John Kerry in Boston discussing his parents. To better understand both Richard and Rosemary Kerry I also interviewed John's siblings Cameron (in Boston), Peggy (in New York), and Diana (in Manchester, Massachusetts). The details regarding Maxwell Air Force Base were drawn from Jerome A. Ennels, *Wisdom of Eagles: A History of Maxwell Air Force Base* (1997).

In his office Kerry keeps photographs of and newspaper clippings about Dick Pershing; these helped me bring his lost friend's life into focus. Research on the 1962 America's Cup race was conducted at the JFK Library in Boston. Also helpful for getting the facts straight from the race was Dennis Conner and Michael Levitt, *The America's Cup: The History of Sailing's Greatest Competition in the Twentieth Century* (1998).

Chapter Two: The Yale Years

It was my good fortune to be able to read the unpublished diary of Blakely Fetridge (Bundy), which is filled with colorful details about Kerry and Yale. She also wrote me an important clarification letter on April 20, 2003. Her husband, Harvey Bundy, whom I met with in Chicago, provided me with correspondence relevant to the period. An excellent work on the Bundy brothers is Kai Bird, *The Color of Truth: McGeorge Bundy and William Bundy; Brothers* in *Arms* (1998).

Kerry keeps the big scrapbook he started at Yale at his home in Boston. The collection contains more than a dozen *Yale Daily News* articles; most helpful to me were Bruce Breimer's "Yale Booters Bomb Crimson 6–3," November 22, 1965; "Triangular Ivy Debaters to Discuss Vietnam Policy," May 13, 1965; and "Yale Debaters Defeat British Team, Defend United Nations," February 13, 1964. Of particular interest was Victor H. Ashe, "Kerry Wins Speech Prize," March 12, 1965.

Correspondence in the scrapbook that helped me better understand Kerry's role at the Yale Political Union included letters to him from U.S. Representative Don Edwards (May 7, 1965), Adlai Stevenson (October 15, 1964), and William P. Bundy (April 16, 1966). Another interesting item was a small blue pamphlet published in 1965, on the history of the Yale Political Union, the last page showing Kerry as the club's president.

I couldn't have written this section without the input of Fred Smith of

Federal Express, whom I interviewed at his office in Memphis, and Danny Barbiero, who lives in Long Island.

The section on the Yale Aviation Club relies largely on information provided by Charles Skelton of Yale Aviation, Inc., now a nonprofit organization devoted to flight training.

Anne Q. Hoy's *Newsday* profile, "Kerry Eyes Revival of Idealism" (March 23, 2003), offers several tidbits about Kerry's time at Yale, including his soccer-team nickname: "the Camel." A first-rate biography of Kerry's political hero while at Yale University is William Chafe, *Never Stop Running: Allard Lowenstein and the Struggle to Save American Liberalism* (1993).

The seminal book on the history of Skull and Bones is Alexandra Robbins's *Secrets of the Tomb: Skull and Bones, the Ivy League, and the Hidden Paths to Power* (2002). Also crucial was Marc Wortman, "Elis Aloft," *Yale*, April 1991.

Two books helped me understand the complex history of the draft better: Lawrence M. Baskir and William A. Strauss, *Chance and Circumstance: The Draft, the War and the Vietnam Generation* (1978), and Sherry Gershon Gottlieb, *Hell, No, We Won't Go: Resisting the Draft During the Vietnam War* (1991).

For understanding of Kennedy administration policies in Vietnam, see David Halberstam, *The Best and the Brightest* (1972), and Fredrick Logevall, *Choosing War* (1999). In general, for capturing the *feel* of the Vietnam generation, I benefited from reading Rick Atkinson's masterpiece, *The Long Gray Line* (1989).

Chapter Three: California Bound

This chapter derives largely from my interviews with John Kerry, Paul Nace, Wade Sanders, David Thorne, and Julia Thorne. For an excellent overview of the U.S. Navy's Pacific Coast operations, see Bruce Linder, *San Diego's Navy* (2001). Retired captain Linder, also the son of a Navy officer, was raised in San Diego and went on to command the Naval Training Center and the Fleet Anti-Submarine Warfare Training Center. He also served as a speechwriter to three vice chiefs of naval operations. My telephone interviews with him proved invaluable. Also useful were Abraham Shragge, "A New Federal City: San Diego During World War II," *Pacific Historical*

Review 63, no. 3 (1994); George W. Mahnke, "The Urban Impact on Coronado of the San Diego–Coronado Bay Bridge" (thesis, San Diego State University, 1970); and Patricia Dibsie, "A Final Drill: NTC Fades Away," *San Diego Union Tribune*, March 22, 1997. Richard F. Pourade's series of books about San Diego—*The Silver Dons* (1963), *The Glory Years* (1964), *Gold in the Sun* (1965), and *The Rising Tide* (1967)—offered delightful color.

Throughout writing this book I drew upon issues of *All Hands*, the Bureau of Navy Personnel Career Publication. The January 1967 issue, which came out when Kerry was first stationed in San Diego, included two articles on the city, "San Diego: A Great Navy Town and One of the Busiest" and "The San Diego Naval Complex."

Books consulted about the Tet Offensive included Peter Braestrup, *Big Story: How the American Press and Television Reported and Interpreted the Crisis of Tet 1968 in Vietnam and Washington* (1977); Phillip B. Davidson, *Vietnam at War: The History, 1946–1975* (1988); Don Oberdorfer, *Tet!* (1971). A recent first-rate essay is Peter Brush's "Reassessing the Viet Cong Role After Tet," *Vietnam* (February 2002).

Tom Wolfe's *The Pump-House Gang* was most helpful in dissecting 1960s California beach culture.

Spencer C. Tucker's article, "The First Tet Offensive" appeared in *Vietnam* in February 2003.

Chapter Four: High Seas Adventures

A fantastic source for this chapter was the U.S.S. *Gridley* 1968 yearbook, on which Kerry served as business and copy editor. The missile-emblazoned yearbook contains capsule biographies of Captains Wyatt E. Harper Jr. and Allen W. Slifer; its section on Danang includes photographs of fishing junks, U.S. gunboats, and a Red Cross ship.

The *Gridley* spanned 533 feet in length and 53 feet across at its widest point. It had a normal displacement of 7,400 tons. The ship's armament consisted of two twin Terrier surface-to-air missiles, anti-submarine rockets, two triple torpedo mounts, and two twin 3"/50 rapid-fire guns.

Officially the *Gridley* was placed in commission on May 25, 1963, with Captain Percy L. Lilly Jr. in command. The ship was named after Captain Charles V. Gridley, who fought gallantly in the Battle of Manila Bay in 1898. Captain Gridley was the commanding officer of Commander

Dewey's flagship *Olympia*. As naval lore has it, Commander Dewey turned to Captain Gridley and said: "You may fire when you are ready, Gridley." This launched the battle in which the U.S. Asiatic Fleet crushed the Spanish navy.

The Naval Historical Center in Washington, D.C., helped me understand that when Kerry was in the Gulf of Tonkin, the *Gridley* assumed duties with an Attack Carrier Task Group. The ship usually stayed only 1,000 yards from a carrier. It would rescue any airman who had trouble. Its number was lucky 21. Kerry saved eleven of his elegant World War II radio addresses he delivered over the *Gridley*'s intercom. They're in his Boston archive.

For the story of Thomas Bennett's Vietnam career, see Edward F. Murphy, "Combat Medic Thomas W. Bennett," *Vietnam*, June 2003, p. 16.

For the story about Black Jack Pershing, see Christine Ammer, "Fighting Words: Terms from Military History," *Military History Quarterly* 15, no. 1 (Autumn 2002), p. 49. *Newsweek*'s article on Dick Pershing's death, "Footsteps," ran March 4, 1968, and an Associated Press story had appeared that February 20 under the headline "General Pershing's Kin Killed in Viet."

For an example of what was shaping Kerry's vision of Vietnam and his infatuation with Senator Robert F. Kennedy of New York, see Kennedy's *To Seek a Newer World* (1967), especially the "Vietnam" chapter, pp. 169–228. Also see Arthur M. Schlesinger Jr., *Robert Kennedy and His Times* (1978).

Interviews with David Simons, Robert Jack, and other *Gridley* veterans were extremely helpful.

Danang is described well in Frances FitzGerald's *Fire in the Lake: The Vietnamese and the Americans in Vietnam* (1972).

Chapter Five: Training Days at Coronado

There are a number of helpful books covering the Navy's involvement in Vietnam, such as Edward J. Marolda's *By Sea, Air, and Land: An Illustrated History of the U.S. Navy and the War in Southeast Asia* (1994); Richard L. Schreadley's *From the Rivers to the Sea: The United States Navy in Vietnam* (1992); and General William C. Westmoreland's memoir *A Soldier Reports* (1976). In addition, there are a few fine works on the Navy's river patrols in Vietnam: Thomas J. Cutler's *Brown Water, Black Berets* (1988); Jimmy R. Bryant's *Man of the River: Memoir of a Brown Water Sailor in Vietnam 1966–1969* (1998); and Gordon L. Rottman's *The Vietnam Brown*

Water Navy: Riverine and Coastal Warfare 1965–1969 (1997). Admiral Horatio Rivero's *Riverine Warfare: The U.S. Navy's Operations on Inland Waters* was published by the Navy itself in 1968. This government document also offers wonderful photographs of U.S. vessels used in Vietnam.

The Naval Institute Oral History Series, U.S. Naval Institute, Annapolis, Maryland, has a fine collection of San Diego reminiscences from former officers, including Rear Admiral Roy S. Benson, Lieutenant Commander Richard A. Harralson, Vice Admiral Paul Stroop, and Rear Admiral Odale D. Waters Jr., which were useful in writing this chapter.

For a comparison of river warfare in the Civil War and Vietnam, see "The River War," *Newsweek*, July 3, 1967, and "Baby Ironclad," *All Hands*, June 1970. For a detailed description of Soviet gunmaker Mikhail T. Kalashnikov's creation of the AK-47, see Gene Gangarosa Jr., "The AK-47 Emerged from the Vietnam War with a Legendary Reputation It Still Enjoys," *Vietnam*, April 2003. A valuable article on how Operating Market Time worked is "Intercept Hot Cargo Below," *All Hands*, October 1967.

Bernard Fall's *Street Without Joy* (1961) and *Hell in a Very Small Place* (1968) had a profound influence on Kerry's thinking about Vietnam.

A conversation with Virgil Jackson of Vidalia, Louisiana, made the Navy's use of helicopters in Vietnam clear.

For more on PT boats, and specifically John F. Kennedy's PT-109, see John Hersey's "Survival" from the August 2, 1944, issue of *The New Yorker*, Robert J. Donovan's *PT-109* (1961), and Robert D. Ballard's *Collision with History: The Search for John F. Kennedy's PT 109* (2003).

A landmark study of small boats in the U.S. Navy is Norman Friedman, *U.S. Small Combatants* (1987). Also see Elmo R. Zumwalt Jr., *On Watch: A Memoir* (1976).

For an in-depth profile of Bob Kerrey's Vietnam experience, see Gregory L. Vistica, *The Education of Lieutenant Kerrey* (2003). Also a must-read is Bob Kerrey's memoir, *When I Was a Young Man* (2002).

San Diego ambience of 1968 is captured in the following *San Diego Union* articles: "City Reports Sharp Boost in Building," August 26, 1968; "Padres, Hahn Reach Stadium Use Accord," August 2, 1968; "Summer's Over? Not in San Diego," September 4, 1968; and "Military Essential to San Diego" September 28, 1968. The San Diego Navy Historical Association helped me locate these articles.

Throughout 1966 Hanson Baldwin wrote on Market Time in the *New York Times*; in the magazine *Shipmate*, his "Spitkits in Tropic Seas," displays the logic behind the operation.

Two John Kerry letters provided key details for his transition from the *Gridley* to Swift school: to Wade Sanders from the *Gridley* May 1, 1968, and to David Thorne from Coronado, September 4, 1968. My understanding of Roy Hoffman comes from interviewing Swift boat officers Mike Bernique, Larry Thurlow, Rich McCann, and Wade Sanders. I also interviewed Hoffman at the 2003 Swift boat reunion in Norfolk, Virginia.

For an account of Colt's manufacturing, including production of M-16s during the Vietnam era, see Dean K. Boorman, *The History of Colt Firearms* (2001).

Various conversations with, and memos from, Skip Barker regarding this period made a more accurate portrayal of it possible. A helpful article on PBRs was "PBR's Make Their Mark," *All Hands*, July 1969.

Once Sewart Seacraft received a U.S. government contract for over 300 boats, it quickly became the leading patrol boat builder in America. In 1969 the company became Swiftships. The smallest boat was 25 feet long, the largest 225 feet. Over the years thirty-eight different countries have purchased vessels from Swiftships. And their client base is growing. Their specialty remains fast crewboats, offshore supply boats, patrol boats, research vessels, and custom motor yachts. They're best known for using aluminum *and* steel in construction. All of these vessels meet U.S. Coast Guard requirements and American Bureau of Shipping classification requirements. The Swiftships Archive at Morgan City, Louisiana, provided me documents that enhanced my understanding of how these vessels were built.

Chapter Six: Trial by Desert

Much of this chapter draws upon Kerry's correspondence with Julia Thorne and his parents from August to October 1968. One letter, written after SERE training in October, was twelve single-spaced typed pages long.

Information about Warner Springs and its naval history came from the San Diego Navy Historical Association. Kerry also spoke about his life in San Diego 1967–68 in his remarks to the California State Democratic Convention, February 16, 2002, and his Jefferson-Jackson Day speech to the San Diego Democratic Party, November 26, 2001 (Kerry Archives, Washington, D.C.).

An extremely helpful and thorough account of the war in Vietnam in 1968–69 is Ronald H. Spector's *After Tet: The Bloodiest Year in Vietnam* (1993). Larry Wasikowski also shared with me his Warner Springs experience.

The personal aspects of the Kerry-Thorne courtship come from interviews with them both.

Chapter Seven: In-Country

For a discussion of Cam Ranh Bay history, see Neil Sheehan, *A Bright Shining Lie: John Paul Vann and America in Vietnam* (1988); Carroll H. Dunn, *Base Development in South Vietnam, 1965–1970* (1972); and Brandan M. Greeley Jr., "Soviets Extend Air, Sea Power with Buildup at Cam Ranh Bay," *Aviation Week*, March 2, 1987. Also see Naval Facilities Engineering Command, *Southeast Asia: Building the Bases* (1975).

Throughout Kerry's diaries he mentions Melvin Laird. To understand Laird's strategic vision, see Juan Cameron, "A Political Pro at the Pentagon," *Fortune*, April 1969; Julius Duscha, "The Political Pro Who Runs Defense," *New York Times Magazine*, June 14, 1971; and Tad Szulc, *The Illusion of Peace: Foreign Policy in the Nixon Years* (1978). For Laird's views on the Cold War, see his two books, *A House Divided: America's Strategy Gap* (1962) and *People, Not Hardware: The Highest Defense Priority* (1980). David Wise's "credibility gap" is discussed in Stanley I. Kutler, *The Wars of Watergate* (1990). My interviews with Charles Horne were invaluable, providing me with background information on his sterling career and answering numerous queries about Operation Sealords. Larry Wasikowski's Web site, http://swiftboats.net, provides the most up-to-date statistics on Swift boats in Vietnam and was an excellent overall source on the riverine forces. For the last months of Lyndon Johnson's presidency, see Robert Dallek *Lyndon B. Johnson: Portrait of a President* (2003). George Butler kindly shared with me his Kerry files.

It should not be forgotten that women played an important part in the Vietnam War; for example, see Karen G. Turner and Phan Thanh Hao, *Even the Women Must Fight: Memories of War from North Vietnam*

(1999), and Lynda Van Devanter with Charles Morgan, *Home Before Morning: The Star of an Army Nurse in Vietnam* (1984).

For the Green Berets in Vietnam—and American popular culture—see Garry Wills, *John Wayne's America: The Politics of Celebrity* (1997), and Shelby Stanton, *Green Berets at War* (1985). Robert Jay Lifton also evaluates the John Wayne myth in *Home from the War* (1973)—see the chapter "From John Wayne to Country Joe and the Fish." For action scenes, I also found useful Charles M. Simpson II, *Inside the Green Berets* (1983). The diaries of Stephen Hayes and the voluminous Don Droz correspondence were exceedingly helpful. Finally, Tedd Peck of Arizona spoke to me frankly concerning what he disliked about John Kerry.

Useful works on the Nixon Doctrine include William Bundy's *A Tangled Web: The Making of Foreign Policy in the Nixon Presidency* (1998) and Robert S. Litwak's *Détente and the Nixon Doctrine: American Foreign Policy and the Pursuit of Stability, 1969–1976* (1984). Sheehan's *A Bright Shining Lie* provides invaluable information about the finer points of America's involvement in Vietnam. I constantly relied on Stanley Karnow's *Vietnam* (1983).

For more on the transfer of PCFs, see "Swift Takeover," *All Hands*, November 1968, and "The Riverine Force," *All Hands*, October 1968.

Kerry's correspondence from Cam Ranh, especially a November 24, 1968, letter to George Butler, explains his attitude about his time there.

A number of distinctive Vietnam details came from novels by Tim O'Brien, including *In the Lake of the Woods* (1994) and *Going After Cacciato* (1978). His ability to capture the emotions of soldiers in-country—and to render the psychological landscape of Vietnam veterans—is unparalleled.

The Bernard Fall articles that Kerry practically memorized are in Fall's *Street Without Joy* (1961).

The Abigail Adams quote is from Evan Thomas, *John Paul Jones: Sailor, Hero, Father of the American Navy* (2003).

Of great help in understanding the geography of the Mekong Delta was Harry G. Summers Jr.'s *Historical Atlas of the Vietnam War* (1995).

Much of Don Droz's profile came from information supplied by his daughter, Tracy Tragos, in Topanga, California. Droz's harsh critique of

Vietnamese sailors comes from his October 23, 1968, letter. Also see Captain John Cantzon Foster, "Co Van My: A Naval Adviser's Story," *Vietnam*, August 2003.

The history of the Purple Heart was provided by the Military Order of the Purple Heart, in Springfield, Virginia (the organization also has chapters throughout the United States). Original Purple Hearts are on display in Washington, D.C., at the Society of the Cincinnati's Anderson House Museum and at Washington's Headquarters State Historic Site in Newburgh, New York.

Michael Kranish's profile of Kerry in Vietnam, "John Kerry, Candidate in the Making, Part 2: Heroism, and Growing Concern About War," *Boston Globe*, June 16, 2003, was helpful for details about how Kerry won his Purple Hearts.

As mentioned in the acknowledgments, Skip Barker sent me numerous memos dealing with Swifts in Vietnam; particularly helpful re An Thoi was the one of October 1, 2003. Stephen Hayes, another lieutenant who skippered a Swift boat, did not know John Kerry personally but provided a number of insights into an officer's life on the Mekong delta.

For more on Oliver Stone, his films and his tour in Vietnam see Frank Beaver, ed., *Oliver Stone: Wakeup Cinema* (1994), and Stone's own reflection, *A Child's Night Dream* (1997).

When one reads letters written from Vietnam by U.S. Swifties there is almost always a mention of monsoons. For an interesting brief discussion of this, see John F. Fuller, *America's Weather Warriors, 1814–1985* (1986). Many veterans explain why and how they thought Vietnamese weren't people in Lifton's *Home from the War.* Sometimes when GIs burned villages, they were called Zippo Squads. Wade Sanders shared with me a first-rate memoir/reflection paper that explained the curfew rules.

Chapter Eight: PCF-44

Kerry's journals and correspondence form the crux of this chapter. Robert Olen Butler A *Good Scent from a Strange Mountain* (1992) and Frances FitzGerald's *Fire in the Lake* (1972) provided key atmospheric details for understanding Vietnamese culture. Kerry reflected on the rivers in his review of Francis Coppola's *Apocalypse Now, Boston Herald American*, October 14, 1979.

Various radio calls and information on Swifts was provided by Larry Wasikowski through his Web site, http://swiftboats.net.

My conversation with Wade Sanders provided the perspective of another young lieutenant during this same period.

In September 2003, Mike Bernique, now a Dallas businessman, also provided me a frank interview about life in the Mekong Delta and the Ca Mau Peninsula. For a discussion of Bernique and how the creek got named after him, I relied on Thomas J. Cutler *Brown Water, Black Berets* (1988); the chapter "Sea Lords" was extremely helpful in writing about Vietnam in 1968–69. Cutler also interviewed Admiral Zumwalt about Bernique.

A truly invaluable unpublished document was Richard L. Schreadley's "The Naval War in Vietnam," which is available in the Naval Historical Center's Operational Archives in Washington, D.C. A condensed version of this study, which Kerry read when it first appeared in the 1971 *Naval Review*, was titled "The Naval War in Vietnam, 1950–1970."

Statistics about the U.S. Navy in Vietnam vary, but I found two reference books extremely readable: the U.S. Naval Institute's *Naval and Maritime Chronology, 1961–1971* (1973) and Jack Sweetman's *American Naval History: An Illustrated Chronology of the U.S. Navy and Marine Corps 1775–Present* (1984).

The Eisenhower quote to his wife comes from my "Introduction" to Andrew Carroll, *War Letters* (2001).

Chapter Nine: Up the Rivers

Again Kerry's war journals form the backbone of this chapter. Kerry's perspective was supplemented by those of the men of PCF-44: James Wasser, William Zaladonis, Drew Whitlow, and Stephen Hatch. I was unable to interview Stephen Gardner. The books I consulted dealing with Agent Orange include Paul F. Cecil, *Herbicidal Warfare: The Ranch Hand Project in Vietnam* (1986); Michael Gough, *Dioxin, Agent Orange: The Facts* (1986); and J. B. Neilands, G. H. Orians, E. W. Pfeiffer, Abje Vennema, and Arthur H. Westing, *Harvest of Death: Chemical Warfare in Vietnam and Cambodia* (1972). An important early report on the environmental impact of U.S. spraying of chemical agents is the Stockholm International Peace Research Institute (SIPRI), "Ecological Consequences of the Second Indochina War" (1976).

For the frequency of friendly-fire accidents in Vietnam see C. D. B. Bryan, *Friendly Fire* (1976); David H. Hackworth, "Killed by Their Commanders" *Newsweek*, November 8, 1991; and Charles Shrader, *Amicide: The Problem of Friendly Fire in Modern War* (1982).

Two books were particularly helpful in my understanding the problems facing ARVN when Kerry was in Vietnam: Jeffrey J. Clarke, *Advice and Support: The Final Years, 1965–1973* (1988); and Andrew F. Krepinevich, *The Army and Vietnam* (1986). Also, my conversations with Ron Drez, my colleague at the Eisenhower Center, were enormously helpful.

Bob Crosby was a greatly loved figure in the Swift boat community. Numerous men I interviewed told me stories about him; Wade Sanders was particularly helpful.

PsyOps was best explained to me by Larry Wasikowski.

For more on the U.S.S. *Liberty*, see James Ennes Jr., *Assault on the Liberty: The True Story of the Israeli Attack on an American Intelligence Ship* (1979).

My conversations with Bob Emory were extremely helpful in understanding the dangers related to maintaining a Swift boat.

Chapter Ten: Death in the Delta

Most of this chapter comes from Kerry's "War Notes." Information about Bob Hope's USO Show comes from Raymond Strait, *Bob Hope: A Tribute* (2003). A useful story about hospitals in Vietnam is Marc Phillip Yablonks, "Doctors in a War Zone," *Vietnam*, February 2002. In conjunction I read two novels on dissent, Norman Mailer, *Why Are We in Vietnam?* (1967) and Tim O'Brien, *In the Lake of the Woods* (1994).

For Operation Rolling Thunder, see W. Hays Parks, "Rolling Thunder and the Law of War," *Air University Review* 33, no. 2 (January–February 1982) and Earl H. Tilford Jr., *Setup: What the Air Force Did in Vietnam and Why* (1991).

James Wasser spent time with me in both Chicago and Charleston discussing the events presented in this chapter.

Sergeant Thomas Oathout's letter to his mother, written on February 14, 1966, was published in Stewart O'Nan, ed., *The Vietnam Reader* (1998).

The descriptions of death in Vietnam come from Philip Caputo, *A Rumor of War* (1977); "copper-jacketed death" is from p. 103.

A truly incredible book is Laura Palmer, *Shrapnel in the Heart: Letters and Remembrances from the Vietnam Veterans Memorial* (1986).

Norodom Sihanouk is quoted in Stanley Karnow, *Vietnam: A History* (1983).

The Chinese game with tiles is discussed in Lawrence H. Clima, M.D., "Unorthodox Practice: An American Doctor in Vietnam," *Vietnam*, April 2003, p. 32.

Statistics on the 1965–1968 U.S. bombing campaigns in North Vietnam come from Michael Maclear, *The Ten Thousand Day War: Vietnam, 1945–1975* (1981), p. 241; those on the 1968 U.S. bombing of North Vietnam and Laos come from William Shawcross, *Sideshow: Kissinger, Nixon, and the Destruction of Cambodia* (1979), p. 93.

Kerry had decided on the name "Nguyen" for the man he witnessed dying because it was both commonplace and popular in Vietnamese history. For example, the prime minister of Vietnam from 1965 to 1967 had been Nguyen Cao Ky; a leading Vietnamese general had been Nguyen Chanh Thi; the North Vietnamese commander of operations in South Vietnam had been Nguyen Chi Thanh; a flamboyant female Vietnamese revolutionary had been Nguyen Thi Dinh; and the president of the Republic of Vietnam from 1967 to 1973 was Nguyen Van Thieu. An interesting report on how U.S. doctors operated in Vietnam is Mark Phillip Yablonka, "Doctors in a War Zone," *Vietnam*, February 2002.

Chapter Eleven: Braving the Bo De River

Once again, Kerry's journals form the crux of this chapter. Kerry's interview with Dorsey Lee comes from the Kerry archive in Boston (this is an exact transcription from the original cassette); in addition, I interviewed Dorsey Lee. Just who "Deuce" was couldn't be ascertained. While there is no book on the Bo De River, it frequently appears in studies of the U.S. Navy in Vietnam as Richard Schreadley's book *From the Rivers to the Sea: The United States Navy in Vietnam* (1972) and, of course, Thomas Cutler's *Brown Water, Black Berets* (1988). When discussing weapons used in Vietnam I consulted three reference works: Ian V. Hoff and John Weeks, *Military Small Arms of the Twentieth Century* (1985); J. I. H. Owen, ed., *Brassey's Infantry Weapons of the World* (1975); and David Rosser-Owen, *Vietnam Weapons Handbook* (1986).

CHAPTER TWELVE: TAKING COMMAND OF PCF-94

The bulk of my discussion on the Saigon summit come from interviews with John Kerry, Wade Sanders, Tedd Peck, Larry Thurlow, Bill Shumadine, Rich McCann, and Bob Hildreth.

Lewis Sorley's interview with Creighton Abrams is part of tape project at the Vietnam Center, Texas Tech University, Lubbock, Texas. Sorley, author of *A Better War: The Unexamined Victories and Final Tragedy of America's Last Year in Vietnam* (1999), plans on publishing important excerpts from the tapes in a book tentatively entitled *Vietnam Chronicles: The Abrams Tapes, 1968–1972*. (It should also be noted that the Vietnam Center helped me fact-check a few questions. My admiration for James Reckner, the director, has no bounds.)

The quote from William Corson comes from his book *The Betrayal* (1968).

I found the descriptions of Zumwalt in Edward J. Marolda and G. Wesley Price, *A Short History of the United States Navy and the Southeast Asia Conflict, 1950–1975* (1984) helpful. For biographical material on Zumwalt I relied on his own writings: *On Watch* (1976); "The War Is Over," *New York Times*, February 7, 1994; and *My Father, My Son* (1987), which he wrote with his son.

Although not a theme in this book, African Americans like David Alston had to deal with racial friction in the navy. A haunting book on this topic is Wallace Terry, *Bloods: An Oral History of the Vietnam War by Black Veterans* (1984). Also helpful was William King, "A White Man's War: Race Issues and Vietnam," *Vietnam Generation*, Spring 1989. For Kerry on Alston, see his Jefferson-Jackson Day address to the South Carolina Democratic Party, May 3, 2002 (Kerry archives, Washington D.C.).

Randolph Forrester's "Perspectives" provides a good analysis of a Vietnam veteran's protest of the war (*Vietnam*, October 2002).

The use of drugs has been a major theme in the best novels written about U.S. soldiers in Vietnam. Kerry believes reports of drug use, however, were exaggerated by the media. Senator Alan Cranston of California did more than anybody to shed light on this issue; especially in his "Legislative Approaches to Addiction Among Veterans," *Journal of Drug Issues* (1974). A good, brief article on the subject is Peter Brush, "Higher and Higher: American Drug Use in Vietnam," *Vietnam*, December 2002.

The topic of atrocities in Vietnam is combustible. Bruce Palmer Jr., who from 1968 to 1972 was vice chief of staff of the Army, deals with the topic head-on in his important book *The 25-Year War: America's Military Role in Vietnam* (1984). The heroic story about rescuing forty-two Vietnamese citizens comes from Skip Barker and John Kerry.

There are many good books and articles on SEALs. A few which I relied on are T. L. Bosiljevac, *SEAL: UDT SEAL Operations in Vietnam* (1990); James Watson and Kevin Dockery, *Point Man* (1993); and Darryl Yound, *The Element of Surprise: Navy SEALs in Vietnam* (1990). The quote from Leonard Waugh comes from Kevin Dockery, *Navy SEALs: A History of the Early Years* (2002). The Dockery quote used comes from the introduction of the book.

To better appreciate the important role the U.S. Coast Guard played in the Vietnam War, see H. R. Kaplan, "Coast Guard Played Vital Role in Viet War," *Navy Magazine*, November 1970; Alex Larzelere, The *Coast Guard at War: Vietnam 1965–1975* (1997); and Eugene N. Tulich, The *United States Coast Guard in South East Asia During the Vietnam Conflict* (1975). They had to patrol 1,200 miles of coastline.

Chapter Thirteen: The Medals

For more on General Creighton W. Abrams Jr., see Lewis Sorley, "Reassessing the ARVN," *Vietnam*, April 2003, and his A *Better War: The Unexamined Victories and Final Tragedy of America's Last Years in Vietnam* (1999).

Kerry's original government-issued award recommendations and citations helped me pin down the exact dates concerned with them.

Because Kerry's river raids became so frequent, he did not keep diaries in these weeks in February and March 1969 when the fighting was most intense. Since this is the only Vietnam chapter not based on Kerry's journals, Michael Medeiros's Swift boat log were an indispensable source. It meticulously—and accurately—documents all of the river runs made by PCF-94. Captain Robert Gormly is quoted in Kevin Dockery, *Navy SEALs: A History of the Early Years.* My interviews with Bob Hildreth, Skip Barker, Del Sandusky, Fred Short, David Alston, and Bill Zaladonis constitute the core of this chapter.

An extremely underappreciated account of what it was like to be a

SEAL in Vietnam is Charles W. Sasser's account of Roy Boehm, "The First SEAL," *Vietnam*, April 2001.

There are two different versions of Kerry's Silver Star and Bronze Star citations. I've included the shorter ones, written in Saigon. I also consulted the more detailed ones in Kerry's military record.

An excellent book on Nixon's policy is Larry Berman, *No Peace, No Honor: Nixon, Kissinger, and the Betrayal of Vietnam* (2001).

Chapter Fourteen: The Homecoming

For a discussion of VC escapees on Phu Quoc, see Michael P. Kelley, *Where We Were in Vietnam* (2002). Wade Sanders's letter was found in Kerry's archive. To complete my understanding of the U.S. war in Cambodia, I consulted Phillip B. Davidson, *Vietnam at War: The History, 1946–1975* (1988); William Shawcross, *Sideshow: Kissinger, Nixon, and the Destruction of Cambodia* (1979); and Tom Wells, *The War Within: America's Battle over Vietnam* (1994). Information on Admiral Walter Schlech's career came courtesy of the Naval Historical Center in Washington, D.C. Tracy Tragos sent me her Vietnam files about her father, Don Droz, which included articles and oral history reminiscences she had collected while making her documentary on children of Vietnam veterans. She also shared twenty-five letters her father wrote home from Vietnam, including the April 11, 1969, one quoted. Peter Upton's Internet article "Death of the 43" was extremely helpful.

I spoke with numerous veterans familiar with Droz's death, including Chuck Mohn and Jim Galvin, and I interviewed George Elliott on the tragedy. Stephen Hayes of Virginia kindly shared with me his personal diary of his time in Vietnam. It's a superb primary source document.

I read a great deal of literature about post-traumatic stress disorder (PTSD). Besides Marilyn B. Young, The *Vietnam Wars: 1945–1990* (1991), I carefully read Myra MacPherson, *Long Time Passing: Vietnam and the Haunted Generation* (1984); Charles R. Figley and Seymour Leventamn, eds., *Strangers at Home: Vietnam Veterans Since the War* (1980); Charles Figley, ed., *Stress Disorders Among Vietnam Veterans Since the War: Theory, Research, and Treatment* (1980); and B. G. Burket and Glenna Whitley Stolen, *Valor: How the Vietnam Generation Was Robbed of Its Heroes and Its History* (1998). A number of important experts have dealt with PTSD, including Robert Jay Lifton, Chaim Shatan, Charles Figley, Arthur Egendorf, and Jack Smith. The John McCain story is from my September 26, 2003, interview with him; also see John McCain with Mark Salter, *Faith of My Fathers* (1999).

For understanding Operation Linebacker II, I relied on Karl J. Eschmann, *Linebacker: The Untold Story of the Raids over Vietnam* (1989).

For the history of the moratoriums, see Jerry Lembcke, *The Spitting Image: Myth, Memory, and the Legacy of Vietnam* (1998). This underappreciated book was also exceedingly helpful in writing about VVAW.

For Nixon's Silent Majority speech, see Stephen E. Ambrose, *Nixon*, vol. 2: *The Triumph of a Politician 1962–1972* (1989), and Richard M. Nixon, *RN: The Memoirs of Richard Nixon* (1978). An interesting article on the U.S. Navy and gradual withdrawal from Vietnam is "ACTOVASAP: Vietnamization," *All Hands*, September 1970, outlines how an operation called Grant Slingshot grew out of Sealords in May 1969. "The Navy is still in the Republic of Vietnam, but every day it becomes more Vietnamese and less American," the article stated. "Ships and boats continue to be transferred to the Republic of Vietnam under the accelerated turnover program." Besides searching sampans and checking ID cards, the Vietnamese navy started a SEAL training program at Cam Ranh Bay.

Besides my interviews with Robert Drinan and David Thorne, various clippings in Kerry's *Boston Globe* file were helpful in understanding his 1970 run for Congress. Samuel Z. Goldhober's February 21, 1970, article in the *Harvard Crimson* was also extremely useful, as were Brian McNiff, "Vietnam Hero Impresses Caucus," *Worcester Evening Gazette*, February 27, 1970, and Robert Healy, "The Caucus Gains Power," *Boston Globe*, March 23, 1970.

My information pertaining to Kerry's marriage comes from Julia Thorne's wedding scrapbook. "Julia Stimson Thorne Bride of John Kerry" from the New York *Sunday News* of May 24, 1970, a fascinating relic of its time, was another excellent source. George Butler shared with me Jamaica stories. The article that had such a profound effect on Kerry during his honeymoon was Charles Childs, "Our Forgotten Wounded: Assignment to Neglect," *Life*, May 22, 1970. I viewed a tape of *The Dick Cavett Show* and interviewed Cavett when he came to New Orleans to participate in a Tennessee Williams festival.

A good article on Special Forces was Russell H. S. Stolfi, "SEAL Assault on a VC Command Center," *Vietnam*, June 2001.

For the impact of moratoriums, see Charles DeBenedetti, *An American Ordeal: The Antiwar Movement of the Vietnam Era* (1990); William Goldman, "The New Mobe," *New York Times Magazine*, November 30, 1969; Francine Du Plessix, "The Moratorium and the New Mobe," *The New Yorker*, January 3, 1970; and Nancy Zaroulis and Gerald Sullivan, *Who Spoke Up? American Protest Against the War in Vietnam, 1963–1975* (1984).

Far and away the best account of Operation RAW is Gerald Nicosia's *Home to War: A History of the Vietnam Veterans' Movement* (2001). Nicosia is particularly good at explaining the role of Al Hubbard in VVAW. The VVAW-Wis archives has wonderful boxes of Operation RAW pamphlets/clippings. See also Karen Burstein, "Vet Doves Hear from Jane Fonda," *Long Island Free Press*, September 9, 1970; and Sheldon Ramsdell "Soldiers of Peace at Valley Forge," *The World Magazine* (Daily World), September 19, 1970. Another interesting book on the veterans' movement is Richard Severo and Lewis Milford *The Wages of War: When America's Soldiers Came Home* (1989), and I also used David Bonoir, Steven M. Champlin, and Timothy Kolly, *The Vietnam Veteran: A History of Neglect* (1986). Also see "River Craft Turned over to South Vietnamese Forces," *All Hands*, May 1969.

CHAPTER FIFTEEN: THE WINTER SOLDIER

For accounts of the My Lai massacre published shortly before the winter soldier investigation, see Seymour Hersh, *My Lai* 4 (1970), and Jack Schwartz, The *My Lai Massacre and Its Cover-Up: Beyond the Reach of Law* (1970). In 1971 Lieutenant William Calley published his own account of what happened with the help of journalist John Sacks. I interviewed Calley in Columbus, Georgia, in October 2003 at the jewelry store he runs. Also see William M. Hammond, *United States Army in Vietnam: Public Affairs; The Military and the Media, 1968–1973* (1995).

An excellent treatment of Nixon's reaction to Dewey Canyon III is Jeffrey Kimball, *Nixon's Vietnam War* (1998). Also helpful was my correspondence with Paul "Pete" McCloskey Jr. throughout 2003.

For a clear and concise breakdown of the organizational structure of

Vietnam Veterans Against the War (VVAW), see the first issue of *Casualty*, August 1971. It's impossible to overstate how seminal Robert Jay Lifton's *Home from the War* was in writing this and the next chapter. Lifton's insight into veterans was astonishing. A more recent personal insight into PTSD, which I also found useful, was Merlene Reynolds, "Hearts Without Homes," *Vietnam*, June 2001.

For the winter soldier investigation in Detroit, see the pamphlet *The Winter Soldier Investigation* published by VVAW, with comments from Timmy Butz, William Crandell, Scott Moore, and Mile Oliver. Also essential were Gerald Nicosia's *Home to War* and Andrew Hunt's *The Turning* (1999). In Turner's footnotes, he quotes VVAW member Mike McKusker, who also testified before Congress: "[Kerry] spoke to the Senate, the House of Lords, he being a clean-shaven Naval Officer—and now he's a member of the House of Lords—and I spoke to the House of Representatives, the House of Commons, and I was an enlisted man, the grunt. And that was our division all the way down the line. John made this speech that he had Xeroxed and sent all over the fucking world. I just had a few remarks that showed up in the Congressional Record and that was it."

Essential to my understanding of the VVAW are the oral histories published in Kerry's *The New Soldier*, especially those of Jim Weber, Skip Roberts, Joe Bangert, Charles Leffler, Steve Noetzel, Michael Hunter, Kevin Byrne, and Franklin Shepard. All Americans should read Myra MacPherson's elegant *Long Time Passing: Vietnam and the Haunted Generation* (1984); her chapter on PTSD is superb.

Some commentators compared the Detroit hearings with the Nuremberg trials after World War II, including Telford Taylor in *Nuremberg and Vietnam* (1971). Kerry appeared on a BBC TV documentary comparing Nuremberg to Vietnam. See, for example, Michael Bukht to John Kerry, June 8, 1971 (VVAW-Wis).

Another incredibly helpful book for this chapter was Tom Wells, *The War Within: America's Battle over Vietnam* (1994).

Thomas Oliphant of the *Boston Globe* wrote a number of seminal articles on Kerry and Dewey Canyon III in April 1971, including "Vets' New War Needs Funding" from April 15, "Arlington Cemetery Gates Barred to Mothers, Vets with Wreaths" from April 20, and three from April 23: "A Demonstration That Gave Lift to a Movement That Was on Wane," "Senators Told 'We Created a Monster,'" and "U.S. Yields, Lifts Ban on Camping Veterans." Mary McGrory, also of the *Boston Globe*, wrote helpful articles on VVAW, including "The Wheel Chair Lobby" (April 20, 1971), "Brothers and Daughters—and Nixon" (April 25, 1971), and "Veterans Wake Up the Nation" (April 26, 1971). Two profiles of Kerry, George C. Wilson, "Viet Veteran Turns Protestor," *Washington Post*, April 18, 1971, and "Angry War Veteran: John Forbes Kerry," *New York Times*, April 22, 1971, also proved quite helpful. Two books edited by Jan Barry—*Peace Is Our Profession: Poems and Passages of War Protest* (1981) and *Winning Hearts and Minds: War Poems by Vietnam Veterans* (1972), with Basil T. Paquet and Larry Rottmann—were indispensable. Joe Klein wrote an outstanding post–Dewey Canyon III article on Kerry for *Boston After Dark*, June 1, 1971.

Many young people wrote Kerry fan letters after his April 22, 1971, Senate testimony. A particularly moving one is Mike Long to John Kerry, August 20, 1971. Kerry wrote back to him on September 16 (VVAW-Wis). William F. Buckley Jr.'s criticisms of Kerry's testimony were published in the *Boston Globe* as "John Kerry's Speech—I," June 14, 1971, and "John Kerry's Speech—II," June 15, 1971. (The Buckley article was syndicated and appeared under various titles.) A major debate erupted in intellectual circles on the unfairness of the draft—see, for instance, James Fallows, "What Did You Do in the Class War, Daddy?," Washington Monthly, October 1975.

William L. Claiborne, "Protest Planned Near Capitol," *Washington Post*, March 17, 1971. Also see John F. Kerry to All U.S. Congressmen, March 27, 1971, explaining the purpose of Dewey Canyon III.

A truly groundbreaking book on the antiwar movement is Fred Halstead, *Out Now!: A Participant's Account of the American Movement Against the Vietnam War* (1978).

For an understanding of Edward Kennedy's views on Vietnam, which closely mirrored John Kerry's, see Adam Clymer, *Edward M. Kennedy* (1999). The story about the Javits dinner comes from Robert Mann, *A Grand Delusion* (2001). The press constantly compared Kerry to a young John F. Kennedy. See, for example, Kay Bartlett, "New JFK Also Appears Destined for White House," Associated Press, July 6, 1971.

Chapter Sixteen: Enemy Number One

Michael Kranish's *Boston Globe* articles, particularly "John Kerry, Candidate in the Making Part III: With Antiwar Role, High Visibility," June 17, 2003, provided much valuable material and proved quite helpful with this chapter. I also made use of the unpublished Nixon Tape transcripts housed at College Park, Maryland, and *The Haldeman Diaries: Inside the Nixon White House*. Some helpful post–Dewey Canyon III articles include Barbara Rabinovitz, "'Winter Soldier' Kerry Plans Long Campaign," *Boston Globe*, May 3, 1971; "John Kerry: A Hot Item," *Boston Globe*, May 6, 1971; and Helen Dudar, "Kerry: Man in Demand," April 30, 1971, *New York Post*. For *Playboy*'s continued role in helping VVAW recruit, see Andrew E. Hunt, *The Turning: A History of Vietnam Veterans Against the War* (1999), which is also the source for the "Master Kerry" story. Rottmann was most helpful, in a series of telephone conversations, helping me better understand VVAW. The vicious anti-Kerry *Detroit News* article was one of many by Jerald F. Horst that were extremely critical of Kerry and VVAW from April through August of 1971. Bill Cardoso's defense of Kerry was "Antiwar Vets Camp at Concord Bridge," *Boston Globe*, May 29, 1971. Morley Safer could not have been more helpful. Not only did he consent to an interview but he helped arrange for CBS to send me a copy of the Kerry *60 Minutes* interview. Also helpful was Safer's *Flashbacks on Returning to Vietnam* (1990). Morley Safer's role as an antiwar reporter is explored in Todd Gitlin, *The Whole World Is Watching* (1980).

For good Vietnam poetry, see W. D. Ehrhart, *Carrying the Darkness: The Poetry of the Vietnam War* (1989), and H. Bruce Franklin, ed., *The Vietnam War in American Stories, Songs and Poems* (1996).

Carol Liston's article on speculating on Kerry's political future, "Kerry Seeking Brooke Seat?," appeared in the *Boston Globe*, April 29, 1971. At the VVAW-Wis archives there is a file that documents Kerry's lecture tour in 1971. A few good Kerry letters can be found here, particularly to Dale Edmondson (September 16, 1971) and Joseph Gleicher (June 26, 1971).

Not much has been written about Operation POW. The *Boston Globe*, however, covered it quite well; for example, William J. Cardoso, "Antiwar Vets Camp at Concord Bridge," May 29, 1971; Bruce McCabe and Joan Mahoney, "Antiwar Vets Defy Ban on Lexington Green," May 30, 1971; John Wood, "Antiwar Veterans March on Bunker Hill," May

31, 1971; two articles without bylines from June 2, 1971, "A Bell Tolls in Lexington" and "Lexington Withdraws Injunction Against Viet Vets"; and a series by Joan Mahoney, "Citizens and Veterans Share Cold Night Vigil" (May 31, 1971), "Lexington Selectmen Hear Irate Critics Next Monday" (June 3, 1971), and "Lexington Prepares for Verbal Battle" (June 7, 1971). Alice Hinkle's "Lexington's Other Battle" in the *Globe* of May 20, 2001, also helped, as did "Lexington Forbids Antiwar Vet Bivouac," *Boston Globe*, May 27, 1971. But it was my conversations with Robert Drinan, Peter Yarrow, Bestor Cramm, Arthur Johnson, and Chris Gregory that were most helpful. The film *The Unfinished Symphony* was also extraordinarily helpful in my understanding what had transpired on the march.

The two journalists out to smear Kerry were J. F. ter Horst and Smith Hempstone. See, for example, ter Horst's "2 Vets with Medals, 1 with Silver Spoon," Morris Plains, NJ, *News-Bee*, May 27, 1971. A good roundup piece on the orchestrated anti-Kerry movement is Sandy Grady, "Anti-war Veteran Meets Hostility," *The Evening Bulletin*, June 15, 1971. Two books that helped me understand Nixon's enemies list are Anthony Summers, *The Arrogance of Power: The Secret World of Richard Nixon* (2000), and Frank Donner, *The Age of Surveillance: The Aims and Methods of America's Political Intelligence System* (1980). My interview with John Dean was also helpful.

The files of Armistead Maupin were most helpful in breaking the story of his involvement with the Nixon administration, particularly his June 3, 1971, letter to Admiral Zumwalt. Maupin also provided the letter from Jesse Helms of February 9, 1971. Also helpful were Brad L. Graham, "For Armistead Maupin There Are Still Tales to Tell," St. *Louis Post-Dispatch*, May 1, 2002, and "A Conversation with Armistead Maupin," *BookBrowser*, July 7, 2003.

The John Lennon story comes from an interview with Danny Barbiero. I could not locate Kerry's November 10 VVAW resignation letter supposedly housed at the Wisconsin archives. The quote I use comes directly from Andrew E. Hunt's essential *The Turning: A History of Vietnam Veterans Against the War* (1999).

I found the following books on Agnew useful: Spiro T. Agnew, *Go Quietly or Else* (1980); Richard Cohen and Jules Witcover, *A Heartbeat*

Away (1974); Paul Hoffman, *Spiro* (1971); and Jules Witcover, *White Knight: The Rise of Spiro Agnew* (1972).

Helpful articles from the *Boston Globe* include Benjamin Kilgore, "Kerry Lauds 'New Temper' in CYO Talk," May 31, 1971; George L. Croft, "Kerry Planning Viet Trip," June 3, 1971; "Mr. Agnew Tees Off Again" June 3, 1971; and "John Kerry: A Hot Item" May 6, 1971. Also see the Veterans Administrations' Report "Myth and Realities: A Study of Attitudes Toward Vietnam Era Veterans" (July 1968).

For the St. Louis VVAW meeting in which Kerry was asked to remain spokesperson, see Connie Rosenbaum, "Antiwar Vietnam Veterans Re-elect Kerry in Move Seen as Confidence Vote," *St. Louis Post-Dispatch*, June 8, 1971.

Robert Jay Lifton's quote comes from my September 21, 2003, interview. Although the FBI was following Lifton around in this period, he never felt threatened. For just how much distance VVAW wanted between themselves and Kerry, see Al Hubbard to George McGovern, Hubert Humphrey, Henry Jackson, and seven other leading Democrats, written on February 8, 1972 (VVAW-Wis).

An important work on understanding conscientious objectors is Stephen Martin Kohn, *Jailed for Peace: The History of American Draft Law Violators, 1658–1985* (1986). From 1965 to 1975, more than 20,000 Americans were indicted for draft violation. When *The New Soldier* was in galleys, Kerry received raves from George McGovern (September 30, 1971) and Ralph David Abernathy (October 1, 1971). George Plimpton explained to me Kerry's role in New York society.

Chapter Seventeen: Duty Continued

A first-rate evaluation of Kerry's political behavior in 1972 is Brian C. Mooney, "John F. Kerry, Candidate in the Making Part 4: First Campaign Ends in Defeat," *Boston Globe*, June 18, 2003.

The *Record American* headline I refer to comes from a March 15, 1972, article by Joe Albano.

A fine analysis of Kerry's 1972 post-election mood appears in Michael C. Pollak, "What Now for John Kerry?," *Worcester Sunday Telegram*, April 8, 1973. A more recent memoir of the Nixon administration's efforts to destroy the veterans' antiwar movement is W.D. Ehrhart's *Busted: A Vietnam Veteran in Nixon's America* (1995). Kerry had read Samuel Eliot Morison, Fredrick Merk, and Frank Freidel's small book *Dissent in Three Wars* (1970) while preparing for his 1972 campaign.

Even as a U.S. senator focused on international affairs Kerry demonstrated a penchant for prosecuting global terrorists and drug lords long before 9/11. During the 1990s, for example, he went after the Russian Mafia, the Chinese triads, the Colombian drug cartels, the Japanese *yakuza*, and the Sicilian Mafia. More than any other U.S. senator, he went after Central American arms merchants and then, Colonel Oliver North's cabal of soldiers of fortune. From the Senate he helped prosecute both the Iran-Contra connection and the illegitimate money laundering of the Bank of Credit and Commerce International. For a discussion of these matters, see John Kerry *The New War: The Web of Crime That Threatens America's Security* (1997). As Douglas Frantz and David McKean make clear in *Friends in High Places: The Rise and Fall of Clark Clifford*, Kerry was the only Democratic senator daring enough to investigate the BCCI in 1992. Taking on Clifford—Lyndon Johnson's last secretary of defense—was akin to heresy.

Thankfully, John Kerry's life and career were well documented by the press—especially in the *Boston Globe* and the *Boston Herald* (and later the *Herald American*)—in the months and years following his Senate Foreign Relations testimony. Among the most helpful articles were Kay Bartlett's "John Kerry: The Fame Came Easy"(Associated Press), July 6, 1971; Al Neenan's "Kerry Weighs Run for Congress," *Boston Globe*, January 10,1972; "Kerry Buying House, Hires Aide in Fourth District," *Boston Globe*, February 4, 1972; Cornelius Dalton, "Kerry-Donohue Race Looms," *Boston Globe*, February 10, 1972; "Dump Nixon Rally Calls for Action," *Boston Globe*, February 11, 1972; "Kerry Calls for Impeachment of President Nixon," *Boston Globe*, February 13, 1972; John B. Wood, "Listless Antiwar Rally on Boston Common," *Boston Globe*, April 30, 1972; "A Party at Plimpton's," *Boston Globe*, May 26, 1972; "Jobs Need '70s Challenge, Kerry Tells Lowell Class," *Boston Globe*, June 12, 1972; "Stiff Congress Battles Loom in State," *Boston Globe*, July 12, 1972; Rachelle Patterson, "$454,200 Already Spent in Fifth District Primary Race," *Boston Globe*,

September 16, 1972; Rachelle Patterson, "Kerry Edges Sheehy in Fifth District Race," *Boston Globe*, September 20, 1972; Jonathan Fuerbringer, "Kerry Wants Viet Pullout Based on POW Return," *Boston Globe*, September 25, 1972; Richard M. Weintraub, "Kerry Candidacy Boosted amid Antiwar Book Criticism," *Boston Globe*, October 19, 1972; "Durkin Suit to Seek Use of Cover on Kerry book," *Boston Globe*, October 20, 1972; Robert Healy, "John Kerry and the Flag," *Boston Globe*, October 27, 1972; William F. Buckley Jr., "Kerry's World," *Boston Globe*, October 28, 1972; Stephen Wemiel, "In Defeat Kerry Calm, No Regrets About Fight," *Boston Globe*, November 8, 1972; John Kerry's own special column " 'Peace with Honor' Phrase Conceals Failure of U.S. Policy," *Boston Globe*, March 25, 1973; Ray Richard, "Lowell's Antiwar Veteran Kerry Asks to Rebut McCord Testimony," *Boston Globe*, June 1, 1973; Gerard Weldmann, "Kerry Led Spending During House Campaign," *Boston Globe*, September 15, 1973; Joe Albano and Bill Duncliffe, "A Decision for Kerry," *Boston Herald*, July 19, 1973; Rachelle Patterson, "Poll to Decide If Kerry Runs," *Boston Globe*, September 13, 1973; John W. Riley, "Kerry Blasts State Fund Handling," *Boston Globe*, July 2, 1974; "John F. Kerry New Director of Mass Action," *Boston Herald*, July 3, 1974; "Mass Action Claims State Losing Millions," *Boston Herald*, July 29, 1974; "Kerry Resigns Mass Action Post," *Boston Herald American*, October 29, 1974; Crocker Snow Jr., "Once a Hot Political Property, Student John Kerry Just Watches," *Boston Globe*, November 3, 1974; John J. Mullins, "It'll Be Law, Politics for John Kerry," *Boston Globe*, October 19, 1975; "Kerry Named Assistant DA," *Boston Globe*, January 25, 1977; "Everett Lawyer Denies Guilt in Fraud Case," *Boston Herald*, December 8, 1977; Kevin R. Convey, "Kerry Is Seen Lt. Gov. Winner," *Boston Herald American*, September 15, 1982; Keith R. Yocum, "The Resurrection of John Kerry," *Boston Herald*, November 15, 1982; and Wayne Woodlief, "Kerry, Markey Lead Senate Donnybrook," *Boston Herald*, February 15, 1984.

John Kerry's staff for the 1972 election included George Butler, Judy Gordon, Mike Grealy, Alice Jelin, Charlie Kenney, Cam Kerry, Tom Kiley, Don McNamee, Ken Opin, Dan Payne, Mike Pollak, David Thorne, Rose Thorne, Tom Vallely, and Harriet Yellin.

Epilogue: September 2, 2003 (Charleston, South Carolina)

Joe Klein's "The Long War of John Kerry"—which ran in *The New Yorker* on December 2, 2002—was particularly helpful in learning about the Doghunters and the Kerry-McCain relationship. Kerry's presidential announcement took place across from Patriot's Point Naval and Maritime Museum. For how the Doghunters helped Kerry get reelected to the U.S. Senate in 1996, see Don Aucoin, "War Journey Bonds Kerry, Key Adviser," *Boston Globe*, October 28, 1996, and Charles M. Sennott, "The '96 Election: Old Friends," *Boston Globe*, November 6, 1996. Interesting essays on how Kerry uses his Navy background to further his political career are James A. Barnes, "War and Politics," *National Review*, July 6, 2002, and Seth Gitell, "John Kerry's Chances for the White House," *Boston Phoenix*, April 20, 2002.

The *Yorktown* had replaced its namesake, which had been lost at the Battle of Midway in June 1942. Kerry had first seen the *Yorktown* in February 1967, when it entered the Long Beach Naval Shipyard for an overhaul. He caught up with it again when he was on the *Gridley* and it was anchored in Subic Bay. Given this, Kerry felt a symbolic attachment to the famous ship. And over the years he kept seeing the *Yorktown* on TV: it served as the platform for the movie *Tora! Tora! Tora!* and was the rescue ship for Apollo 8. In 1974, the Navy Department donated it to the Patriot's Point Development Authority. It was opened to the public on October 13, 1975—the two hundredth anniversary of the U.S. Navy.

Useful information about Congressman Jim Kolbe's Swift boat experience came from his oral history from the Veterans History Project, Washington, D.C.

Columnist David Kusnet of Salon.com wrote an extensive article covering Kerry's announcement in Charleston entitled "Can John Kerry Turn It Around?"

For Ross Perot's view on the POW/MIA issue, see Todd Mason, *Perot: An Unauthorized Biography* (1990). Also see Ellen Chris Fanizzi, "North Vietnam's POW Propaganda War," *Vietnam*, August 2003.

Back in November 2001 Kerry had attended a Swift boat reunion in San Diego. See Mark Sauer, "Senator Kerry Visits to Honor Vietnam Swift

Boat Sailors," *San Diego Union-Tribune*, November 27, 2001.

In addition to the Swift boat veterans group, there is also a Mobile Riverine Force Association based in Conover, North Carolina.

The article Kerry had read was John McCain III, "How the POWs Fought Back," *U.S. News & World Report*, May 14, 1973. A wonderful essay on the Kerry-McCain collaboration is James Carroll, "A Friendship That Ended the War," *The New Yorker*, October 21 and 28, 1996.

As Kerry traveled around Vietnam looking for POW/MIAs, he was amazed at how much had changed since 1969. Wherever he went in Hanoi, Danang, Haiphong, and Ho Chi Minh City, he saw signs of *doi moi*—the Vietnamese term for economic reforms. During these trips he carefully read such books as Henry Kamm's *Dragon Ascending: Vietnam and the Vietnamese* (1996) and Neil Sheehan's *After the War Was Over* (1992). He also studied Tom Mangold's *The Tunnels of Cu Chi* (1985), which detailed the Viet Cong's underground fortifications, hoping to find clues as to where the remains of U.S. servicemen might be.

Kerry's Charleston speech closely resembled an essay he wrote for Andrew Cuomo, ed., *Crossroads: The Future of American Politics* (2003) titled "Renewing the Democratic Party: A New Patriotism for America." For Kerry's defense of veterans' rights, I read his October 10, 2002, letter to President George W. Bush, at his Washington, D.C., archive.

There is important literature on the road to normalization of relations between the United States and Vietnam, see R. W. Apple Jr. and Fredrick Z. Brown, *Second Chance: The United States and Indochina in the 1990s* (1989); Matthew Cooper, "Give Trade a Chance," *U.S. News & World Report*, February 14, 1994; Douglas Jehl, "Clinton Drops Nineteen-Year Ban on U.S. Trade with Vietnam," *New York Times*, February 4, 1994; and Keith Richburg, "Back to Vietnam," *Foreign Affairs*, Fall 1991.

For Kerry's views on the importance of normalization, see his op-ed piece "Vietnam Finally More than a War," *USA Today*, November 16, 2000.

Although I didn't use it in the text Swift boat veteran Michael Solhaug sent me "self inflections" about Vietnam, which he had delivered at St. Olaf College in Minnesota. He also helped me get in touch with other Swifties.

A good brief summary of Bob Kerrey's Vietnam woes is Michael

Elliott, "A Tale of Innocence Lost," *Newsweek*, June 10, 2002. Kerry's defense of his Senate colleague was published April 29, 2001, in the *Boston Globe* under the headline "Civilians Underscored How Little We Understand the Vietnam Experience."

John Kerry's remarks at the twentieth anniversary of the Vietnam Veterans Memorial on December 2, 2002, is his most famous speech besides his 1971 Senate testimony. Also, of interest are Kerry's opening statement at the Senate Finance Committee Hearings on the U.S.-Vietnam Bilateral Trade Agreement from June 26, 2001, and a public letter to Donald Rumsfeld from Kerry urging full benefits to veterans exposed to Agent Orange, Gulf War Syndrome, and post-traumatic stress disorder, dated April 17, 2003.

Selected Bibliography

Books

Adair, Gilbert. *Vietnam on Film: From "The Green Berets" to "Apocalypse Now."* New York: Proteus, 1981.

Aiken, E. E. *The Secret Society System*. New Haven: Tuttle, Moorehouse & Taylor, 1882.

Aitken, Jonathan. *Nixon vs. Nixon: An Emotional Tragedy*. New York: Farrar Straus Giroux, 1977.

Alexander, Paul. *Man of the People: The Life of John McCain*. Hoboken, N.J.: John Wiley, 2003.

Ambrose, Stephen E. *Nixon*. Vol. 2, *The Triumph of a Politician, 1962–1972*. New York: Simon & Schuster, 1989.

——*Nixon*. Vol. 3, *Ruin and Recovery, 1973–1990*. New York: Simon & Schuster, 1991.

Anderson, David L., ed. *Shadow on the White House: Presidents and the Vietnam War, 1945–1975*. Lawrence: University Press of Kansas, 1993.

Andrade, Dale. *Ashes to Ashes: The Phoenix Program and the Vietnam War*. Lexington, Mass.: Lexington Books, 1990.

Atkinson, Rick. *The Long Gray Line*. Boston: Houghton Mifflin, 1989.

Auster, Albert, and Leonard Quart. *How the War was Remembered: Hollywood and Vietnam*. New York: Praeger, 1988.

Ballard, Robert D. *Collision with History: The Search for John F. Kennedy's PT 109*. Washington, D.C.: National Geographic, 2003.

Barone, Michael, and Grant Ujifusa. *The Almanac of American Politics 2000*. Washington, D.C.: National Journal, 1999.

Barry, Jan, ed. *Peace Is Our Profession: Poems and Passages of War Protest.* Montclair, N.J.: East River Anthology, 1981.

Barry, Jan, Basil T. Paquet, and Larry Rottmann, eds. *Winning Hearts and Minds: War Poems by Vietnam Veterans.* Brooklyn, N.Y.: First Casualty Press, 1972.

Beaver, Frank, ed. *Oliver Stone: Wakeup Cinema.* New York: Twayne, 1994.

Berrigan, Daniel. *American Is Hard to Find.* New York: Doubleday, 1972.

Bird, Kai. *The Color of Truth: McGeorge Bundy and William Bundy; Brothers in Arms.* New York: Simon & Schuster, 1998.

Boettcher, Thomas D. *Vietnam: The Valor and the Sorrow.* Boston: Little, Brown, 1985.

Bryant, Jimmy R. *Man of the River: Memoir of a Brown Water Sailor in Vietnam, 1966–1969.* Fredericksburg, Va.: Sergeant Kirkland's Museum and Historical Society, 1998.

Bundy, William. *A Tangled Web: The Making of Foreign Policy in the Nixon Presidency.* New York: Hill & Wang, 1998.

——, ed. *Two Hundred Years of American Foreign Policy.* New York: New York University Press, 1977.

The Bunker Papers: Reports to the President from Vietnam. 1967–1973. Vol. 3. Berkeley: Institute of East Asian Studies, University of California, 1990.

Burdick, Eugene, and Harvey Wheeler. *Fail Safe.* New York: McGraw-Hill, 1962.

Butler, James. *River of Death: Song Vam Sat.* Reseda, Calif.: Mojave, 1979.

Buttinger, Joseph. *Vietnam: A Political History.* New York: Praeger, 1968.

Buzzanco, Robert. *Masters of War: Military Dissent and Politics in the Vietnam Era.* New York: Cambridge University Press, 1996.

Camil, Scott. "Undercover Agents' War on Vietnam Veterans." In *It Did Happen Here*, edited by Bud Scultz and Ruth Scultz. Berkeley: University of California Press, 1989.

Caputo, Philip. *A Rumor of War.* New York: Holt, Rinehart and Winston, 1977.

Carroll, Andrew, ed. *War Letters.* New York: Scribner, 2001.

Carson, Clayborne. *In Struggle: SNCC and the Black Awakening of the 1960s.* Cambridge, Mass.: Harvard University Press, 1981.

Chafe, William. *Never Stop Running: Allard Lowenstein and the Struggle to Save American Liberalism*. New York: Basic Books, 1993.

Charlton, Michael, and Anthony Moncrieff. *Many Reasons Why: The American Involvement in Vietnam*. New York: Hill & Wang, 1978.

Chong, Dennis. *The Girl in the Picture: The Story of Kim Phuc, the Photograph, and the Vietnam War*. New York: Penguin, 2000.

Clifford, Clark, with Richard Holbrooke. *Counsel to the President*. New York: Random House, 1991.

Clodfelter, Mark. *The Limits of Air Power: The American Bombing of North Vietnam*. New York: Free Press, 1989.

Collier, Peter. *The Fondas: A Hollywood Dynasty*. New York: G.P. Putnam's Sons, 1991.

Colson, Charles W. *Born Again*. London: Hodder & Stoughton, 1980.

Commager, Henry Steele. *Freedom and Order*. Indianapolis: Bobbs–Merrill, 1966.

Conner, Dennis, and Michael Levitt. *The America's Cup: The History of Sailing's Greatest Competition in the Twentieth Century*. New York: St. Martin's, 1998.

Corson, William R. *The Betrayal*. New York: W.W. Norton, 1968.

Crandell, William. "They Moved the Town." In *Give Peace a Chance: Exploring the Vietnam Antiwar Movement*, edited by Melvin Small and William D. Hoover. Syracuse, N.Y.: Syracuse University Press, 1992.

Croizat, Victor. *The Brown Water Navy: The River and Coastal War in Indo-China and Vietnam, 1948–1972*. Dorset, Eng.: Blandford Press, 1984.

Crowley, Monica. *Nixon in Winter*. New York: Random House, 1998.

——. *Nixon Off the Record*. New York: Random House, 1996.

Cutler, Thomas J. *Brown Water, Black Berets: Coastal and Riverine Warfare in Vietnam*. 1988. Reprint, Annapolis, Md.: Naval Institute Press, 2000.

Davidson, Phillip B. *Vietnam at War: The History, 1946–1975*. New York: Oxford University Press, 1988.

Dean, John W., III. *Blind Ambition: The White House Years*. New York: Simon & Schuster, 1976.

DeBenedetti, Charles, with Charles Chatfield. *An American Ordeal: The Antiwar Movement of the Vietnam Era*. Syracuse, N.Y.: Syracuse University Press, 1990.

Donovan, Robert J. *PT-109: John F. Kennedy in World War II*. New York: McGraw-Hill, 1961.

Ehrhart, W. D. *Busted: A Vietnam Veteran in Nixon's America*. Amherst: University of Massachusetts Press, 1995.

Ehrlichman, John. *Witness to Power*. New York: Simon & Schuster, 1982.

Ellsberg, Daniel. *Papers on the War*. New York: Simon & Schuster, 1972.

——. *Secrets: A Memoir of Vietnam and the Pentagon Papers*. New York: Viking, 2002.

Eschmann, Karl J. *Linebacker: The Untold Story of the Air Raids over Vietnam*. New York: Ivy Books, 1989.

Fall, Bernard B. *Last Reflections on a War*. New York: Doubleday, 1964.

——. *Street Without Joy*. New York: Schocken, 1961.

Frost, David. *"I Gave Them a Sword": Behind the Scenes of the Nixon Interviews*. New York: William Morrow, 1978.

Frost, Robert. *The Complete Poems of Robert Frost*. New York: Holt, Rinehart & Winston, 1949.

Fulton, William B. *Vietnam Studies: Riverine Operations, 1966–1969*. Washington, D.C.: Department of the Army, 1973.

Gitlin, Todd. *The Sixties: Years of Hope, Days of Rage*. New York: Bantam, 1987.

——. *The Whole World Is Watching: Mass Media in the Making and Unmaking of the New Left*. Berkeley: University of California Press, 1980.

Greene, Graham. *The Quiet American*. New York: Viking, 1956.

Haig, Alexander M., Jr. *Inner Circles: How America Changed the World*. New York: Warner, 1992.

Haldeman, H. R. *The Haldeman Diaries: Inside the Nixon White House*. New York: G.P. Putnam's Sons, 1994.

Haldeman, H. R., and Joseph DiMona. *The Ends of Power*. New York: Times Books, 1978.

Haley, Edward P. *Congress and the Fall of South Vietnam and Cambodia*. Rutherford, N.J.: Fairleigh Dickinson University Press, 1982.

Hallock, Daniel. *Hell, Healing and Resistance: Veterans Speak*. Farmington, Pa.: Plough Publishing, 1998.

Halstead, Fred. *Out Now: A Participant's Account of the American Movement Against the Vietnam War*. New York: Monad Press, 1978.

Hammond, William M. *United States Army in Vietnam: Public Affairs; The Military and the Media, 1968–1973*. Washington, D.C.: Center of Military History, 1996.

Heckscher, August. *St. Paul's: The Life of a New England School*. New York: Charles Scribner's Sons, 1980.

Helmer, John. *Bringing the War Home: The American Soldier in Vietnam and After*. New York: Free Press, 1974.

Herring, George C. *America's Longest War: The United States and Vietnam, 1950–1975*. 2nd ed. Philadelphia: Temple University Press, 1986.

——. *LBJ and Vietnam: A Different Kind of War*. Austin: University of Texas Press, 1994.

——, ed. *The Secret Diplomacy of the Vietnam War: The Negotiating Volumes of the Pentagon Papers*. Austin: University of Texas Press, 1983.

Hersh, Seymour. *The Price of Power: Kissinger in the Nixon White House*. New York: Summit, 1983.

Hershberg, James. *James B. Conant: Harvard to Hiroshima and the Making of the Nuclear Age*. New York: Alfred A. Knopf, 1993.

Hodgson, Godfrey. *The Colonel: The Life and Wars of Henry Stinson, 1867–1950*. New York: Alfred A. Knopf, 1990.

Hoff, Joan. *Nixon Reconsidered*. New York: Basic Books, 1994.

Hoffman, Stanley. *Primacy or World Order: American Foreign Policy Since the Cold War*. New York: McGraw-Hill, 1978.

Hogan, Michael J., and Thomas G. Paterson, eds. *Explaining the History of American Foreign Relations*. Cambridge: Cambridge University Press, 1991.

Hooper, Edwin Bickford. *Mobility, Support, Endurance: A Story of Naval Operational Logistics in the Vietnam War, 1965–1968*. Washington, D.C.: Naval Historical Center, Department of the Navy, 1972.

Hunt, Andrew E. *The Turning: A History of Vietnam Veterans Against the War*. New York: New York University Press, 1999.

Isaacs, Arnold R. *Without Honor: Defeat in Vietnam and Cambodia*. New York: Vintage, 1984.

Isaacson, Walter. *Kissinger*. New York: Simon & Schuster, 1992.

Isaacson, Walter, and Evan Thomas. *The Wise Men: Six Friends and the World They Made*. New York: Simon & Schuster, 1986.

Jackson, Joe. *A Furnace Afloat*. New York: Free Press, 2003.

Johnson, Lyndon Baines. *The Vantage Point: Perspectives of the Presidency, 1963–1969*. New York: Popular Library, 1971.

Johnson, Raymond W. *Postmark: Mekong Delta*. Westwood, N.J.: Fleming H. Revell, 1968.

Jones, Howard. *Death of a Generation: How the Assassinations of Diem and JFK Prolonged the Vietnam War*. New York: Oxford University Press, 2003.

Kalb, Marvin. *The Nixon Memo: Political Respectability, Russia, and the Peace*. Chicago: University of Chicago Press, 1994.

Kalb, Marvin, and Bernard Kalb. *Kissinger*. Boston: Little, Brown, 1974.

Kane, Rod. *Veteran's Day: A Combat Odyssey*. New York: Simon & Schuster, 1990.

Karlin, Wayne, Basil T. Paquet, and Larry Rottmann, eds. *Free Fire Zone: Short Stories by Vietnam Veterans*. New York: McGraw-Hill/First Casualty Press, 1973.

Karnow, Stanley. Introduction. In *Historical Atlas of the Vietnam War*. New York: Houghton Mifflin, 1995.

——. *Vietnam: A History*. New York: Viking, 1983.

Kelley, Michael P. *Where We Were in Vietnam: A Comprehensive Guide to the Firebases, Military Installations and Naval Vessels of the Vietnam War, 1945–75*. Central Point, Ore.: Hellgate Press, 2002.

Kerrey, Bob. *When I Was a Young Man*. New York: Harcourt, 2002.

Kerry, John. *The New War: The Web of Crime That Threatens America's Security*. New York: Simon & Schuster, 1997.

Kerry, John, and the Vietnam Veterans Against the War. *The New Soldier*. New York: Collier/Macmillan, 1971.

Kerry, Richard J. *The Star-Spangled Mirror*. New York: Rowman & Littlefield, 1990.

Kimball, Jeffrey. *Nixon's Vietnam War*. Lawrence: University Press of Kansas, 1998.

Kissinger, Henry A. *White House Years*. Boston: Little, Brown, 1979.

Kovic, Ron. *Born on the Fourth of July*. New York: Simon & Schuster, 1976.

Kutler, Stanley L., ed. *Abuse of Power: The New Nixon Tapes*. New York: Free Press, 1997.

——, ed. *Encyclopedia of the Vietnam War*. New York: Scribner, 1996.

——. *The Wars of Watergate: The Last Crisis of Richard Nixon.* New York: Alfred A. Knopf, 1990.

Larzelere, Alex. *The Coast Guard at War: Vietnam 1965–1975.* Annapolis, Md.: Naval Institute Press, 1997.

Lederer, William J., and Eugene Burdick. *The Ugly American.* New York: W.W. Norton, 1958.

Leepson, Marc, with Helen Hannaford, eds. *Webster's New World Dictionary of the Vietnam War.* New York: Simon & Schuster, 1999.

Lembcke, Jerry. *The Spitting Image: Myth, Memory, and the Legacy of Vietnam.* New York: New York University Press, 1998.

Lewy, Guenter. *America in Vietnam.* New York: Oxford University Press, 1978.

Lifton, Robert Jay. *Home from the War: Vietnam Veterans; Neither Victims nor Executioners.* New York: Simon & Schuster, 1973.

——. *The Spoils of War.* Boston: Houghton Mifflin, 1974.

Lind, Michael. *Vietnam: The Necessary War.* New York: Simon & Schuster, 1999.

Linder, Bruce. *San Diego's Navy.* Annapolis, Md.: Naval Institute Press, 2001.

Litwak, Robert S. *Détente and the Nixon Doctrine: American Foreign Policy and the Pursuit of Stability, 1969–1976.* New York: Cambridge University Press, 1984.

Lodge, Henry Cabot. *The Storm Has Many Eyes.* New York: W.W. Norton, 1973.

Love, Robert William, Jr., ed. *Changing Interpretations and New Sources in Naval History.* New York: Garland, 1980.

Maclear, Michael, *The Ten Thousand Day War: Vietnam, 1945–1975.* New York: St. Martin's, 1981.

Mailer, Norman. *The Armies of the Night.* New York: Signet, 1968.

——. *Why Are We in Vietnam?* New York: Putnam, 1967.

Makower, Joel, *Boom! Talkin' About Our Generation.* Chicago: Contemporary, 1985.

Marolda, Edward J. *The U.S. Navy in the Vietnam War: An Illustrated History.* Washington, D.C.: Brassey's, 2002.

Marolda, Edward J., and G. Wesley Price III. *A Short History of the United States Navy and the Southeast Asian Conflict, 1950–1975.*

Washington, D.C.: Naval Historical Center, Department of the Navy, 1984.

Marolda, Edward J. and Oscar P. Fitzgerald. *The United States Navy and the Vietnam Conflict.* Vol. 2, *From Military Assistance to Combat, 1959–1965.* Washington, D.C.: Naval Historical Center, 1986.

Mason, Bobbie Ann. *In Country.* New York: Harper & Row, 1985.

Mason, Robert. *Chickenhawk.* New York: Viking, 1983.

Matusow, Allen J. *Nixon's Economy: Booms, Busts, Dollars, and Votes.* Lawrence: University Press of Kansas, 1998.

Maupin, Armistead. *The Night Listener.* New York: HarperCollins, 2000.

McCain, John, with Mark Salter. *Faith of My Fathers.* New York: Random House, 1999.

McCarthy, Eugene, *The Year of the People.* Garden City, N.Y.: Doubleday, 1969.

McNamara, Robert S., with Brian VanDeMark. *In Retrospect: The Tragedy and Lessons of Vietnam.* New York: Times Books, 1995.

MacPherson, Myra. *Long Time Passing: Vietnam and the Haunted Generation.* Garden City, N.Y.: Doubleday, 1984.

Millett, Allan R., and Peter Maslowski. *For the Common Defense: A Military History of the United States of America.* New York: Free Press, 1984.

Naval and Maritime Chronology, 1961–1971. Annapolis, Md.: U.S. Naval Institute, 1973.

Nguyen Quang Truoung. *The Easter Offensive of 1972.* Washington, D.C.: U.S. Army Center of Military History, 1980.

Nicosia, Gerald. *Home to War: A History of the Vietnam Veterans' Movement.* New York: Crown, 2001.

Nixon, Richard. *In the Arena: A Memoir of Victory, Defeat, and Renewal.* New York: Simon & Schuster, 1990.

——. *The Real War.* New York: Warner, 1981.

——. *RN: The Memoirs of Richard Nixon.* 1978. Reprint, New York: Simon & Schuster, 1990.

Oberdorfer, Don. *Tet!: The Turning Point in the Vietnam War.* Garden City, N.Y.: Doubleday, 1971.

Palmer, Bruce, Jr. *The 25-year War: America's Military Role in Vietnam.* Lexington: University of Kentucky Press, 1984.

Parmet, Herbert S. *Richard Nixon and His America.* Boston: Little, Brown, 1990.

Peace in Vietnam: A New Approach in Southeast Asia; A Report Prepared for the American Friends Service Committee. New York: Hill & Wang, 1966.

The Pentagon Papers: The Defense Department History of United States Decision-making on Vietnam. [Senator Gravel Edition.] Boston: Beacon Press, 1971.

Pike, Douglas. *Viet Cong: The Organization and Techniques of the National Liberation Front of South Vietnam.* Cambridge, Mass.: MIT Press, 1966.

——. *Vietnam and the Soviet Union: Anatomy of an Alliance.* Boulder, Colo.: Westview Press, 1987.

Pourade, Richard F. *City of the Dream.* La Jolla, Calif.: Copley, 1977.

——. *The Glory Years.* San Diego: Union-Tribune Publishing, 1964.

——. *Gold in the Sun.* San Diego: Union-Tribune Publishing, 1965.

——. *The Rising Tide.* San Diego: Union-Tribune Publishing, 1967.

——. *The Silver Dons.* San Diego: Union-Tribune Publishing, 1963.

Powers, Thomas. *The Man Who Kept the Secrets: Richard Helms and the CIA.* New York: Alfred A. Knopf, 1979.

——. *The War at Home: Vietnam War and the American People.* New York: Grossman, 1973.

Prados, John. *The Sky Would Fall: Operation Vulture, The Secret U.S. Bombing Mission to Vietnam, 1954.* New York: Dial, 1983.

Puller, Lewis B., Jr. *Fortunate Son.* New York: Grove Weidenfeld, 1991.

Rhodes, Richard. *Dark Sun: The Making of the Hydrogen Bomb.* New York: Simon & Schuster, 1995.

Riverine Warfare: The U.S. Navy's Operations on Inland Waters. Rev. ed. Washington, D.C.: Naval Historical Center, Department of the Navy, 1969.

Safire, William. *Before the Fall: An Inside View of the Pre-Watergate White House.* New York: Doubleday, 1975.

Santoli, Al, *Everything We Had: An Oral History of the Vietnam War by Thirty-three American Soldiers Who Fought It.* New York: Random House, 1981.

Schell, Jonathan. *The Time of Illusion.* New York: Alfred A. Knopf, 1975.

Schiller, Alec C. "Congressman Bob Wilson's Contribution to the Navy and San Diego, 1952–62." Master's thesis, San Diego State University, 1990.

Schlesinger, Arthur M., Jr. *The Bitter Heritage: Vietnam and America Democracy, 1941–1966*. Boston: Houghton Mifflin, 1967.

——. *The Imperial Presidency*. Boston: Houghton Mifflin, 1973.

Schreadley, Richard L. *From the Rivers to the Sea: The United States Navy in Vietnam*. Annapolis, Md.: Naval Institute Press, 1992.

Schulzinger, Robert D. *Henry Kissinger: Doctor of Diplomacy*. New York: Columbia University Press, 1989.

Schurnam, Franz. *The Foreign Politics of Richard Nixon: The Grand Design*. Berkeley: Institute of International Studies, University of California, 1987.

Scott, Peter Dale. *The War Conspiracy: The Secret Road to the Second Indochina War*. Indianapolis: Bobbs-Merrill, 1972.

Severo, Richard, and Lewis Milford. *The Wages of War: When American Soldiers Came Home—From Valley Forge to Vietnam*. New York: Simon & Schuster, 1989.

Shaplen, Robert. *The Road from War: Vietnam, 1965–1970*. New York: Harper & Row, 1970.

Shawcross, William. *Sideshow: Kissinger, Nixon and the Destruction of Cambodia*. Rev. ed. New York: Simon & Schuster, 1987.

Sheehan, Neil. *A Bright Shining Lie: John Paul Vann and America in Vietnam*. New York: Random House, 1988.

Sheehan, Neil, et al., eds. *The Pentagon Papers as Published by the* New York Times. New York: Bantam, 1971.

Sherry, Michael S. *The Rise of American Air Power: The Creation of Armageddon*. New Haven: Yale University Press, 1987.

Showalter, Dennis E., and John G. Albert, eds. *An American Dilemma: Vietnam, 1964–1973*. Chicago: Imprint Publications, 1993.

——. *Johnson, Nixon, and the Doves*. New Brunswick, N.J.: Rutgers University Press, 1988.

Small, Melvin, and William D. Hoover, eds. *Give Peace a Chance: Exploring the Vietnam Antiwar Movement*. Syracuse, N.Y.: Syracuse University Press, 1992.

Spector, Ronald H. *After Tet: The Bloodiest Year in Vietnam*. New York: Free Press, 1993.

——. *United States Army in Vietnam: Advice and Support; The Early Years, 1941–1960.* Washington, D.C.: Center of Military History, 1983.

Strait, Raymond. *Bob Hope: A Tribute.* New York: Kensington, 2003.

Summers, Harry G., Jr. *On Strategy: A Critical Analysis of the Vietnam War.* Novato, Calif.: Presidio Press, 1982.

Sweetman, Jack. *American Naval History: An Illustrated Chronology of the U.S. Navy and Marine Corps, 1775–Present.* Annapolis, Md.: Naval Institute Press, 1984.

Szulc, Tad. *The Illusion of Peace: Foreign Policy in the Nixon Years.* New York: Viking, 1978.

Terry, Wallace, II. *Bloods: An Oral History of the Vietnam War by Black Veterans.* New York: Random House, 1984.

Thompson, Kenneth W., ed. *Portraits of American Presidents.* Vol. 6, *The Nixon Presidency: Twenty-two Intimate Perspectives of Richard M. Nixon.* New York: University Press of America, 1987.

Tollefson, James W. *The Strength Not to Fight: An Oral History of Conscientious Objectors of the Vietnam War.* Boston: Little, Brown, 1993.

Toplin, Robert Brent. *Reel History: In Defense of Hollywood.* Lawrence: University of Kansas Press, 2002.

Van Tien Dung. *Our Great Spring Victory: An Account of the Liberation of South Vietnam.* Translated by John Spragens Jr. New York: Monthly Review Press, 1977.

Vietnam Documents and Research Notes Series: Translation and Analysis of Significant Viet Cong/North Vietnamese Documents. Bethesda, Md.: University Publications of America, 1991. Microfilm.

Vietnam Veterans Against the War. *Twenty-five Years Fighting for Veterans Peace and Justice.* Chicago: Vietnam Veterans Against the War, 1992.

——. *The Winter Soldier Investigation: An Inquiry into American War Crimes.* Boston: Boston Press, 1972.

Wells, Tom. *The War Within: America's Battle over Vietnam.* Berkeley: University of California Press, 1994.

Westmoreland, General William C. *A Soldier Reports.* Garden City, N.Y.: Doubleday, 1976.

Wicker, Tom. *One of Us: Richard Nixon and the American Dream.* New York: Random House, 1991.

Williams, William Appleton, Thomas McCormick, Lloyd Gardner, and Walter LaFeber. *America in Vietnam: A Documentary History.* Garden City, N.Y.: Anchor Press/Doubleday, 1985.

Wills, Garry. *John Wayne's America: The Politics of Celebrity.* New York: Simon & Schuster, 1997.

——. *Nixon Agonistes: The Crisis of the Self-Made Man.* Boston: Houghton Mifflin, 1969.

Wittner, Lawrence S. *Resisting the Bomb: A History of the World Nuclear Disarmament Movement, 1954–1970.* Stanford, Calif.: Stanford University Press, 1997.

Wolfe, Tom. *The Right Stuff.* New York: Farrar Straus Giroux, 1979.

Young, Marilyn B. *The Vietnam Wars, 1945–1990.* New York: HarperCollins, 1991.

Zabecki, David T., ed. *Vietnam: A Reader.* New York: Ibooks, 2002.

Zaroulis, Nancy, and Gerald Sullivan. *Who Spoke Up?: American Protest Against the War in Vietnam, 1963–1975.* Garden City, N.Y.: Doubleday, 1984.

Zinn, Howard. *Vietnam: The Logic of Withdrawal.* Boston: Beacon Press, 1967.

Zumwalt, Elmo R., Jr. *On Watch.* New York: Quadrangle, 1976.

ARTICLES

Alexander, Paul. "John Kerry: Ready for His Close-Up." *Rolling Stone*, April 11, 2002.

"Antiwar Veterans Going to Court on Washington Mall Campground issue." *Boston Globe*, April 19, 1971.

"Antiwar Vet Hits the Trail, Dogged by Apathy." *National Observer*, August 9, 1971.

"Antiwar Vet Says Issue Is Fading." *Boston Globe*, December 10, 1971.

"Antiwar Vets Begin Week of Protest." *Boston Globe*, April 15, 1971.

Baker, John W., and Lee C. Dickson. "Army Forces in Riverine Operations." *Military Review*, August 1967, 64–74.

Baldwin, Hanson W. "Spitkits in Tropic Seas." *Shipmate*, August–September 1966, 8–12.

Barker, Karlyn. "Vietnam Veterans Return to Help." *Washington Post*, September 23, 1971.

Barnhart, Aaron. "She Found Her Father's Story and Herself." *Kansas City Star*, May 23, 2003.

Baxter, Tom, and Jim Galloway. "Max Returns, with Fire in His Eyes." *Atlanta Journal-Constitution*, June 16, 2003.

Bearden, Bill. "River Craft Turned Over to South Vietnamese Forces." *All Hands*, May 1969.

Beatty, Jack. "A Race Too Far?" *Atlantic Monthly*, August 1996.

Beaumont, Thomas. "Creativity, Single-minded Ambition Buoyed Kerry Through Tough Times." *Des Moines Register*, August 17, 2003.

Beisner, Robert L. "History and Henry Kissinger." *Diplomatic History* 14 (Fall 1990): 511–28.

Benjamin, Dick. "Shuttle Run." *All Hands*, July 1968.

Beschloss, Michael R. "How Nixon Came in from the Cold." *Vanity Fair*, June 1992, 114–19, 148–52.

Blais, Madeline. "Viet Veterans Seek Drug Center from Army." *Boston Globe*, August 11, 1971.

Blumenfeld, Laura. "John Kerry: Hunter, Dreamer, Realist." *Washington Post*, June 1, 2003.

"Boston Whaler: The Unsinkable Legend," bostonwhaler.com, February 27, 2003.

Brands, H. W. "Fractal History, or Clio and the Chaotics." *Diplomatic History* 16 (Fall 1992): 495–510.

Briggs, Glenn H. "Baby Ironclad." *All Hands*, June 1970.

Brinkley, Douglas. "A Second Act for the Work of Dos Passos." *New York Times*, August 30, 2003.

Bucklew, Phil H. "Navy Small Craft in Market Time." *Naval Engineers Journal*, June 1966, 395–402.

Buckley, William F., Jr. "John Kerry's Speech—I." *Boston Globe*, June 14, 1971.

———. "John Kerry's Speech—II." *Boston Globe*, June 15, 1971.

"Candidate Charges Harassment: Kerry's Brother and Helper Arrested." *Boston Globe*, September 15, 1972.

Cardoso, William J. "Antiwar Vets Camp at Concord Bridge." *Boston Globe*, May 29, 1971.

Carlson, Peter. "Political Veteran." *Washington Post*, July 3, 2003.

Carroll, James. "A Friendship That Ended the War." *The New Yorker*, October 21 and 23, 1996.

Chapelle, Dickey. "Water War in Viet Nam." *National Geographic*, February 1966, 270–96.

"Court Lashes Justice Dept.: U.S. Yields, Lifts Ban on Camping Veterans." *Boston Globe*, April 23, 1971.

Dalton, Cornelius. "Kerry-Donohue Race Looms." *Boston Globe*, February 10, 1972.

DeBenedetti, Charles. "On the Significance of Peace Activism: America, 1961–1975." *Peace and Change: A Journal of Peace Research* 9, nos. 2/3 (Summer 1983): 6–20.

Delawala, Imtiyaz. "McCain, Kerry Lauded for Vietnam Service." *Harvard Crimson*, September 11, 2001.

Dudar, Helen. "Kerry: Man in Demand." *New York Post*, April 30, 1971.

"Editorial Comments on Vets Week." *Boston Globe*, May 3, 1971.

Edwards, James. "Antiwar Vets Plan Protest as Ex-GI Faces Drug Charge." *Boston Globe*, September 20, 1971.

Farrell, John Aloysius. "John F. Kerry, Candidate in the Making, Part 6: With Probes, Making His Mark." *Boston Globe*, June 20, 2003.

——. "John F. Kerry, Candidate in the Making, Part 7: At the Center of Power, Seeking the Summit." *Boston Globe*, June 21, 2003.

Gallagher, Thomas C. "A Rising Political Star." *Boston Globe*, May 7, 1971.

Garment, Leonard. "The Annals of Law." *The New Yorker*, April 17, 1989, 90–110.

Gelzinis, Peter. "John Kerry: War Hero and Former Vets Spokesman Says He's Political Activist . . . Not Pol." *Boston Herald*, January 16, 1978.

"Gen. Pershing's Kin Killed in Viet." *Boston Globe*, February 20, 1968.

"Gesture Ends Week of Protest: Antiwar Veterans Discard Medals." *Boston Globe*, April 24, 1971.

Gilgore, Benjamin. "Kerry Lauds 'New Temper' in CYO Talk." *Boston Globe*, May 31, 1971.

Glad, Betty, and Matthew W. Link. "President Nixon's Inner Circle of Advisers." *Presidential Studies Quarterly* 26 (Winter 1996): 13–40.

"Group Doesn't Worry Kerry." *Boston Globe*, December 12, 1971.

Halbfinger, David M. "Eyes on the White House, Kerry Keeps Focus on Vietnam." *New York Times*, August 26, 2003.

Hayes, Stephen Douglas. "Broken Promise." *Northern Virginian*, July–August 1979.

Healy, Robert. "Veterans' Protest Reaches Different Segment of Society." *Boston Globe*, June 6, 1971.

Hennessey, Gregg. "San Diego, the U.S. Navy, and Urban Development." *California History* 72, no. 2 (Summer 1993).

Hinkle, Alice. "Lexington's Other Battle." *Boston Globe*, May 20, 2001.

Hoff, Joan. "Researchers' Nightmare: Studying the Nixon Presidency." *Presidential Studies Quarterly* 26 (Winter 1996): 250–66.

——. "A Revisionist View of Nixon's Foreign Policy." *Presidential Studies Quarterly* 26 (Winter 1996): 107–30.

Holcomb, Will H. "Getting a Naval Training School." *San Diego Magazine*, April 1930, 50.

Janofsky, Michael. "Rocking the Vote, John Kerry Style." *New York Times*, September 14, 2003.

Jensen, Holger. "'All Vietnam Veterans Aren't Potheads, Smack Freaks and Radical Peaceniks.'" *Charleston* (S.C.) *News and Courier*, September 2, 1971.

"John Kerry: A Hot Item." *Boston Globe*, May 6, 1971.

"John Kerry: A Navy Dove Returns." *Boston Globe*, February 24, 1970.

"John Kerry: The Fame Came Easy." *Boston Globe*, July 6, 1971.

Johnson, Arthur, and Bestor Cramm. "To Glorify Life." *Boston Globe*, April 23, 1971.

Johnson, Brooks. "Viet Vets Throw Away Medals at War Protest on Capitol Steps." *Boston Globe*, April 23, 1971.

Kaplan, H. R. "The Coast Guard in Viet Nam." *Navy Magazine*, June 1966.

——. "Coast Guard Played Vital Role in Viet Nam." *Navy Magazine*, November 1970.

Kerry, John. "'Peace with Honor' Phrase Conceals Failure of U.S. Policy." *Boston Globe*, March 25, 1973.

——. "Vietnam Then Was Worlds Apart from 'Apocalypse Now.'" *Boston Herald*, October 19, 1979.

"Kerry Agrees to Debate on War." *Boston Globe*, June 3, 1971.

"Kerry Avoids a Debate." *Boston Globe*, July 22, 1971.

"Kerry Buying House, Hires Aide in Fourth District." *Boston Globe*, February 4, 1972.

"Kerry Calls for Impeachment of President Nixon." *Boston Globe*, February 13, 1972.

"Kerry Clash Shows Rift Among POW Kin." *Boston Globe*, July 24, 1971.

"Kerry Denies Plan to Seek Public Office." *Boston Globe*, May 14, 1971.

"Kerry Hit, Defended by POW Relatives." *Boston Globe*, July 23, 1971.

"Kerry in Nation-wide Focus." *Boston Globe*, April 26, 1971.

"Kerry, O'Neill Present Rhetorical De-escalation." *Boston Globe*, August 3, 1971.

"Kerry Plans Return Visit to Viet." *Boston Globe*, April 28, 1971.

"Kerry 'Running for Something.'" *Boston Globe*, March 15, 1972.

"Kerry Says Administration Is Neglecting War Veterans." *Boston Globe*, December 6, 1971.

"Kerry's Brother Arraigned: Row over Arrests Clouds Fifth Voting." *Boston Globe*, September 19, 1972.

"Kerry, Vets Know Facts." *Boston Globe*, May 3, 1971.

"Kerry Weighs Run for Congress." *Boston Globe*, January 10, 1972.

Kissinger, Henry A. "The Viet Nam Negotiations." *Foreign Affairs* 47 (January 1969): 211–34.

Klein, Joe. "The Long War of John Kerry." *The New Yorker*, December 2, 2002.

Kranish, Michael. "John Kerry, Candidate in the Making Part 1: A Privileged Youth, a Taste for Risk." *Boston Globe*, June 15, 2003.

———. "John Kerry, Candidate in the Making Part 2: Heroism, and Growing Concern About War." *Boston Globe*, June 16, 2003.

———. "John Kerry, Candidate in the Making Part 3: With Antiwar Role, High Visibility." *Boston Globe*, June 17, 2003.

———. "Sailors' Reunion Takes Kerry down Memory Lane." *Boston Globe*, March 24, 2003.

Kurtz, Howard. "When the White House Feared Kerry." *Washington Post*, June 17, 2003.

Kusnet, David. "Can John Kerry turn it around?" Salon.com, September 19, 2003.

"Let's Try and Glorify the Living." *Time*, May 3, 1971.

Liston, Carol. "Kerry Seeking Brooke seat?" *Boston Globe*, April 29, 1971.

Mahoney, Joan. "Citizens and Veterans Share Cold Night Vigil." *Boston Globe*, May 31, 1971.

——. "Lexington Prepares for Verbal Battle." *Boston Globe*, June 7, 1971.

——. "Lexington Selectmen Hear Irate Critics Next Monday." *Boston Globe*, June 3, 1971.

Mailin, Brendan. "The Liabilities of Vietnam war." *Boston Globe*, May 27, 1971.

"Mass. Officer Asks: 'Why?'" *Boston Globe*, April 27, 1971.

Maupin, Armistead, Jr. "The Ten Vets Who Went Back." *National Review*, December 31, 1971.

McCabe, Bruce, and Joan Mahoney. "Antiwar Vets Defy Ban on Lexington Green." *Boston Globe*, May 30, 1971.

McGrory, Mary. "Brothers and Daughters—and Nixon." *Boston Globe*, April 25, 1971.

——. "John Kerry Now a Household Word." *Boston Globe*, June 6, 1971.

——. "Major Fought Lonely Battle Against War." *Boston Globe*, September 28, 1971.

——. "Veterans Wake Up the Nation." *Boston Globe*, April 26, 1971.

——. "Veterans Win Mall Battle." *Boston Globe*, April, 24, 1971.

Miller, Richards T. "Fighting Boats of the United States." *Naval Review* (1968): 297–329.

Mooney, Brian C. "John F. Kerry, Candidate in the Making, Part 4: First Campaign Ends in Defeat." *Boston Globe*, June 18, 2003.

——. "John F. Kerry, Candidate in the Making, Part 5: Taking One Prize, Then a Bigger One." *Boston Globe*, June 19, 2003.

Mulligan, John E. "Hawk and Dove—Sen. John F. Kerry Has Been Both." *Providence Journal*, March 23, 2003.

Murphy, Jeremiah V. "15,000 Traveling from Boston Area." *Boston Globe*, April 24, 1971.

Nixon, Richard. "Asia After Viet Nam." *Foreign Affairs* 46 (October 1967): 109–25.

Nolan, Martin F. "D.C. Braces as Thousands Pour in for Massive March." *Boston Globe*, April 24, 1971.

"Observances Over, Protest Points Made." *Boston Globe,* June 1, 1971.

Oliphant, Thomas. "Antiwar Veterans Discard Veterans." *Boston Globe,* April 24, 1971.

———."A Demonstration That Gave Lift to a Movement That Was on Wane." *Boston Globe,* April 23, 1971.

———. "Scene: Looking for Fellow New Englanders in Sea of Marchers." *Boston Globe,* April 25, 1971.

———. "Senators Told 'We Created a Monster.'" *Boston Globe,* April 23, 1971.

———. "U.S. Yields, Lifts Ban on Camping Veterans." *Boston Globe,* April 23, 1971.

———. "Vets Defy Supreme Court, Stay on Mall." *Boston Globe,* April 22, 1971.

"PBRs Make Their Mark." *All Hands,* July 1969.

Pilati, Joe. "An Anti-war Notebook: The Vietnam Veterans' March on Washington." *Boston Globe,* May 30, 1971.

Pollak, Michael C. "What Now for John Kerry?" *Worcester Sunday Telegram,* April 8, 1973.

Powers, Robert C. "Beans and Bullets for Sea Lords." *U.S. Naval Institute Proceedings,* December 1970, 95–97.

"POW Wives Rap Their Spokesman." *Boston Globe,* July 23, 1971.

"Pro-Nixon Vets Challenge Kerry to Debate on War." *Boston Globe,* June 2, 1971.

"Protest: A Week Against the War." *Time,* May 3, 1971.

Rainwater, Tom. "Coastal Group Advisor." *Navy Magazine,* December 1971, 39–40.

"The Riverine Force." *All Hands,* October 1968.

Rosenbaum, Connie. "Antiwar Vietnam Veterans Re-elect Kerry in Move Seen as Confidence Vote." *St. Louis Post-Dispatch,* June 8, 1971.

Ruskin, Jesse. "Unfinished Symphony." *Mass Humanities,* Spring 2001.

Sales, Robert J. "Antiwar Vets End Protest on Common." *Boston Globe,* June 1, 1971.

Schreadley, Richard L. "The Naval War in Vietnam,1950–1970." *Naval Review* (1971): 180–209.

———. "'Nothing to Report': A Day on the Vam Co Tay." *U.S. Naval Institute Proceedings,* December 1970, 23–27.

——. "SEA LORDS." *U.S. Naval Institute Proceedings,* August 1970, 22–31.

——. "Swift Raiders." *U.S. Naval Institute Proceedings,* June 1984, 53–56.

"Sen. Kerry Defends Kerrey's Vietnam Record." *USA Today,* April 25, 2001.

Shain, Percy. "Kerry Says He's Hopeful, Depressed." *Boston Globe,* May 26, 1971.

Shields, Jim. "ACTOV ASAP: Vietnamization." *All Hands,* September 1970.

"Shot Himself in Copley-Plaza." *Boston Globe,* November 23, 1921.

Simon, Roger. "Eyes on the Prize." *U.S. News & World Report,* March 17, 2003.

Simpson, Thomas H., and David La Boissiere. "Fires Support in Riverine Operations." *Marine Corps Gazette,* August 1969.

"Skull and Spare Ribs." *Economist,* November 2, 1991.

Small, Melvin. "Influencing the Decision Makers: The Vietnam Experience." *Journal of Peace Research* 24, no. 2 (1987): 185–98.

"Small Craft Operation." *All Hands,* February 1967.

"Spokesman Stuns Senators." *Boston Globe,* April 23, 1971.

"Swift Takeover." *All Hands,* November 1968.

"Text of Speech by Waltham Veteran." *Boston Globe,* April 25, 1971.

Trachtenberg, Marc. "A 'Wasting Asset': American Strategy and the Shift in Nuclear Balance, 1949–1954." *International Security* 13 (Winter 1988–89): 5–49.

"Two Major Protest Missiles to Blast Off from Boston." *Boston Globe,* October 5, 1971.

VandeHei, Jim. "Kerry Opens Campaign on War Theme." *Washington Post,* September 3, 2003.

Vennochi, Joan. "Who is John Kerry?" *Boston Globe,* February 3, 2003.

"Veterans' Protest Reaches Different Segment of Society." *Boston Globe,* June 6, 1971.

"Veterans Versus War." *Boston Globe,* March 28, 1971.

"Vet Protestors Seized by Police After Court Defy." *Boston Globe,* April 22, 1971.

"Vets' Boston Mission a Warm-up for Washington." *Boston Globe,* April 15, 1971.

"Vets' New War Needs Funding." *Boston Globe,* April 15, 1971.

"Vet's Spokesman Kerry: 'He Does the Leg Work.'" *Boston Globe*, April 25, 1971.

"Vietnam Veterans Learn of Hypocrisy." *Boston Globe*, April 25, 1971.

"Vietnam: Yale Men Buckley and Kerry Go at It." *Boston Globe*, November 15, 1971.

"Viet Veteran Turns Protestor." *Boston Globe*, April 18, 1971.

"Viet Veterans Endorse War Policy. Assail Kerry." *Boston Globe*, June 7, 1971.

"Walinsky Denies Writing Veteran's Speech." *Boston Globe*, June 4, 1971.

Welch, Donald P. "Rung Sat—Dangerous Mission." *All Hands*, December 1969.

Wells, W. C. "The Riverine Force in Action, 1966–1967." *Naval Review* (1969): 46–83.

White, Peter T. "Behind the Headlines in Viet Nam." *National Geographic*, February 1967, 149–93.

——. "The Mekong: River of Terror and Hope." *National Geographic*, December 1968, 737–87.

Wood, John. "Antiwar Veterans March on Bunker Hill," *Boston Globe*, May 31, 1971.

Archives

Armistead Maupin personal archive, Navy files, San Francisco, California

Boudinot family portraits, from the collection of Mr. and Mrs. Landon K. Thorne, Bay Shore, New York

Don Droz archive, Topanga, California

Elliott "Skip" Barker personal archive, Selma, Alabama

Harvey and Blakely Bundy personal archive (includes diary of Blakely Fetridge), Chicago, Illinois

John Kerry, personal archive (includes journals, diaries, correspondence, etc.), Boston, Massachusetts; and John Kerry, U.S. Senate archive (includes speeches and correspondence pertaining to Vietnam since 1984), Washington, D.C.

Julia Thorne personal archive, Bozeman, Montana

Lexington Oral History Projects, Lexington, Massachusetts

Michael Medeiros personal log (1969), San Leandro, California

Mike Solhaug personal archive (unpublished reminiscences), St. Olaf, Minnesota

National Security Archive, Washington, D.C.

Nixon Presidential Materials, National Archives and Records Administration, College Park, Maryland. Of special interest was Conversation 602-9.

Sewart Seacraft company archive, Berwick, Louisiana

Social Action Collection, State Historical Society of Wisconsin, Madison

Social Protest Project, The Bancroft Library, University of California, Berkeley

Stephen D. Hayes personal archive, Alexandria, Virginia. Of special interest was his Vietnam diary.

U.S. Air Force Historical Research Agency, Maxwell Air Force Base, Montgomery, Alabama

U.S. Department of the Navy, Warner Springs File, Naval Historical Center, San Diego, California

Veterans History Project, American Folklife Center of the Library of Congress, Washington, D.C.

Vietnam Veterans Against the War National Office Papers and Film Archive, Chicago, Illinois; and Vietnam Veterans Against the War Papers, State Historical Society of Wisconsin, Madison

Wade Sanders personal archive, San Diego, California

FILMS

Born on the Fourth of July (director, Oliver Stone, 1989)

Different Sons (director, Jack Oldfield, 1971)

I'll Never Do That Again (documentary, 1987)

Only the Beginning (documentary, produced by Vietnam Veterans Against the War, 1972)

The Unfinished Symphony (documentary, Northern Light Productions, 2001)

Winter Soldier (Winterfilm in association with Vietnam Veterans Against the War, 1971)

Acknowledgments

Writing *Tour of Duty* would have been impossible without the full cooperation of Senator John Kerry. Beginning in 1965, while he was an undergraduate at Yale University, he sporadically kept a journal, which became a far more regular diary in 1968 and 1969, when he was in Vietnam. The middle chapters of this book contain extensive quotes and other material adapted from these descriptive and detailed journal entries. They are extremely well written, especially coming from a Navy officer in his early- to midtwenties, portraying a consciously literary effort on Kerry's part to preserve a record of his experiences as an ensign aboard the U.S.S. *Gridley* in the Pacific and as a lieutenant (j.g.) in charge of PCFs-44 and -94 in Vietnam. His journals provide, to my knowledge, the most riveting first-person accounts that exist of U.S. Navy Swift boat raids on the waterways of the Mekong Delta and Ca Mau Peninsula.

In addition to his journals, Kerry's personal archive at his house in Boston also contains files of his correspondence from 1963 to 1973. Particularly during his time in Vietnam, he regularly typed long, artful missives to his friends and family from Danang, Cam Ranh Bay, and An Thoi, which further illuminate his actions in and observations of Southeast Asia. While skippering PCFs-44 and -94, with typewriters often unavailable, he wrote many of his journal entries by hand. In November 1971, after he resigned from Vietnam Veterans Against the War, Kerry toyed with writing a war memoir based on his journals and letters from the period, but eventually gave up the idea. Reliving those bloody, death-filled days in Vietnam proved simply too painful. The long book outline he put together at the time, however, turned out to be very useful for the book at hand. Many of the river-action sequences in *Tour of Duty* are, in fact, paraphrased directly from Kerry's unfinished book proposal. He also shared with me various other handwritten notes, random jottings, poems, interview transcripts, scrapbooks, photographs, and home movies pertaining to the Vietnam era and kindly let me borrow helpful items such as obscure pamphlets and out-of-print books from his memorabilia collection. The five or so hours of movie film Kerry shot while in Vietnam are simultaneously unremarkable yet fascinating. Along with mundane footage of Swift boats churning up the coastal waters of Southeast Asia, he also captured B-24 bombers flying overhead, a helicopter lifting a small boat off a bay, Vietnamese children waving from ashore, and coils of ammo rounds primed for firing. Time has muted the color images into dull, dated browns, blues, and grays.

Like many of his fellow brown water navy veterans, Kerry wished that the story of what the Swift boats did in Vietnam was better known, and toward that end granted me permission to quote from what he collectively called his "War Notes" with only one string attached: that I write any book or article drawn from them within two years. When I started writing *Tour of Duty* in 2002, Kerry had not yet announced he was going to run for the Democratic presidential nomination in 2004. Once he declared, his schedule intensified. But he still allowed me to catch up with him on the campaign trail at such events as a Swift boat reunion in Virginia, a public forum in Illinois, and a fund-raiser in Louisiana. Also with Kerry's permission, I obtained his Navy records and have used them as a reliable source. Finally, the senator granted me interviews on nine different occasions, in Boston, New Orleans, and Washington, D.C. On any number of evenings he personally answered fact-checking queries by telephone. But Kerry had *no* editorial control over this project whatsoever. This is my book, not his.

Tour of Duty traces the tale of one American serviceman's journey through the Vietnam War and then the movement against it. To further illuminate his character, it was important to interview Kerry's family, friends, associates, and foes. The most prodigious fount of information on his early career was Kerry's first wife, Julia Thorne. Now remarried to a prominent Montana architect, Thorne granted me an interview at her Bozeman home in early August 2003, as well as several more hours by telephone. In addition to her time, she also gave me leave to rummage through her scrapbooks, newspaper clippings, and photograph box. This book owes much to Julia Thorne's thoughtful generosity. She is a wonderful woman.

The main highlight of researching *Tour of Duty* was the chance to get to know the Swift boat community. It would be difficult to exaggerate how helpful Larry Wasikowski of Omaha, Nebraska, was in teaching me about patrol craft fast in Vietnam. Still in active service in the U.S. Navy, Wasikowski served on PCF-58 in 1968 and 1969, spending time in Danang, Chu Lai, Cua Viet, and Hue. In 1995 he created a Web site devoted to the history of PCFs in Vietnam, earning Wasikowski the moniker "keeper of the Swiftie flame." Whenever I had a question related to the men of Coastal Divisions 11 and 13, I turned to Larry, who also helped me track down telephone numbers and addresses for many other Swift boat veterans. He proofread many of the chapters. He truly is a gentleman and a scholar.

Two lieutenants who served alongside Kerry in Vietnam—and attended the Swift skippers' Saigon summit of January 22, 1969, hosted by Admiral Zumwalt and General Abrams—also proved particularly helpful: Wade Sanders of San Diego, California, and Elliott "Skip" Barker of Selma, Alabama. Both sent me detailed letters about Swift boat operations, read over chapters, and tolerated numerous telephone interviews. Someday they should write their own memoirs of the Vietnam War as they saw it.

I also interviewed most of the enlisted men who served under Kerry on PCFs-44 and -94, including James Wasser, Eugene Thorson, William Zaladonis, Stephen Hatch, Michael Medeiros, Fred Short, David Alston, Drew Whitlow, and Del Sandusky. Each of these veterans shared with me his memories, letters, and photographs. They're all great Americans.

Early in the process of writing *Tour of Duty*, my friend Cullen Murphy, editor of the *Atlantic Monthly*, served as a great encourager. He is one of the smartest, nicest, funniest and hardworking individuals I know. Twice he flew down to New Orleans to discuss the book with the idea of acquiring first serial rights. The result of those trips led to the *Atlantic*'s December 2003 cover story, "Tour of Duty: John Kerry in Vietnam."

Special thanks are due to San Francisco novelist Armistead Maupin, who let me use his personal archive's Navy files. For providing information on Kerry's family history, I owe thanks to the Essex County Trail Association,

Massachusetts Historical Society, New England Genealogical Society, and Groton House Farm. Jeffery M. Perin, Captain, USN (Ret.), provided access to his Swift boat files in Morgan City, Louisiana. (Unfortunately, a fire destroyed most of the documents on the building of PCFs between 1965 and 1970.) Others whose assistance deserves special mention include Nova M. Seals and James Shattuck of the John Fitzgerald Kennedy Library in Boston; Mark Hopkins, George Butler, Laura Rollison, and Sarah Scully of White Mountain Films; George Tobia of Burns and Levinson; Ben Healy and Yvonne Rolzhausen of the *Atlantic Monthly*; Mike Walker and the staff of the U.S. Naval Historical Center; Carl Mehler and Matt Chwastyk from the National Geographic Society for their cartographic expertise; and Dimitri Reavis of the Nixon Presidential Materials staff at the National Archives, who sifted through hours of the thirty-seventh president's White House tapes on my behalf.

Among the toughest aspects of writing *Tour of Duty* was getting on John Kerry's hectic schedule, particularly as he prepared to run for president. No words could ever express how helpful his chief of staff, David McKean—himself the author of books on Washington, D.C., insiders Tommy "the Cork" Corcoran and Clark Clifford—proved in making sure my requests were brought to the senator's immediate attention. Other especially helpful members of Kerry's team included David Wade, Kelly Benander, and the always gracious Tricia Ferrone of the senator's Washington office. In Senator John McCain's office, Chief of Staff Mark Salter was extraordinarily helpful. And special thanks to Robert Rosen, Beth Haugland, and Marybeth Cahill in Senator Edward Kennedy's office.

At the 2003 Swift crews reunion in Norfolk, Virginia, I crossed paths with *Boston Globe* reporter Michael Kranish, whose profiles of Kerry in his newspaper's June 2003 "Candidate in the Making" series broke new ground in illuminating the history behind the presidential contender. Much of the material pertaining to the Nixon White House's anti-Kerry campaign was unearthed by Kranish among the Nixon tapes housed at the National Archive facility in College Park, Maryland. Others at the *Boston Globe* who were extremely helpful included editor Matthew Storin and my friends David Beard and Jim Concannon. Columnist Alex Beam's piece on Julia Thorne, "Divorced from Politics," which appeared in the *Globe* on March 6, 2003, was very informative. The *Globe*'s editors graciously let me look at all of their paper library files on Kerry; it would have been near impossible to have written the antiwar segments of *Tour of Duty* without their help. Political writer Joe Klein of *Time* also deserves a salute for his superb *New Yorker* profile, "The Long War of John Kerry," from December 2, 2002.

Since 1983, the Eisenhower Center at the University of New Orleans has been compiling oral histories from veterans of D-Day, the Battle of the Bulge, Iwo Jima, and other World War II hot spots. It is known as the Peter Kalikow World War Two Oral History Collection. In recent years our center has turned its attention to Vietnam as well. Under the direction of Ron Drez, a captain in the U.S Marine Corps during Vietnam, we have collected more than 150 interviews with survivors of the Battle of Khe Sanh—in which Drez took part. The students in my Spring 2003 Vietnam History class at UNO indulged me in my John Kerry stories throughout the semester. Eventually, my *Tour of Duty* interviews relating to Swift boats will become part of the Eisenhower Center collection.

My staff at the Eisenhower Center worked tirelessly to help make *Tour of Duty* a reality. Lisa Weisdorffer toiled far above and beyond the call of duty to prepare the manuscript for publication while constantly managing my hectic calendar and various other operations at the same time. Andrew Travers also put in long hours transcribing interviews and fact-checking myriad queries. Kevin Willey and Michael Edwards kept the day-to-day operations of the center going while I was writing the book. This book is a tribute to their allegiance. Shawn King also helped transcribe interviews—including one with her own father—and offered constant support.

My dear friend Shelby Sadler once again helped out in innumerable ways. She is a tremendously gifted line editor, an astute political analyst, and takes pride in an obsessive knowledge of the Nixon administration that saved me from potentially embarrassing errors. She provided many smart and important recommendations that greatly enhanced the book. Her knowledge of American politics and tireless wit are legendary. So is the rock 'n' roll expertise of her consort, Tom Beach, who helped with fact-checking through their massive combined archives and the naval library of retired Chief Petty Officer Robert M. Beach. They're the best.

More than a decade ago, I worked with Claire Wachtel of William Morrow on my book *The Majic Bus: An American Odyssey*. It was terrific to once again have her as my editor. She did an incredible job of cutting my manuscript down to size and getting me to ax out tangential material; she also did a first-rate job of bringing *Tour of Duty* to press with breakneck speed. Her assistant Jennifer Pooley was a joy to work with throughout the editorial process. Without her there would be no book. Others at Morrow who helped make *Tour of Duty* a reality include Cathy Hemming, Michael Morrison, Lisa Gallagher, Kim Lewis, Lorie Young, Sharyn Rosenblum, and Dee Dee DeBartlo. And once again the indomitable Trent Duffy, working against tight deadlines, copyedited the final manuscript and ferreted out numerous errors and inconsistencies.

Three books were invaluable to *Tour of Duty*'s chapters on Kerry's antiwar activism: Andrew E. Hunt's *The Turning: A History of Vietnam Veterans Against the War* (1999); Gerald Nicosia's *To War for All* (2000), and Robert Mann's *A Grand Delusion: America's Descent into Vietnam* (2001). Also helpful, in addition to being a fascinating artifact of the era, is John Kerry's own *The New Soldier* (1971), edited by his friends David Thorne and George Butler. Historian Stanley I. Kutler's *Encyclopedia of the Vietnam War* (1996) also proved an invaluable reference tool. My bible on Swift boats was Thomas J. Cutler's *Brown Water, Black Berets* (1988). A truly generous man, Cutler read my manuscript, catching some glaring omissions and factual mistakes. Special thanks also to Lynne and Jerome Goldman, Anne and Ed Brinkley, Frances and Calvin Fayard, Elaine and Frank Sadler, and Rini and Wally Marcus.

Finally, this book is dedicated to my beautiful wife Anne, who chose the title. She has become my life, my reason for pressing on.

Index